ENCYCLOPEDIA OF
WEATHER AND CLIMATE
VOLUME II

ENCYCLOPEDIA OF
WEATHER AND CLIMATE
REVISED EDITION

VOLUME II

P–Z

MICHAEL ALLABY

Facts On File
An imprint of Infobase Publishing

ENCYCLOPEDIA OF WEATHER AND CLIMATE, Revised Edition

Facts On File, Inc.
An Imprint of Infobase Publishing
132 West 31st Street
New York NY 10001

ISBN-10: 0-8160-6350-8 (set)
ISBN-13: 978-0-8160-6350-5 (set)

Library of Congress Cataloging-in-Publication Data
Allaby, Michael.
Encyclopedia of weather and climate / [Michael Allaby].—Rev. ed.
p. cm.
Includes bibliographical references and index.
ISBN 0-8160-6350-8 (set)—
ISBN 0-8160-6348-6 (v.1)—ISBN 0-8160-6349-4 (v. 2) 1. Meteorology—
Encyclopedias. 2. Climatology—Encyclopedias. I. Title.
QC854.A452007
551.603—dc22   2006018295

Text design by Joan M. McEvoy
Cover design by Salvatore Luongo
Illustrations by Richard Garratt
Photo research by Tobi Zausner, Ph.D.

Printed in the United States of America

VB Hermitage 10 9 8 7 6 5 4 3 2 1

This book is printed on acid-free paper.

CONTENTS

ENTRIES P–Z

P

Pacific- and Indian-Ocean Common Water (PIOCW)
The deep waters of the Pacific and Indian Oceans, which are usually considered together, as a single mass of water. Their characteristics are very similar. Both have a temperature of 35.3°F (1.5°C) and a salinity of 34.7 per mil.

Pacific Decadal Oscillation (PDO) A change that occurs over a period of several decades in the ocean-atmosphere system in the Pacific basin. It affects the temperature of the lower atmosphere, passing through alternating warm and cold phases.

The PDO is similar to El Niño (*see* ENSO), but differs in timescale. ENSO events last for 6–18 months; PDO events last for 20–30 years. The two also differ in that ENSO affects mainly the TROPICS, with secondary effects elsewhere, whereas the PDO affects the North Pacific Ocean with secondary effects in the Tropics. Emphasizing the similarity with El Niño and La Niña, warm PDO phases have been nicknamed El Viejo (the old man) and cool phases La Vieja (the old woman).

During warm PDO eras, biological productivity increases in the coastal seas off Alaska, but is suppressed off the North American coast farther south. During cold PDO eras the opposite occurs.

Cold PDO eras predominated from 1890 through 1924 and 1947–76. Warm eras predominated from 1925 through 1946 and from 1977 through about 1995. There were full PDO cold phases from about 1900–1925 and 1950–1975, and full warm phases from about 1925–1950 and 1975 until the mid 1990s.

Cold and warm eras lasting just a few years have alternated since the middle 1990s. Some climate scientists suspect that the abrupt shift from a cool phase to a warm phase that occurred in 1976–1977 was responsible for most of the warming that occurred in Alaska since about that time.

Further Reading
Mantua, Nate. "The Pacific Decadal Oscillation (PDO)." Available online. URL: http://tao.atmos.washington.edu/pdo/. Accessed July 14, 2005.

Pacific high An ANTICYCLONE that covers a large part of the subtropical North Pacific Ocean. There is a similar anticyclone over the South Pacific. The Pacific highs are source regions for maritime AIR MASSES. The North Pacific high and North ATLANTIC HIGH together cover one-quarter of the Northern Hemisphere and for six months of each year they cover almost 60 percent of it.

Paleoarchean An era of Earth history that began 3,600 million years ago and ended 3,200 million years ago. The Paleoarchean is the second oldest of the four eras that comprise the ARCHEAN eon. It was a time when the Earth's crust was still forming. (*See* APPENDIX V: GEOLOGIC TIMESCALE.)

Paleocene The epoch of geologic time that began 65.5 million years ago and ended 55.8 million years ago. It is the earliest epoch of the PALEOGENE period. "Paleocene" means "ancient, recent life."

The continents had separated by the start of the Paleocene. Sea levels fell, exposing large areas of land that had previously lain beneath water. Climates were warm, the epoch ending with a rise in TEMPERATURE known as the PALEOCENE–EOCENE THERMAL MAXIMUM. Dense forests covered most of North America and Europe. It was a time of rapid mammal evolution as species radiated to occupy the niches formerly held by dinosaurs, which had become extinct at the end of the CRETACEOUS. (*See* APPENDIX V: GEOLOGIC TIMESCALE.)

Paleocene-Eocene thermal maximum (PETM, Initial Eocene Thermal Maximum, IETM) A CLIMATIC OPTIMUM that occurred approximately 55 million years ago and lasted for less than 100,000 years, at the boundary between the PALEOCENE and EOCENE epochs. At the onset of the climatic optimum, the global mean TEMPERATURE rose rapidly by between 9°F and 14°F (5°C–8°C) and the oceans warmed throughout their depths, not simply at the surface. A second brief temperature rise occurred about 2 million years later.

Scientists are uncertain what triggered the rise in temperature. Some suggested that it followed the release into the atmosphere of 1,500–3,300 billion tons (1,400–2,800 billion tonnes) of METHANE, which triggered the warming through a GREENHOUSE EFFECT. The methane had been held in methane hydrates in sedimentary rocks on the ocean floor. Scientists have proposed various mechanisms by which the methane hydrates might have been induced to release their methane. The most likely cause was astronomical and linked to the cyclical change in the ECCENTRICITY of the Earth's orbit (*see* MILANKOVITCH CYCLES). An alternative explanation is that a surge in volcanic activity released large amounts of CARBON DIOXIDE into the air.

Paleogene The period of geologic time that began 65.5 million years ago and ended 23.03 million years ago. It was the first period of the CENOZOIC ERA.

The climate was warm everywhere throughout the Paleogene. The Indian continental shelf collided with the Asian continental shelf prior to the start of the Paleogene and the Himalayan mountain ranges were already rising, but a land bridge became established for the first time during the EOCENE (*see* PLATE TECTONICS). Shallow seas separated the continents allowing different groups of mammals and birds to evolve on each continent. (*See* APPENDIX V: GEOLOGIC TIMESCALE.)

Paleoproterozoic The first era of the PROTEROZOIC eon of the Earth's history, which began 2,500 million years ago and ended 1,600 million years ago. Complex single-celled organisms appeared during the Paleoproterozoic, including bacteria and cyanobacteria, which formed mats called stromatolites.

These organisms performed PHOTOSYNTHESIS, releasing oxygen as a by-product, and in 2005 Robert E. Kopp and his colleagues at the Division of Geological and Planetary Sciences of the California Institute of Technology suggested that the release of OXYGEN into the atmosphere had dramatic consequences. During the early Paleoproterozoic abundant atmospheric METHANE sustained a GREENHOUSE EFFECT that maintained global TEMPERATURES similar to those of today. The release of oxygen led to the oxidation of the methane, the consequent breakdown of the greenhouse effect, and a rapid drop in temperature. There were three GLACIAL PERIODS between 2,450 million years ago and 2,220 million years ago. The final glacial period, from 2,300 million years ago until 2,200 million years ago was global in extent. Known as the Makganeyene glacial from the South African site where evidence for it was found, it produced a SNOWBALL EARTH. (*See* APPENDIX V: GEOLOGIC TIMESCALE.)

Paleozoic The first era of the PHANEROZOIC eon of the Earth's history, which began 542 million years ago and ended 251 million years ago. By the end of the Paleozoic life had established itself on land, vascular plants had appeared, and most invertebrate groups, including insects, as well as fish, amphibians, and reptiles had evolved.

The supercontinent of Pangaea formed during the Paleozoic. There were TWO GLACIAL PERIODS. Earth rotated faster than it does now, so the days were shorter, and the TIDES were bigger, because the Moon was closer.

At first the climate was mild, then it became warmer, but the continental shelves, where most organisms lived, grew steadily cooler. By the ORDOVICIAN period most of western Gondwana (what are now Africa and South America) lay at the South Pole. Baltica (Russia and northern Europe) and Laurentia (eastern North America and Greenland) lay in the TROPICS, and Australia and China were in temperate latitudes. The Ordovician ended with a GLACIAL PERIOD. The middle part of the Paleozoic was a time of climatic stability. During the late Paleozoic atmospheric OXYGEN and CARBON DIOXIDE levels fluctuated, and there was one and possibly two glacial periods during the CARBONIFEROUS. The climate deteriorated and the Paleozoic ended with a mass extinction of living organisms at the end of the PERMIAN. (*See* APPENDIX V: GEOLOGIC TIMESCALE.)

parameterization The physical laws governing many processes are well known and can be applied in a general way. If the equations describing such processes are then stored as programming subroutines, they can be called on as required during the construction of a computer MODEL. The use of subroutines in this way is called parameterization.

CLIMATE MODELS impose a grid over the world, typically with grid points 78–155 miles (125–250 km) apart horizontally, and with 10–30 levels 650–1,300 feet (200–400 m) apart vertically. Running the model produces "snapshots" of atmospheric conditions at 30-minute intervals.

Many atmospheric processes take place on a smaller scale than the model grid. These processes include CONVECTION, CLOUD FORMATION, and the transport of heat through the PLANETARY BOUNDARY LAYER. Their development cannot be calculated directly, because the grid is too coarse to accommodate them. Consequently, they can be included in the model only by making the general assumptions stored as subroutines. This is parameterization.

Further Reading

Houghton, J. T., Y. Ding, D. J. Griggs, M. Noguer, P. J. van der Linden. X. Dai, K. Maskell, and C. A. Johnson (eds.). "Modelling and Projection of Anthropogenic Climate Change," in *Climate Change 2001: The Scientific Basis.* Cambridge: Cambridge University Press with the Intergovernmental Panel on Climate Change, 2001, p. 94.

parcel of air (air parcel) A volume of air that can be considered separately from the air surrounding it and from which it is assumed to be physically isolated. The concept is theoretical, since no volume of air can be completely isolated from the air around it, but it is useful in calculating the behavior of the atmosphere. It is used in calculating adiabatic (*see* ADIABAT) changes and in calculations involving the equation of state.

The rate at which the volume of a given mass of air changes is called its mass flux. Air is very elastic, which means that a parcel of air expands, is compressed, and its shape is deformed when it moves against bodies that are denser or less dense than it is. Nevertheless, the total amount of gas (its mass) remains constant. Consequently, when the parcel of air expands, contracts, or is deformed some of its mass flows out of or into the original volume. The rate at which this happens is the mass flux.

An atmospheric cell is a parcel of air inside which the air is moving. In the GENERAL CIRCULATION of the atmosphere, Hadley cells, Ferrel cells, and polar cells are atmospheric cells.

Convective inhibition occurs when the rise of a parcel of air by CONVECTION is checked by an INVERSION. In order to reach its level of free convection (*see* STABILITY OF AIR) the parcel of air must possess sufficient energy to overcome the inhibition.

The TEMPERATURE a parcel of air would have if it were decompressed at the saturated adiabatic LAPSE RATE almost to zero pressure and then recompressed at the dry adiabatic lapse rate to 1000 mb (sea-level pressure) is called its equivalent potential temperature. As the saturated air is decompressed, its temperature falls and its WATER VAPOR condenses. This releases LATENT HEAT of CONDENSATION, warming the air and the warmth is retained, so the equivalent potential temperature is higher than the actual temperature of saturated air at the 1000-mb level. The difference can be estimated from:

$$\theta_e - \theta_w = 2.5 q_s$$

where θ_e is the equivalent potential temperature, θ_w is the temperature of saturated air at the 1000-mb level, and q_s is the specific HUMIDITY of the saturated air at the 1000-mb level expressed in grams of water vapor per kilogram of air including the water vapor.

The temperature a parcel of air would have if all the water vapor it contained were condensed from it, the latent heat of condensation were allowed to warm the air, and the pressure remained constant is called its equivalent temperature, also known as its isobaric equivalent temperature. "Equivalent temperature" is also sometimes used as a synonym of equivalent potential temperature.

passive instrument An instrument that measures radiation falling on it (so it is passive), rather than sending out a signal that returns to it.

past weather   The weather that has prevailed over the period, usually of six hours, since a weather station last submitted a report. The past weather is reported as a series of numbers between 0 and 9.

0   Cloud covered half of the sky or less throughout the period
1   Cloud covered more than half the sky for part of the period and less than half for the remainder of the period
2   Cloud covered more than half the sky throughout the period
3   Sandstorm, dust storm, or blowing snow
4   Visibility was less than 0.6 mile (1 km) due to fog, ice fog, or haze
5   Drizzle
6   Rain
7   Snow or a mixture of rain and snow (in Britain, called sleet)
8   Showers
9   Thunderstorms, with or without precipitation

path length   The distance that incoming solar radiation must travel between the top of the atmosphere and the land or sea surface. The path length has a value of 1 when the Sun is directly overhead, or 90° above the horizon, and it increases as the angle of the Sun decreases. The path length can therefore be calculated as 1/cosA, where A is equal to 90° minus the angle of the Sun. If the Sun is at 45°, the path length is 1.4, and if the Sun is at 30°, the path length is 2 (90 - 30 = 60; 1/cos60 = 2). Solar radiation travels twice as far through the air when the Sun is 30° above the horizon than it does when the Sun is directly overhead.

pentad   A period of five days. Meteorologists and climatologists often use the pentad in preference to the week, because there is an exact number of pentads in a 365-day year (73).

penumbra   The less deeply shaded area that lies near the edge of a shadow. The central, totally shaded area is called the umbra. Although they refer to any shadow, the terms are most often applied to ECLIPSES of the Moon or Sun and to SUNSPOTS.

If the source of light is a point, the shadow is sharply defined and there is no penumbra. If the source covers a large area, such as a substantial portion of

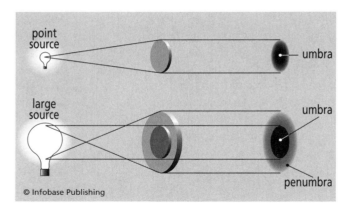

The dark center of a shadow is called the umbra. An area adjacent to the umbra that is only partly shaded is called the penumbra.

the sky, then some of the light will illuminate the area around the umbra, producing a penumbra.

perihelion   The point in the eccentric solar orbit (*see* ECCENTRICITY) of a planet or other body when it is at its closest to the Sun. At present, Earth is at perihelion on about January 4 each year, but the dates of APHELION and perihelion change over a cycle of about 21,000 years (*see* MILANKOVITCH CYCLES and PRECESSION OF THE EQUINOXES). The Earth receives 7 percent less solar radiation at aphelion than it does at perihelion.

permafrost (**pergelisol**)   Ground that remains frozen throughout the year. In order for permafrost to form, the TEMPERATURE must remain below freezing for at least two consecutive winters and the whole of the intervening summer. Permafrost covers approximate-

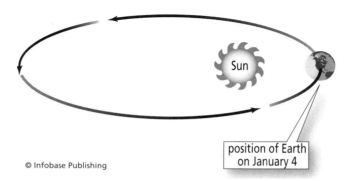

Perihelion is the point in its elliptical orbit at which a body is closest to the Sun.

ly 26 percent of the land area of the Earth, including more than half of the land area of Canada.

The depth of the permafrost layer varies. In Canada it is about 6.5 feet (2 m) thick along the southern edge of the permafrost zone and it is up to 1,000 feet (300 m) thick in the far north. On the North Slope of Alaska the permafrost is 2,000 feet (600 m) thick in places and in parts of Siberia it is up to 4,600 feet (1,400 m) thick.

The soil that lies above a layer of permafrost and that thaws during the summer and freezes again in winter is called the active layer. A layer of frozen ground, or ground containing frozen water, that lies at the base of the active layer in a permafrost region is known as

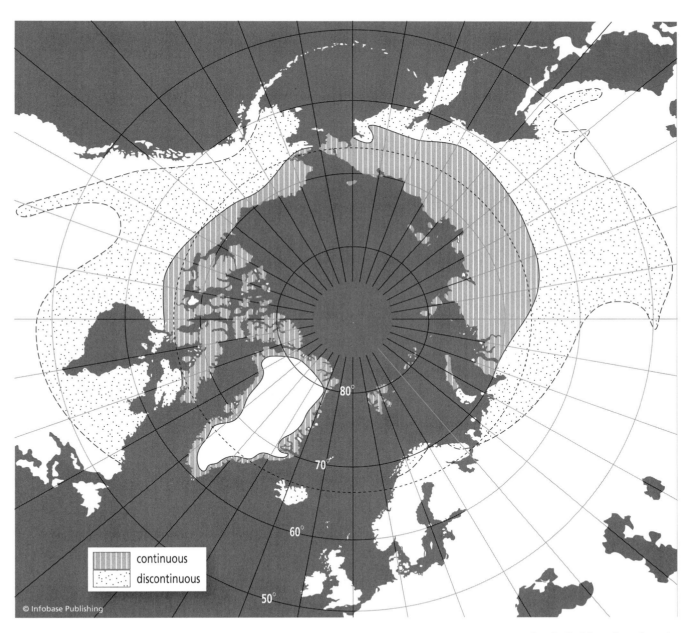

continuous
discontinuous

© Infobase Publishing

Together, the regions of continuous and discontinuous permafrost in the Northern Hemisphere cover more than half of Canada and much of Siberia.

the tjaele, or frost table. The tjaele moves downward during the summer thaw in the active layer.

Soil is a poor conductor of heat, and changes in the average air temperature take many years to penetrate. After about 100 years the change will affect material about 500 feet (150 m) below the surface, and it will be 1,000 years before it is felt at 1,640 feet (500 m). Material deep below the surface is so well insulated from temperature changes above the surface that scientists working out past climatic conditions sometimes use it. They take long, vertical cores of soil from the ground and measure the temperature at intervals along them. Some of the Canadian, Alaskan, and Siberian permafrost areas have remained frozen for several thousand years, since the end of the most recent (Wisconsinian or Devensian) GLACIAL PERIOD.

In the northern part of the Arctic, adjacent to the Arctic Circle and to the north of it, the permafrost is continuous. This means that all of the ground is frozen all of the time. Farther south, and covering a much larger land area, the permafrost is discontinuous. It occurs in patches and is found where the average temperature locally is lower than that of the surrounding area, on north-facing slopes that are never exposed to direct sunshine, and on land that is wet for most of the time.

Except for the northernmost tip of the Antarctic Peninsula, all of Antarctica lies within the Antarctic Circle. Ground that is not covered by the ICE SHEET is permanently frozen. Apart from the continent and its offshore islands, there are no other regions of permafrost in the Southern Hemisphere, because there is no large expanse of land in a sufficiently high latitude.

Even the southern islands, such as the Falkland Islands (Las Malvinas), have no permafrost.

Permafrost may include areas that remain unfrozen. These are called talik. Where summer temperatures remain above freezing for several weeks or months the upper part of the soil will thaw. This produces a thin active layer in which a few plants can grow. Over the years as plants die and shed leaves, plant material accumulates to form a mat of dead vegetation that insulates the ground below, preventing the summer warmth from penetrating. The existence of an active layer therefore tends to perpetuate the permafrost layer beneath it.

A permafrost layer impedes drainage and, because it is hard as rock, it presents an impenetrable barrier to plant roots.

permeability (hydraulic conductivity) The ability of a soil, sediment, or rock to allow water or air to pass through it. Permeability is measured as the volume of water that flows through a unit cross-sectional area in a specified time. Soil permeability is categorized as slow, moderate, or rapid according to the rate of flow and is reported as the distance traveled in one hour. If the water is moving vertically downward through the soil, the process is called percolation. Percolation is reported as the time taken for the water to move a specified distance. The classification system comprises seven classes for both permeability and percolation.

Permeability and the percolation rate depend on the structure of the material through which air or water moves. Sand does not hold water well. Pour water onto a sandy beach on a fine day and the sand is dry again

Permeability and percolation classes

| Class | Permeability | | Percolation | |
	inches per hour	cm	minutes per inch	per cm
1. very slow	less than 0.05	less than 0.1	more than 1200	more than 470
2. slow	0.05–0.20	0.1–0.5	300–1200	118–470
3. moderately slow	0.20–0.80	0.5–2.0	75–300	30–118
4. moderate	0.80–2.50	2.0–6.4	24–75	9–30
5. moderately rapid	2.50–5.00	6.4–12.7	12–24	5–9
6. rapid	5.00–10.00	12.7–25.4	6–12	2–5
7. very rapid	more than 10.00	more than 25.4	less than 6	less than 2

within at most a few minutes. Clay becomes sodden, because water does not pass through it easily. The difference is due to the size and shape of the particles from which the soil is made.

The spaces between soil particles are called pores, and if all the particles are spherical and the same size, the pore space will amount to about 45 percent of the total volume filled by the particles. This proportion is the same whatever the size of the particles themselves. In a soil, however, the particles are not all of the same size or shape.

If the particles are large, the pore spaces between them will also be large. This will allow water and air to move through the soil rapidly. Gravel and sand do not retain water for very long because they consist of large particles. Sand grains have a granular texture and gravel consists of single stones, or grains. Water permeates and percolates rapidly through them.

If the particles are very small, the amount of pore space is the same, but the individual pores are small. Water tends to be held to the surfaces of the grains by molecular attraction and the distance between one particle surface and the next is so small that there is little room for water to flow past the water adhering to the particles. Some very small particles tend to pack together to form a solid mass. This squeezes out the pore

spaces and produces a massive texture. Clay particles pack together, but in layers. The particles themselves are flat, like flakes, but microscopically small. They arrange themselves with their flat sides adjacent. This produces a platy structure that prevents water passing. Water permeates and percolates slowly through soils of these types.

Between the two extremes there are soils made from particles that arrange themselves into prismatic columns. Water is able to move between the columns. Other soils consist of big, irregular particles. These form a blocky structure with channels along which water can move. Water permeates and percolates moderately quickly through soils of these types.

Permeability affects the way soils respond to the weather. Areas where the permeability is rapid will not be susceptible to flooding, because water leaves them quickly. Dry weather can parch the ground just as quickly, however, so there is a relatively high risk of DROUGHT. Where permeability is slow, drought is less of a risk, but flooding is more likely. Also, soils that retain water tend to be cold in early spring, because of the high HEAT CAPACITY of water. This can delay the sowing of spring crops.

Further Reading

Allaby, Michael. *Temperate Forests*. Rev. ed. Ecosystem. New York: Facts On File, 2007.

Ashman, M. R., and G. Puri. *Essential Soil Science*. Oxford: Blackwell Science, 2002.

Permian The final period of the PALEOZOIC era of the Earth's history, which began 299 million years ago and ended 251 million years ago. The Permian was named in 1841 by the British geologist Roderick Impey Murchison (1792–1871) after the ancient kingdom of Permia and the city of Perm in the Ural Mountains of Russia, where he found vast deposits of rocks characteristic of the period.

During the Permian all the continents moved together, finally joining to form the supercontinent Pangaea, which was surrounded by the world ocean, Panthalassa. A salty inland sea, called the Zechstein Sea, covered part of Europe, and much of southern and central Europe lay beneath a large bay in the Panthalassa Ocean, called the Tethys Sea.

At the start of the Permian, the Earth was gripped by a GLACIAL PERIOD. ICE SHEETS covered both polar

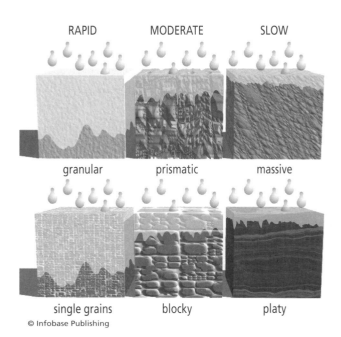

RAPID   MODERATE   SLOW

granular   prismatic   massive

single grains   blocky   platy

© Infobase Publishing

The rate at which water flows through a soil depends on the size and arrangement of the soil particles.

regions and there were glaciers in many parts of Gondwana, the southern continent. Swamp forests extended over much of the Tropics. By the middle of the Permian, the climate was becoming warmer. The ice sheets and glaciers retreated, and the interiors of the continents became drier. Once the continents had merged, the interior of Pangaea probably had a very dry climate, with marked wet and dry seasons. As the climate became generally drier the swamp forests disappeared.

The predominant plants were seed ferns and ferns, together with coniferous trees and ginkgos. The seas teemed with life, including fishes. On land there were ancestors of the reptiles, some resembling crocodiles and turtles, and the ancestors of the mammals. A mass extinction marked the end of the Permian. *See* Appendix V: Geologic Timescale.

Phanerozoic The present eon of the Earth's history, which began 542 million years ago and continues to the present day. The name is derived from the Greek words *phaneros* meaning "visible" and *zoion* meaning "animal." The Phanerozoic is the time of "visible life," although paleontologists now know that living organisms existed long before the start of the Cambrian period, which marks the base of the Phanerozoic. *See* Appendix V: Geologic Timescale.

phase The word *phase* has two meanings. The first describes a part of a system that is of the same composition throughout and that is clearly distinct from all other parts of the same system. A block of ice is a system consisting of one phase (solid), as is a mass of liquid water (liquid) or water vapor (gas). A mixture of water and ice comprises a two-phase system and a mixture of ice, liquid water, and water vapor comprises a three-phase system. A solution of salt or sugar in water is a single-phase system, because the solution is entirely liquid.

When water changes between gas, liquid, and solid, it is said to change phase. At a pressure of 611.2 Pa (6.112 mb) and a temperature of 32°F (0°C) water can exist as gas, liquid, and solid simultaneously. This is known as its triple point (*see* boiling) and the triple point of water forms the basis of the definition of the kelvin (0°C = 273.16K; 0K = -273.16°C; *see* temperature scales and units of measurement).

The stage that a regularly repeating motion has reached is also called a phase. Usually the phase of a periodic motion (*see* wave characteristics) is

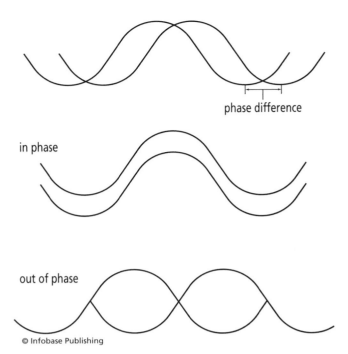

© Infobase Publishing

The distance between the troughs and peaks of two waves of the same wavelength is the phase difference between them. The waves are in phase when their peaks and troughs coincide and out of phase when the peaks of one coincide with the troughs of the other.

described by comparison with another motion having the same wavelength. If the peaks and troughs of two or more waves coincide, the waves are said to be in phase. If the peaks of one wave coincide with the troughs of the other, the two are said to be out of phase. If the peaks and troughs are between the two extremes of being in phase or out of phase, the distance between them is called the phase difference.

phenology The scientific study of periodic events in the lives of plants and animals that are related to the climate. Such events include the dates on which deciduous trees come into leaf, crops germinate, flower, ripen, and are harvested, and the dates of arrival and departure of migratory birds.

Phenological events comprise all the familiar signs of the changing seasons. The dates of particular events at places some distance apart can be used to compile a phenological gradient marking the geographic movement of the seasons across a continent or, on a much smaller scale, to measure the effects of such influences as exposure or shade within a garden.

Phenological data collected in Europe over a period of 30 years from the INTERNATIONAL PHENOLOGICAL GARDENS has shown that during this time spring events have advanced to occur six days earlier and autumn events now occur 4.8 days later. The dates of European grape harvests have been of particular value to historians tracing the history of climate and was used extensively by the French historian Emmanuel Le Roy Ladurie.

Further Reading

Ladurie, Emmanuel Le Roy. *Times of Feast, Times of Famine: A History of Climate Since the Year 1000.* New York: Doubleday, 1971.

photochemical smog A form of AIR POLLUTION that occurs in strong sunlight when ULTRAVIOLET RADIATION acts upon hydrocarbon compounds emitted by vehicle exhausts. It is quite different from SMOG of the London type.

The branch of chemistry that studies the chemical effects of electromagnetic radiation, including visible light, is called photochemistry. Many chemical reactions take place only when radiation supplies the energy that is needed to drive them. Photochemical smog is the product of a series of reactions that take place

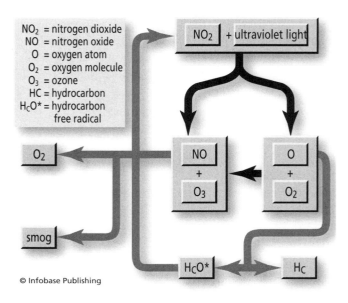

NO₂ = nitrogen dioxide
NO = nitrogen oxide
O = oxygen atom
O₂ = oxygen molecule
O₃ = ozone
HC = hydrocarbon
HcO* = hydrocarbon free radical

© Infobase Publishing

Pollution occurs when some of the atomic oxygen produced by the photolytic cycle combines with unburned hydrocarbons to form hydrocarbon free radicals.

in strong sunlight. OZONE is formed in the stratosphere (*see* ATMOSPHERIC STRUCTURE) by the action of ULTRAVIOLET RADIATION on oxygen (*see* PHOTODISSOCIATION). These are photochemical processes, but the best known and most important photochemical reactions are those in PHOTOSYNTHESIS.

Photochemical smog contains compounds such as aldehydes (compounds containing the -COOH group joined directly to another CARBON atom), ketones (compounds containing the C.CO.C group), and formaldehyde (or methanal, HCHO), which impart a characteristic odor, and nitrogen dioxide (NO_2) and solid particles that cause a brownish haze. Ozone (O_3), aldehydes, and peroxyacetyl nitrates (PAN) cause irritation to the eyes and throats of persons exposed to smog, and plants are damaged by NITROGEN OXIDES (NO_x), O_3, PAN, and ethene ($CH_2=CH_2$, also called ethylene).

The formation of photochemical smog begins with the photolytic cycle, in which NO_2 is broken down and reformed. Atomic oxygen (O) is produced in the first stage of the photolytic cycle. This is highly reactive and will oxidize hydrocarbons (H_c) to hydrocarbon free radicals (H_cO^*). Free radicals are atoms or molecules that have unpaired electrons as a result of which they are extremely reactive. These react further to reform NO_2, allowing the cycle to continue, and to yield O_2 and the hydrocarbon ingredients of smog. Using oxygen from the photolytic cycle, the reactions are:

$$O + H_c \rightarrow H_cO^* \qquad (1)$$

$$H_cO^* + O_2 \rightarrow H_cO_3^* \qquad (2)$$

$$H_cO_3^* + NO \rightarrow H_cO_2^* + NO_2 \qquad (3)$$

$$H_cO_3^* + H_c \rightarrow \text{aldehydes, ketones, etc.} \qquad (4)$$

$$H_cO_3^* + O_2 \rightarrow O_3 + H_cO_2^* \qquad (5)$$

$$H_cO_x^* + NO_2 \rightarrow PAN \qquad (6)$$

photodissociation The splitting of a molecule into smaller molecules or single atoms using light as a source of energy. Photodissociation occurs when an atom or molecule absorbs a photon, which is a unit (quantum) of electromagnetic radiation, possessing precisely the energy needed to allow it to break free from the group to which it is attached.

In the upper stratosphere (*see* ATMOSPHERIC STRUCTURE) oxygen molecules (O_2) absorb solar ULTRAVIOLET

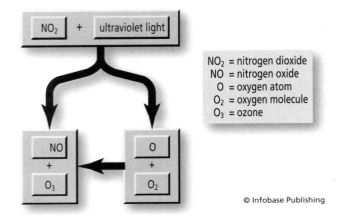

Ultraviolet light splits nitrogen dioxide into nitrogen oxide and atomic oxygen. Oxygen atoms then combine with oxygen molecules to form ozone.

RADIATION (UV) at wavelengths (*see* WAVE CHARACTERISTICS) below about 0.3 μm and are photodissociated into single oxygen atoms (O + O). UV radiation at less than 0.23 μm causes the photodissociation of chlorofluorocarbon (CFC) compounds in the stratosphere. In the troposphere, UV radiation at 0.37–0.42 μm photodissociates nitrogen dioxide (NO_2) into nitrogen oxide (NO) and oxygen (O). This photodissociation is followed by the reformation of NO_2 in a process known as the photolytic cycle.

The chemical reactions that comprise the photolytic cycle are:

$$NO_2 + UV \rightarrow NO + O \qquad (1)$$

$$O + O_2 \rightarrow O_3 \qquad (2)$$

$$O_3 + NO \rightarrow NO_2 + O_2 \qquad (3)$$

If hydrocarbons from vehicle exhausts are also present in the air, the natural cycle is disrupted and a range of other compounds are produced, causing PHOTOCHEMICAL SMOG.

photoperiod The number of hours of daylight that occur during a 24-hour period. Except at the equator, this varies with the season. Many plants respond physiologically to changes in the photoperiod. Responding to changes in the photoperiod is called photoperiodism.

The most common responses are those affecting the time of flowering. Some plants will not flower if the daily cycle includes long periods of darkness. These are known as long-day plants (but are really short-night plants), and they flower in late spring and early summer, when the days are lengthening and the nights are growing shorter. Lettuce, wheat, and barley are long-day plants. Strawberries and chrysanthemums are among the plants that flower between late summer and early spring, when days are short and nights are long. They are called short-day plants.

Not all plants are affected by the photoperiod. Tomatoes and cucumbers are among the plants in which the time of flowering is not determined by the duration of daylight. These are called day-neutral plants.

photosynthesis The series of chemical reactions by which green plants (as well as certain bacteria and cyanobacteria) synthesize (construct) sugars, using CARBON DIOXIDE and WATER as the raw materials. The first stage in the process depends on light as a source of energy and is called the light-dependent stage or light stage. The second stage also takes place in light, but it does not use light energy, and so it is called the light-independent stage or dark stage.

The overall set of reactions can be summarized as:

$$6CO_2 + 6H_2O + [\text{light energy}] \rightarrow C_6H_{12}O_6 + 6O_2 \uparrow$$

The OXYGEN is released into the air, indicated by the arrow pointing upward. $C_6H_{12}O_6$ is a simple sugar from which more complex sugars and starches can be made.

The process begins with chlorophyll, the green pigment that is held in bodies called chloroplasts in the cells of plant leaves and some stems. Chlorophyll is a very complex substance with the property of absorbing light energy. When a photon of light possessing exactly the right amount of energy strikes a chlorophyll molecule, one electron in the chlorophyll molecule absorbs that energy and becomes excited. It jumps from its ground state to its excited state and escapes from the molecule, but is immediately captured by a neighboring molecule. An electron is then passed from molecule to molecule along an electron-transport chain until it is used to split water (*see* PHOTODISSOCIATION) into HYDROGEN and oxygen:

$$H_2O \rightarrow H^+ + OH^-$$

Having lost an electron, the chlorophyll molecule carries a positive charge. The hydroxyl ion (OH) produced by the photodissociation of water carries a negative charge in the form of an extra electron. The hydroxyl ion passes this electron to the chlorophyll. Both hydroxyl and chlorophyll are then neutral and hydroxyls combine to form water:

$$4OH \rightarrow 2H_2O + O_2 \uparrow$$

The oxygen is released into the air. Free hydrogen atoms attach themselves to molecules of nicotinamide adenine dinucleotide (NADP), converting it to NADPH. This completes the light-dependent stage.

NADP loses its hydrogen again during the light-independent stage, and the NADP then returns to the light-dependent stage, ready to be used again.

The light-independent stage begins when a molecule of carbon dioxide (CO_2) becomes attached to a molecule of ribulose biphosphate (RuBP) in the presence of an enzyme, RuBP carboxylase (usually called rubisco for short). The captured carbon then enters a sequence of reactions that end with the construction of sugar molecules and also of the RuBP with which the process began. Because the RuBP is reconstructed so it can be used again, the sequence of reactions forms a cycle. It is known as the Calvin cycle, because its details

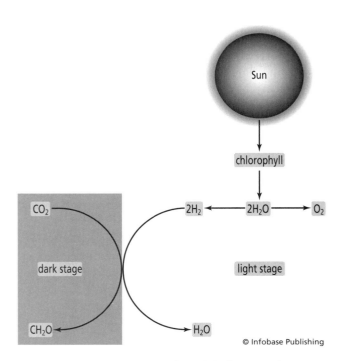

The series of chemical reactions by which plants use the energy of sunlight to manufacture sugars has two stages. Water is broken down into hydrogen and oxygen in the light-dependent stage, and in the light-independent stage carbon from carbon dioxide is used to make sugar.

were discovered by the American biochemist Melvin Calvin (1911–97). Calvin was awarded the 1961 Nobel Prize in chemistry for this work.

There are variations on the photosynthetic pathways. If the first product of the light-independent stage, made when CO_2 joins RuBP, is 3-phosphoglycerate, the pathway is called C3, because 3-phosphoglycerate has three carbon atoms. Most plants, including all trees, are C3 plants.

Other plants use the C4 pathway, in which CO_2 combines with phosphoenol pyruvic acid (PEP) rather than RuBP, and the first product is oxaloacetic acid, which has four carbon atoms. This pathway uses more energy than the C3 pathway, but it produces more sugar for a given leaf area, making C4 plants grow faster than C3 plants. C4 plants can also tolerate higher light intensities and lower CO_2 concentrations than C3 plants. Most C4 plants are either grasses or plants that grow in desert. The C4 grasses include sugarcane and corn (maize).

Some desert plants, including cacti and the pineapple, have evolved a third pathway. Plants exchange gases through their STOMATA, but if the stomata are open in bright sunshine, when the rate of EVAPORATION is high, water vapor also passes through them. To minimize the loss of water, these plants keep their stomata closed during the day and open them for gas exchange at night. CO_2 enters at night, combines with PEP to produce oxaloacetic acid, and this is then converted into malic acid ($C_4H_6O_5$, also called hydroxysuccinic acid). The carbon is stored as malic acid until the next day when it is broken down, releasing CO_2, which enters the Calvin cycle.

This method of photosynthesis was first observed in plants belonging to the family Crassulaceae, and it is known as the crassulacean acid metabolism (CAM). The Crassulaceae are succulent herbs and small shrubs that are found mainly in warm, dry regions. The family includes the stonecrops and houseleeks.

Physical Oceanography Distributed Active Archive Center (PO.DAAC) The branch of the Data Information System of the EARTH OBSERVING SYSTEM that is responsible for storing and distributing information about the physical state of the ocean. Most of the data held at the PO.DAAC was obtained from satellites. It is technical and intended for research and educational use. The data is available to anyone free of charge, but must not then be sold. The center is part of NASA and is located at the Jet Propulsion Laboratory, California Institute of Technology.

pilot report   A description of the current weather conditions that is radioed to air traffic control or to a meteorological center by the pilot or other crew member of an aircraft. The report may consist of nothing more than the height of the CLOUD BASE or cloud top (*see* FLYING CONDITIONS) or the presence of CLEAR AIR TURBULENCE. A full pilot report should contain, in this order: the extent or location of the reported conditions; the time they were observed; a description of the conditions; the altitude of the conditions; and in the case of a report of clear air turbulence or icing, the type of aircraft.

Pine Island Glacier   An outlet glacier in the West Antarctic Ice Sheet that is one of the largest outlet glaciers in Antarctica and the glacier with the greatest rate of discharge. The center of the glacier (called the trunk) moves at about 650 feet a year (200 m/yr). The Pine Island Glacier drains 82.5 million tons (75 million tonnes) of ice from the ICE SHEET each year. Between

1992 and 1996 the glacier retreated inland by 3 miles (5 km), and between 1992 and 1999 the interior of the region it drains grew thinner at a rate of 39 inches (100 cm) a year. Near the grounding line, marking the boundary between that part of the glacier which rests on solid rock and that which floats on the sea, the glacier thinned by approximately 5 feet (1.6 m) a year. If the Pine Island Glacier should continue to lose ice at this rate, the glacier could be completely afloat within 600 years and a substantial part of the West Antarctic Ice Sheet might disappear within the next few centuries.

Scientists believe the retreat and thinning of the Pine Island Glacier is caused by internal changes within the glacier itself. It is losing mass at a rate that cannot be explained by present climate change, and the retreat and thinning may be the result of climate changes in the distant past that are only now affecting the glacier and its drainage basin. Tributary glaciers converge about 110 miles (175 km) inland and feed ice into the Pine Island Glacier. The trunk, which is losing about 0.4 percent of its mass a year, sits in a trough about 1,650 feet (500 m) deeper than the bed beneath some of the tributaries. The thinning affects mainly the trunk of the main glacier and not the tributaries.

pitot tube (**pitot-static tube**)   A device that is used to sample air in order to measure the speed with which the air is moving (or the speed with which the pitot tube is moving in relation to the air surrounding it) and the atmospheric pressure. Pitot tubes are used at weather stations, but their most widespread use is on aircraft. Every aircraft carries a pitot tube to supply air to its ALTIMETER, airspeed indicator, and vertical speed indicator (which shows the rate at which the aircraft is climbing or descending). The device was invented by the French physicist Henri Pitot (1695–1771).

A pitot tube comprises two thin-walled tubes, one enclosing the other, and mounted so that the ends of the tubes face into the airstream. Strictly, it is only the tube used to measure the speed of the airflow that should be called a pitot tube. The second tube is a static tube. The end of the pitot tube that faces into the airstream is open. The end of the static tube is closed, but there is a belt of small holes around its circumference. These are located a distance from the forward end of the tube that is equal to not less than five times the diameter of the tube. Usually, the pitot tube is housed inside the static tube.

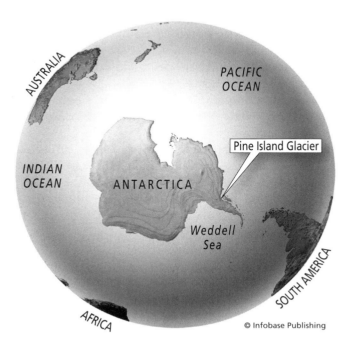

Pine Island glacier is one of the largest outlet glaciers in Antarctica. It is retreating rapidly, but not because of the present global warming.

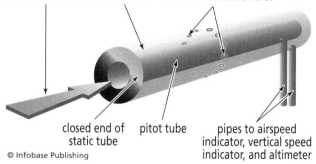

The outer, static tube encloses the inner, pitot tube. Air from both tubes is fed to the instruments that indicate wind speed or, in an aircraft, airspeed, altitude, and rate of climb or descent (vertical speed).

Air enters the pitot tube through the open end and enters the static tube through the belt of small holes. Pipes conduct the sampled air from both tubes to the instruments.

When air enters the pitot tube, it is brought to rest, exerting a pressure of ½ ρv^2, where ρ is the density of the air and v is its speed. The air pressure in the static tube is exactly equal to the external atmospheric pressure.

The WIND SPEED or airspeed is calculated from the difference in pressure between the two tubes. In an aircraft this is shown on the airspeed indicator. Changes in the pressure in the static tube indicate that the tube is changing its altitude and are used to calculate the vertical speed, which is shown on the vertical speed indicator. Comparing the pressure in the static tube with the sea-level pressure, recorded in the instrument prior to takeoff, gives the height above sea level and is shown as altitude on the altimeter.

Planck's law A description of the relationship between the wavelength of electromagnetic radiation and the TEMPERATURE of the body emitting it that was first stated in 1900 by the German physicist Max Planck (1858–1947). The law states that the intensity of radiation emitted at a given wavelength is determined by the temperature of the emitting body. This can be written as:

$$E_\lambda = c_1/[\lambda^5(\exp(c_2/\lambda T) - 1)]$$

where E_λ is the amount of energy (expressed in watts per square meter per micrometer of wavelength), λ is the wavelength (in µm), T is the temperature in kelvins,

c_1 is the first radiation constant, with the value 3.74 × 10^{16} W/m², and c_2 is the second radiation constant, with the value 1.44 × 10^{-2} m/K.

plane of the ecliptic An imaginary disk, the circumference of which is defined by the path the Earth follows in its ORBIT about the Sun. Each day of the year, the noonday Sun is at a slightly different position in the sky from the position it occupied on the preceding day. If its position is plotted for every day of the year on a picture of the landscape that shows a clear view in the direction of the equator, the varying positions of the Sun will appear to follow a path across the sky. This path marks the ecliptic, and the imaginary disk it encloses is the plane of the ecliptic.

If the rotational axis of the Earth were normal (at right angles) to the plane of the ecliptic, the position of the noonday Sun would be the same on every day of the year and, therefore, there would be no SEASONS. In fact, however, the axis is tilted. At present, its angle to the ecliptic is 66.55°, so it is tilted 23.45° from the normal (90 - 66.55 = 23.45), but this angle changes in the course of a cycle with a period of about 41,000 years (*see* MILANKOVITCH CYCLES). Latitudes 23.5° N and S mark the tropics of Cancer and Capricorn and latitudes 66.5° N and S mark the Arctic and Antarctic Circles.

An imaginary circle with a radius drawn from the center of the Earth, through the equator, and projected to the outermost limit of the universe (the celestial

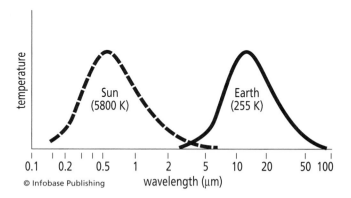

The graph shows the intensity of the radiation emitted at different wavelengths by a body at the temperature of the surface of the Sun compared with that emitted by a body at the temperature of the surface of the Earth. Obviously, the Sun emits far more radiation than does the Earth, but for ease of comparison the graph assumes the same amount for both bodies.

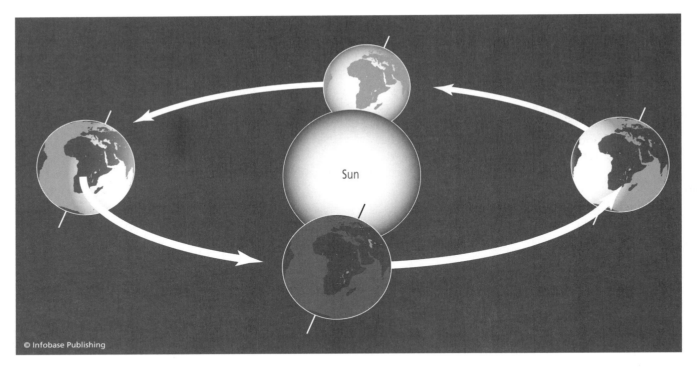

The plane of the ecliptic is an imaginary disk, the edge of which marks the Earth's orbital path around the Sun.

sphere) is called the celestial equator. Because of the Earth's AXIAL TILT, the celestial equator is inclined with respect to the plane of the ecliptic. The latitude of an astronomical body, such as the Sun, measured in relation to the celestial equator, is known as the declination. Declination is the angle north (above, designated +) or south (below, -) of the celestial equator. *See also* MAGNETIC DECLINATION.

planetary boundary layer (atmospheric boundary layer) The lowest part of the atmosphere, in which the movement of the air is strongly influenced by the land or sea surface. FRICTION between moving air and the land or sea surface causes EDDIES, making the flow of air turbulent (*see* TURBULENT FLOW). The depth of the planetary boundary layer varies from place to place and from time to time, but it is usually less than about 1,700 feet (519 m).

The lowest approximately 10 percent of the total depth of the planetary boundary layer—about 170 feet (52 m)—is called the friction layer, or surface boundary layer. Turbulent flow within the friction layer ensures thorough mixing of the air. Consequently, the characteristics of the air in this layer are fairly constant throughout.

The air above the planetary boundary layer, comprising about 95 percent of the total mass of the atmosphere, is called the free atmosphere. Air in the free atmosphere is not directly influenced by friction with the land or sea surface. The planetary boundary layer is sometimes called the shielding layer, referring to the fact that the boundary layer shields the surface from events in the free atmosphere.

The Ekman layer, discovered by the Swedish oceanographer and physicist Vagn Walfrid Ekman (1874–1954; *see* APPENDIX I: BIOGRAPHICAL ENTRIES) is the part of the planetary boundary layer in which the wind blows at an angle across the isobar (*see* ISO-). Within this layer the movement of air is balanced between the PRESSURE GRADIENT FORCE, CORIOLIS EFFECT, and friction. The balance is maintained by the air flowing inward toward a center of low pressure and outward from a center of high pressure. Wind accelerates as it moves toward a center of low pressure due to the conservation of its angular MOMENTUM as it spirals inward. For the same reason, it slows as it spirals outward from a center of high pressure. That is why winds are always stronger around a CYCLONE than they are around an ANTICYCLONE. *See also* UPWELLING.

plasma One of the four states of matter (the other three are gas, liquid, and solid) in which a gas is ionized (*see* ION). The KINETIC ENERGY of particles in a plasma exceeds the energy of attraction (POTENTIAL ENERGY) between particles that are close together. Electrons move rapidly among the particles, neutralizing any net charge, so that each charged particle is surrounded by a cloud of particles with an opposite charge and the electric forces within the plasma are low. These clouds overlap and consequently each particle is linked to many others. Particles rarely collide.

Plasmas occur naturally in the atmospheres of stars, including the Sun, and constant bombardment by the charged particles of the SOLAR WIND continuously creates a plasma in the region of space immediately surrounding Earth. The MAGNETOSPHERE consists entirely of plasma.

plate tectonics The theory that describes the surface of the Earth as a number of solid sections that are able to move in relation to one another, thereby causing the deformation of rocks and the production of new structures. The sections are called "plates," and "tectonics" (from the Greek *tektonikos*, meaning "carpenter") is a geologic term referring to rock structures resulting from deformation and the forces that produce them.

The theory of plate tectonics explains the presence of features that must have formed under conditions very different from those of today. Limestone rocks, for example, are abundant in many parts of the world. Limestone forms only by the heating and compression of sediments on the floor of a shallow sea. Consequently, limestone regions must once have been covered by sea. Coal measures can form only in mud beneath shallow coastal waters in the TROPICS. Areas where coal is found, such as the northern United States, northern Europe, Russia, and China, must once have lain close to the equator. Other places, such as parts of Devon in southwestern England, have rocks of a type that forms only in hot, dry deserts.

The theory developed slowly over a long period. Geographers had speculated about how it could be that the shape of the continent of Africa looks as though it would fit snugly against Central and South America, but assumed this was mere coincidence. Then, in 1879, Sir George Darwin (1845–1912, a son of Charles Darwin) suggested that the Moon might have formed by breaking away from the Earth. Geologists then believed that the mantle, beneath the solid rocks of the Earth's crust, is liquid, and in 1882 and 1889 the Rev. Osmond Fisher (1817–1914) proposed that the Pacific Ocean might fill a basin caused by the removal of the rocks that formed the Moon. Osmond thought the continents to either side of the Pacific might have moved together to close the gap, and that this movement might have caused a split that widened to form the Atlantic. The most comprehensive proposal for the motion of continents was made some years later, however, by Alfred Wegener (1880–1930; *see* APPENDIX I: BIOGRAPHICAL ENTRIES). But by then opinions had changed. Geologists believed the mantle to be solid and Wegener's idea was dismissed.

Support for Wegener's idea began to grow in the middle 1940s, when scientists first acquired the technological means to study the floor of the deep oceans. The most important development was the discovery that rocks to either side of the ridges running across all the major ocean basins formed distinct bands of reversed magnetic polarity. Scientists knew that from time to time the Earth's magnetic field reverses its polarity, so that north becomes south and south north. Mineral grains in molten rock align themselves with the magnetic field, and as the rock solidifies, their magnetic orientation becomes fixed. The magnetic bands in the rocks of the ocean floor matched each other on either side of the ridges. This suggested that over a long period new, molten rock had emerged from the ridges and solidified on either side, and that the ocean floor had move away from the ridges to accommodate the new rock. In 1963, an American naval oceanographer, Robert Sinclair Dietz (1914–95) called this "seafloor spreading." It was in 1967 that Professor Dan McKenzie (born 1942) of the University of Cambridge brought all the different strands of evidence together and proposed the theory of plate tectonics.

The theory proposes that the Earth's crust is formed from a number of major and minor plates. Some of the plates are relatively thin and made from dense rock. These form the floors of the oceans. Other plates are much thicker and made from rocks that are less dense. These form the continents, rising above the sea.

Geologists differ in the way they classify some of the plates, but most now consider that there are eight major plates: the African, Eurasian, Pacific, Indian, North American, South American, Antarctic, and Nazca plates.

The Cocos, Caribbean, Somali, Arabian, Philippine, and Scotia plates are smaller and are classed as lesser plates. In addition, there are minor plates such as the Juan de Fuca and Gorda plates, as well as microplates and fragments of former plates. The plates are made from solid rock, and they move because of convection currents in the hot rock of the mantle. In some places where the oceanic crust is thin there are hot spots, where VOLCANOES are especially active. Earthquakes and volcanism are also common in the vicinity of plate margins.

Plate margins are of several types. At divergent, or constructive, margins plates are moving apart and new crustal rock is being added. At convergent, or destructive, margins plates are moving toward each other. The denser oceanic rock sinks beneath the lighter continental rock, and dense rock that has already descended below the crust drags the rest of the plate behind it.

This process is called subduction. As the oceanic plate sinks, the sedimentary rock and loose sediment on the surface of the subsiding rock is scraped off, forming mountains along the coast of the continent. When two continental plates collide, both are of equal density and neither can sink below the other, so the rocks of both are crumpled upward to form mountains. The Himalayas are the most recent mountains to be produced in this way, by the collision between the Indian and Eurasian plates that began about 40 million years ago. The Indian plate is still moving northward, so the mountains are still rising, although they are eroding almost as quickly owing to the extreme weather conditions to which they are exposed. At other margins, the plates are moving parallel to each other but in opposite directions, a process that produces a type of rock fracture called transform faulting.

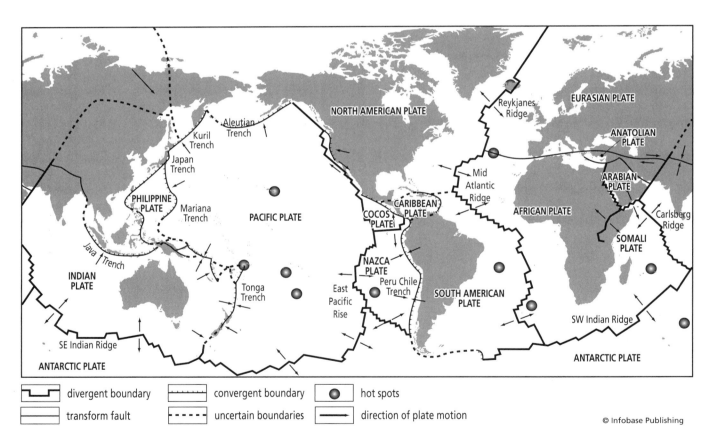

The major plates into which the Earth's crust is divided, with ridges at which plates are separating and new rock is being added, and trenches at which one plate is being subducted beneath its neighbor. At transform faults the plates are moving past one another in opposite directions.

Plate movements cause the continents to change their positions. About 200 million years ago, for example, there was just one "supercontinent" called Pangaea (from the Greek *pan*, meaning "all," and *gi*, meaning "Earth") and one ocean called Panthalassa (*thalassa* means "sea"). An arm of the sea, called Tethys, penetrated deeply into Pangaea, partly separating it into a northern part, called Laurasia, that contained all the present northern continents, and a southern part, called Gondwana, containing the southern continents. About 180 million years ago, Pangaea began to break apart. North America broke away first from Africa and about 150 million years ago from Europe.

Pleistocene The epoch of geologic time during which the most recent sequence of GLACIAL PERIODS occurred. The Pleistocene epoch began about 1.81 million years ago and ended about 11,000 years ago, which is when the Wisconsinian glacial ended. The Pleistocene epoch was followed by the HOLOCENE EPOCH, which continues to the present day. Together these two epochs comprise the PLEISTOGENE period. Technically, the start of the Pleistocene is defined by a change in the Earth's magnetic polarity and the extinction of certain microscopic marine organisms and first appearance of others. The end is defined as exactly 10,000 radiocarbon years ago (*see* RADIOCARBON DATING), which is equivalent to about 11,000 calendar years ago.

When the Pleistocene was first identified, its climate was thought to have been uniformly cold throughout. It was equated with the "Great Ice Age," the existence of which was discovered by Louis Agassiz (*see* APPENDIX I: BIOGRAPHICAL ENTRIES). In fact, the climate was much more complex. About 3 million years ago, the continents of North and South America joined, closing a seaway that had previously separated them. This created conditions that allowed glaciations to develop in the Northern Hemisphere. The first of these occurred about 2.36 million years ago (during the Pliocene epoch of the NEOGENE period). During the Pleistocene, glaciations occurred at intervals of about 100,000 years, and their onset and ending was driven by the astronomical events of the MILANKOVITCH CYCLES. During each glaciation, ICE SHEETS extended approximately to a line running from Seattle to New York and from London to Berlin and Moscow. Sea levels fell to about 395 feet (120 m) lower than they are today, because of the large amount of water held as ice.

Between glaciations there were INTERGLACIALS. These were episodes of warmer climates lasting an average of 10,000 years. During some interglacials, including the Sangamonian interglacial, which was the one prior to the present, Flandrian interglacial, TEMPERATURES were markedly higher than those of today. About 528,000 cubic miles (2.2 million km³) of ice melted from the West Antarctic Ice Sheet during the Sangamonian, raising the sea level to about 20 feet (6 m) above its present level, but the East Antarctic and Greenland ice sheets remained intact. *See* APPENDIX V: GEOLOGIC TIMESCALE.

Pleistogene The period of geologic time that began about 1.81 million years ago and that continues to the present day. Because this is the period during which human beings evolved, some people have suggested the period be called the Anthropogene. The Pleistogene is the third period of the CENOZOIC era and comprises the PLEISTOCENE and HOLOCENE epochs. *See* APPENDIX V: GEOLOGIC TIMESCALE.

Climatostratigraphy is the study of traces of soil and living organisms found in sedimentary rocks that were formed during the Pleistogene. These rocks can be dated (*see* RADIOMETRIC DATING) and the fossils and other materials found in them provide clues to the climatic conditions at the time the sediments were deposited.

Pliocene The epoch of geologic time that began about 5.3 million years ago and ended about 1.81 million years ago. It comprises the most recent epoch of the NEOGENE period.

During the Pliocene, North America and South America became joined by a land bridge. Otherwise a map of the Pliocene world is very similar to a map of the modern world. Climates everywhere began to grow cooler during the preceding MIOCENE epoch, and this trend continued throughout the Pliocene. The Antarctic ice cap began to form, but Antarctica was not yet completely frozen. Marine invertebrates typical of arctic waters appeared as far south as Britain. With falling temperatures, grasslands spread and forests retreated. Consequently, grazing mammals such as cattle, sheep, antelopes, and gazelles became more numerous and more widespread. Saber-toothed cats hunted them. *See* APPENDIX V: GEOLOGIC TIMESCALE.

pluvial A prolonged period of increased PRECIPITATION that affects a large region. The increase in precipitation is caused by increased EVAPORATION from the ocean and is associated with generally warmer conditions. Pluvial periods are separated by drier interpluvial periods.

Increased rainfall during a pluvial may produce lakes that subsequently disappear when drier conditions return. These are known as pluvial lakes. Lakes that formed during the PLEISTOCENE and of which only traces now remain are the best known examples of pluvial lakes. The lakes formed during warmer INTERGLACIAL episodes, when rainfall increased, glaciers retreated, ICE SHEETS melted, and rivers flowed, carrying vast quantities of water that accumulated in depressions hollowed out by the ice. In North America, lakes in the Great Basin expanded to form large inland seas, such as LAKE BONNEVILLE and Lake Lahontan, and extensive lakes also formed in East Africa. All of these lakes reduced greatly in size during the interpluvial periods that separated the pluvials.

Poisson's equation An equation from which it is possible to calculate the TEMPERATURE of a PARCEL OF AIR at any height, provided the air is on a dry ADIABAT. The equation is:

$$(T \div \Phi) \times c_{\mathrm{p}} \div R = p \div p_{\mathrm{o}}$$

where T is the actual temperature, Φ is the POTENTIAL TEMPERATURE, R is the specific gas constant of air, c_{p} is the specific heat at constant pressure, p is the pressure at the position of the air parcel, and p_{o} is the surface pressure. The equation was devised by the French mathematician Siméon-Denis Poisson (1781–1840).

polar high The persistent region of high surface atmospheric pressure that covers the Arctic Basin and Antarctica. In winter it consists of continental arctic (cA) air (*see* AIR MASS) over both polar regions. In summer the cA air continues to cover Antarctica, but maritime arctic (mA) air covers the arctic. The polar highs are the source of the polar easterlies (*see* GENERAL CIRCULATION).

polar low (polar hurricane) A small, intense CYCLONE that forms during winter in the cold AIR MASS on the side of the polar FRONT that is nearest the pole. Polar lows usually bring heavy hail and snow, and winds of up to gale force.

Polar lows are 125–500 miles (200–800 km) in diameter and develop when large amounts of very cold air spill out from the ice-covered continents across a markedly warmer sea. A small, upper-level TROUGH may also need to be present to trigger the disturbance that grows into the cyclone.

Once it has formed, a polar low is similar to a TROPICAL CYCLONE in many respects, although it is much smaller. Like a tropical cyclone, it is circular, generates strong winds, and has a cloud-free center surrounded by towering cumuliform cloud (*see* CLOUD TYPES) sustained by CONVECTION and extending to the tropopause (*see* ATMOSPHERIC STRUCTURE). Air flows outward from the cyclone at high level, producing cirrus cloud. A polar low dissipates quickly when it crosses land.

In other respects a polar low is unlike a tropical cyclone. From its first appearance, it reaches its full strength within 24 hours or less. It travels at up to 30 knots (34.5 MPH, 55.5 km/h), which is much faster than a tropical cyclone, and it lasts for no longer than 48 hours before it reaches land and dies.

Polar lows occur in many parts of the North Pacific and North Atlantic Oceans and also over the Tasman Sea, close to New Zealand. So far as is known, they rarely form over the Southern Ocean. They are most common in the Greenland, Norwegian, and Barents Seas, but sometimes they also form on the western side of Greenland and in the Beaufort Sea, to the north of Alaska.

polar molecule A molecule in which the electromagnetic charge is separated, so that one end of the molecule carries a positive charge and the other end carries a negative charge, although the molecule as a whole is neutral. Because it carries charge at its ends, the molecule is a dipole.

The WATER molecule is polar. This is because its two HYDROGEN atoms share their single electrons with the OXYGEN atom. Lines drawn from the two hydrogen atoms to the center of the oxygen atom meet at an angle of 104.5°, so both hydrogen atoms are positioned on the same side of the oxygen atom. This gives the oxygen side of the molecule a negative charge and the hydrogen side a positive charge, and it is this char-

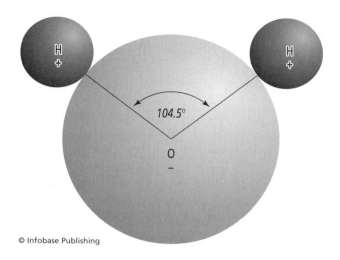

© Infobase Publishing

Water molecules (H₂O) are polar because their hydrogen atoms share electrons with the oxygen atom, and both hydrogens, bearing positive charge, are on the same side of the oxygen, which carries negative charge.

acteristic that allows water molecules to form hydrogen bonds (*see* CHEMICAL BONDS) and liquid water to act as a very efficient solvent.

Polar Pathfinder Program   A program that uses sensors on orbiting satellites to monitor the polar regions. The program was initiated by the EARTH OBSERVING SYSTEM program.

Climatic changes on a global scale are likely to be amplified in the Arctic and Antarctic, because ICE SHEETS and the area of SEA ICE are especially sensitive to changes in temperature. There is a second reason why changes due to the GREENHOUSE EFFECT are expected to appear first and most strongly at the poles. Because it is so cold, polar air is extremely dry. WATER VAPOR is the most important greenhouse gas, but its atmospheric concentration varies widely from place to place and from hour to hour, so it tends to mask signals from other greenhouse gases. Those signals, for example from a rising concentration of CARBON DIOXIDE, will therefore be detectable in very dry air some time before they are evident in the moister air of lower latitudes. These considerations make it highly desirable to maintain a close watch on the polar regions.

The advanced very high resolution radiometer (AVHRR, *see* SATELLITE INSTRUMENTS) and the TIROS operational vertical sounder (TOVS) carried on the TELEVISION INFRARED OPERATIONAL SATELLITE have

been acquiring data for many years. There is a continuous record since 1982 from the AVHRR and from 1978 from the TOVS. In addition, there are data from microwave sensors from 1978.

Little use has been made of these data, because they are so voluminous that individual laboratories found them too costly to acquire and process, and the data contained errors. The Polar Pathfinder Program aims to remedy that situation by supplying raw data that are consistently calibrated to allow data from different instruments to be compared.

Further Reading

Maiden, M., R. G. Barry, R. L. Armstrong, J. Maslanik, T. Scambos, A. Brenner, S. T. X. Hughes, D. J. Cavalier, A. C. Fowler, J. Francis, and K. Jezek. "The Polar Pathfinders: Data Products and Science Plans." American Geophysical Union. Available online. URL: www.agu.org/eos_elec/96149e.html. Accessed December 15, 2005.

polar stratospheric clouds (PSCs)   Clouds that form in winter over Antarctica and, less commonly, over the Arctic. They occur in the stratosphere (*see* ATMOSPHERIC STRUCTURE) at a height of 9–15.5 miles (15–25 km). Usually they are too thin to be visible, but when the Sun is about 5° below the horizon they can sometimes be seen as nacreous clouds (*see* CLOUD TYPES). What is known about them has been discovered by means of LIGHT DETECTION AND RANGING (LIDAR).

There are two principal types of PSC, known as Type 1 and Type 2, and there are two or possibly three varieties of Type 1 PSCs, known as Types 1a, 1b, and 1c. All Type 1 PSCs form at about 9 miles (15 km) altitude in air that is just above the frost point. At this height the frost point temperature is about -109°F (195K, -78°C). As the temperature falls below the frost point, Type 1 clouds begin to form rapidly, with very small ICE CRYSTALS acting as FREEZING NUCLEI. The source of the WATER VAPOR to produce the ice crystals is not known, but it may result from the oxidation of methane (CH_4) to carbon dioxide (CO_2) and water (H_2O). The Type 1 particles then grow rapidly as nitric acid (HNO_3) and more water vapor condense onto them. The resulting solution may be either liquid or solid, depending on the conditions around it.

Type 1a clouds are believed to consist of irregularly shaped, liquid particles about 0.004 inch (0.1 mm) in diameter made from approximately one molecule of HNO_3 to three molecules of H_2O.

Type 1b clouds are made from much smaller (about 0.00004 inch, 0.001 mm) liquid particles that are spherical and probably made from a mixture of sulfuric acid (H_2SO_4) and HNO_3 dissolved in water. There may also be Type 1c clouds, made from solid crystals of HNO_3 and water.

Type 2 PSCs are made from ice crystals. These PSCs form at lower temperatures and, therefore, at a greater altitude, most commonly at around 15.5 miles (25 km), where the temperature is about -121°F (188K, -85°C). Type 2 PSCs occur over Antarctica, but are very rarely observed over the Arctic, where winter temperatures seldom fall low enough for them to form.

The chemical reactions involved in the removal of OZONE from the OZONE LAYER take place on the surface of PSC ice crystals. Chlorine (Cl) is present in the stratosphere in two forms that are fairly inert chemically. These are hydrochloric acid (HCl) and chlorine nitrate ($ClONO_2$). On the surface of PSC particles they are converted into the much more reactive forms Cl_2 and HOCl by the reactions:

$$HCl + ClONO_2 \rightarrow HNO_3 + Cl_2$$

$$ClONO_2 + H_2O \rightarrow HNO_3 + HOCl$$

In both reactions, the HNO_3 remains inside the cloud particles. HOCl then reacts further to release free atomic chlorine (Cl).

Chlorine then reacts to remove ozone (O_3) in a series of steps.

$$Cl + O_3 \rightarrow ClO + O_2$$

This reaction takes place twice, to yield two molecules of ClO, which combine and in two subsequent reactions release 2 chlorine atoms.

$$ClO + ClO + M \rightarrow Cl_2O_2 + M$$

$$Cl_2O_2 + h\upsilon \rightarrow Cl + ClO_2$$

$$ClO_2 + M \rightarrow Cl + O_2 + M$$

where M is any air molecule and $h\upsilon$ is a quantum of solar energy of near-UV wavelength.

NITROGEN OXIDES (NO_x) are also removed by reactions on the surface of PSC particles. The process is called denoxification, and it is important because nitrogen dioxide (NO_2) removes chlorine oxide (ClO),

which is a key ingredient in the ozone-depletion process, by the reaction:

$$ClO + NO_2 + M \rightarrow ClONO_2 + M$$

NO_2 changes back and forth into gaseous N_2O_5:

$$4NO_2 + O_2 \leftrightarrow 2N_2O_5$$

Denoxification then removes gaseous nitrogen oxides by the reactions:

$$N_2O_5 + H_2O \rightarrow 2HNO_3$$

$$N_2O_5 + HCl \rightarrow ClNO_2 + HNO_3$$

Nitric acid is also removed, because as the PSC particles grow bigger their weight increases and they start to settle out of the stratosphere.

Further Reading

Carver, Glenn. "The Ozone Hole Tour." Centre for Atmospheric Science, University of Cambridge. Available online. URL: www.atm.ch.cam.ac.uk/tour/psc.html. Accessed January 20, 2006.

Salawitch, Ross J. "Polar Stratospheric Clouds." Jet Propulsion Laboratory. Available online. URL: http://remus.jpl.nasa.gov/info.htm. Accessed January 20, 2006.

Pole of Inaccessibility The point in the Arctic Ocean that is farthest from land. It lies between the North Pole and Wrangel Island, off the coast of eastern Siberia, and is sometimes taken to be the center of the Arctic.

pollen Pollen grains are the male reproductive cells of seed plants. Seed plants are plants that reproduce by producing seeds, rather than SPORES. The group includes the coniferous plants (gymnosperms) and the flowering plants (angiosperms). Pollen consists of individual pollen grains, which plants produce in vast numbers. Many insect-pollinated plants produce pollen grains that are sticky or barbed, to make them adhere to the pollinator. Pollen from wind-pollinated plants is usually smooth.

Most pollen lives for only a very short time (a few hours in the case of grasses), but the protective outer coating, called the exine, resists decay. Under ideal conditions, the exteriors of pollen grains can be preserved for many thousands of years.

A pollen grain is contained in a coat with two layers. The inside layer, called the intine, is soft, but the outer layer, the exine, is very tough and often sculptured. Grooves, called colpi (singular colpus), form patterns on the exine that are characteristic for particular plant families and in some cases for genera and even species. The pollen grains of some plants have pores in the exine through which the intine protrudes, and this adds to the markings on the exine. These sculptured shapes and markings are visible under a powerful microscope (usually at a magnification of at least ×300 and more often ×400 or ×1000). The markings on pollen grains allow palynologists to identify the family of the plant that produced the pollen (for example, the birch family Betulaceae). In some cases it is possible to go further and identify the genus (for example, an alder, *Alnus,* which is a genus of trees belonging to the birch family) or even the species (for example, green alder, *A. crispa*).

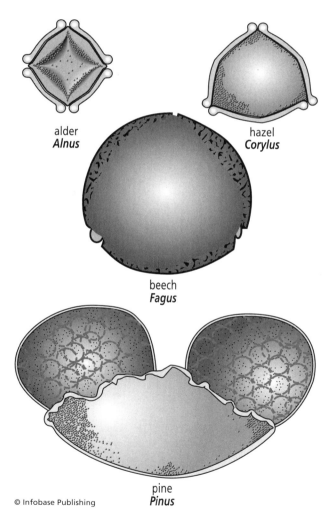

alder
Alnus

hazel
Corylus

beech
Fagus

pine
Pinus

© Infobase Publishing

Pollen grains from four different tree genera, all drawn to the same scale. Note the air sacs on the pine pollen.

The science of identifying and classifying pollen grains is called palynology, and the interpretation of pollen and spores that are found in sediments is called pollen analysis, although the two terms are often used synonymously.

Pollen analysis is the reconstruction of past climates and environments through the study of pollen grains and plant spores that are recovered from sediments. Pollen analysis began early in the 20th century. The first scientist to apply it was the Swedish geologist Lennart von Post. Palynology is the study of pollen grains, spores, and the shells of some aquatic organisms in order to classify them and discover their distributions. Palynology developed from pollen analysis and is used in many fields, including archaeology, petroleum exploration, and paleoclimatology (*see* CLIMATOLOGY).

Once the pollen has been identified, its presence can be interpreted. The Betulaceae is a family that grows in cool or cold climates, so its pollen indicates that at one time the area had a climate like that of northern Canada or Eurasia, regardless of what the climate is like today. Alders grow near water, and so their pollen indicates that the ground was wet. It might have been a riverbank or the shore of a lake.

An interpretation as simple as this would be unreliable, however. It reads too much into a very small amount of evidence. Pollen can travel, stuck to the skins or coats of animals, or blown by the wind, and so its presence in a particular place does not necessarily mean that that is where the plant actually grew. Conifers are wind-pollinated and produce pollen grains that have two air sacs to increase their BUOYANCY. This pollen can travel very long distances.

Plant species are never alone, however. There is always a community of them, so where pollen is found it usually represents several species. That makes interpretation much more reliable, because of the improbability of pollen from a group of species being brought together by chance. It is much more likely that the plants grew close to each other. They then constituted a life assemblage. After they all died, their pollen became a death assemblage. Its composition is different from that of the life assemblage, because the amount of pollen produced varies greatly from one species to another, the pollen grains themselves are distributed differently, and not all pollen survives equally well. The scientist analyzing the pollen allows for these factors.

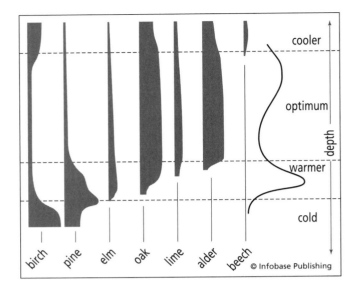

A pollen diagram shows how the amount of pollen of particular plant genera changes with depth through a section cut vertically through the soil. This indicates the relative abundance of each genus at different times, from which climatic changes can be inferred.

The pollen is preserved by being buried in soil or lake sediment. It is best preserved under anaerobic, acid conditions, such as those of a peat bog or the bed of a lake. Samples are recovered by drilling vertically through the sediment to extract a core. Organic material found at carefully marked depths can then be dated by RADIOCARBON DATING. This gives a date for the pollen found at those depths. Several samples are needed from each site, and it is usually necessary to count at least 200 pollen grains in each sample.

The absolute pollen frequency (APF) is the actual number of pollen grains that are counted in a unit volume of a sediment and, where the rate of deposition is known, the number per unit time. The pollen grains are those of a particular species, genus, or family of plants. The relative pollen frequency (RPF) is the number of pollen grains belonging to a particular species, genus, or family of plants that can be counted in a given volume of sediment. If the rate of deposition is known in a unit of time, this is expressed as a percentage of either the total amount of pollen present or of the total amount of tree pollen present.

RPF is the most widely used method for comparing pollen diagrams, but APF is more useful than counting the relative pollen frequency where several sites are to

be compared and different plants at each site produce the most abundant pollen. The amount and type of pollen present in a sediment that can be dated, for example by radiometric dating, allows the vegetation at that time to be identified. This gives a reliable indication of the type of climate at the time.

A pollen diagram is used to illustrate the pattern of vegetation that occupied a site at particular times in the past. It is compiled from counts of the absolute or relative pollen frequency for certain species, genera, or families of plants. The pollen counts are made from samples of soil that are taken from different depths. Often, the samples are obtained from a cliff face or exposed soil profile in a ditch or other cutting. In the diagram, pollen from each plant is shown as a vertical bar, plotted against depth with the ground surface at the top. The bar varies in thickness according to the amount of that pollen present at each level.

The resulting representation of vegetation types reveals the climatic conditions at different levels, because plants are sensitive to climatic change. If birch (*Betula*) and pine (*Pinus*) predominate, for example, the climate was cold and similar to that of northern Canada and Siberia today. The appearance of elm (*Ulmus*) and oak (*Quercus*) at a higher level, accompanied by a decline in birch and pine, indicates that the climate was growing warmer. Basswoods, also called lime and

Godwin's pollen zones

Years ago	Zone	Plants	Climate
14,000	I	creeping willow	Older Dryas; cold
12,000	II	birch	Allerød; milder
	III	creeping willow	Younger Dryas; cold
10,000	IV	birch, pine	Pre-Boreal; dry
	V	hazel, birch	Boreal; cool, dry
8,800	VIa	hazel, pine	Boreal; cool, dry
8,000	VIb	hazel, pine	Boreal; warmer, dry
	VIc	hazel, pine	Boreal; warmer, dry
7,500	VIIa	alder, oak, elm, lime	Atlantic; warm, moist
5,000	VIIb	alder, oak, lime	Sub-Boreal; cooler, drier
2,800	VIII	alder, birch, oak, beech	Sub-Atlantic; cool, wet

linden, (*Tilia*) grow only in warm temperate climates so their presence indicates a period of warmth. A decline in the abundance of these trees, accompanied by an increase in birch and the appearance of beech (*Fagus*) indicates cooler conditions.

A pollen zone, also called a pollen-assemblage zone, is an assemblage of pollen grains and spores that is considered to be characteristic of a particular climate that was once the climate of a large region. The concept was introduced in 1940 by the English botanist Sir Harry Godwin, and although the zones he proposed are now known to overlap locally and have largely fallen from use, many older textbooks include them.

Godwin proposed eight zones, identifying them with Roman numerals. They extend from the latter part of the Devensian GLACIAL PERIOD until the present day. The eight zones are summarized in the table (bottom, right) on page 364.

pollution   Pollution is the act of causing any direct or indirect alteration of the properties of any part of the environment in such a way as to present an immediate or potential risk to the health, safety, or well-being of any living species. The alteration may be chemical, biological, radioactive, or thermal.

Pollution is usually associated with human activities, but it can also occur naturally. For example, volcanic eruptions release large quantities of pollutants, and many plants emit ISOPRENES and terpenes that contribute to the formation of OZONE and PHOTOCHEMICAL SMOG. Certain plants emit METHANE, possibly amounting to 10–30 percent of the total quantity of methane entering the atmosphere, and methane is a greenhouse gas (*see* GREENHOUSE EFFECT).

Chemical pollution occurs when substances are released which are toxic, such as tetraethyl lead, or which are harmless in themselves but engage in reactions that yield toxic products, such as peroxyacetyl nitrate, which one of the chemical compounds formed during the sequence of reactions that result in photochemical smog. Biological pollution involves the release of harmful bacteria, viruses, or fungal spores. These can be carried long distances in the air. Radioactive pollution is caused by the release of radioactive substances that may emit IONIZING RADIATION.

Thermal pollution occurs when gases or liquids are released at a markedly different (almost always higher) temperature than that of the medium into which they

discharge. Injecting relatively warmer or cooler gases into the atmosphere can change the way air moves. Discharging hot air can produce a HEAT ISLAND effect that radically alters the local climate. Raising the TEMPERATURE of WATER can harm aquatic organisms, because they depend for RESPIRATION on OXYGEN that is dissolved in water, and the amount of dissolved oxygen water can hold is inversely proportional to the water temperature. *See* AIR POLLUTION.

pollution control   The recognition, many years ago, that AIR POLLUTION is a serious problem that must be addressed led to steps to bring it under control. The first requirement was technical. Reducing the emission of pollutants had to be made physically possible, so physical devices were invented that would prevent polluting substances from entering the atmosphere. There is a range of such devices.

Pollutants enter the air from a flue, which is a passage designed to remove the hot waste gases and other by-products of COMBUSTION from an incinerator or other combustion site. It must be built from materials that are capable of withstanding the temperatures to which the combustion products will subject them and arranged in such a way as to produce a steady flow of air to carry the products. Gases passing through the flue are known as flue gases, and most pollution control aims to trap particles or gases before they leave the flue.

A scrubber is a device that removes solid and liquid particles and some gases from a stream of gas. It consists of a space into which water is sprayed or that contains wet packing material. The particles and gas molecules adhere to water molecules or dissolve in the liquid water. Scrubbers are used to take samples of gas streams and to remove potential pollutants from waste gases.

A wet scrubber consists of an ABSORPTION TOWER in which the gas stream is brought into contact with a liquid that absorbs the pollutants. Different liquids are used, depending on the pollutants to be removed.

Caustic scrubbing is a process for the removal of sulfur dioxide (SO_2) from flue gases. It involves passing the gas stream through a solution of caustic soda (sodium hydroxide, NaOH). The SO_2 and NaOH react to form sodium sulfite (Na_2SO_3) and sodium hydrogen sulfite ($NaHSO_3$). The addition of calcium carbonate ($CaCO_3$) then causes the precipitation of insoluble gypsum (calcium sulfate, $CaSO_4$) and leaves the water enriched in sodium carbonate (Na_2CO_3), a harmless substance that is present in most mineral waters. After it has been diluted, the solution can be safely discharged into surface waters.

A bag filter is used to remove small particles from industrial waste gases. It consists of a tube-shaped bag made from woven or felted fabric, up to 33 feet (10 m) long and 3 feet (1 m) wide, and closed at one end. The open end is attached to the pipe carrying the gases. Provided the gas is traveling fairly slowly, the filter can trap more than 99 percent of the particles carried in the waste stream. The material from which the filter is made must be suitable for the temperature of the gas. Natural fibers can be used at temperatures up to 194°F (90°C), nylon up to 392°F (200°C), and glass fibers up to 500°F (260°C).

Bringing two or more substances or objects into contact so that one adheres to the other is called impingement. For example, DUST is made to impinge onto a dust collector. Dry impingement is a technique that is used to remove particulate matter from a stream of waste gases. The gases are blown against a surface to which the particles adhere.

Impingement can be made more efficient if an electrostatic charge (see STATIC ELECTRICITY) is applied to the filter, which is then known as an electrostatic filter. Particles bearing a charge are attracted to the opposite charge on the filter.

An electrostatic precipitator is used to remove solid and liquid particles from a gas. The gas is made to flow between two electrodes across which a high voltage is applied. This produces an ELECTRIC FIELD. As the particles pass through the electric field they acquire a charge and move to the electrode bearing the opposite charge, where they are held. Electrostatic precipitators are extremely efficient at collecting small particles and they are used widely to clean industrial emissions.

The second prong in the approach to pollution control aimed to encourage the installation of the available antipollution devices. Companies must purchase these devices and pay for their installation and maintenance. They must also find alternative means for disposing of their unwanted by-products. In a few cases these by-products may have industrial uses and can be sold, but in most cases the cost of disposal must be borne by the company that produces and collects them. A company that accepts all of these costs will find itself at a commercial disadvantage if it competes with other companies that continue to emit pollutants as before.

Responsible organizations will recognize the desirability of reducing pollution and will willingly play their part in achieving this, but legislation is needed to compel less responsible organizations to comply, and thereby eliminate their unfair advantage. Regulations are also required to list the forms of pollution that are to be reduced and to stipulate the actions required.

Legislation therefore accompanied the introduction of devices to reduce pollution. Environmental quality standards were introduced to establish the maximum limits or concentrations of polluting substances that are permitted in air, water, or soil. In the United States, there are primary and secondary standards. With an allowance to provide a safety margin, primary standards represent those maxima that present no threat to human health. Secondary standards are those maxima that present no threat to public welfare.

The threshold limit value (TLV) is the greatest concentration of a specified airborne pollutant to which workers may be legally exposed day after day. The TLV is calculated to produce no adverse effect on persons experiencing that level of exposure.

Emissions trading, or trading in permitted pollutant emissions, is an approach to pollution control that has proved highly successful. A political agreement is reached on the total concentration of a particular pollutant that is considered safe and acceptable. Measurements of rates of emission, accumulation, and RESIDENCE TIMES allow scientists to calculate the amount of the pollutant that can be emitted annually if the total environmental burden is not to be exceeded. Scientists also identify the principal sources of the emissions. It is then possible to share the total acceptable emission amounts among the principal sources in the form of emission permits. These stipulate that a particular factory is permitted to emit a specified quantity of the named pollutant in the course of a year. Exceeding emission permits incurs penalties, ranging from a scale of fines to the enforced closure of operations that refuse or are unable to comply.

Every organization with an emission permit may then trade. If a factory's emissions fall below the amount allowed, the company can sell the surplus to another company that wishes to exceed its permitted emissions. Trading quickly establishes a price structure for emission permits, which are then traded much like any other commodity.

The bubble policy introduced in the Clean Air Act (see APPENDIX IV: LAWS, REGULATIONS, AND INTERNA-

TIONAL AGREEMENTS) had a similar effect in the United States. This policy allowed companies to agree with the Environmental Protection Agency, which is the regulatory authority, to exceed their allotted emission limit at certain of their installations in return for equivalent reductions in emissions at other installations.

postglacial climatic revertence A period during which the climate became cooler and wetter than it had been previously. The change began abruptly about 2,500 years ago, and the cool, wet conditions have continued to the present day, marking the sub-Atlantic period (*see* HOLOCENE EPOCH).

The onset of the deterioration was marked by a decline in the number of lime trees (*Tilia* species) in England and Wales. Lime trees demand warm conditions. There was also a decline in the number of pine trees (*Pinus* species) in northern Britain, indicating that summers were cool.

Further Reading

White, Iain. "The Flandrian: The Case for an Interglacial Cycle." University of Portsmouth. Available online. URL: www.envf.port.ac.uk/geog/teaching/quatgern/q8b.htm. Last updated July 2002.

potato blight Two diseases of potatoes that are most likely to cause damage under certain weather conditions. The less serious of the two is early blight, caused by the fungus *Alternaria solani*. It occurs in hot, dry weather. If the temperature remains below about 81°F (27°C), early blight produces brown marks on the leaves of the potato plant, but at higher temperatures it may destroy the foliage.

Late blight is caused by the *Phytophthora infestans* and occurs when the weather is cool and wet, with daytime temperatures between about 50°F and 78°F (10–25°C). *P. infestans* is a water mold, also known as a downy mildew, a funguslike organism classified as an oomycete (Oomycota). Late blight can rapidly destroy the foliage. The tubers then start rotting and soon turn into an inedible brown pulp. It was late blight that caused the failure of the potato crop over Britain in 1845 and 1846 and the potato famine in Ireland.

potential energy The energy that is stored in a body by virtue of its position or state. Energy is stored in a ball that is stationary at the top of an incline and will be converted into KINETIC ENERGY if the ball should start rolling. Gravitational, chemical, nuclear, and electrical energy are all forms of potential energy.

potential temperature The TEMPERATURE a volume of a fluid would have if the pressure under which it is held were adjusted to sea-level pressure, of 1,000 mb (100 kPa), and its temperature were to change adiabatically (*see* ADIABAT). Potential temperature depends only on the actual temperature and pressure of the fluid. It is conventionally represented by the Greek letter Phi (Φ), which is the "phi" in TEPHIGRAM. In METEOROLOGY, the concept of potential temperature is used to calculate the STABILITY OF AIR—the likelihood that air will move vertically with consequent condensation or evaporation of moisture.

The concept also explains why cold air does not sink from the upper atmosphere to the surface. Air temperature decreases with height and according to the GAS LAWS, as the temperature of a mass of gas decreases so does its volume. If a given mass contracts, its DENSITY increases. It would seem, then, that cold air high above the ground is denser than the warmer air close to the ground. If that is so, why is it that the very cold air near the tropopause (*see* ATMOSPHERIC STRUCTURE) remains there?

Suppose that the air is fairly dry, with no clouds in the sky, and that the temperature near to ground level is 80°F (27°C). Up near the tropopause, at a height of 33,000 feet (10 km), suppose that the air temperature is -65°F (-54°C). This is very much colder than the air temperature near the ground. Increase the atmospheric pressure to its sea-level value, however, and as the air is compressed, it will warm at the dry adiabatic LAPSE RATE (DALR) for air, of 5.4°F per 1,000 feet (9.8°C km^{-1}). The consequence of adjusting the pressure is to convert the actual temperature to the potential temperature. At once the reason the cold air does not sink becomes evident.

$$\Phi = (\text{DALR} \times A) + t_t$$

where A is the altitude of the cold air (in this example 33,000 feet) and t_t is its temperature (-65°F). Therefore:

$$\Phi = (5.4 \times 33) - 65$$

$$\Phi = 113.2°F$$

The potential temperature of the air at the tropopause, of 113.2°F (45°C), is much higher than the actual temperature of air at ground level, which is 80°F (27°C).

The difference between the actual and potential temperature is the reason why the cold air remains aloft.

The wet-bulb temperature that a parcel of saturated air would have if it were taken adiabatically to the 1,000-mb level is called the wet-bulb potential temperature.

The potential temperature of any PARCEL OF AIR is said to be conserved. This means it does not change as a consequence of the vertical movement of the air. Consequently, knowledge of the potential temperature makes it possible to study the thermodynamic characteristics of the air and to represent them in diagrams. The concept is therefore of great importance to meteorologists and is widely used in WEATHER FORECASTING.

A line joining points of equal potential temperature is called an isentrope, and a surface of equal potential temperature is known as an isentropic surface. Entropy is conserved in all adiabatic processes (*see* THERMODYNAMICS, LAWS OF), so an isentropic surface is one over which entropy is everywhere the same. The mixing of air that takes place across an isentropic surface is called isentropic mixing. Isentropic analysis is a procedure in which winds, pressures, temperatures, and humidities across several isentropic surfaces are extracted from radiosonde data (*see* WEATHER BALLOON).

power-law profile A mathematical expression that describes the variation of the wind with height. Many attempts have been made to devise such a formula, but they tend to fail when the air is very stable. The most successful is probably:

$$u = (u_*/k)[(\ln\{z/Z_o + b/4L'\})(z - z_o)]$$

where u is the wind speed, u_* is the friction velocity, k is the von Kármán constant (*see* WIND PROFILE), ln means the natural logarithm, z is the height, z_o is the roughness length (*see* AERODYNAMIC ROUGHNESS), b is a coefficient, and L' is the length of the gradient, and the gradient is assumed to remain constant with height.

precession of the equinoxes A change in the dates at which the Earth reaches APHELION and PERIHELION and therefore in the position of the Earth in its ORBIT at the EQUINOXES and SOLSTICES. Precession is a property of gyroscopes. When a force is applied to the rotational axis of a gyroscope, the axis moves at right angles to the force in the direction of rotation. Because it is spinning, the Earth behaves as a gyroscope and forces on its

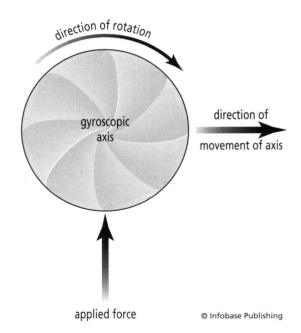

Precession causes the axis of a gyroscope to move at 90° to the direction of a force applied to it.

axis (more strictly on its equatorial bulge) are exerted by the gravitational attraction of the Moon, Sun, and, to a lesser extent, Jupiter. This causes the axis to wobble, like that of a toy gyroscope or spinning top (*see* MILANKOVITCH CYCLES).

If the equator is extended to the edge of the universe, the circle it forms is called the celestial equator. This is at an angle to the PLANE OF THE ECLIPTIC, because of the tilt in the Earth's rotational axis (*see* AXIAL TILT). The equinoxes occur when the Earth reaches the two positions in its orbit at which the celestial equator intersects the ecliptic. The axial wobble causes the celestial equator to change its position, and this causes the orbital positions of the equinoxes to change. The result is that the orbital positions of the Earth at the equinoxes move westward by 50.27″ (arcseconds) every year and complete a circuit of the ecliptic in about 25,800 years (360° ÷ 50.27″ = 25,800).

At present, Earth is at perihelion in early January and at aphelion in early June. In about 12,000 years, it will be at perihelion in June and at aphelion in January.

precipitation WATER that falls from the sky to the surface in either liquid or solid form is called precipitation, a word derived from the Latin *praecipitatio*, which means "I fall headlong." Precipitation physics is the branch of physical METEOROLOGY that is concerned with the physical processes that are involved in the formation of CLOUD DROPLETS, ICE CRYSTALS, and the resulting precipitation.

The average depth of the precipitation that falls during a specified time over a specified area is known as the depth-duration-area value (DDA value). The precipitation intensity is the amount of precipitation that falls to the surface within a specified period. It is measured in inches or millimeters per hour or day.

The zone of maximum precipitation is the elevation at which the precipitation that falls on a mountainside is greatest. Spillover is precipitation due to OROGRAPHIC lifting that is blown over the top of the mountain by the wind, so that it falls inside the area that is ordinarily in the rain shadow.

The probable maximum precipitation is an estimate of the greatest amount of precipitation that could conceivably fall on a given drainage area over a given period. It is calculated from records of the worst storms ever known in the area. The amount of precipitation is then converted into the amount of water that will flow through streams and rivers as a result. This figure is used to calculate the probable maximum flood, a figure that engineers use when designing dams. Although this calculation provides an estimate of the magnitude of the most severe flood, it does not calculate the probability that such a flood will occur within any stated period (such as 50 or 100 years).

Precipitable WATER VAPOR is the total amount of water vapor that is present in a column of air above a point on the surface of the Earth, or in a column of air within a layer of the atmosphere that is defined by the atmospheric pressure at its base and top. Precipitable water vapor is measured as the mass of water vapor in a unit area (such as pounds per square yard or kilograms per square meter). It is the amount of water that would fall as precipitation if all of it were to condense, but it is also the amount of water vapor that will react, as vapor, with outgoing radiation, thereby affecting the rate of atmospheric heating (see GREENHOUSE EFFECT). The proportion of the precipitable water that can fall as precipitation is known as the effective precipitable water.

When the relative HUMIDITY of the air exceeds 100 percent, water vapor will condense in the presence of CLOUD CONDENSATION NUCLEI to form cloud droplets or freeze in the presence of FREEZING NUCLEI to form ice crystals. Depending on the TEMPERATURE, the droplets or crystals then grow either by collision or by the Bergeron–Findeisen mechanism. They fall from the cloud when their weight exceeds the ability of vertical air currents to support them.

The effective precipitation is a value for the aridity of a climate that is important in many schemes for CLIMATE CLASSIFICATION, because the aridity of a climate determines its suitability for agriculture and the type of natural vegetation it will support. Aridity refers to the amount of water that is available to plants and it is equal to the difference between precipitation and EVAPORATION, or in other words to the effective precipitation. Effective precipitation is calculated as r/t, where r is the mean annual precipitation in millimeters and t is the mean annual temperature in °C. Precipitation that falls as SNOW must first be converted to its rainfall equivalent. If r/t is less than 40, the climate is considered to be arid, and if r/t is greater than 160, the climate is perhumid (extremely wet).

Between these extremes, the boundary between steppe grassland and desert, and the boundary between forest and steppe can be determined by values for r/t, depending how the precipitation is distributed. If precipitation falls mainly in winter, the two boundaries lie where $r/t = 1$ and $r/t = 2$, respectively. If precipitation is distributed evenly through the year, they fall where $r/(t + 7) = 1$ and $r/(t + 7) = 2$. If precipitation falls mainly in summer, they fall where $r/(t + 14) = 1$ and $r/(t + 14) = 2$.

The humidity coefficient is a measure of the effectiveness for plant growth of the precipitation falling over a region during a specified period. It relates the amount of precipitation to the temperature and is calculated as $P/1.07^t$, where P is the amount of precipitation in centimeters and t is the mean temperature for the period in question in degrees Celsius.

The moisture factor is also a measure of the effectiveness of precipitation. It is calculated by dividing the amount of precipitation in centimeters by the temperature in degrees Celsius for the period under consideration.

Water droplets, ice crystals, hailstones, and SNOWFLAKES that fall from clouds are called hydrometeors. Hydrometeors include drizzle, RAIN, freezing rain, HAIL, graupel, sleet, snow, and ice pellets. The fall of charged hydrometeors causes a downward flow of electric charge. This is known as a precipitation current. Meteoric water is water that falls from the sky as precipitation and then moves downward through the soil until it joins the GROUNDWATER (*see* JUVENILE WATER).

Not all precipitation falls from clouds. The term also includes DEW, white dew, hoar FROST, rime frost, glaze, FOG, and freezing fog. Precipitation that falls from a cloud but evaporates before reaching the ground is called virga (*see* CLOUD TYPES).

Precipitation that is wholly caused by the distribution of temperature and moisture within an AIR MASS and not to orographic lifting or the lifting of air in a frontal system (*see* FRONT), is called air mass precipitation.

Convectional precipitation results from thermal CONVECTION in moist air. The precipitation usually takes the form of rain, snow, or hail SHOWERS, which are sometimes heavy.

Precipitation that falls from clouds associated with a weather front is called frontal precipitation. Cyclonic rain is associated with a CYCLONE or DEPRESSION. Steady, persistent rain or snow falls from stratiform clouds along the warm front and cumuliform clouds along the cold front produce showers, which are sometimes heavy.

Mist is liquid precipitation in which the droplets are 0.0002–0.002 inch (0.005–0.05 mm) in diameter. These are large enough to be felt on the face of a person walking slowly through the mist. VISIBILITY in mist is greater than 1,094 yards (1 km). Mist is usually associated with stratus cloud. If the droplets increase in size, mist becomes drizzle.

Drizzle is liquid precipitation in which the droplets are smaller than 0.02 inch (0.5 mm) in diameter, are all of approximately similar size, and are very close together. They form by the coalescence of smaller droplets near the base of stratus cloud. As soon as a sufficient number have coalesced to reach a size that is just heavy enough to fall, droplets start to sink slowly. There are no upcurrents in stratus, so the droplets have no opportunity to coalesce to a larger size before they leave the CLOUD BASE and the base is low enough for them to reach the surface before evaporating.

Freezing drizzle is drizzle made of supercooled droplets (*see* SUPERCOOLING) that freeze on contact with the ground. Inside the stratus cloud from which the precipitation falls, the temperature is a little above freezing. Droplets falling through the cloud do not freeze, but when they leave the base of the cloud, their temperature is close to freezing. If the air temperature between the cloud base and the ground is below freez-

ing, the droplets are chilled further as they fall, but because the cloud base is low they do not remain airborne long enough to freeze. Instead, they freeze immediately on contact with the ground, the temperature of which is also below freezing.

Freezing rain forms in the same way as freezing drizzle. Supercooled raindrops fall quickly from the cloud and are not exposed to the cold air long enough to freeze, but when they strike the ground they freeze on contact. Other types of frozen precipitation include freezing fog, frozen fog, frost smoke, graupel, hail, sleet, and snow.

Ice pellets are ice particles that are transparent or translucent, spherical or irregular in shape, and less than 0.2 inch (5 mm) in diameter. They are snowflakes that have partly melted and then refrozen, frozen raindrops or drizzle droplets, or SNOW PELLETS that are enclosed in a thin coating of ice.

Sleet consists of small raindrops that freeze in cold air beneath the base of the cloud in which they formed. They fall as small particles of ice of various shapes. Heavy sleet can reduce visibility. In Britain, sleet is a mixture of rain and snow falling together. All of it falls as snow from the base of the cloud and enters air that is below freezing temperature, but close to the ground there is a layer of warmer air in which some, but not all, of the snowflakes melt.

An outburst is a sudden, very heavy fall of precipitation from a cumuliform cloud that is caused by the strong downcurrents in the cloud. Outbursts often accompany THUNDERSTORMS.

Throughfall is the proportion of the total precipitation falling over a forest that reaches the ground. This consists of the sum of the amount that is not intercepted by plants and reaches the ground surface directly, the amount that drips from leaves, and the amount of STEM FLOW. It is equal to the total amount of precipitation minus the amount lost by evaporation from plant surfaces.

present weather Current weather conditions are included in a report from a weather station. Known as "present weather," these are represented by a series of two-digit numbers, from 00 to 99. Most of the categories are listed below. Categories 30–39 and 40–49 are not given in detail to avoid repetition. The descriptions are also simplified from the official versions, which are worded very carefully to avoid ambiguity and are therefore not easy to understand. When the present weather is reported, the number that is used is the highest that is applicable to the situation. In the following descriptions "freezing" means freezing on impact:

00 No cloud developing during the past hour
01 Cloud dissolving during the past hour
02 Cloud generally unchanged during the past hour
03 Cloud developing during the past hour
04 Visibility reduced by smoke
05 Haze
06 Dust widespread
07 Dust or sand raised by local wind, but not by dust storms, sandstorms, or whirls (devils)
08 Dust or sand whirls seen in the past hour, but no dust storms or sandstorms
09 Dust storm or sandstorm seen nearby during the past hour
10 Mist
11 Shallow, patchy fog or ice fog
12 Shallow continuous fog or ice fog
13 Lightning but no thunder
14 Precipitation seen, but not reaching the surface
15 Precipitation seen reaching the surface in the distance
16 Precipitation seen reaching the surface nearby, but not at the station
17 Thunderstorm but no precipitation seen
18 Squalls at the time of observation or during the past hour
19 Funnel cloud seen at the time of observation or during the past hour
20 Precipitation, fog, or thunderstorm during the past hour but not at the time of observation
21 Drizzle (not freezing) or snow grains, but not in showers
22 Rain (not freezing), but not in showers
23 Rain and snow or ice pellets, but not in showers
24 Freezing drizzle or freezing rain
25 Rain showers
26 Showers of rain and snow (British sleet) or snow
27 Showers of hail and rain or hail
28 Fog or ice fog in the past hour
29 Thunderstorm
30–39 Dust storms, sandstorms, drifting snow, or blowing snow
40–49 Fog or ice fog at the time of observation
50 Drizzle (not freezing) that is intermittent and slight at the time of observation

51 Drizzle (not freezing) that is continuous at the time of observation
52 Drizzle (not freezing) that is intermittent and moderate at the time of observation
53 Drizzle (not freezing) that is continuous and moderate at the time of observation
54 Drizzle (not freezing) that is intermittent and heavy at the time of observation
55 Drizzle (not freezing) that is continuous and heavy at the time of observation
56 Slight freezing drizzle
57 Moderate or heavy freezing drizzle
58 Slight drizzle and rain
59 Moderate or heavy drizzle and rain
60–69 The same as 50–59, but with rain instead of drizzle and in 58 and 59 snow instead of rain
70 Snowflakes, intermittent and slight at the time of observation
71–75 The same as 51–55, but with snow instead of drizzle
76 Ice prisms with or without fog
77 Snow grains with or without fog
78 Isolated, starlike, snow crystals with or without fog
79 Ice pellets
80 Rain showers, slight
81 Rain showers, moderate or heavy
82 Rain showers, violent
83 Rain and snow showers, slight
84 Rain and snow showers, moderate or heavy
85 Snow showers, slight
86 Snow showers, moderate or heavy
87 Slight showers of snow pellets, encased in ice or not, with or without rain or rain and snow (British sleet) showers
88 Moderate showers of snow pellets, encased in ice or not, with or without rain or rain and snow (British sleet) showers
89 Slight hail showers, without thunder, with or without rain or rain and snow (British sleet)
90 Moderate or heavy hail showers, without thunder, with or without rain or rain and snow (British sleet)
91 Slight rain
92 Moderate or heavy rain
93 Slight snow, or rain and snow (British sleet), or hail
94 Moderate or heavy snow, or rain and snow (British sleet), or hail
95 Slight or moderate storm with rain and/or snow, but no hail

96 Slight or moderate storm with hail
97 Heavy storm with rain and/or snow, but no hail
98 Storm with sandstorm or dust storm
99 Heavy storm with hail

pressure gradient (isobaric slope) Atmospheric pressure changes over a horizontal distance and the rate of change constitutes a gradient of pressure, or pressure gradient. The isobars (*see* ISO-) that join points of equal pressure on a surface resemble the contour lines on a physical map, and a line drawn at right angles to them shows the direction of the gradient, or slope, that inclines from a region of high pressure to one of low pressure.

The distance between isobars indicates the steepness of the gradient, just as does the distance between contour lines. That distance is proportional to the difference in pressure between the high and low regions, just as the distance between contour lines is proportional to the elevation of high and low areas of ground. Air moves because there is a pressure gradient, but it does not move directly down the gradient (*see also* gradient wind).

In synoptic METEOROLOGY, an increase in the pressure gradient that takes place over hours or days, leading to a strengthening of the winds, is called intensification. Weakening is a decrease in the pressure gradient that takes place over hours or days, causing a reduction in WIND SPEED.

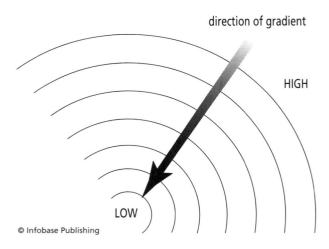

direction of gradient

HIGH

LOW

© Infobase Publishing

The rate at which air pressure changes between centers of high and low pressure varies according to the distance between the centers and the intensity of the high and low pressure. The rate of change therefore constitutes a gradient, like the slope of a hillside, with the high pressure at the top of the hill and low pressure in the valley. The isobars resemble contours.

The area between two centers of high or low pressure where the pressure gradient is low is called a col. A col is the region of highest pressure between two CYCLONES and of lowest pressure between two ANTICYCLONES. On a constant-pressure (*see* WEATHER MAP) chart a col is shaped like a saddle.

pressure gradient force (PGF) It is the pressure gradient force that accelerates air horizontally across the surface of the Earth. It is produced by the PRESSURE GRADIENT, and its magnitude is proportional to the steepness of the gradient. The PGF has both vertical and horizontal components, but the vertical component, which tends to make the air rise, is balanced by the force of gravity and therefore can be ignored.

The pressure-gradient force always acts in the same direction as the pressure gradient. Consequently, that is the direction in which it tends to move the air—by the most direct route from a region of high pressure to a region of low pressure until the two are equal and the pressure gradient disappears. This is not what happens in fact, however, because of the CORIOLIS EFFECT, but it is the direction in which the PGF acts.

A force that continues to act causes the body on which it acts to accelerate, so the PGF is a force of ACCELERATION. Its magnitude can be calculated provided the relevant factors are known. These are the air DENSITY, the distance between the two points over which the PGF is being calculated, and the difference between the pressures at those two points. These values must be in compatible SI units (*see* APPENDIX VIII: SI UNITS AND CONVERSIONS). Distances must be converted to meters and pressures from millibars to pascals (1 mb = 100 Pa, *see* UNITS OF MEASUREMENT). One pascal is the pressure that will impart an acceleration of 1 meter per second per second, per kilogram, per square meter. The equation is:

$$F_{PG} = (1 \div d) \times (\Delta_p \div \Delta_n)$$

where F_{PG} is the PGF, d is the density of the air (in kilograms per cubic meter), Δ_p is the difference in pressure (in pascals), and Δ_n is the distance between the two places (in meters).

Suppose the sea-level air pressure is 1004 mb at Boston and 980 mb at New York, about 200 miles away.

1004 mb = 100,400 Pa;

980 mb = 98,000 Pa;

Δ_n (distance of 200 miles) = 321,800 m;

d (density of air) at sea level = approximately 1 kg/m^3;

Δ_p (pressure difference) = 100,400 - 98,000 Pa = 2,400 Pa.

Applying the equation:

$F_{\mathrm{PG}} = (1 \div 1) \times (2,400 \div 321,800) = 0.00746 \text{ m/s}^2$

This is a very small acceleration, but it applies to only one kilogram of air, the mass of air that occupies one cubic meter, and it is an acceleration, not a speed. Although the sea-level air density is 1 kg/m^3, so the first term in the equation is 1 ÷ 1, the PGF is usually measured at some height above sea level, where the density of air is lower and the first term does not cancel.

pressure system An atmospheric feature that is characterized by AIR PRESSURE. The term is usually applied to a CYCLONE or ANTICYCLONE, but it may also refer to a RIDGE or TROUGH.

If a pressure system is embedded in the general westerly airflow of middle latitudes, the entire pressure system will travel from west to east. Such a pressure system is said to be migratory.

Program for Regional Observing and Forecasting Services (PROFS) A program that began in 1980–81 under the auspices of the NATIONAL OCEANIC AND ATMOSPHERIC ADMINISTRATION (NOAA). Its aim was to test and apply new scientific knowledge and technological innovations in order to improve operational weather services. The program was later renamed the Forecast Systems Laboratory (FSL). The FSL is now called the GLOBAL SYSTEMS DIVISION (GSD) of the Earth System Research Laboratory (ESRL).

Proterozoic The immensely long eon of the Earth's history in which multicellular organisms first appeared. The eon began 2,500 million years ago and ended 542 million years ago.

PLATE TECTONICS became the dominant force in shaping the Earth's crust, and by the middle of the Proterozoic, approximately 1,100 million years ago, all the continents were joined together to form a supercontinent called Rodinia. By about 900 million years ago Rodinia was starting to break apart. *See* APPENDIX V: GEOLOGIC TIMESCALE.

proxy data Climatologists cannot rely wholly on direct measurements of the state of the atmosphere in order to build up a picture of climate, because such measurements may not exist. The scientists are not left helpless, however, because there are other measurements that do not refer to the climate directly, but which can be interpreted to yield information about climate. These are called proxy data.

TREE RINGS, ICE CORES, POLLEN, and animal remains (*see* BEETLE ANALYSIS) are among the sources of proxy data. Proxy data is the only data available for the reconstruction of prehistoric climates.

psychrometry The Greek word *psukhros* means "cold" and *metron* means "measure," so psychrometry is literally the measurement of coldness. In fact, psychrometry is the operation of calculating the relative HUMIDITY and DEW point TEMPERATURE from the extent to which the reading given by a THERMOMETER with a bulb that is kept wet is lower than the reading from a thermometer with a dry bulb. In other words, the calculation is based on a measurement of the coldness of the wet-bulb thermometer. The instrument used to make the measurement is a type of HYGROMETER called a psychrometer.

A psychrometer comprises two thermometers mounted parallel to each other and held a short distance apart. One, known as the dry-bulb thermometer, measures the air temperature. The other, known as the wet-bulb thermometer, indirectly measures the rate of EVAPORATION. The bulb of this thermometer is wrapped in a wick, usually made from muslin, which is partly immersed in a reservoir of distilled water.

Water is drawn by CAPILLARITY into the wick from where it evaporates. The LATENT HEAT of vaporization is taken from the thermometer bulb, thus depressing its temperature. The rate at which water evaporates, and therefore the amount of latent heat supplied by the bulb, varies according to the relative humidity of the air.

To ensure an accurate reading, it is necessary to provide a flow of air over the thermometers and to ensure that the wick remains wet at all times, but is not allowed to become sodden. If the wick is sodden, water will evaporate from its surface, but the layer of wick in contact with the bulb will remain unaffected. Neither relative humidity nor the dew point temperature can be

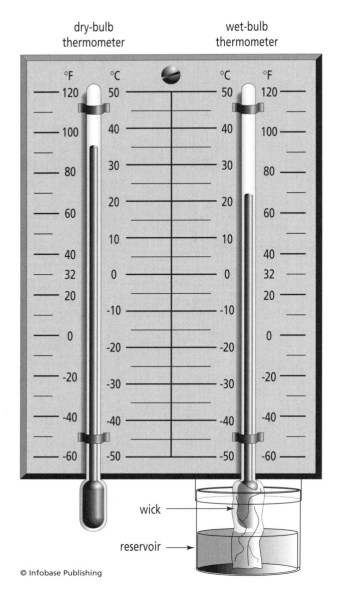

© Infobase Publishing

A psychrometer is an instrument for measuring wet-bulb and dry-bulb temperatures. Two identical thermometers are mounted side by side on a board. The wet bulb is wrapped in a wick that is partly immersed in a reservoir of distilled water.

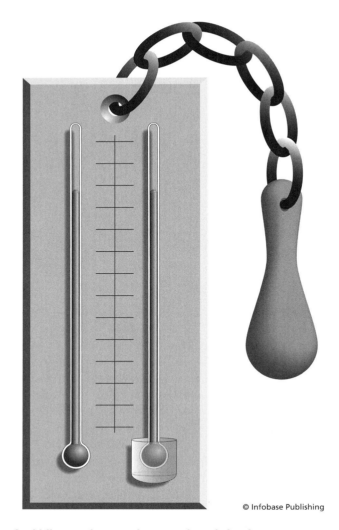

© Infobase Publishing

A whirling psychrometer is swung through the air to ensure an evenly distributed flow of air over both bulbs. This process increases the accuracy of the instrument.

read directly from the instrument, but both can be calculated from its readings.

An aspirated hygrometer is a psychrometer in which the necessary flow of air over the thermometers is provided by placing them inside a tube through which air is blown.

In a whirling psychrometer, also called a sling psychrometer, the flow of air across the bulbs of the dry-bulb and wet-bulb thermometers is produced by manually whirling the instrument through the air. In one version, a chain is attached at the top of the board to which both thermometers are securely fixed and there is a handle on the other end of the chain. In another version, the board forms the horizontal arm of a device resembling a rattle. The top of the board is fixed to a rod, one end of which is attached to a handle in which it is free to rotate.

quasi-biennial oscillation (QBO)   An alternation of easterly and westerly winds that occurs in the stratosphere (*see* ATMOSPHERIC STRUCTURE) above the TROPICS, between about 20°S and 20°N. The change happens on average every 27 months, but the period varies from less than two years to more than three years.

Easterly winds become established at a height of about 12 miles (20 km) usually between about May and September. Then westerly winds in the upper stratosphere, above about 19 miles (30 km), begin to extend downward until by about January westerlies predominate.

The QBO causes a meridional circulation (*see* GENERAL CIRCULATION) that affects the distribution of traces gases, including OZONE, in the stratosphere over the Tropics and subtropics, and some scientists suspect that in winter the QBO may be linked to the arctic polar VORTEX and to sudden periods of warm weather. This happens because sudden warming in high latitudes is associated with ROSSBY WAVES that start in the troposphere and propagate upward into the stratosphere. Rossby waves are able to propagate in this way if the stratospheric winds are westerlies, but not if they are easterlies.

KELVIN WAVES are also influenced by the QBO. The severity of winters in the Northern Hemisphere is linked to both the QBO and the SUNSPOT cycle. Winters are generally warmer when the stratospheric winds are westerlies than they are when the winds are easterlies. The phase of the QBO affects the frequency and intensity of Atlantic and Pacific TROPICAL CYCLONES, and the QBO is one of the factors influencing the MONSOON. It also affects rainfall patterns in the SAHEL.

Scientists do not fully understand the QBO or the ways in which it affects the weather, however.

Further Reading
Heaps, Andrew, William Lahoz, and Alan O'Neill. "The Quasi-Biennial zonal wind Oscillation (QBO)." University of Reading. Available online. URL: http://ugamp.nerc.ac.uk/hot/ajh/qbo.htm. Accessed January 24, 2006.

Quaternary   The Quaternary sub-era of the Earth's history began approximately 1.81 million years ago and continues to the present day. Although the name is still widely used, it is likely to be dropped from formal use within the next few years. The Quaternary sub-era covers the same span of time as the PLEISTOGENE period. *See* APPENDIX V: GEOLOGIC TIMESCALE.

QuikScat   A NASA satellite that was launched on June 19, 1999, from Vandenberg Air Force Base, California, on a Titan II rocket of the U.S. Air Force. It entered an elliptical orbit at a height of about 500 miles (800 km).

QuickScat, short for Quick Scatterometer, replaced the NASA Scatterometer (NSCAT) that was lost in June 1997, when the satellite carrying it lost power. It carries the SEAWINDS RADAR instrument.

radar An electromagnetic device that is used to detect and measure the direction, distance, and motion of objects that are otherwise invisible due to darkness or because they are obscured by a medium the radar can penetrate. The name is from "*ra*dio *d*etection *a*nd *r*anging."

Radar devices transmit an electromagnetic wave as a beam from an antenna. If the beam is interrupted by striking an object, a part of it is reflected and the device receives the reflection. Information is obtained by comparing the signal and its reflection.

Electromagnetic waves travel at the speed of light, so the time that elapses between the transmission and the reception of its reflection indicates the distance to the object (about 500 feet (150 m) for every microsecond of delay). This type of measurement is made by pulse radar, in which short, intense bursts of radiation are transmitted, with a fairly long interval between bursts. Continuous-wave radar transmits a continuous beam. In its basic form, continuous-wave radar cannot be used to measure distance, but it can measure the speed at which the target object is moving, as the DOPPLER EFFECT. Frequency-modulated radar is a more advanced version of continuous-wave radar that can measure distance, because each part of the signal is tagged to make both it and its reflection recognizable.

A cloud echo is a radar signal that is reflected by CLOUD DROPLETS. The wavelength (*see* WAVE CHARACTERISTICS) of the radar transmission determines the size of the objects that will reflect it. The longer the wavelength, the bigger the objects it will detect. Conse-quently, very short-wave radar beams are used to detect cloud droplets, which are extremely small. It is not possible to tell from a cloud echo whether the cloud is producing PRECIPITATION.

If the cloud is producing precipitation, the radar echo will produce a precipitation echo, which is a characteristic image on the radar screen. A precipitation cell is an area indicated by radar within which precipitation is fairly continuous. A spiral band is a pattern on a radar screen made by echoes from the center of a TROPICAL CYCLONE. The radar reflections are from areas of heavy rainfall, and the pattern forms a roughly spiral shape that curves inward toward the storm center and merges in the wall cloud (*see* CLOUD TYPES).

A bright band is an enhanced radar echo that is seen in an image of a cloud where SNOWFLAKES are melting into RAINDROPS. The melting band is a region in certain clouds, especially nimbostratus, where melting snowflakes become coated in a layer of water. The snow then reflects radar waves more strongly than the ICE CRYSTALS and SNOWFLAKES above or the raindrops below, so the region appears on radar screens as a bright, horizontal band. The existence of melting bands confirms that in middle and high latitudes most of the rain that reaches the ground is melted snow. The height at which melting bands occur is known as the melting level. It is the level at which the temperature is slightly above freezing, so that falling ice crystals and snowflakes begin to melt.

A blob is a signal on a radar screen that indicates a small-scale difference in TEMPERATURE and HUMIDITY.

It is produced by atmospheric turbulence (*see* TURBULENT FLOW).

A wind that is observed by radar is called a radar wind. The radar tracks a radiosonde WEATHER BALLOON or a balloon carrying a radar reflector in order to determine the speed and direction of the wind at a particular height. Knowing the distance to the target from the time lapse between transmission and echo and its angle of elevation allows the height of the balloon or reflector to be calculated by trigonometry.

Doppler radar uses two radar devices to measure the Doppler effect on water droplets and, from that, to determine the speed at which the droplets are rotating about a vertical axis. Doppler radar is employed to study air movements and measure the wind speed inside TORNADOES.

WEATHER RADAR is widely used in the study of clouds, precipitation, and storms. A tornado funnel is visible because WATER VAPOR condenses in the low atmospheric pressure prevailing inside the funnel, producing water droplets that are detectable by radar. The funnel also extends upward, through the wall cloud and into the MESOCYCLONE. These regions cannot be seen, because they are surrounded by cloud that obscures them, but they are visible to radar, allowing the entire structure of a tornadic storm to be examined.

If two radar transmitters are set some distance apart horizontally, the two beams they transmit will enter the cloud at different points. If there is a horizontal component to the movement of water droplets at these points toward or away from the transmitter, there will be a measurable Doppler effect on the reflected beams. This will reveal whether the air inside the cloud is rotating, because if it is rotating the reflection from one side of the center of rotation will be red-shifted and that from the other side will be blue-shifted. If the air is rotating, the extent of the red and blue shifts will indicate the speed of rotation.

The Doppler radar devices used to study tornadoes and other severe storms produce real-time color displays, allowing storms to be studied and measured while they are still active. This has made it possible to issue increasingly accurate severe storm and tornado warnings.

The use of Doppler radar for meteorological research was developed during the 1970s, principally at the National Severe Storms Laboratory, at Norman, Oklahoma. Its first success was in 1973, with the study of a SUPERCELL storm at Union City, Oklahoma. It was by means of Doppler radar that meteorologists discovered the way a mesocyclone forms high in the storm cloud and then extends downward shortly before a tornado emerges beneath the cloud. A device called Doppler on Wheels (DOW), developed by a team led by Joshua Wurman at the Center for Severe Weather Research, was first deployed in 1995; since then it has collected data on approximately 100 tornadoes, as well as eight tropical cyclones and several forest fires. On May 3, 1999, DOW also measured the fastest wind speed ever recorded, of 301 MPH (484 km/h) near ground level in an Oklahoma tornado. Most Doppler radars transmit a single beam and take about five minutes to complete a vertical and horizontal scan of a tornadic storm. Rapid-Scan Doppler on Wheels, also developed by the Wurman team and capable of scanning a tornadic storm at 5–10 second intervals, was deployed in 2005.

Until Doppler radar became available, there was no way to measure the wind speed around the center of a tornado accurately. People guessed at the speeds, and some scientists suggested the winds might occasionally reach or even exceed the speed of sound. At 68°F (20°C) this is about 770 MPH (1,239 km/h). The radar detected winds approaching 300 MPH (483 km/h) and meteorologists now believe this is close to the highest speed they ever reach.

These successes led to the establishment of the Joint Doppler Operational Project (JDOP), based at Norman, Oklahoma, which ran from 1976 until 1979. This led in turn to the NEXT GENERATION WEATHER RADAR program, which began in 1980.

Portable Doppler radars are sometimes used to study tornadoes. These can be set up within a few miles of a tornado. They transmit at a relatively short wavelength. This allows them to emit a narrow beam with a resolution high enough to measure conditions at different parts of the storm, and therefore wind speeds, but their reflections can sometimes be difficult to interpret. Bigger radars study storms from a distance of 100 miles (160 km) or more, but at long distances the curvature of the Earth places the lowest part of a storm below the horizon and out of the reach of a radar beam. Two sets of Doppler radars are used to study storms forming along SQUALL lines, where the storms themselves are moving across the field of view.

Radar will also scan solid surfaces. Radar altimetry is a technique for measuring the topography of a land surface by means of a radar ALTIMETER carried by an aircraft or space vehicle. The altimeter measures the distance between the vehicle carrying it and the surface vertically below it. Provided the vehicle remains at a constant height in relation to a DATUM level (not the ground surface), its distance to the surface will vary according to changes in the ground elevation. With a series of passes the physical features of the landscape can be measured and plotted, and the plots used to compile a map.

Radar interferometry is used to measure very small changes in the shape of features on the solid surface of the Earth. These can give warning of an impending volcanic eruption. An instrument called an interferometer mounted on an observational satellite transmits radar waves that interact with each other to produce characteristic patterns known as interference fringes. The appearance of the fringes varies with very small changes in the distance traveled by the radar pulse and its echo.

Synthetic aperture radar (SAR) is a type of radar in which the instrument moves on board an aircraft or satellite. The SAR transmits a continuous-wave signal at a precisely controlled frequency, so the signal is coherent, and it stores the reflected signals in a memory. This allows it to process a large number of echoes at a time. The effect is similar to that of an instrument with a large antenna, although only a small antenna is used. SAR is used to map the surface of Earth or other planets in great detail.

radiation balance   When the amount of energy that the Earth receives from the Sun is compared to the amount that is reflected and radiated from the Earth into space, the difference between incoming and outgoing radiation represents a radiation balance. This is the energy budget of the Earth, in which the amount of energy the Earth receives from the Sun is set against the way that energy is distributed and eventually returned by radiating it back into space. It can be summarized as:

$$R = (Q + q)(1 - a) - I$$

where R is the radiation balance, Q is the direct sunlight reaching the surface, q is the diffuse sunlight reaching the surface, a is the ALBEDO of the surface, and I is the outgoing long-wave radiation from the surface. On average, the surface of the Earth absorbs about 124 kly (kilolangleys, *see* UNITS OF MEASUREMENT) of radiation each year and radiates about 52 kly

into space. This means that the surface absorbs 72 kly more than it radiates.

The same equation can also be applied to the atmosphere. It absorbs about 45 kly of energy a year and radiates about 117 kly into space. The atmosphere therefore radiates 72 kly more than it absorbs. Consequently, the figures for the surface and atmosphere are in balance over the year. If this were not the case, the world would be growing either warmer or cooler (but *see* GLOBAL WARMING).

The balance is achieved by the transfer of energy from the surface to the atmosphere. Some of this transfer occurs directly. As the surface is warmed by the Sun, the air in contact with it is also warmed by CONDUCTION, and its warmth is transferred upwards by CONVECTION. Most of the transfer is due to the EVAPORATION of water from the surface and its CONDENSATION in the atmosphere. Evaporation absorbs LATENT

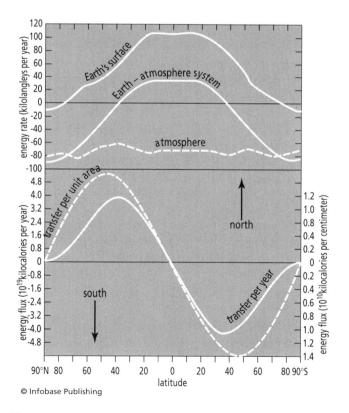

© Infobase Publishing

The upper section of the graph shows the distribution of solar energy per year for the surface of the Earth, the atmosphere, and the surface and atmosphere considered together. The lower section shows the horizontal transfer of energy from low to high latitudes.

HEAT from the surface and condensation releases latent heat into the air.

Solar energy is not received equally in all areas of the Earth. The Sun shines more intensely on the equator than it does on the poles. The resulting difference in available energy increases with latitude. This is because in latitudes higher than 40°, the atmosphere loses more energy in a year than the surplus that is absorbed by the surface, and in latitudes lower than 40° the reverse is true and the surface absorbs more energy than the atmosphere loses. Despite this, the poles do not grow colder year by year, nor the TROPICS warmer. The balance is maintained by the horizontal transfer of heat. Without that transfer the radiation budget would balance at each latitude only if the mean temperature at the equator were 25°F (14°C) higher than it is and the mean temperature at the poles 45°F (25°C) lower.

Ocean currents (*see* GYRES) and the GENERAL CIRCULATION of the atmosphere transfer heat away from the equator. ADVECTION through the atmosphere accounts for a little more than half of the total transfer. Evaporation and condensation also increase the rate of this horizontal transfer of heat from low to high latitudes.

The total amount of annual heat transfer reaches a maximum at about latitudes 40°N and S. The maximum amount per unit area of the surface is transferred at about 45°N and S. It is slightly greater in the Southern Hemisphere than in the Northern Hemisphere. Much of the transfer in the Tropics and subtropics occurs in the upper troposphere (*see* ATMOSPHERIC STRUCTURE) through the Hadley cell circulation. The transfer in middle latitudes occurs mainly through the CYCLONES and frontal systems (*see* FRONT) that move with the prevailing westerlies (*see* WIND SYSTEMS). The rate of meridional transport varies with the temperature gradient. The gradient is steepest in winter, and that is when the atmospheric circulation is most vigorous.

radiation cooling At night the Earth's surface radiates heat, thereby reducing the surface temperature. This is radiation cooling.

During the day the Earth's surface absorbs solar radiation. Some of the absorbed radiation is in the form of heat, and absorbed light (*see* SOLAR SPECTRUM) is converted to heat. As heat absorption causes its temperature to rise, the surface also emits infrared radiation (*see* BLACKBODY). The amount of energy radiated in this way is proportional to the temperature of the surface (*see* STEFAN-BOLTZMANN LAW), but during the day it is always less than the amount of solar radiation that is being absorbed. Consequently, the surface temperature rises through the day, reaching a peak in the middle of the afternoon. At night the blackbody radiation from the surface continues, but the absorption of solar radiation ceases. Consequently, there is a net loss of energy from the surface and therefore the surface cools. Its temperature continues to fall through the night, then starts rising again at dawn.

Most of the blackbody infrared radiation is absorbed by gases in the atmosphere (*see* GREENHOUSE EFFECT). These then reradiate it. On clear nights atmospheric absorption occurs throughout the troposphere (*see* ATMOSPHERIC STRUCTURE). It warms the air, but allows the ground surface to cool. On cloudy nights the clouds form a barrier to the radiation. They reflect some of it and absorb some of it, reradiating a proportion of it downward. This greatly reduces the rate at which the surface cools and explains why in spring and fall frosts are more likely on clear nights than on cloudy nights.

A night when the sky is clear is known as a radiation night. The absence of cloud allows the surface to cool rapidly. The drop in temperature close to the ground is especially marked when the air is still, because air movements would mix warmer air with the cold air at ground level.

radiative dissipation When fingers of warm air penetrate cooler air, they cool by radiating away, or dissipating, their surplus heat. This is radiative dissipation.

While the air is at a higher temperature than its surroundings, the temperature difference represents available POTENTIAL ENERGY that could be converted into KINETIC ENERGY. By radiating its heat, however, the available potential energy is dissipated before it can be converted.

radioactive decay Unstable atomic nuclei lose elementary particles and continue to lose them until the nuclei become stable. This represents the decay of those nuclei. ISOTOPES of chemical elements that have nuclei subject to this type of decay are said to be radioactive, since the decay results from the emission of electromagnetic radiation and the loss of particles. It is radioactive decay.

For example, naturally occurring uranium (U) is a mixture of three isotopes. ^{238}U accounts for 99.28 percent of the total mass, ^{235}U for 0.71 percent, and ^{234}U for 0.006 percent. All three isotopes are radioactive. ^{234}U and ^{238}U both decay in the same way and eventually become ^{206}Pb (lead-206), which is stable. ^{238}U has a HALF-LIFE of 4,510 million years and the half-life of ^{234}U is 2.48 million years. These isotopes undergo ALPHA DECAY eight times and BETA DECAY six times. ^{235}U, with a half-life of 713 million years, undergoes seven alpha decay steps and four beta decay steps, and finally it also becomes a stable isotope of lead, ^{207}Pb.

It is impossible to predict when an individual nucleus will undergo decay, but the rate at which a particular isotope decays is very constant and can be used to determine the half-life for that isotope. The constancy of radioactive decay means it can be used for RADIOMETRIC DATING, the best known version of which is RADIOCARBON DATING.

radiocarbon dating   A method that exploits the regularity of RADIOACTIVE DECAY to calculate the age of a sample taken from what was once a living organism by measuring the proportions of two carbon ISOTOPES that the sample contains. The two ISOTOPES are carbon-12 (^{12}C) and carbon-14 (^{14}C).

^{12}C is the stable isotope. ^{14}C is formed in the atmosphere by the action of COSMIC RADIATION on nitrogen (^{14}N). Cosmic radiation bombards the air with neutrons. Occasionally a neutron strikes the nucleus of a nitrogen atom and replaces one of its protons. That replacement leaves the mass of the atom unchanged, but alters it from $^{14}_{7}N$ to $^{14}_{6}C$. The ^{14}C is then oxidized to carbon dioxide (CO_2).

Plants absorb CO_2 and incorporate its carbon in sugars by the process of PHOTOSYNTHESIS. The plants do not discriminate between ^{12}C and ^{14}C, so the two isotopes are present in plants in the same proportion as they are in the atmosphere. The carbon then passes to animals that eat the plants and their bodies also contain both isotopes.

^{14}C is unstable and undergoes radioactive decay with a HALF-LIFE of 5,730 years. While the plant or animal is alive, the ^{14}C in its tissues is constantly being replenished, but after its death the organism ceases to absorb carbon and replenishment ceases. The ^{14}C continues to decay at a steady and known rate. This alters the ratio of $^{14}C:^{12}C$. Measuring the ratio in the sample and comparing it to that in the atmosphere therefore make it possible to calculate the time that has elapsed

since the organism died. Obviously the technique can be used only with material that once formed part of a living organism. It can date wood, linen, cotton, wool, hair, or bone, but not stone or metal.

When the technique was first introduced in the late 1940s by the American chemist Willard Frank Libby (1908–80) the half-life of ^{14}C was thought to be 5,568 years. The year 1950 was set as the base (the "present") against which ages were to be reported. That date was chosen because it was about then that the atmospheric testing of nuclear weapons altered the $^{14}C:^{12}C$ ratio. Before long, material was being examined and then dated as having been formed so many years BP (before the present). Some years later the ^{14}C half-life was recalculated and all the previously announced dates had to be revised.

Then another difficulty was discovered. Radiocarbon dating was based on the assumption that the ratio of $^{14}C:^{12}C$ in the atmosphere had remained constant for tens of thousands of years. This is now known to be untrue. The ratio varies a little. This means that ages measured by radiocarbon dating do not correspond directly to historical ages. The variation is not constant, the extent of it fluctuating in an irregular fashion, which makes dating far from simple.

The difference can cause confusion. A human skeleton that was found in California in 2000 was reported as being 13,000 years old. At Monte Verde, a site in southern Chile, there are objects made by people and dated at 12,500 years. It sounds as though the California skeleton is older than the Monte Verde site, but this is not so. The California date was in calendar years, the Monte Verde date in radiocarbon years, and 12,500 radiocarbon years are equal to 14,700 calendar years.

It is important, therefore, to state what kind of years are being reported. Radiocarbon dates should always be described as "radiocarbon years" BP, or as "percent modern," which means the proportion of ^{14}C in the sample compared with that in samples that were formed in 1950, such as a piece of wood cut from a tree in that year.

Radiocarbon years are calibrated to convert them into calendar years. The calibration compares radiocarbon years and TREE RING years. The $^{14}C:^{12}C$ ratio is measured in the sample. Then a tree ring with a similar ratio is found by searching through published data. The age of the tree ring is known, and because it contains the same carbon isotope ratio the tree ring

Calendar and radiocarbon ages

Calendar	Radiocarbon
11	9.6
12	10.2
13	11.0
14	12.0
15	12.7
16	13.3
17	14.2
18	15.0
19	15.9
20	16.8

must be the same age as the sample. The calibrated age is then reported as CalBC, CalAD, or CalBP, to indicate whether the age is measured as B.C. (B.C.E.), A.D. (C.E.), or BP.

The table above compares some radiocarbon and calendar ages (in thousands of years).

Radiocarbon dating is widely used in paleoclimatology (see CLIMATOLOGY). It makes it possible for scientists to date plant material that they can link to particular climatic conditions.

Further Reading

Guilderson, Tom P., Paula J. Reimer, and Tom A. Brown. "The Boon and Bane of Radiocarbon Dating." *Science*, 307, 362–364. January 21, 2005.

Higham, Thomas. "Radiocarbon Dating." Available online. URL: www.c14dating.com/int.html. Accessed January 25, 2006.

radiometric dating Any method that exploits the very regular rate of RADIOACTIVE DECAY in order to determine the age of rocks and organic material can be called radiometric dating. This regularity means that the proportions of parent (original, radioactive) and daughters (decay products) ISOTOPES present in the material can be used to calculate the time that has elapsed since the material was formed. At that time the material either contained the isotopes in different proportions, or contained only the parent isotope if the daughters are produced only by the decay of the parent.

Several elements are used in radiometric dating. Organic material that is thousands of years old can be dated by RADIOCARBON DATING. Other decay series are used to date rocks, which are much older. The first to be introduced was the decay of uranium (HALF-LIFE 4.51 billion years or 713 million years, depending on the isotope) to lead (Pb). ^{232}Th (thorium-232), with a half-life of 13.9 billion years, decays to ^{207}Pb and this decay is also used for dating.

Potassium (^{40}K) decays to argon (^{40}Ar) with a half-life of 1.5 billion years. Rubidium (^{87}Rb) decays to strontium (^{87}Sr) by a single BETA DECAY, but there is uncertainty about the half-life, which may be 48.8 billion years or 50 billion years. Samarium (^{147}Sm), with a half-life of 250 billion years, undergoes ALPHA DECAY to become neodymium (^{143}Nd). All of these decays are used to date rocks.

radon (Rn) A radioactive element that is one of the NOBLE GASES. Radon is colorless, tasteless, and odorless. It has atomic number 86, relative atomic mass 222, and its DENSITY is 0.09 ounces per cubic inch (0.97 g/cm^3). It melts at -95.8°F (-71°C) and boils at -79.24°F (-61.8°C).

Radon is formed by the RADIOACTIVE DECAY of radium-226 and undergoes ALPHA DECAY. There are at least 20 ISOTOPES of radon, the most stable of which is ^{222}Rn, which has a HALF-LIFE of 3.8 days.

Air contains very small but variable amounts of radon. Radium, the source of radon, is a rare metal that occurs mainly in granites. Consequently, the concentration of radon is highest where the underlying rock is granite.

As they decay, radon atoms ionize (see ION) nearby atoms of atmospheric gases. This process increases the electrical conductivity of air, but the effect is extremely small. Most radon decay products, called radon daughters, also emit alpha radiation. In large doses, radon and its daughters are known to cause lung cancer and so measures are taken to prevent the accumulation of radon inside buildings where the natural radon level is high.

rain Rain is liquid PRECIPITATION consisting of droplets that are between 0.02 inch and 0.2 inch (0.5–5.0 mm) in diameter. In heavy rain the droplets vary in size considerably. Small RAINDROPS are spherical, but those approaching 0.2 inch (5 mm) in size are variable in shape. Fine rain falls from nimbostratus and stratocumulus (see CLOUD TYPES) when the cloud base is low. Droplets more than about 0.4 inch (1 mm) in size require to fall through several thousand feet of cloud in order to grow to this size. These are usually associated with cumulus or cumulonimbus clouds and they fall as SHOWERS or in rainstorms.

Most raindrops are SNOWFLAKES that have melted in the lower part of the cloud or between the CLOUD BASE and the surface. Rain falling from a warm cloud— a cloud consisting entirely of liquid droplets, with no ICE CRYSTALS—is known as warm rain. Warm rain is at a higher temperature than rain that falls from cold clouds—consisting entirely of ice crystals—or mixed clouds—containing both ice crystals and liquid droplets.

Rain that contains a quantity of fine soil particles that are large enough to discolor the rainwater and leave a mudlike deposit on surfaces is called mud rain.

Rainfall frequency is a measure of how often rain falls in a particular place. This is of vital agricultural importance, and it also determines the type of natural vegetation a region supports. Two places may both receive the same amount of rain in the course of a year, but the place where rainfall is distributed evenly throughout the year will have a quite different type of natural vegetation from the place where all the rain falls in one short season. The number and distribution of rain days reveal the effectiveness of rainfall for agriculture and natural plant growth. Rainfall frequencies are also used to define RAINFALL REGIMES.

As the name suggests, a rain day is a day on which rain falls. This is usually taken to mean a period of 24 hours, beginning at 0900 Z (see UNIVERSAL TIME), during which at least 0.08 inch (0.2 mm) of rain falls. The number of rain days in one month and in one year indicate the distribution of rainfall through the year and provide a useful measure for comparing the climates of two places.

Seattle, Washington, for example, has an average of 51 rain days between April and September and 99 rain days between October and March, indicating that its climate is rainier in winter than in summer. Las Vegas, Nevada, on the other hand, is dry throughout the year, with no more than one or two rain days in each month and a total of 18 rain days in the year. Taken together, the number of rain days and total annual rainfall give a clear picture of the seasonality of a climate. Seattle receives a total of 33.4 inches (848 mm) of rain a year on 150 rain days. Mumbai, India, with a strongly seasonal MONSOON climate, receives 71.3 inches (1,811 mm) a year but has only 72.7 rain days. Dividing the amount of rainfall by the number of rain days gives an idea of the intensity of the rainfall.

A period each year when the amount of precipitation is much higher than it is at other times is known as a rainy season. The rainy season may occur in winter or in summer. Rainy winters occur around the Mediterranean Sea and on the western sides of continents in latitudes 30–45°N and S. San Francisco, California, receives an average 22 inches (561 mm) of rain a year, of which 21 inches (534 mm) fall between October and April. Gibraltar, at the southern tip of Spain, has a very similar climate. Its average annual rainfall is 30 inches (770 mm), of which 28 inches (711 mm) fall between October and April. In regions with a monsoon climate it is the summer that is wet. Mumbai receives an average 71.3 inches (1,811 mm) of rain a year. Of this total, 67 inches (1,707 mm) falls between June and September.

Zenithal rain is rain that falls every year, or every other year, during the season when the Sun is most nearly overhead—at its zenith. This type of rainfall distribution occurs in parts of the TROPICS and subtropics.

A hyetogram is a chart that records the amount and duration of rainfall at a particular place. Hyetography is the scientific study of the annual geographic distribution of precipitation and variations in it.

raindrops RAIN is PRECIPITATION that reaches the ground as drops of WATER larger than those of drizzle. The drops of water that fall from a cloud to reach the surface as rain vary in size, but there is a minimum size below which they are unable to survive their passage through the air below the CLOUD BASE.

Mist and FOG consist of droplets about 100 μm (0.004 inch) in diameter, but these are really very large CLOUD DROPLETS. They form near the surface, and they are too small to fall because air resistance and turbulence are sufficient to keep them aloft. Raindrops are at least about 1 mm (0.04 inch) across. The size limitation is due to the rate at which a drop of water falls in relation to the rate at which it evaporates.

The TERMINAL VELOCITY (V) of a drop of water falling through still air is equal to 8,000 times its radius ($V = 8 \times 10^3 r$), where r is the radius measured in SI units (see APPENDIX VIII: SI UNITS AND CONVERSIONS). A drop that is 1 mm (0.04 inch) in diameter therefore falls at about 4 m/s (157 inches per second). The base of a cloud marks the boundary between saturated and unsaturated air. If the base is at 60 m (200 feet), the drop will spend 15 seconds falling through the dry air. Suppose, though, that the drop is half that size. It will then spend 30 seconds in the dry air. If the air is turbulent, as it is inside cumuliform clouds (*see* CLOUD TYPES) and around drops that are very much bigger, the terminal velocity is much slower ($V = 250r^{1/2}$). A raindrop 5 mm (0.2 inch) across falls at about 0.4 m/s (16 inches per second) and would take 150 seconds to fall 60 m (200 ft).

The rate of EVAPORATION depends on the TEMPERATURE and relative HUMIDITY (RH) of the air beneath the cloud. RH is variable, but even if it is about 95 percent, so the air is almost saturated, a drop that is less than about 30 µm (0.0012 inch) will have evaporated completely by the time it has fallen a few inches. No drop smaller than about 100 µm (0.004 inch) in diameter is likely to reach the surface, because the base of most clouds is higher than 500 feet (150 m). That is the maximum distance such a drop can fall through air at 40°F (5°C) and RH 90 percent before it evaporates.

Bigger drops survive much better. A droplet 1 mm (0.04 inch) across can fall 42 km (26 miles) through air at this temperature and RH before it evaporates completely, and one 2.5 mm (0.1 inch) can fall 280 km (174 miles). In fact it is not possible for raindrops to fall this distance, because above about 7 miles (11 km) the air is too dry for drops to form.

Consequently, bigger drops reach the surface as raindrops. Typical raindrops are between 0.08 inch (2 mm) and 0.2 inch (5 mm) in diameter, and they fall at 14–20 MPH (23–33 km/h). Drops smaller than these that reach the ground are classed as drizzle. Raindrops larger than about 0.2 inch (5 mm) usually break apart into two or more smaller drops.

Raindrops form from cloud droplets. These are typically about 20 µm (0.0008 inch) across. Depending on the temperature inside the cloud, they grow by collision or by the Bergeron–Findeisen mechanism. In order to grow to the size of a drizzle droplet, about 300 µm (0.1 inch) across, the cloud droplet must increase its volume more than 3,000 times and to attain the size of a raindrop 2 mm (0.08 inch) across it must grow almost 1 million times bigger. (The volume of a sphere is equal to $4/3\ \pi r^3$, where r is the radius.)

The Bergeron–Findeisen mechanism is a theory to explain how cloud droplets grow into raindrops in cold clouds (clouds in which the ambient temperature is below freezing). It is sometimes called the Wegener–Bergeron–Findeisen mechanism, because the German meteorologist Alfred Wegener suggested the first stage in the process in 1911. The Norwegian meteorologist Tor Bergeron was the first scientist to propose it in full in 1935 (*see* APPENDIX I: BIOGRAPHICAL ENTRIES for information about Wegener and Bergeron), and later the German meteorologist Walter Findeisen demonstrated it in large CLOUD CHAMBERS.

Cold clouds contain both ICE CRYSTALS and supercooled (*see* SUPERCOOLING) water droplets. The SATURATION VAPOR PRESSURE over an ice surface is lower than that over a liquid water surface. This is especially true at temperatures between about 5°F and -13°F (-15°C and -25°C), when the difference amounts to about 0.2 mb (20 Pa). Consequently, water will evaporate from the droplets and accumulate on the crystals by direct deposition from WATER VAPOR. The crystals grow, collide with one another, and stick together to form aggregations—SNOWFLAKES. This continues until the snowflakes are heavy enough to start falling. As they fall, the snowflakes collide with more supercooled droplets. The water freezes onto them, so the flakes keep on growing. Some are broken apart by air currents, and the splinters of ice produced in this way act as FREEZING NUCLEI for the formation of more ice crystals that join together into more snowflakes. Those snowflakes that remain intact fall from the base of the cloud. If the temperature in the air beneath the cloud is above freezing, the snowflakes will start to melt and some or all of them will reach the ground as rain. In middle latitudes, most rain SHOWERS consist of melted snow, even in the middle of summer.

A small region inside a cloud where ice crystals are growing more rapidly than they are elsewhere in the cloud at the expense of a concentration of supercooled droplets is called a precipitation-generating element. When they exceed a certain size, the ice crystals will fall through the cloud, generating precipitation by the Bergeron-Findeisen mechanism.

The alternative mechanism for raindrop formation is called collision theory. This theory that describes the way raindrops form in warm clouds, where no ice crystals are available and therefore the Bergeron-Findeisen mechanism does not apply. Warm clouds contain droplets of varying sizes. The larger ones fall through the cloud at a terminal velocity that varies with their size,

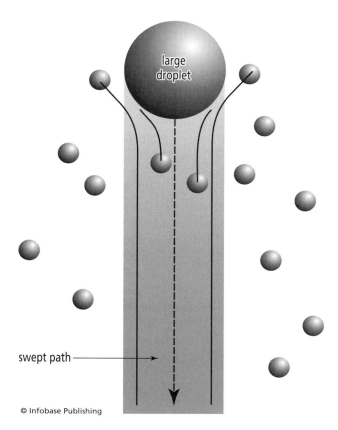

© Infobase Publishing

As it falls, the large droplet collides with the smaller droplets in its path. Only those small droplets that are close to the center of the path swept by the large droplet collide with it. Those farther away are carried to the sides by the airflow.

so large droplets fall faster than smaller ones. As they fall, the large droplets collide and coalesce with smaller droplets in their path. The rate at which collisions and coalescence occurs is measured by the collision efficiency and coalescence efficiency.

The collision efficiency is measured by the proportion of droplets in a cloud that collide with other droplets. Although it might seem that water droplets are crowded closely together inside a cloud, in fact they are widely separated in relation to their own size and so collisions are by no means inevitable. A large droplet has a higher terminal velocity than a small one and therefore falls faster. A warm cloud therefore consists of relatively large droplets falling through smaller ones. In order to collide, the small droplets must be very close to the center of the path followed by the large ones. If they are not, the displacement of air by the large droplets will sweep the small ones to the sides and

away and collisions will not occur. Collision efficiency increases with the size of the large droplets. Droplets smaller than 20 μm in diameter are swept aside without colliding. Droplets more than about 40 μm across collide with most of the small droplets in their path. The higher the collision efficiency, the more rain the cloud will produce. Collision does not necessarily mean the droplets will coalesce.

Sweeping is a mechanism by which raindrops are believed to grow. It is a variant of the collision and coalescence theory that is based on the fact that the terminal velocity of a raindrop is proportional to its size. Large drops therefore fall faster than small drops. They will collide with some slower, smaller drops and these will merge with them (collision and coalescence). Other small droplets will be swept into the wake of the bigger drop and absorbed by them in that way.

Coalescence is the merging of two or more cloud droplets into a single, larger droplet. Colliding droplets may bounce away from each other. Droplets of similar size usually coalesce temporarily, oscillate, and then separate into two or more smaller droplets. But droplets of widely different sizes will coalesce to form a stable droplet.

The proportion of colliding cloud droplets that merge to form larger drops is known as the coalescence efficiency. Coalescence efficiency is greatest where there is the greatest difference in size between the large and small droplets. It also varies with the relative velocities of the droplets and the angle at which they collide. The higher the coalescence efficiency, the more rain the cloud will produce. Atmospheric electricity increases coalescence efficiency by placing opposite charges on the surfaces of droplets, so they are attracted to one another. The merging of cloud droplets that carry opposite electrostatic charges (*see* STATIC ELECTRICITY) on their surfaces is called electrostatic coalescence.

Falling raindrops engulf solid particles with which they collide, and the rain then carries the particles to the surface. This process is called washout. Particles are also removed from the air by FALLOUT, impaction (*see* AIR POLLUTION), and rainout (*see* CLOUD CONDENSATION NUCLEI).

rainfall regime   A CLIMATE CLASSIFICATION that is based on rainfall frequency (*see* RAIN). It was devised by the British climatologist W. G. Kendrew and first published in his book *The Climates of the Continents.* Kendrew proposed six regimes.

(i) Equatorial regime. This comprises two seasons of heaviest rain, at or about the time the Sun is directly overhead, with drier periods between, but no dry season.

(ii) Tropical regime. There are two types of tropical regime. In the inner tropical, between the equator and latitudes 10°N and S, maximum rainfall occurs twice in the year, separated by a long dry season. In the outer tropical regime, between 10° and the TROPICS, the two wet seasons merge and the dry season is longer.

(iii) Monsoon regime. This has a marked summer maximum and a dry winter (*see* MONSOON). It occurs both inside and outside the Tropics.

(iv) Mediterranean regime. Most of the rain falls in winter, with either a single maximum in the middle of winter or two maxima, in spring and fall, and the summer is dry.

(v) Continental interior regime. This regime occurs in temperate latitudes. Most rain falls in summer. Winter rainfall is much less, but the winter is not completely dry.

(vi) West coastal regime. This regime occurs in temperate latitudes. Rain is abundant in all seasons, but heavier in the fall or winter than it is at other times.

Further Reading
Kendrew, W. G. *The Climates of the Continents.* Oxford: Oxford University Press, 1st ed. 1922, 5th ed., 1961.

rain forest A forest that grows where the rainfall is heavy and spread fairly evenly through the year. Rain forests grow in both the TROPICS and in temperate latitudes.

Tropical rain forest was first defined in 1898 by the German botanist and ecologist Andreas Franz Wilhelm Schimper (1856–1901), who also coined the term (in German, as *tropische Regenwald*). Schimper said the forest comprises evergreen trees that are at least 100 feet (30 m) tall, rich in thick-stemmed lianas (creepers), and with many woody and herbaceous (nonwoody) epiphytes growing on them. An epiphyte is a plant that grows on the surface of another plant, but that is not a parasite. This definition still stands. The forest trees form a continuous canopy. There are also trees up to 100 feet (60 m) tall that stand high above the canopy.

Temperate rain forest develops outside the Tropics, wherever the annual rainfall exceeds 60–120 inches (1,500–3,000 mm). It consists of broad-leaved evergreen trees, often with coniferous species, and with abundant climbers and epiphytes. This type of forest is found in coastal areas of the southeastern United States, northwestern North America, southern Chile, parts of Australia and New Zealand, and in southern China and Japan.

rain gauge The instrument that is used to measure the amount of rain that falls in a particular place during a given period of time, usually one day, is known as a rain gauge. The amount of rain that falls over a specified period into a particular rain gauge, or the amount that is estimated to have fallen during that period at a particular place, is known as the point rainfall. The point rainfall often refers to the amount of rain that fell during a single storm, or during a period of unusually wet or dry weather.

There are several ways to measure rainfall. The simplest is to leave an open-topped container exposed to the rain. This will collect rain, but water will also evaporate from it, especially after the rain has ceased falling, so it will not give an accurate reading. It is impossible to correct for this, because the rate of EVAPORATION varies with the air TEMPERATURE and wind speed.

The standard rain gauge is designed to minimize evaporation losses. It is called "standard" because it is the design that is approved internationally for use at WEATHER STATIONS. All standard rain gauges are made to the same specification and dimensions. When standard gauges are used, the scientists who use the data to prepare forecasts know that all the measurements reported to a meteorological center have been made in exactly the same way. If the staff at each station could decide for themselves how to make their measurements and what instruments to use, the data from one station would not be strictly comparable to that from another. What one station called 1.01 inches (25.65 mm) of rain, for example, another might call 1.00 inch (25.4 mm) and a third 1.02 inches (25.91 mm). The difference amounts to no more than a tiny one-hundredth of an inch (0.25 mm), but it would introduce an uncertainty into the data that is easily avoided by standardization.

The exterior of a standard gauge is a cylinder 7.9 inches (20 cm) in diameter that is mounted vertically with its top 39.4 inches (1 meter) above ground level. Rain falling into the cylinder enters a funnel that guides the water into a measuring tube inside the cylinder. The diameter of the measuring tube is 2.49 inches (6.32 cm), therefore the cross-sectional area of the measuring tube is one-tenth that of the mouth of the funnel. Consequently, a column of water 10 inches (or millimeters) high in the measuring tube represents one inch (or millimeter) of rain entering the funnel. Multiplying by 10 makes it easier to read the rainfall accurately. The measuring tube may be calibrated, or the height of water in it may be measured with a ruler or other measuring stick.

A standard gauge has to be visited at regular intervals, but it is also possible to measure and record rainfall automatically. The instrument most often used for this purpose is the tipping-bucket gauge.

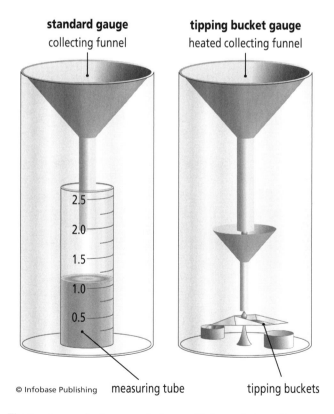

standard gauge
collecting funnel

tipping bucket gauge
heated collecting funnel

2.5
2.0
1.5
1.0
0.5

© Infobase Publishing   measuring tube   tipping buckets

The two types of rain gauge that are widely used at weather stations. The standard gauge (left) has to be visited at regular intervals to be read and reset. The tipping-bucket gauge (right) allows the rainfall to be recorded automatically.

Like the standard gauge, a tipping-bucket gauge is contained in a cylinder with a diameter of 7.9 inches (20 cm), but it has no measuring tube. Instead, rainwater entering the funnel is guided to a second, smaller funnel and from there into one of two buckets. Each bucket holds 0.01 inch (0.25 mm) of water. The two buckets are mounted on a rocker, like a seesaw. When 0.01 inch (0.25 mm) of water has flowed into a bucket, its weight tips the bucket downward. The bucket then makes an electrical contact that is transmitted to a recording pen which moves on a graph mounted on a rotating drum. At the same time, as the buckets tip the second bucket is positioned to collect water and the first bucket is emptied.

There are also gauges that automatically record the height of the water in the measuring tube. One design uses a float valve to do this and another measures the flow of an electric current through the water column, from which the height of the column is calculated.

A hyetograph is an instrument that measures and automatically records rainfall. It consists of a reservoir in which rainwater collects. The reservoir also contains a float that is connected mechanically to a pen. As the water accumulates the float rises, moving the pen, which traces a line on a chart fixed to a rotating drum.

Rainfall can also be measured automatically by a weighing gauge. In this device the collected water is fed into a cylinder that rests on a balance. Nowadays this is usually an electronic balance, but earlier instruments used a spring balance. The weight of the water is recorded on a graph.

All of these instruments are subject to errors. Although the standard gauge minimizes evaporation losses by enclosing the collecting funnel and measuring tube inside the outer cylinder, it cannot eliminate them entirely. Some water evaporates through the tube of the collecting funnel. The first rain to fall into the gauge wets the funnel and the film of water coating the funnel does not reach the measuring tube, bucket, or weighing cylinder. This is called wetting, and although the amount is no more than 0.04–0.08 inch (0.1–0.2 mm), this can be significant where rainfall is extremely light. Heavy rain can overflow tipping buckets, so they rock rapidly back and forth but under-record the amount of rainfall.

Wind causes even greater inaccuracies. EDDIES that form around the gauge carry a variable proportion of

RAINDROPS across the mouth of the funnel. Gauges are often placed inside a shield in order to minimize wind losses. The shield consists of a horizontal circular hoop, 20–40 inches (50–100 cm) in diameter, with baffles hanging vertically from it.

Trees, buildings, and other obstructions can shelter a rain gauge from rain that is falling obliquely. To avoid this, the distance between the gauge and the nearest obstruction should be at least equal to the height of the obstruction.

An optical rain gauge measures the intensity of rainfall. It transmits an infrared beam along a horizontal path 20–40 inches (50–100 cm) long to a detector. When the beam strikes a raindrop, the radiation is scattered forward. The detector records the SCATTERING, from which the number of raindrops per second and in a unit volume of air can be counted. This provides a continuous record of rainfall intensity, but it does not measure the amount of rain directly.

rainmaking Any attempt to induce PRECIPITATION to fall from a cloud that otherwise might not have released it can be called rainmaking. Several methods have been tried to achieve this. All them aim to induce the formation of CLOUD DROPLETS and then to stimulate their growth. This can be done by injecting suitable material into the cloud. The process is called CLOUD SEEDING.

raised beach A strip of land that contains rounded pebbles, sand, and seashells by which it can be recognized as having once been a beach, although it is now some distance above the sea shore. The location of the shore has changed because either the land has risen as a result of movements in the Earth's crust, such as GLACIOISOSTASY, or because the sea level has fallen. The accumulation of water in ICE SHEETS and GLACIERS is the most likely explanation for a fall in sea level (as opposed to a rise in land level).

Raoult's law A law that was formulated by the French physical chemist François-Marie Raoult (1830–1901) in 1886. It states that when one substance (the solute) is dissolved in another (the solvent), the partial pressure (see AIR PRESSURE) of the solvent vapor that is in equilibrium with the solution is directly proportional to the ratio of the number of solvent molecules to solute molecules.

The law means that the equilibrium vapor density (see BOUNDARY LAYER) above the surface of a solution is lower than that above the surface of pure solvent and the more concentrated the solution, the greater is the difference. Consequently, more water will enter the solution. CLOUD DROPLETS that form on hygroscopic nuclei (see CLOUD CONDENSATION NUCLEI) are solutions that are often quite concentrated. Raoult's law shows that such droplets will then grow rapidly by the CONDENSATION of more WATER VAPOR.

Rapid Climate Change (RAPID) A British program launched in 2001 and funded by the UK government through the Natural Environment Research Council that aims to quantify the likelihood and magnitude of rapid climate change in the future. The program cost £20 million (about $36 million) and runs from 2001 until 2007. It studies all aspects of rapid climate change, but concentrates especially on changes affecting the THERMOHALINE CIRCULATION in the Atlantic Ocean.

RAPID monitors change by means of moored devices called profilers. These are sensors that move vertically up and down wires connected to buoys on the sea surface and moorings on the seafloor. As they move, the sensors acquire data that they transmit to an orbiting satellite.

Further Reading

Natural Environment Research Council. "Welcome to the Rapid Climate Change Home Page." NERC. Available online. URL: www.noc.soton.ac.uk/rapid/rapid.php. Last modified January 19, 2006.

Rayleigh number A DIMENSIONLESS NUMBER, calculated by Lord Rayleigh (see APPENDIX I: BIOGRAPHICAL ENTRIES), that describes the amount of TURBULENT FLOW in air that is being heated from below by CONVECTION. The Rayleigh number (Ra) is calculated from the equation:

$$Ra = [(g\Delta\theta)/kv\theta](\Delta z)^3$$

where g is the gravitational acceleration, $\Delta\theta$ is the POTENTIAL TEMPERATURE lapse in a layer of air with a depth Δz, k is the thermal conductivity of the air, and v is the kinematic VISCOSITY of the air.

Convection will always produce turbulence when the amount of heating is too great for energy to be

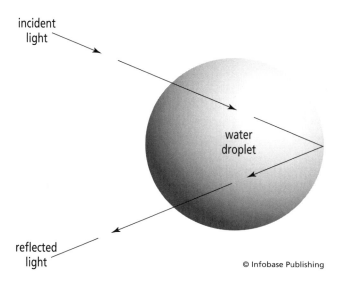

In internal reflection, incident light rays pass through the water droplet to its far side from where they are reflected.

transferred by conduction between molecules and when the resulting air motion is too vigorous for LAMINAR FLOW to be sustained by viscosity. Turbulence will occur when the Rayleigh number exceeds approximately 50,000.

reflection The "bouncing" of light when it strikes an opaque surface. The angle at which light strikes a surface is known as the ANGLE OF INCIDENCE, and the angle at which it is reflected is the angle of reflection. The angle of reflection is always equal to the angle of incidence.

Objects that are not themselves sources of light are made visible by the light reflected from them. If the surface is smooth, then all the light will strike it and be reflected from it at the same angle. If the surface is uneven, the angle at which light strikes it will vary from place to place, as will the angle at which it is reflected. Such uneven reflection scatters the light, producing multiple or distorted images.

Light that has passed through a transparent body may be reflected from the inside surface of the body on the far side. This is called internal reflection. The internal reflection from water droplets is partly responsible for such OPTICAL PHENOMENA as rainbows.

refraction When light passes obliquely from one transparent medium to another transparent medium through which it travels at a different speed, the light is bent, or refracted. The extent of the refraction is proportional to the difference in the speed of light in the two media and to the angle at which the light enters.

The ratio of the speed of light in air to the speed of light in the medium is a constant for that medium, known as its refractive index. Air has a refractive index of 1.0003, the refractive index of ice is 1.31, that of water is 1.33, and that of window glass is 1.5.

The angle through which light is refracted is known as the angle of refraction. This angle varies according to the angle at which the light strikes the boundary between the two media and also to the difference in their refractive indices. The smaller the incident angle, the greater will be the angle of refraction, and when the incident angle is 90°, light is not refracted at all, although its speed changes. The greater the difference in the refractive indices of the two media, the greater will be the angle of refraction.

When light passes from a medium with a low refractive index into one with a high refractive index, it is bent toward the vertical in relation to the surface

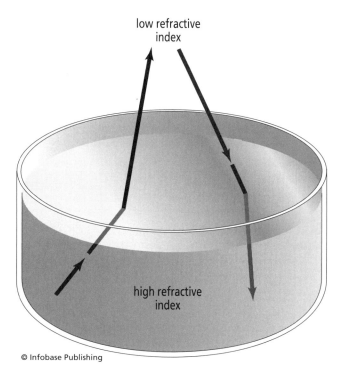

Light rays bend as they pass across the boundary between two transparent substances, such as air and water, through which light travels at different speeds.

between the two media. Light passing from a high to a low refractive index is bent in the opposite direction, away from a line vertical to the boundary.

remote sensing   Remote sensing is the acquisition of information about an object without being in direct physical contact with that object. Instruments carried by a WEATHER BALLOON are in contact with the air they are monitoring, so although the meteorologists who receive data from them are not in contact with that air, the sensors are, so the sensing itself is direct, not remote. An orbiting satellite, in contrast, transmits atmospheric data that it acquires from outside the atmosphere. This sensing is remote and works by acquiring images in various parts of the electromagnetic spectrum, such as the microwave, infrared, visible light, and ultraviolet wavebands. Scientists on the ground are able to interpret these images to obtain information about various aspects of climate, such as TEMPERATURES, clouds, and HUMIDITY.

A photopolarimeter–radiometer (PPR) is an instrument used in remote sensing. It supplies data from which the temperature and cloud formation in the atmosphere of a planet or satellite can be determined, as well as some surface detail. The PPR measures the intensity and polarization of sunlight in the visible part of the SOLAR SPECTRUM.

residence time   The atmosphere is a dynamic system. DUST particles, mineral grains, POLLEN, SPORES, as well as molecules of WATER VAPOR and the other gases that constitute the air are constantly entering the atmosphere and leaving it. The residence time of any individual atom, molecule, or particle is the length of time that it remains in the air.

Residence time (R) is calculated as the total mass (M) of the substance divided by the FLUX (F), which is the rate at which it is entering and leaving the air ($R = M/F$). The larger the total mass the longer the residence time will be. NITROGEN is the principal atmospheric gas. Although a very large amount of nitrogen leaves the air every day due to nitrogen fixation and enters it by denitrification, the atmosphere contains approximately 4,466 million million tons (4.466×10^{15}, or 4.06×10^{18} kg). The flux therefore represents only a minute fraction of the total mass of nitrogen in the air. Consequently, the average time a nitrogen molecule spends in the air between being released by denitrification—the bacterial

process that releases gaseous nitrogen from the breakdown of nitrogen compounds—and removed again by nitrogen fixation is about 42 million years.

The atmosphere contains about 1,200 million million tons (1.2×10^{15}, or 1.09×10^{18} kg) of OXYGEN. Oxygen is removed from the air mainly by respiration and is returned to it by the process of PHOTOSYNTHESIS. An oxygen molecule remains in the air for an average of about 1,000 years.

CARBON DIOXIDE is present in trace amounts and is cycled rapidly by photosynthesis and RESPIRATION. A CO_2 molecule remains airborne for an average of about 55 years. METHANE has a residence time of about 11 years. Methyl chloride (CH_3Cl), which is produced by chemical reactions in seawater and is the largest natural source of atmospheric chlorine, has a residence time of 1.5 years. Halons (*see* OZONE LAYER) have residence times of 20 years for H-1211 (CF_2ClBr) and 65 years for H-1301 (CF_3Br). CFC-11 and CFC-12, which are the commonest CFCs, have residence times of 50 years and 100 years respectively.

There is even more WATER on the Earth than there is nitrogen. The total amount of water is about 1.54×10^{18} tons ($1,400 \times 10^{18}$ kg). The residence time of a water molecule is shorter than that of a nitrogen molecule, however, because the flux is much greater. Much more water is moving through the HYDROLOGICAL CYCLE than there is nitrogen moving through the nitrogen cycle. A water molecule spends an average 4,000 years in the ocean, about 400 years in lakes, rivers, GROUNDWATER, and as ice, and only about 10 days in the atmosphere.

Solid particles remain in the air for quite short periods. They obey the same law as gas molecules, but their total mass is much smaller than that of the atmospheric gases and their flux rate is greater. Some particles act as CLOUD CONDENSATION NUCLEI and are removed by rainout or snowout (*see* SNOW). Others are swept from clouds and from the air beneath clouds by PRECIPITATION and so are removed from the air by washout. Washout is much more efficient than rainout at removing particles. In addition, solid particles settle by gravity and the rate at which they do so must also be taken into account. Most particles bigger than 0.00004 inch (1 μm) in diameter are removed from the air by settling, and those more than 0.0004 inch (10 μm) usually remain in the air for only a few minutes. Smoke particles usually remain airborne for a few hours.

Residence times

Substance	Symbol	Atmospheric residence time
nitrogen	N_2	42 million years
oxygen	O_2	1,000 years
CFC-114	$C_2F_4Cl_2$	300 years
HCFC-23	CHF_3	250 years
CFC-12	CF_2Cl_3	100 years
CFC-113	$C_2F_3Cl_3$	85 years
H-1301	CF_3Br	65 years
carbon dioxide	CO_2	55 years
CFC-11	$CFCl_3$ 5	50 years
carbon tetrachloride	CCl_4	42 years
H-1211	CF_2ClBr	20 years
HCFC-22	CHF_2Cl	13.3 years
methane	CH_4	11 years
HCFC-124	C_2HF_4	5.9 years
methyl choride	CH_3CL	1.5 years
HCFC-123	$C_2HF_3Cl_2$	1.4 years
water	H_2O	10 days
ammonia	NH_3	7 days
smoke		hours
large particles		minutes

The table above lists the atmospheric residence times of a range of substances.

respiration Respiration is the process by which living organisms release energy by the oxidation of carbon. Breathing is the pumping action by which vertebrate land animals draw air into their lungs to obtain OXYGEN and expel CARBON DIOXIDE.

Oxidation was once thought to be a chemical reaction in which a substance acquires oxygen. The reverse process, in which a substance loses oxygen, and also a reaction in which a substance acquires hydrogen, was known as reduction. Nowadays these terms are defined more generally. Oxidation is a reaction in involving a loss of electrons and reduction is a reaction involving the acquisition of electrons. Consequently, it is possible for an oxidation reaction to take place in the absence of oxygen.

The carbon oxidized in respiration is in the form of glucose ($C_6H_{12}O_6$), a sugar that is produced in green plants and some BACTERIA and cyanobacteria by the process of PHOTOSYNTHESIS. In Archaea, some bacteria, and some fungi respiration is anaerobic (takes place in the absence of oxygen). The anaerobic reaction in the case of yeast, which is a fungus, is known as fermentation and it can be summarized as:

$$C_6H_{12}O_6 \rightarrow 2C_2H_5OH + 2CO_2$$

C_2H_5OH is ethanol, which is also called ethyl alcohol or just alcohol.

Aerobic respiration, which requires the presence of oxygen, can be summarized as:

$$C_6H_{12}O_6 + 6O_2 \rightarrow 6CO_2 + 6H_2O$$

Plants and all animals practice aerobic respiration. Species that dwell on land obtain their oxygen directly from the atmosphere. Aquatic organisms rely on oxygen that is dissolved in the water (they do not obtain oxygen by splitting water molecules).

In fact, the reaction in both cases takes place as a sequence of steps and is a great deal more complicated than this summary makes it appear. The energy that is released is used to attach a phosphate group to adenosine diphosphate (ADP), making it into adenosine triphosphate (ATP). ATP is transported through the organism, and wherever energy is required a phosphate group is discarded (ATP becomes ADP) with a release of energy. All the energy used by living organisms from bacteria to trees to people is transported and released by the ADP ↔ ATP reaction.

So far as the atmosphere is concerned, the process of respiration returns to the air the CO_2 that is removed by photosynthesis and aerobic respiration removes from the air the oxygen that is released by photosynthesis. Between them, photosynthesis and respiration maintain a constant atmospheric concentration of these two gases.

response time The time that elapses between a change in the amount of energy that is available in one part of the climate system and the effect that energy produces. This varies greatly according to the type of surface.

For example, in early spring there is an increase in the amount of solar heat that reaches the ground. If snow and ice cover the ground, however, most of that heat is reflected. This greatly reduces the warming effect of the spring sunshine and the extent to which it warms the air in contact with the surface. The climate responds slowly to the increase in energy. Where the ground is free from ice and snow, it warms rapidly and the response time of the climate is short.

The oceans absorb a large amount of heat before the water TEMPERATURE increases. This is because of the high HEAT CAPACITY of water. Air in contact with the sea surface warms eventually, but the absorption of heat delays that response. Ocean currents also transport the warmed water over long distances, and it may also be carried deep below the surface and held there for years before returning to the surface and warming the air. In this case the response time is very long.

return period The frequency with which a rare natural phenomenon may be expected to occur. It is based on recorded occurrences in the past and is then expressed as a range of values. These values also represent the statistical probability that the phenomenon will occur in any particular year. For example, records and calculations may indicate that a certain area is liable to be flooded once every 10 years. There is therefore a 10 percent chance that it will be flooded in any particular year, and the flood that affects it is known as a 10-year flood. Ten years is then said to be the return period for that event. Using the same method, there can be 50-year, 100-year, or 1,000-year floods.

The less frequent the event, the more severe it is when it happens. In 1952, the English village of Lynmouth was severely damaged by flooding. This was an event so unlikely that it was classed as a 50,000-year flood. Wind storms, BLIZZARDS, DROUGHTS, and other types of hazardous weather can be assigned probabilities in the same way.

Further Reading
Allaby, Michael. *Floods*. Rev. ed. Dangerous Weather. New York: Facts On File, 2003.

Reynolds number (*Re*) A DIMENSIONLESS NUMBER that is used to measure the extent to which a fluid flows smoothly (*see* LAMINAR FLOW) or turbulently (*see* TURBULENT FLOW). For a fluid flowing through a pipe it is calculated by:

$$Re = v\rho l/\eta$$

where v is the flow VELOCITY, p is the DENSITY of the fluid, l is the radius of the pipe, and η is the VISCOSITY of the fluid. For air, *Re* can be calculated by

$$Re = LV/v$$

where L is the distance over which the air is moving, V is its velocity, and v is its kinematic viscosity, which is equal to approximately 16×10^{-5} ft^2/s (1.5×10^{-5} m^2/s). If *Re* is less than about 1,000, the flow is dominated by viscosity. If *Re* is greater than about 1,000, the flow is dominated by turbulence.

Except on a very small scale, such as the flow around a spherical ball, *Re* is usually much larger than 10^3. Typically it is 10^6 or 10^7.

The number was discovered by the physicist Osborne Reynolds (*see* APPENDIX I: BIOGRAPHICAL ENTRIES).

Rhyacian A period during the PALEOPROTEROZOIC era of the Earth's history that began 2,300 million years ago and ended 2,050 million years ago. At this time the continents were smaller than those existing today, but they moved faster (*see* PLATE TECTONICS) because the magma beneath the Earth's crust was hotter and less viscous than it is now. (*See* APPENDIX V: GEOLOGIC TIMESCALE.)

ridge A long, tonguelike protrusion of high AIR PRESSURE into an area of lower pressure. The waves in the polar front JET STREAM associated with the index cycle (*see* ZONAL INDEX) that extend toward the North Pole are also called ridges.

A ridge in which the isobars (*see* ISO-) make a V-shaped point is called a wedge.

root-mean-square (**RMS**) Calculating the RMS is a technique for determining the value of a quantity that is fluctuating. RMS is calculated by sampling the values, squaring them, averaging them, and then calculating the square root of the mean.

Rossby number (*Ro*) A DIMENSIONLESS NUMBER that was discovered by the Swedish-American meteorologist Carl-Gustav Rossby (*see* APPENDIX I: BIOGRAPHICAL ENTRIES). The Rossby number is the ratio of the ACCELERATION of moving air due to the PRESSURE GRADIENT and the CORIOLIS EFFECT: *Ro* = (relative acceleration)/(Coriolis effect). It is given by:

$$Ro = U/\Omega L$$

where U is the horizontal wind VELOCITY, Ω is the angular velocity of the Earth, and L is the horizontal distance over which the wind travels.

Rossby waves (long waves, planetary waves) Waves that develop in moving air in the middle and upper troposphere. They have wavelengths (*see* WAVE CHARACTERISTICS) of 2,485–3,728 miles (4,000–6,000 km) and are named after Carl-Gustav Rossby (*see* APPENDIX I: BIOGRAPHICAL ENTRIES), the Swedish-American meteorologist who discovered them.

The angular wave number, also called the hemispheric wave number, is the circumference of the Earth measured at a specified latitude divided by the wavelength of the Rossby waves associated with a particular weather pattern. It gives the number of waves of that wavelength that are required to encircle the Earth and, therefore, the number of times the weather pattern repeats around the world.

The irregular change that takes place in the Rossby waves surrounding each hemisphere as their amplitude oscillates between a maximum and a minimum is called vacillation. The series of steps by which the oscillation occurs makes up the index cycle (*see* ZONAL INDEX).

runoff Runoff is the movement of water that falls to the ground as PRECIPITATION, including melting SNOW, FROST, DEW, and FOG droplets, and that then flows across the ground surface directly into rivers or lakes. As it crosses the ground, some of the water filters into the soil, eventually joining the GROUNDWATER. The remainder, or net runoff (symbolized by Δr), is not available to plants. Measurement of the amount of runoff is used in calculating the WATER BALANCE for an area.

Saffir/Simpson hurricane scale For more than a century, wind force was reported using the scale that had been devised by Admiral Beaufort (*see* APPENDIX I: BIOGRAPHICAL ENTRIES). The BEAUFORT WIND SCALE is still in use, but it has one major disadvantage: It is designed for temperate regions, where winds stronger than 75 MPH (120.6 km/h) are very uncommon. All such winds are classed on the Beaufort scale as being of hurricane force.

This is inadequate for those parts of the world that experience real TROPICAL CYCLONES. All of these cyclones produce winds of greater force than the 75 MPH, force 12, which is the highest value on the Beaufort scale, but they vary considerably in the winds they generate and, therefore, in the damage they are capable of inflicting.

To address this difficulty, meteorologists at the U.S. Weather Bureau (now part of the NATIONAL WEATHER SERVICE) introduced in 1955 an extension to the Beaufort scale: the Saffir/Simpson hurricane scale, named after the scientists who devised it. It adds five more points to the wind scale, but it also conveys more information than does the Beaufort scale. As well as wind speed and a general description of the type of possible wind damage, it includes the surface AIR PRESSURE at the center of the storm and the height of the STORM SURGE. The pressure indicates the intensity of the storm—the lower the pressure the more violent the storm will be—and information about the anticipated storm surge is vital, because tropical cyclones begin at sea and affect mainly coastal areas. Tropical cyclones everywhere are now classified according to the Saffir/Simpson scale, and their values are determined by their core pressure and not by the wind speeds they sustain.

Sahel Sahel is the name of the region in northern Africa that lies along the southern margin of the Sahara. The region extends from Senegal in the west to Sudan and Ethiopia in the east and includes parts of Senegal, Guinea Bissau, Mauritania, Mali, Burkina Faso, Niger, Nigeria, Chad, Central African Republic, Sudan, Eritrea, and Ethiopia.

The Sahel forms a transitional zone between the desert to the north and the humid tropical grasslands to the south. The vegetation comprises short grasses, tall herbs, and thorn scrub with species such as acacias (*Acacia* species) and baobab trees (*Adansonia digitata*), but the plants are scattered and there are few places where the vegetation cover is continuous.

The climate is semi-arid and strongly seasonal, with a short rainy season in summer. Niamey, Niger, receives an average 22 inches (554 mm) of rain a year, but no rain at all falls between the end of October and the beginning of March. Most of the rain, about 20 inches (495 mm), falls in June, July, August, and September. N'Djamena, the capital of Chad, receives no rain from the end of October until the beginning of April, and of the 29 inches (744 mm) it receives in an average year, 24 inches (610 mm) fall in July, August, and September. August is the wettest month.

Temperatures change little through the year, and the climate is hot. April is the warmest month at

Saffir/Simpson Hurricane Scale

Category	Pressure at center mb in. of mercury cm. of mercury	Wind speed mph km h⁻¹	Storm surge feet meters	Damage
1	980 28.94 73.5	74–95 119–153	4–5 1.2–1.5	Trees and shrubs lose leaves and twigs. Mobile homes destroyed.
2	965–979 28.5–28.91 72.39–73.43	96–110 154.4–177	6–8 1.8–2.4	Small trees blown down. Exposed mobile homes severely damaged. Chimneys and tiles blown from roofs.
3	945–964 27.91–28.47 70.9–72.31	111–130 178.5–209	9–12 2.7–3.6	Leaves stripped from trees. Large trees blown down. Mobile homes demolished Small buildings damaged structurally.
4	920–944 27.17–27.88 69.01–70.82	131–155 210.8–249.4	13–18 3.9–5.4	Extensive damage to windows, roofs, and doors. Mobile homes destroyed completely. Flooding to 6 miles (10 km) inland. Severe damage to lower parts of buildings near exposed coasts.
5	920 or lower below 17.17 below 69	more than 155 more than 250	more than 18 more than 5.4	Catastrophic. All buildings damaged, small buildings destroyed. Major damage to lower parts of buildings less than 15 ft (4.6 m) above sea level to 0.3 mile (0.5 km) inland.

N'Djamena, when the average daytime temperature is 107°F (42°C) and has been known to reach 114°F (46°C). In December, the coldest month, the average daytime temperature is 92°F (33°C) and 101°F (38°C) has been recorded. It is much cooler at night, but the lowest nighttime temperature recorded is 47°F (8°C), and the average temperature at night in December and January is 57°F (14°C). Niamey experiences almost identical temperatures.

Averages can be misleading, however. It is the northward movement of the INTERTROPICAL CONVERGENCE ZONE (ITCZ) that brings the tropical rain belt to the Sahel and causes the summer rains, but occasion-ally the ITCZ remains to the south. When this happens, the rains are lighter than usual or, if the ITCZ remains a long way to the south, they fail altogether. During the late 1960s, the Sahel experienced a sequence of years when the summer rains were light. This produced DROUGHT. Then, in 1972 and 1973, the rains failed completely, and the rainfall did not return to normal until the 1980s.

It was not the first time the Sahel had been afflict-ed with severe drought. Several droughts occurred in the 17th century and caused serious famines. Those droughts were associated with the coldest part of the LITTLE ICE AGE. No one knows what caused the

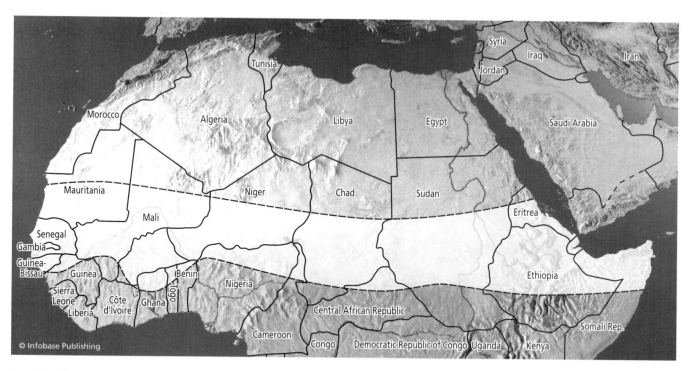

The Sahel forms a belt along the southern margin of the Sahara, extending from Senegal to Sudan, Eritrea, and Ethiopia.

drought in the 20th century, although it did coincide with the latter part of a period when temperatures were falling sharply throughout the Northern Hemisphere.

Some of the people of the Sahel grow crops around the oases. Others live a seminomadic life, taking their herds and flocks of cattle, sheep, goats, and camels to traditional seasonal pastures. The drought that peaked in the 1970s proved devastating. It is estimated that between 100,000 and 200,000 people and up to 4 million cattle died. Countless more people were forced to migrate south across the national frontiers that are a legacy of European colonialism. It was the Sahel drought that alerted the international community to the difficulties facing people who live along the borders of deserts.

It is sometimes asserted that overgrazing caused the drought, but this is untrue. It was an entirely natural event and was not the fault of the people living in the region. Overgrazing did exacerbate its effects, however. As the pastures failed, livestock was crowded into increasingly smaller areas, where they did destroy the sparse vegetation. Governments also encouraged nomadic people to settle in permanent villages. This gave them access to medical care and schools, but it also placed excessive pressure on the grazing around the villages.

Further Reading

Allaby, Michael. *Droughts.* Rev. ed. Dangerous Weather. New York: Facts On File, 2003.

salt crystal The solid form that common salt (sodium chloride, NaCl) takes when it comes out of solution. The basic crystal is cubic in shape, about 0.0004 inch (10 µm) along each side, and it grows by the addition of more cubes.

Salt enters the air when drops of sea spray evaporate. The tiny crystals are then carried by air currents. Salt is hygroscopic, which means it dissolves in water that its crystals absorb from the atmosphere and airborne salt crystals act as hygroscopic nuclei. These are the most efficient of all CLOUD CONDENSATION NUCLEI. Water will condense onto a salt crystal at a relative HUMIDITY as low as 75 percent. Salt crystals are sometimes used in CLOUD SEEDING, where their effect is to increase the range of size of CLOUD DROPLETS. This increases the likelihood of PRECIPITATION, because some of the droplets grow large enough to fall and continue growing by collision and coalescence (*see* RAINDROPS) as they do so.

The efficiency of using salt crystals for cloud seeding is initially due to the readiness with which they dissolve in water. When a crystal dissolves, the resulting droplet is a fairly strong saline solution. The vapor pressure (*see* WATER VAPOR) is always lower over a solution than it is over pure water, so water evaporates more slowly from the solution. Once they form, therefore, saline cloud droplets resist EVAPORATION. More water condenses onto them, and as the droplets grow, the salt solution is diluted. This increases the vapor pressure.

The accumulation of more water would also increase the rate of evaporation if it were not for a counteracting effect that is due to the size of a droplet. Water molecules are linked by weak hydrogen bonds (*see* CHEMICAL BONDS). Molecules are held less firmly at the surface than they are in other parts of the body of liquid, because above the surface there are no molecules to which they can be linked. The attraction between molecules at the surface is strongest if the sur-

face is flat. It is weaker over a curved surface by an amount that is proportional to the degree of curvature. Consequently, small droplets evaporate faster than large droplets.

As the saline droplet begins to grow and the vapor pressure over it increases, it also grows larger and its surface becomes less curved. This reduces the rate of evaporation from it.

The overall result is that salt crystals readily cause water vapor to condense, and the resulting droplets tend to survive long enough to grow. They continue to grow until they attain a size that is in equilibrium with the amount of moisture present in the cloud.

sand Sand is a granular material that results from the WEATHERING of rock. To a geologist, sand is defined only by the size of its grains. There are several classifications of particle size. In the Udden–Wentworth scale, which is widely used by geologists, sand grains are 0.0025–0.079 inch (0.0625–2.0 mm) across. Soil scientists use a different scale, with sand comprising grains 0.02–0.079 inch (0.5–2.0 mm) in the U.S. scale and 0.0008–0.079 inch (0.02–2.0 mm) in the international scale.

Composition is much less important, and sand can be made from the crushed remains of seashells, skeletons, fragments of volcanic rock, or almost any mineral. Most beach sand contains calcium carbonate ($CaCO_3$) fragments from seashells. In fact, though, most sand is made from silicate minerals, of which quartz (silica, or silicon dioxide, SiO_2) is by far the most common. Desert sand consists almost entirely of quartz.

Sand grains that have been lifted from the ground by wind and are transported through the air are known as blowing sand. A wind of 12 MPH (20 km/h) will raise sand grains smaller than 0.01 inch (0.25 mm) in diameter provided they are dry. Strong local convergence (*see* STREAMLINE) over a sandy desert will produce WHIRLWINDS, and a strong wind blowing over a wider area will produce a SANDSTORM. *See also* AQUIFER.

sand dune Dry sand is easily transported by wind, and it accumulates in particular places, where the wind may pile it into large heaps, called sand dunes. As well as building sand dunes, the wind also shapes them.

If there is abundant sand, a wind that blows almost always from a single direction will build transverse dunes. These are long, with the gradual slope on

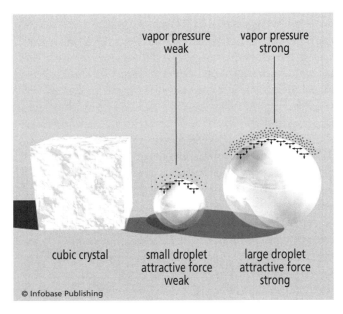

vapor pressure weak

vapor pressure strong

cubic crystal

small droplet attractive force weak

large droplet attractive force strong

© Infobase Publishing

Salt crystallizes into very small cubes that dissolve to form tiny droplets of salt water. The vapor pressure over the solution is weak, reducing the rate of evaporation. As the droplets grow, the solution becomes more dilute and the vapor pressure increases, tending to increase the rate of evaporation. At the same time, however, their surface curvature decreases. This increases the intermolecular forces at the surface and reduces the rate of evaporation.

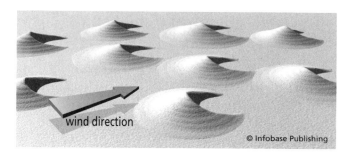

A barchan dune is crescent-shaped and formed by wind that blows predominantly from a single direction.

the side facing the wind, and the line of the dunes is at right angles to the wind direction. Transverse dunes can be up to 60 miles (96 km) long and up to 300 feet (90 m) high, and they move downwind at up to about 80 feet (25 m) a year.

Transverse dunes sometimes develop wavy crests, so the face of the dune faces alternately into and away from the wind direction. Dunes of this type are called aklé dunes.

If the wind direction varies to either side of an average direction, it will build longitudinal dunes. These are aligned with the average wind direction. Longitudinal dunes have long, sharp, sinuous crests. In some places, especially in the western Sahara, the biggest of them are known as seif dunes, which can stretch for hundreds of miles. Their curved shapes resemble sword blades, and seif is from *sayf*, the Arabic word for sword.

Where the supply of sand is limited, winds blowing from either side of an average direction will blow the sand into a crescent shape, with the horns pointing downwind. If the crescent is narrow, so the horns are close together, this is called a parabolic dune— the shape is that of a parabola. If the horns are fairly wide apart, so the crescent is open, it is a barchan dune. Barchans are up to about 100 feet (30 m) high, and the tips of their horns are up to 1,200 feet (370 m) apart. Where there is enough sand, adjacent barchans may join to form an aklé dune, and sometimes one limb of a barchan can be blown away altogether, leaving the remaining limb as an isolated seif dune.

If there is no predominant wind direction, the sand may form star dunes, also called stellar dunes. A star dune consists of a series of ridges that radiate from a central point, making a shape resembling a star. Dunes of this shape can also form where other dunes intersect. A dune of this, intersecting, type is called a rhourd.

The largest sand formation of all is called a draa. A draa is a ridge or chain of sand dunes, sometimes more than 1,000 feet (300 m) high, and 0.3–3 miles (0.5–5 km) from its nearest neighbor. Draas are found in the sand seas, or ergs, of the Sahara. Intersecting draas form rhourds.

Sand dunes form from sand grains that are blown up the shallow windward slope and fall from its crest down the steeper slope, which is at an angle of about 32°. The constant movement of sand from one side of the dune to the other causes the dune to move in the direction of the wind. Draas move 0.8–2 inches (2–5 cm) a year. Barchan dunes move an average 30–65 feet (10–20 m) a year.

Seif dunes are long, tapering, and slightly curved.

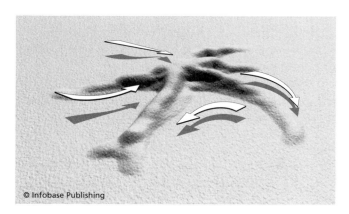

A star dune consists of ridges that radiate from a central point. The arrows indicate the direction of the winds that formed the dune.

A sandstorm approaching a town in Kansas, from "Effect of Dust Storms on Health," U.S. Public Health Service, Reprint No. 1707 from *Public Health Reports,* 50, No. 40, October 4, 1935 *(Historic NWS Collection/NOAA)*

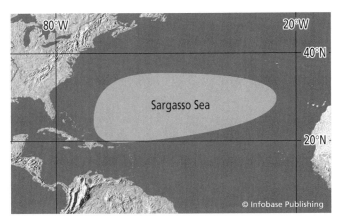

The roughly elliptical Sargasso Sea lies in the western North Atlantic and is enclosed by ocean currents that flow clockwise around it.

sandstorm   A wind storm becomes a SAND storm if the wind lifts sand grains into the air and transports them. Sandstorms often travel long distances. The threshold velocity (*see* WIND SPEED) for dry, medium-sized sand grains about 0.01 inch (0.25 mm) in diameter is 12 MPH (19 km/h). Winds of more than 15 MPH (24 km/h) will raise enough sand to cause a sandstorm provided the air is unstable (*see* STABILITY OF AIR).

Wherever there is loose, dry sand, a wind of this speed will raise sand high enough to greatly reduce visibility and drive it with enough force to make exposure to it extremely uncomfortable. The wind blows horizontally, however, and although collisions between sand grains and the turbulence of the air (*see* TURBULENT FLOW) can raise the sand a short distance above the ground, it cannot lift it high enough for it to travel far. In unstable air, however, upcurrents can lift the sand to a considerable height. A sandstorm is produced in the same way as a dust storm, from which it differs only in the size of particles involved.

Sargasso Sea   An area in the western North Atlantic Ocean that lies approximately between latitudes 20°N and 35°N and longitudes 30°W and 70°W. The sea is roughly elliptical in shape and occurs inside a system of ocean currents that rotate clockwise. The Gulf Stream, Canary Current, and North Atlantic Drift flow around its edges (*see* APPENDIX VI: OCEAN CURRENTS).

The waters of the Sargasso Sea are relatively calm. Winds are light, the EVAPORATION rate is high, and rainfall is low. The water is very clear and warm, with an average temperature of 64°F (18°C). The high rate of evaporation combined with low PRECIPITATION produce water with a salinity of 36.5–37.0 per mil. The average salinity of seawater is about 35.0 per mil. The sea is famous for its abundance of gulf weed, a brown, floating seaweed (several *Sargassum* species). It is not true that ships have ever been caught and trapped by the weed.

The Sargasso Sea is the breeding ground of the American and European eels (*Anguilla rostrata* and *A. anguilla* respectively). The larvae of European eels drift with the Gulf Stream, taking about three years to reach the cool, shallow waters off the coast of Europe, where they turn into elvers and migrate into rivers. American eels breed in the western part of the Sargasso Sea, and their larvae take only one or two years to reach the mouths of rivers.

satellite instruments   Orbiting satellites are able to acquire data that is inaccessible to observers on the ground. Satellites have a wide field of view. Those in geostationary ORBITS monitor the surface from the equator almost to the pole, while satellites in polar or Sun-synchronous orbits scan the entire surface of the Earth every day. Their field of view allows satellites to produce mesoscale images, which are images that extend horizontally from 0.6–60 miles (1–100 km). METEOROLOGY at this scale, called mesometeorology, became feasible once satellite images showed the clouds associated with entire weather systems.

Having acquired data, satellites transmit them to a surface center, often by a method known as automatic picture transmission (APT). Usually, APT images have a resolution of 2.5 miles (4 km) and are transmitted at two lines per second. Images are transmitted as soon as they have been taken, rather than having to be stored for transmission later, as was the case with earlier equipment.

The images themselves are in digital form. A digital image is a picture compiled from a continuously varying stream of data received from an orbiting satellite or other remote source. The continuous variation is converted into discrete variation, in which the data change in small steps. Each discrete change is then given a numerical value as a picture element (conventionally abbreviated to "pixel") and assigned a location defined by coordinates.

High-resolution picture transmission (HRPT) is a method used to transmit images with a resolution of 0.7 mile (1.1 km), allowing them to depict features 0.5 mile (0.8 km) across. HRPT transmits two visible and three infrared channels.

False colors are used in many satellite images. These are colors that differ from the actual colors of the surfaces they represent. Instruments carried on satellites are able to detect radiation at any wavelength (*see* WAVE CHARACTERISTICS), including wavelengths outside the waveband to which the human eye is sensitive. The different wavelengths all convey useful information and the allocation of fairly arbitrary colors to them makes the areas emitting them clearly visible. In many false-color images infrared radiation appears as visible red. Vegetation is highly reflective to infrared wavelengths, and so it often appears red in false color pictures.

The use of satellite images to identify areas of the Earth that are experiencing deforestation, DROUGHT, or desert encroachment is known as vegetation index mapping. The greenness of plant cover can be measured from images transmitted by satellites in polar orbit and the health of vegetation inferred by comparing the color to a series that have been compiled into an index.

The operational linescan system (OLS) is the primary imaging system used on some of the satellites in the DEFENSE METEOROLOGICAL SATELLITE PROGRAM. Because the background to the images is dark, it is possible to increase the gain on the OLS photomultiplier tube at night. This allows the OLS to detect LIGHTNING discharges, and the system has produced the longest set of data for lightning, dating from 1973. It also detects waste gas flares at oil wells and has revealed the large extent of this practice. The OLS scans the whole Earth once every day at visible and infrared wavelengths with a resolution of 1.74 and 0.37 miles (2.8 and 0.6 km).

Microwave images of SNOW- and ICE-covered areas that are received from satellites are allotted colors according to a code based on a unit called the brightness temperature. Brightness temperatures correspond closely to the intensity of the microwave radiation, but they are given values in kelvins (*see* UNITS OF MEASUREMENT) to reflect differences in EMISSIVITIES from different surfaces. At a microwave wavelength of 1.55 cm, for example, ice has a brightness temperature of 190K or more, but water has a lower brightness temperature, of less than 160K. The boundary between water and SEA ICE shows clearly on the resulting color-coded image. Brightness temperatures vary with the wavelength of the microwave radiation, and by comparing brightness temperatures at two different wavelengths it is possible to calculate the depth of the snow covering an area.

A particular problem occurs with satellite images near to coasts. It is called land contamination, and it is the effect of the footprint, about 30 miles (50 km) in diameter, which occurs near coasts in satellite images. Land contamination occurs because some of the radiation received at the satellite comes from land and some from the ocean, making the data imprecise near coasts. This is especially important where the land is covered with ice, but the sea is not, because it can make it appear that ice covers coastal waters.

Satellites monitoring the Earth's atmosphere and surface carry a variety of instruments to capture the data they transmit to the ground. Some of these are briefly described here.

A radiometer measures electromagnetic radiation. It may be passive or active. A PASSIVE INSTRUMENT measures the radiation falling upon it. An ACTIVE INSTRUMENT emits a signal that is reflected back to it and compares the emission with its reflection. A radiometer may be designed to respond to any wavelength.

The advanced very high resolution radiometer (AVHRR) senses clouds and surface temperatures. It stores its data on magnetic tape and transmits them on command to surface receiving stations. It also transmits both low- and high-resolution images in real time. The first AVHRR was launched in October 1978 on the *TIROS-N* (TELEVISION AND INFRARED OBSERVATION SATELLITE) satellite. It transmitted on four channels. Other four-channel AVHRRs were carried on *NOAA 6* and other even-numbered satellites in the NOAA (*see* NATIONAL OCEANIC AND ATMOSPHERIC ADMINISTRATION) series. The first five-channel AVHRR was launched in June 1981 on *NOAA 7* and others on subsequent odd-numbered NOAA satellites. The five channels are: 0.58–0.68 μm (visible part of the spectrum); 0.725–1.10 μm (near-infrared); 3.55–3.93 μm (intermediate infrared); 10.3–11.3 μm and 11.5–12.5 μm (thermal infrared on *NOAA 7* and 9); and 10.5–11.5 μm and 11.5–12.5 μm (thermal infrared on *NOAA 11*).

The scanning multichannel microwave radiometer (SMMR) was carried on the SEASAT and *Nimbus-7* satellites (*see* NIMBUS SATELLITES). It first went into service in 1978 and continues to transmit valuable data from Nimbus. It carries six radiometers with 10 channels delivering measurements at five microwave wavelengths (0.81, 1.36, 1.66, 2.8, and 4.54 cm). The SMMR measures SEA-SURFACE TEMPERATURE, WIND SPEED, WATER VAPOR, clouds and cloud content, snow cover, the type of SNOW, rainfall rates, and different types of ice. It also measures the concentration and extent of sea ice. It has provided detailed information about El Niño events (*see* ENSO) since the 1982–83 El Niño, and it is used to monitor changes in sea ice.

Multichannel sea surface temperature (MCSST) is a procedure in which sea-surface temperatures are calculated from data received from an advanced very high resolution radiometer. First the data are checked to identify points referring to clouds or AEROSOLS; these are removed. Using the remaining data, sea-surface temperatures are calculated by an ALGORITHM such as:

$$SST = a_0 + a_1 T_1 + a_2 T_2$$

where T_1 is the AVHRR brightness temperature at the waveband 3.55–3.93 μm, T2 is the brightness temperature at the waveband 10.3–11.3 μm or 10.5–11.5 μm depending on the channel used, and a_0, a_1, and a_2 are coefficients that convert the T values into sea-surface temperatures.

The cryogenic limb array etalon spectrometer (CLAES), carried by the UPPER ATMOSPHERE RESEARCH SATELLITE, measures infrared radiation emitted from the atmosphere. Its etalon (*see* INTERFEROMETER) is kept chilled (cryogenic). The instrument measures the temperature in the stratosphere and lower mesosphere (*see* ATMOSPHERIC STRUCTURE) and the trace constituents of the atmosphere: OZONE (O_3), nitric oxide (NO), nitrogen dioxide (NO_2), nitrous oxide (N_2O), nitric acid (HNO_3), dinitrogen pentoxide (N_2O_5, *see* NITROGEN OXIDES), METHANE (CH_4), CFC-11 (CCl_3F), CFC-12 (CCl_2F_2), and $ClONO_2$.

The wind imaging interferometer (WINDII) carried on the Upper Atmosphere Research Satellite measures the DOPPLER EFFECT on the spectral lines emitted by airglow emissions and aurorae (*see* OPTICAL PHENOMENA), from which it calculates temperatures and winds in the thermosphere (*see* ATMOSPHERIC STRUCTURE).

The electrically scanning microwave radiometer (ESMR) is a radiometer transmitting in the microwave waveband that was launched on the *NIMBUS 5* satellite in December 1972 and continued to function until the end of 1976. Its purpose was to monitor sea ice. It operated on a single channel, collecting data at a wavelength of 1.55 cm. At this wavelength WATER has an emissivity of about 0.44, but the emissivity of ice is between 0.80 and 0.97, so the rate of emission is much greater for sea ice than for water.

The geostationary Earth radiation budget (GERB) is an instrument carried on MSG (*see* METEOSAT) satellites that will observe and measure the radiation reflected and emitted by the Earth.

The high resolution Doppler interferometer (HRDI) is an interferometer carried on the Upper Atmosphere Research Satellite. It measures the Doppler effect on sunlight that is scattered in the stratosphere and on airglow emissions in the mesosphere. From this it provides data on the temperature and winds throughout much of the stratosphere (but with a gap in the upper stratosphere) and most of the mesosphere.

The improved stratospheric and mesospheric sounder (ISAMS), carried on the Upper Atmosphere Research Satellite, measures the temperature in the stratosphere and mesosphere. It also measures concentrations of the atmospheric gases: ozone (O_3), nitric oxide (NO), nitrogen dioxide (NO_2), nitrous oxide (N_2O), nitric acid (HNO_3), dinitrogen pentoxide (N_2O_5), water vapor (H_2O), methane (CH_4), and carbon monoxide (CO).

The microwave limb sounder (MLS) is carried on the Upper Atmosphere Research Satellite. It is a spectrometer that measures in the microwave waveband and detects concentrations of ozone (O_3), water vapor (H_2O), and chlorine monoxide (ClO) in the stratosphere, and O_3 and H_2O in the mesosphere.

The microwave sounding unit (MSU) is carried on the *TIROS-N* series of NOAA satellites. An MSU measures the emissions of microwave radiation from molecular oxygen in the troposphere. The resultant readings are used to calculate atmospheric temperature with an estimated accuracy of ±0.01°C (±0.02°F). The satellites are in polar orbits that pass over every part of the surface of the Earth several times every day. The continuous record of atmospheric temperature measured by the MSUs began in January 1979.

For many years the MSU data showed no significant rise in the temperature of the atmosphere, in contrast to the rise detected by surface stations. It was eventually found that certain features of the satellite orbit had introduced errors into the readings. These were finally resolved, and when the MSU and surface measurements were reconciled, they showed that a steady rise in atmospheric temperature was occurring at a rate of 2.16°F–3.42°F (1.2°C–1.9°C) per century. This rate of warming agrees with the estimate by the Intergovernmental Panel on Climate Change of 3.15°F (1.75°C) per century.

The multispectral scanner carried on Landsat satellites obtains images of the Earth's surface with a spatial resolution of 24 feet (80 m). It is used to monitor surface changes, for example in vegetation, coastlines, ice sheets, glaciers, and volcanoes.

The near-infrared mapping spectrometer is an instrument carried on some satellites that takes readings in the near-infrared part of the solar spectrum. The chemical composition, structure, and temperature of the atmospheres of planets and satellites can be calculated from the data that are produced, as well as details of the mineral and geochemical composition of the surface.

A scatterometer is an instrument that measures the scattering of radar waves by the small capillary waves (*see* wave characteristics) on the ocean surface. The speed and direction of the surface wind can be calculated from these measurements.

The special sensor microwave imager (SSM/I) is carried on the satellites *DMSP F-8, F-10, F-11, F-12,* and *F-13* belonging to the Defense Meteorological Satellite Program. The SSM/I is a passive microwave radiometer with seven channels and operating at four frequencies (19.35 GHz, 22.235 GHz, 37.0 GHz, and 85.5 GHz). It collects linearly polarized microwave radiation and measures the surface brightness over land and sea. This provides data on clouds and other meteorological phenomena, primarily in support of U.S. military operations, but declassified so they are available to meteorologists. The instruments are carried in a nearly circular, Sun-synchronous, nearly polar orbit with a period of 102 minutes, at a height of 534 miles (860 km) and an inclination of 98.8°.

The stratospheric aerosol and gas experiment (SAGE) comprises a set of instruments carried on the Earth Radiation Budget Satellite that are used to measure the material injected into the stratosphere by volcanoes. The second set of instruments (SAGE-II) was launched in October 1984. SAGE-III was launched in 1999 on the Earth Observing System satellite. As well as aerosols and volcanic gases such as sulfur dioxide, these versions also measure ozone, nitrogen dioxide, and water vapor.

The thematic mapper (TM) is a sensing device carried on the *Landsat 4* and *Landsat 5* satellites in Sun-synchronous orbits. It detects reflected visible and near-infrared radiation and obtains information about the surface of the Earth with a resolution of 100 feet (30 m), producing images that are detailed enough to show individual fields. All TM transmissions from *Landsat 4* ended in August 1993 due to failure of the equipment and some *Landsat 5* transmissions ended in February 1987, also due to failure.

The total ozone mapping spectrometer (TOMS) is the instrument that provided the first satellite evidence of the depletion of stratospheric ozone over Antarctica. It was launched on October 24, 1978, on board the *Nimbus-7* satellite and transmitted daily maps of ozone distribution until 1993. A second TOMS, TOMS-METEOR, was launched in 1991 on the Russian Meteor satellite and a third, TOMS-ADEOS, in 1996 on a Japanese satellite. Although satellites carry other instruments that measure ozone concentrations, the data from TOMS are the most detailed.

TOMS consists of an instrument directed vertically downward that measures the intensity of radiation being reflected upward from the ground or ocean surface or from cloud tops. It samples radiation at six

ULTRAVIOLET (UV) (*see* ULTRAVIOLET INDEX) wavelengths between 312.5 nanometers (nm) and 380 nm in a repeating sequence. UV absorption varies according to the wavelength, so by comparing the amount reflected at each of the wavelengths it is possible to calculate the amount of UV that is being absorbed in the atmosphere. Since it is ozone that absorbs UV at UV-B wavelengths, the density of ozone in the column of air directly beneath the TOMS can be inferred from the amount of UV absorbed.

Wefax is an abbreviation for "weather facsimile," which is a system for transmitting by radio such material as graphic reproductions of weather maps, summaries of temperatures, and cloud analyses. Most Wefax transmissions are from GEOSTATIONARY OPERATIONAL ENVIRONMENTAL (GOES) satellites. Schools and individual enthusiasts can receive Wefax data provided they have suitable equipment.

saturated adiabatic reference process An idealized representation of the way moist air behaves that is used as a standard, or reference, with which events in the real atmosphere can be compared. In fact, it is a fairly accurate description of what usually happens.

The reference process assumes that rising air that is cooling adiabatically (*see* ADIABAT) and is saturated remains very close to SATURATION. WATER VAPOR begins to condense when the temperature of the rising air falls to its DEW point temperature. CONDENSATION releases LATENT HEAT, which sustains the BUOYANCY of the air. As the air continues to rise, its temperature also continues to fall. This chills the air between CLOUD DROPLETS sufficiently to cause the condensation of excess water vapor, maintaining the air at saturation. It is assumed that the moisture condenses as liquid, not ice. This is realistic. In real clouds, water droplets often remain liquid until the temperature falls below about -13°F (-25°C, *see* SUPERCOOLING).

The process is reversible. As the temperature of subsiding air rises, the resulting EVAPORATION of droplets adds enough water vapor to the air to maintain it at saturation. This is close to the process that has been observed in the downdrafts of cumulonimbus clouds (*see* CLOUD TYPES).

saturation The condition in which the moisture content of the air is at a maximum. If additional WATER molecules enter saturated air, then an equal number must leave it, by condensing into liquid water or being deposited as solid ICE (but *see* SUPERSATURATION).

Over the surface of water, water molecules are constantly escaping into the air by evaporation. Once in the air, they add to the vapor pressure (*see* WATER VAPOR). As they move through the air, a proportion of the water molecules will strike the surface of the liquid water. Their energy of motion (KINETIC ENERGY) will be absorbed by the relatively denser mass of water molecules composing the liquid, and the impacting molecules will no longer possess the energy needed to escape into the air. They will enter the water mass. Molecules also escape from and merge with an ice surface in the same way (*see also* SUBLIMATION and DEPOSITION).

As evaporation continues, the vapor pressure increases in the air above the water surface. Eventually, however, a point will be reached at which the number of water molecules evaporating from the liquid surface every second precisely balances the number recondensing into it over the same period. The amount of water vapor present in the air cannot increase beyond this point. The vapor pressure at which this occurs is known as the SATURATION VAPOR PRESSURE, and at the saturation vapor pressure the air is said to be saturated.

Strictly speaking, it is not the air that is saturated, but the water vapor. This is easier to understand if the air is likened to a dry sponge onto which water is sprinkled. As they fall, the water drops disappear into the sponge, which grows steadily moister. Continue with this for long enough and all the tiny air spaces in the sponge will be filled with water. The sponge can then hold no more, and water will start to drip out of it. This has no effect whatever on the material from which the sponge is made—it is the millions of air spaces that are filled with water. These lie between the cells or inside the bubbles that make up the foam, but they are not the solid matter of the sponge itself. Consequently, it is the water that is saturated, and not the sponge. Similarly, atmospheric water vapor comprises molecules that move among the other molecules of the air and the dry air is entirely unaffected by their presence. It is usual, however, to think of the air as being saturated.

Raising the TEMPERATURE of the water molecules increases their kinetic energy. In practice, collisions between air molecules and molecules of water vapor ensure they all have much the same kinetic energy. Water molecules therefore acquire more energy as the air temperature increases. This increases the ease with which they are able to escape from the liquid surface, so the quantity of water vapor increases until a new balance is reached at the higher temperature, with a raised saturation vapor pressure. The result is that the quantity of water vapor that air can hold increases with temperature. In fact, it does so extremely rapidly. Air at 80°F (27°C) holds more than four times the amount of water vapor air at 40°F (4°C) can hold.

The difference between the actual vapor pressure and the saturation vapor pressure at the same temperature is called the saturation deficit, also known as the vapor-pressure deficit. This is also the amount of water vapor, usually measured in grams per cubic meter (g/m^3), that must be added to the air to bring it to saturation at the existing temperature and pressure.

saturation vapor pressure The vapor pressure at which the WATER VAPOR in the layer of air immediately above the surface of liquid water is saturated at a given temperature. The table top right gives a number of representative values. The table shows that as the air temperature increases, the amount of water vapor needed to saturate the air also increases, demonstrating the relationship between the temperature of air and its capacity to hold water vapor. Saturation vapor pressure reaches sea-level atmospheric pressure (1013 mb; 101.3 kPa) at 212°F (100°C).

The saturation vapor pressure over an ICE surface is lower than that over the surface of liquid water supercooled (see SUPERCOOLING) to the same temperature, because the stronger bonds between water molecules in ice reduce the rate at which they enter the air by SUBLIMATION. As a result, air over an ice surface holds less water vapor than air over a liquid surface at the same temperature.

The Clausius–Clapeyron equation relates the saturation vapor pressure (e_s) to the absolute temperature (T). The equation is:

$$de_s/dT = L/T(\alpha_2 - \alpha_1)$$

where L is the LATENT HEAT of vaporization, α_2 is the SPECIFIC VOLUME of water vapor and α_1 is the specific volume of liquid water. Since α_2 is usually very much larger than α_1, α_1 can be ignored.

Saturation Vapor Pressure

Temperature °F (°C)	Pressure mb (Pa)
-58 (-50)	0.039 (3.94)
-40 (-40)	0.128 (12.83)
-22 (-30)	0.380 (37.98)
-4 (-20)	1.032 (103.2)
14 (-10)	2.597 (259.7)
32 (0)	6.108 (610.78)
50 (10)	12.272 (1227.2)
68 (20)	23.373 (2337.3)
86 (30)	42.430 (4243.0)
104 (40)	73.777 (7377.7)

scalar quantity A physical quantity that either does not act in a particular direction, as in the case of TEMPERATURE, or for which the direction of action is unimportant or not specified, as in the case of speed. This is contrasted with a VECTOR QUANTITY, in which the direction of action must be specified.

scale height The thickness the atmosphere would have if its DENSITY were constant throughout at its sea-level value of 1.23 kg/m^3. The scale height is 8.4 km (5.2 miles).

scattering The result of the reaction that occurs when visible light (see SOLAR SPECTRUM) passes through the atmosphere and collides with air molecules and AEROSOL particles. The light changes direction repeatedly due to the combined effects of DIFFRACTION, REFLECTION, and REFRACTION. Molecules and very small aerosol particles, with sizes smaller than the wavelength of the light, also absorb the radiation. They are excited by it and reradiate it in all directions.

The size of molecules and particles in relation to the wavelength of light is known as the size parameter (X) and is given by:

$$X = \pi d/\lambda$$

where d is the diameter of the molecule or particle and λ is the wavelength of the light. Air molecules and the smallest aerosol particles are much smaller than the wavelength of light (d is smaller than λ), so X is less than 1. The smaller they are, the less efficiently bodies scatter radiation, and their efficiency decreases as the difference between their size and the wavelength increases. The efficiency of scattering is inversely proportional to the fourth power of the wavelength (λ^{-4}). This means that the shorter wavelengths are scattered most efficiently.

As light passes through the upper atmosphere, the shortest visible wavelengths of 0.4–0.44 µm, which correspond to violet and indigo light, are scattered first. Each time violet or indigo radiation strikes a molecule or particle it rebounds in a random direction. This happens repeatedly, the amount of scattering increasing with the distance the radiation travels through the air. Violet and indigo radiation is scattered so thoroughly that it contributes very little to the sky color. The sky is not violet or indigo.

Blue light is scattered next. Because the efficiency of scattering is proportional to λ^{-4}, blue light (0.44–0.49 µm) is approximately nine times more likely to be scattered than red light (0.64–0.7 µm). By the time sunlight reaches an observer at the surface, the blue light has been scattered in all directions so that the clear sky appears blue in all directions.

At sunrise and shortly before sunset, the Sun is low in the sky, and its light travels through a much thicker layer of atmosphere before reaching the surface. If the sky is clear, the distance is sufficient for all the blue and green light to be scattered, so it disappears, but the longer distance also means there is a greater chance for light at longer wavelengths to be scattered, because the light encounters more air molecules. The blue and green light having been removed, the scattered light that penetrates to the surface is orange and red. This accounts for the colors of the sky at dawn and sunset.

At night the sky is conventionally described as being black. In fact, scattering of starlight continues and the sky is really a very deep shade of blue, aptly named midnight blue.

Bigger particles, for which X is greater than 1 (d is greater than λ), scatter light much more efficiently. When the relative HUMIDITY is high, small aerosol particles absorb water vapor and expand. As they grow larger, the amount of light they scatter increases. They scatter all wavelengths equally, and so the scattered light is not separated into its constituent colors. The scattered light is white, and as scattering increases the sky becomes whiter and hazier. HAZE reduces VISIBILITY, and it can turn into FOG if the relative humidity reaches 100 percent and water starts to condense onto the particles.

Scattering by molecules and very small aerosol particles was first observed experimentally by John Tyndall (see APPENDIX I: BIOGRAPHICAL ENTRIES). Lord Rayleigh (see APPENDIX I: BIOGRAPHICAL ENTRIES), however, discovered the reason and showed that radiation is scattered most when the molecules and particles are smaller than the wavelengths of radiation being scattered. Consequently, this is known as Rayleigh scattering. A Rayleigh atmosphere is an idealized atmosphere that consists only of molecules and particles that are smaller than about one-tenth the wavelength of the solar radiation passing through it. In such an atmosphere, the radiation would be subject only to Rayleigh scattering.

Scattering by larger particles is known as Mie scattering. This occurs when the radiation interacts with particles of a size similar to the wavelength of the radiation. Diffraction, reflection, and refraction combine to cause the change in direction. Mie scattering is predominantly in a forward direction, and all wavelengths are affected. The German physicist Gustav Mie (1868–1957, see APPENDIX I: BIOGRAPHICAL ENTRIES) was the first person to describe this process in detail, in 1908.

The scattering of all wavelengths of radiation equally as the radiation passes through the atmosphere is called nonselective scattering. It is caused by particles that are bigger than the wavelength of the radiation. Because all wavelengths are scattered equally, nonselective scattering produces a white sky.

Afterglow is a bright arch that appears in the west above the highest clouds just after sunset. It is caused by the scattering of light by DUST particles suspended in the upper troposphere.

Airlight is light that is scattered toward an observer by aerosols or air molecules lying between the observer and more distant objects. This light is visual NOISE that makes the more distant objects less clearly visible, thereby reducing visibility. The appearance of a cloudless sky in daytime is due entirely to airlight. At dawn the amount of airlight increases, and it is this that obscures the stars, rendering them invisible. At sunset, as the amount of airlight diminishes, the stars reappear.

Alpine glow is a series of colors that are sometimes seen over mountains in the east, especially if the mountains are covered with snow, as the Sun is setting in the west, and over mountains in the west as the Sun is rising in the east. The phenomenon is caused by the scattering of light reflected from the mountains. The colors change from yellow to orange to pink to purple at sunset and in the reverse order at sunrise.

scavenging The removal of particulate matter (*see* AIR POLLUTION) from the air by the action of PRECIPITATION. The natural processes involved are rainout (*see* CLOUD CONDENSATION NUCLEI), snowout (*see* SNOW), and washout (*see* RAINDROPS).

These processes remove most particles within hours or at most days from the time they enter the air (*see* RESIDENCE TIME). The term is also applied to the removal of gaseous pollutants as a result of chemical reactions. These most commonly involve free radicals, such as hydroxyl (OH).

sclerophyllous plant A sclerophyllous plant is one that is adapted to prolonged periods of hot, dry weather. The Greek word *skleros* means "hard" and *phullon* means "leaf." Sclerophyllous plants have leaves that are small, thick, hard, and leathery. They are also evergreen, which means they are not all shed at the same time and the plant retains leaves through the year.

Holly (*Ilex* species), holm (or holly) oak (*Quercus ilex*), and California lilac (*Ceanothus* species) are sclerophyllous plants with broad leaves, as are the gum trees (*Eucalyptus* species) native to Australia. Many pine trees (*Pinus* species) are also sclerophyllous.

Further Reading
Allaby, Michael. *Temperate Forests.* Rev. ed. Ecosystem. New York: Facts On File, 2007.

sea ice Ice that forms by the freezing of sea water is called sea ice. When the TEMPERATURE at the sea surface is below 32°F (0°C) and the sea is calm, SNOW falling on the sea may settle and accumulate. It is able to do so because snow consists of freshwater, which has a higher freezing temperature than salt water. Ice may also be carried to the sea by rivers or reach it by breaking away from GLACIERS to form ICEBERGS. Although these may float on the sea, accumulated snow and ice that originated on land are not counted as sea ice. That term is reserved for ice resulting from the freezing of the sea itself.

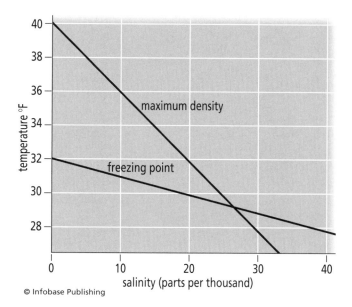

© Infobase Publishing

The freezing point of seawater varies according to the salinity of the water. As seawater freezes, the salinity of the surrounding water increases, raising its density.

WATER freezes at 32°F (0°C) at average sea-level atmospheric pressure, but only if it is pure H_2O. If other substances are dissolved in it, the freezing temperature is lower by an amount proportional to the strength of the solution. The average salinity of seawater, measured as all the dissolved salts but consisting mainly of sodium chloride (NaCl), is 35 grams per kilogram. This is the same as 35 parts per thousand (because 1 g = 1/1,000 kg) and is written as 35‰ (pronounced "per mil"). At 35‰ salinity, the freezing point of water is 28.56°F (-1.91°C). If the sea-surface temperature is between 32°F and 29°F, the seawater will not freeze, but SNOW falling onto it will not melt.

When the temperature falls below freezing, ICE CRYSTALS will start to form. These will consist of pure water. The dissolved salt will be excluded from the crystals. This will increase the salinity of the water adjacent to each crystal, lowering its freezing point still more, but also increasing the DENSITY of the water by an amount equal to that of the salt molecules that are added to it. The denser water will sink, and less dense water will rise from below to replace it, so freezing at the surface increases the rate at which water mixes in the uppermost layer of the sea (*see* THERMOHALINE CIRCULATION).

Provided the sea temperature remains below freezing, ice crystals will continue to form until they cover large areas of the surface. This is known as frazil ice. Frazil ice dampens down small wave movements, which makes the water appear oily, and as freezing continues and more ice crystals form, the sea becomes covered with slush.

As the process continues, the frazil ice thickens, becoming grease ice, and then breaks into pieces due to the motion of the sea. The pieces jostle against one another, which gives them a rounded shape. They are then known as pancake ice. Pancake ice consists of fairly thin patches of ice. These constantly collide with one another, and their circular shape results from the collisions.

As pancake ice forms, salt water becomes trapped between ice crystals, so that although the ice itself consists of only pure water, the pancake ice has salt within it. How much salt becomes trapped in this way depends on the speed with which the ice forms and, therefore, on the air temperature. When the air temperature remains at about 3°F (-16°C) while the ice is forming, the salinity of pack ice is approximately one-fifth that of the seawater (7‰) and at -40°F (-40°C) it is roughly one-third (12‰).

Pack ice comprises large blocks of ice that cover the surface of the sea. In winter, pack ice covers about 50 percent of Antarctic water and about 90 percent of Arctic water, although the area covered by sea ice in the Arctic has been decreasing, probably due mainly to the phase of the ARCTIC OSCILLATION that prevailed through much of the 1990s. In summer, ice covers about 10 percent of the sea in the Antarctic and about 80 percent in the Arctic. The percentage of an area of ocean surface that is covered by ice is called the ice concentration. An ice concentration of 0 percent means there is no ice; an ice concentration of 50 percent means half the area is covered; an ice concentration of 100 percent means the area is fully covered. Close pack ice has an ice concentration of 70–80 percent, with most of the floes in contact. Open pack ice has an ice concentration of 40–60 percent.

Ships can usually move through pack ice if it covers less than 75 percent of the surface and there is open water between blocks. A stretch of open water big enough for a ship to pass is called a lead. An irregular area of open water that is surrounded by sea ice is called a polyn'ya (plural polynyi). Polynyi may contain brash ice or young ice The presence of pack ice can be detected from a distance by the appearance of ice blink (*see* OPTICAL PHENOMENA).

When the blocks of pack ice unite to form a complete ice cover, they are known first as young ice, then as winter ice, and eventually as polar ice. Young ice is a layer of ice that is less than one year old and that forms a complete covering on a large area of the surface of the sea with a thickness of more than about 2 inches (5 cm). Winter ice is a layer less than one year old that forms a complete covering on a large area of the surface of the sea with a thickness of more than 8 inches (20 cm). Polar ice is a layer that forms a complete covering on a large area of the surface of the sea and that is more than one year old.

Floes are flat pieces of ice. A floe that is 6.5–330 feet (2–100 m) across is called a small floe. A medium floe is 330–1,600 feet (100–500 m) across, a big floe is 0.3–1.2 miles (0.5–2 km) across, and a giant floe is more than 6 miles (10 km) across. A vast floe is 1.3–6 miles (2–10 km) across.

Polar ice is less saline than winter ice and young ice, because in summer, when the ice partly melts, the spaces in which salt water are trapped open, allowing salt to drain away. Salinity is lowered further as a consequence of the low thermal conductivity of ice. This insulates the water below the ice, preventing its temperature from falling and so decreasing the rate at which ice accumulates on the underside of the surface layer. At the same time, snow falling onto the surface of the ice remains there, diluting the salt content of the ice as a whole.

Submerged ice that is firmly attached to the seabed is called anchor ice. Brash ice comprises fragments of broken ice not more than about 6 feet (1.8 m) across.

Seasat   The first satellite that used imaging RADAR to study the Earth. It was equipped with a scatterometer and with a scanning multichannel microwave radiometer (*see* SATELLITE INSTRUMENTS). *Seasat-A* was launched on an Atlas–Agena rocket from Vandenberg Air Force Base, California, on June 26, 1978, into a slightly elliptical polar ORBIT at a height of 482–496 miles (775–798 km). Its orbit carried it over nearly 96 percent of the surface of the Earth every 36 hours. On October 10, after transmitting data for 106 days, a short circuit drained all the power from its batteries and the satellite ceased to function.

seasons   In summer, the days are long, the nights short, and the weather is relatively warm. In winter, it is the opposite. Days are short, nights are long, and the weather is relatively cold. These variations in weather conditions define the seasons—the four periods of equal length we know as spring, summer, fall or autumn, and winter.

This description is true only in latitudes outside the TROPICS, however, and even there in some places the difference in temperature between one season and another is much less important—or marked—than the difference in rainfall. In these regions, the names "summer" and "winter" are replaced by "rainy season" and "dry season." In low latitudes, spring and fall are short, or barely happen at all. Nor does the change in day length affect all places equally. At the summer SOLSTICE, for example, people at the equator experience 12 hours of daylight, while those at the North and South Poles experience 24 hours. At the winter solstice, people at the equator still experience 12 hours of daylight, but for those at the Poles the Sun does not rise at all and they have 24 hours of darkness, or to be more precise, of twilight. Although the Sun remains below the horizon, the atmosphere refracts and reflects some of its light over the horizon.

There are few places on Earth where no seasonal changes at all occur in day length, mean TEMPERATURE, or rainfall, but the seasons become more strongly differentiated with increasing distance from the equator. That is because the amount of solar radiation received at the surface changes through the year as a consequence of the tilt in the rotational axis of the Earth with respect to the PLANE OF THE ECLIPTIC.

Instead of being normal (at right angles) to the plane of the ecliptic, the Earth's axis is at an angle of 66.55° to it, so it is tilted 23.45° from the normal. As the Earth moves along its orbital path, first one hemisphere and then the other is tilted toward the Sun. This produces four clearly defined positions. In two of them, known as the solstices, one hemisphere receives maximum exposure to sunlight and the other receives minimum exposure. In the others, known as the EQUINOXES, both hemispheres are equally exposed. Seen from a position on the equator, at the equinoxes the Sun is directly over head at noon and at the solstices it is at an elevation of 66.55° above the horizon at noon, or 23.45° from the vertical, displaced either to the north or to the south. At the solstices, the Sun is directly overhead at one or other of the Tropics.

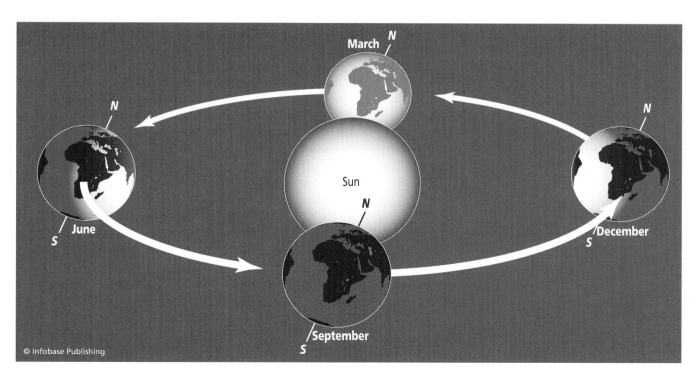

Because the rotational axis of the Earth is tilted in respect to the plane of its solar orbit, in June the Northern Hemisphere receives more solar radiation than the Southern Hemisphere and in December the situation is reversed. This produces the seasons.

The change in angle alters the ANGLE OF INCIDENCE of solar radiation, and this in turn alters the intensity of the radiation that is received at each unit area of the surface. Because the Earth is almost spherical in shape, the angle of incidence increases with latitude and so, therefore, does the intensity of radiation per unit area at the surface. It is this change that causes mean temperatures to be higher in summer than in winter and higher in low latitudes than in high ones.

sea-surface temperature (SST) The TEMPERATURE of the WATER at the surface of the sea, or sea-surface temperature, is routinely measured by drifting buoys, ships, and orbiting satellites. Of these, the satellite observations are the most extensive in their coverage and also the most accurate. Buoys can measure only the temperature of the water around them, which may not be representative of the ocean as a whole, and ships measure the temperature of the water they take on board to cool their engines. Water intakes are located about 16 feet (5 m) below the surface, and so ship measurements must be corrected to give the temperature at the surface.

Sea-surface temperature is climatically important because it affects the temperature of the air immediately above the surface and the EVAPORATION rate of water. This in turn affects air temperature, because WATER VAPOR is the most important greenhouse gas (see GREENHOUSE EFFECT), as well as cloud formation, ALBEDO, and PRECIPITATION.

Sea-surface temperatures change with the SEASONS, but they are also subject to other influences. Latitudinally, they change because of the presence of warm and cold ocean currents, and they also rise and fall in fairly regular cycles. Some cyclical variations operate with a period of a few years, others of decades or centuries. There is still much to be learned about these cycles.

EDDIES in ocean currents can produce local variations in sea-surface temperatures. These are similar to atmospheric CYCLONES and ANTICYCLONES, but they persist for months rather than days.

seawater Seawater is the water found in the seas and oceans. It contains an average of 35 parts per thousand (per mil) of dissolved compounds, known collectively as salts.

The proportion of salts determines the salinity of the water, so if water contains 35 per mil of salts,

Major Constituents of Seawater

	Parts per thousand	Percentage by weight
Chloride*	19.35	55.07
Sodium	10.76	30.62
Sulfate*	2.71	7.72
Magnesium	1.29	3.68
Calcium	0.41	1.17
Potassium	0.39	1.10
Bicarbonate*	0.14	0.40
Bromide*	0.067	0.19
Strontium	0.008	0.02
Boron	0.004	0.01
Fluoride*	0.001	0.01
Total		99.99

*Chlorine (Cl), bromine (Br), and fluorine (F) are present as compounds with other elements and so are measured as chlorides, bromides, and fluorides. Sulfur is present as sulfate (SO_4^{2-}) compounds. Bicarbonate is HCO_3^-, which is a salt of carbonic acid (H_2CO_3).

its salinity is 35 per mil. Salinity ranges from 34 per mil to 37 per mil in coastal areas, but may be close to 0 per mil where rivers discharge large volumes of FRESHWATER, or as high as 40 per mil where a large amount of water is lost by EVAPORATION from a partially enclosed body of water, such as the Persian Gulf and Red Sea.

Chlorine (Cl), sodium (Na), sulfur as sulfate (SO_4), magnesium (Mg), calcium (Ca), and potassium (K) together account for more than 99 percent of the dissolved matter present in seawater. The table above lists the major constituents of seawater.

seaweed A plant belonging to any one of several thousand species of multicellular marine algae may be described as seaweed. Some seaweeds are large. Certain species of *Macrocystis* and *Nereocystis,* found in the Pacific and Southern Oceans, grow to more than 100 feet (30 m) in length.

Seaweeds comprise three plant phyla (or divisions), the Rhaeophyta (brown seaweeds), Rhodophyta (red seaweeds), and Chlorophyta (green seaweeds). Seaweeds grow in coastal waters throughout the world, from the uppermost part of the shore reached by spring TIDES to where the water is about 165 feet (50 m) deep.

At greater depths there is insufficient light for PHOTO-SYNTHESIS.

Seaweeds have many uses. Some are eaten, others are used to make fertilizer, and some were traditionally used to foretell the weather. The "meteorological" species are adapted to survive the very harsh environment of the upper shore, where they are alternately submerged in SEAWATER and exposed to the air and warm sunshine. They survive the dry conditions by shriveling and becoming brittle, but as soon as they detect moisture they begin to absorb water and revive. The wracks are the weeds that demonstrate this capacity best. These are brown seaweeds of the genera *Fucus* and *Ascophyllum*, such as bladder wrack (*F. vesiculosus*) and knotted wrack (*A. nodosum*).

People used to bring these seaweeds home from the shore and hang them outside the door. If the seaweed was dry and shriveled the weather would be fine, but if it became flexible and rubbery it meant rain was likely. The method worked, but only up to a point, because by the time the seaweed responded to the rise in HUMIDITY the change in the weather was usually self-evident.

SeaWinds   A RADAR instrument that is carried on board the QUIKSCAT satellite. SeaWinds was launched on June 19, 1999, and collects data from a continuous band, 1,118 miles (1,800 km) wide, covering 90 percent of the Earth's surface and making approximately 400,000 measurements in a day.

SeaWinds has a rotating dish antenna with two spot beams that sweep in a circular pattern radiating microwave pulses at 13.4 gigahertz. It gathers data on low-level wind speed and direction over the oceans and also tracks the movement of Antarctic ICEBERGS. The data is used in scientific studies of global climate change and weather patterns, interactions between the atmosphere and the ocean surface, to track changes in tropical rain forests, and to monitor movements at the edge of the Antarctic SEA ICE and pack ice.

Further Reading
NASA, "SeaWinds Wind Report." Available online. URL: http://haifung.jpl.nasa.gov/. Accessed January 27, 2006.

sector   A horizontal plane that is bounded on two sides by the radii of a circle and on the third side by an arc that forms part of the circumference of the same circle. A warm sector (*see* FRONT) is an area that is

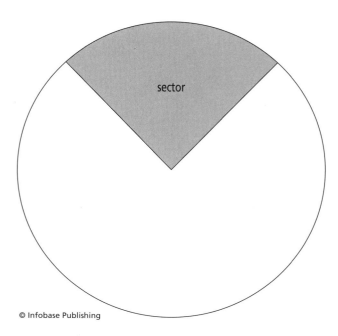

© Infobase Publishing

A sector is a plane surface bounded by two radii and an arc of a circle.

approximately of this shape, and bounded on two sides by a warm front and a cold front, although there is no arc bounding it on the third side.

serein   Rain that falls from a clear sky is called serein. The word is derived from the Latin *serum*, which means "evening," and serein is a fine rain that falls at evening in the TROPICS.

There are several possible explanations for this very rare phenomenon. CLOUD DROPLETS may evaporate after very small RAINDROPS have started to fall. Because of their size, the drops take several seconds to reach the ground, and by the time they do so the cloud has dissipated. Alternatively, the cloud may move away while the raindrops are falling, so that by the time they reach the ground the cloud is no longer overhead. It may also happen that the wind blows fine rain so that it reaches the ground at a point that is not beneath the cloud. In this case, however, the rain arrives at an angle and its source is fairly obvious.

Severe Local Storms unit (SELS)   A meteorological center, located in Kansas City, Missouri, where severe weather is monitored and forecasts of storms are issued for up to 6 hours ahead.

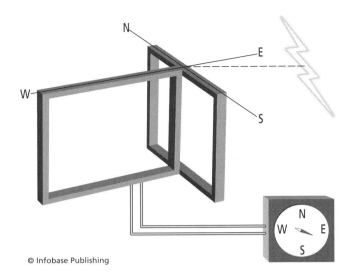

© Infobase Publishing

A sferics receiver has two square aerials at right angles to each other. This arrangement allows the receiver to detect the direction from which a disturbance caused by lightning is coming. The location of the storm can be identified by using two or more sferics receivers.

sferics   A word that is derived from "atmo*spherics*," sferics are the electromagnetic disturbances caused by natural electrical phenomena. They interfere with radio transmissions and can sometimes be heard on a radio as crackling or whistling noises.

LIGHTNING causes sferics, and this is used to locate the source of the THUNDERSTORM. The device uses two square radio receiver aerials mounted at right angles to each other so that one is aligned north–south and the other east–west. The strength of the sferics signal varies according to the angle at which it approaches the aerials. The signal strength is converted to a direction that is displayed on a screen. A sferics receiver can detect a thunderstorm up to a distance of about 1,000 miles (1,600 km).

When two or more widely separated sferics receivers detect the same thunderstorm, they can reveal its location. This will be at the intersection of lines drawn on a map from the position of each receiver in the direction it has measured. The position of a storm that is identified in this way is known as a sferics fix, and a report of a storm that is based on measurements by a sferics receiver is called a sferics observation.

shear   Shear is a force that acts parallel to a plane, rather than at right angles to it. If two plane surfaces experience a shearing force, one surface is being pushed one way and the other in another (not necessarily opposite) direction.

Fluids as well as solids can experience shear. If two bodies of fluid are in motion, the shearing force acting on them may result in a change in their relative speeds rather than direction. When crossing from one body to the other, the shear is evident as an abrupt change in either the direction or speed of movement, or both. A shear line is a line or narrow belt that marks an abrupt change, or shear, in the direction or speed of the wind.

A shear wave may form where there is strong horizontal wind shear in stable air (*see* STABILITY OF AIR). The difference in wind speed across a boundary between two layers of air causes TURBULENT FLOW. Air from the lower layer rises, but its stability causes it to sink again, establishing a wave pattern. In a vertical rather than horizontal plane, this is the mechanism that causes a flag to flap in the wind. Air is moving at different speeds on either side of the flag. The wind shear generates a wave pattern that is prevented from breaking down by the cloth of the flag, which acts in the same way as the inherent stability of the air in a shear wave. Where the Richardson number falls below the critical value of 0.25, the stability of the air is insufficient to sustain the wave pattern, which breaks down into general turbulence. The breakdown of shear waves plays an important part in transferring energy and transporting materials such as WATER VAPOR to and from the ground surface.

The force exerted on a surface by air that passes across it is called surface shearing stress. Because the force acts on a surface area, surface shearing stress is expressed as a pressure, measured in pounds per square

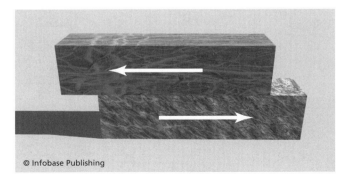

© Infobase Publishing

A shearing force acts parallel to a plane rather than at right angles to it. The arrows represent the direction of shear.

inch or, in SI units (*see* Appendix VIII: SI Units and Conversions) in pascals. The pressure exerted as surface shearing stress is equal to the opposing force of drag that is exerted by the surface on the moving air. Drag slows the air, but its immediate effect is felt only in the layer of air in immediate contact with the surface.

shelter belt A line of trees that is grown at right angles to the prevailing wind (*see* wind systems) in order to reduce the wind speed on the lee side as a means of protecting ground or crops that might be damaged by strong winds. Air approaching the trees is forced to rise. This squeezes the streamlines together, accelerating the air as it passes over the tops of the trees, but the moving air decelerates as soon as it has crossed the barrier. The air then separates from the surface of the trees and forms large eddies that

The trees form a barrier that slows the wind. If the barrier is dense, the wind is slowed greatly, but soon recovers. A low-density barrier slows the wind less, but the effect extends farther.

gradually become smaller in the wake (*see* turbulent flow) of the shelter belt. Finally, the air resumes its former movement and speed. A shelter belt (or wall, fence, embankment or other obstacle used for the same purpose) affects the flow of air above and in front of the barrier for a distance equal to three times the height of the barrier.

The effect on the downwind side of the barrier depends on the density of the barrier. If the barrier is very dense, the wind speed is greatly reduced in the large eddy that forms immediately downwind of it, because air cannot penetrate the barrier. The wind speed quickly recovers to about 90 percent of its former value, however, so the effect is limited to a distance about 10–15 times the height of the barrier. If the barrier is less dense, so it allows some of the air to pass through it, no eddy forms. The reduction in wind speed is smaller immediately behind the barrier, but it extends downwind for a distance that is equal to 15–20 times the height of the barrier. A medium-density barrier performs even better. The wind recovers to 90 percent of its original speed about 20–25 times the barrier height downwind, and some shelter belts provide this amount of protection for a distance equal to 40 times their height.

shock wave A shock wave is a traveling wave that moves through a fluid as a narrow band across which the pressure and/or temperature increase abruptly. Any object that moves through air or water generates a disturbance that propagates as a series of shock waves. Any sudden expansion or movement, such as an explosion or earthquake, generates shock waves. Sound waves are also shock waves.

shower A shower is a short period of precipitation that falls from a convective cloud such as cumulus or, more commonly, cumulonimbus (*see* cloud types). A shower is produced when moist, cool, unstable air (*see* stability of air) crosses a warmer surface. Precipitation starts and ends abruptly, varies in intensity but can be very heavy, and is often followed by sunshine and a blue sky. A shower of this type is sometimes called an air mass shower.

When the mechanism sustaining a cumulonimbus cloud fails and the cloud starts to dissipate, there can be a sudden, very intense shower of rain. This is a cloudburst. Individual cloud bursts are usually of brief duration, but they may be repeated, because as one cloud dissipates another forms along a SQUALL line.

The vertical currents inside a cumuliform cloud carry water droplets upward, and in a large cumulonimbus the updrafts are often strong enough to keep a large amount of water airborne. Eventually, downdrafts inside the cloud overlap and suppress the updrafts. When this happens, the water droplets are no longer supported and the cloud loses all its water at once in the form of a cloudburst. A big, fully developed cumulonimbus may hold 300,000 tons (275,000 tonnes) of water. If that amount of water falls on an area of 10 square miles (26 km²) it will deliver about 4 inches (10 cm) of rain. Rainfall as intense as this may cause a FLASH FLOOD.

A fog shower is quite different. This is a type of precipitation that can occur on mountains at elevations that are higher than the lifting condensation level (*see* CONDENSATION) when the mountain is engulfed by a passing cumulonimbus or large cumulus cloud that contains supercooled (*see* SUPERCOOLING) water droplets. Supercooled droplets freeze on contact with small objects, producing a coating of rime frost or glaze (*see* ICE). The cloud appears as fog to an observer on the mountainside, and the impact of the very cold droplets feels like a shower of rain.

Siberian high   The Siberian high is a region of high surface AIR PRESSURE that forms over Siberia in winter. Centered to the south of Lake Baikal, its influence covers all of Asia north of the Himalayas and extends westward, centered on latitude 50°N, across southern Russia and central Europe as far as the Atlantic.

The high produces a wind divide. The mid-latitude westerlies (*see* WIND SYSTEMS) prevail to its north and northeasterlies prevail to its south, across the steppes from the Ukraine eastward. The northeasterlies bring very dry, continental polar air (*see* AIR MASS).

The Siberian high produces the highest pressures known anywhere on Earth. On December 31, 1968,

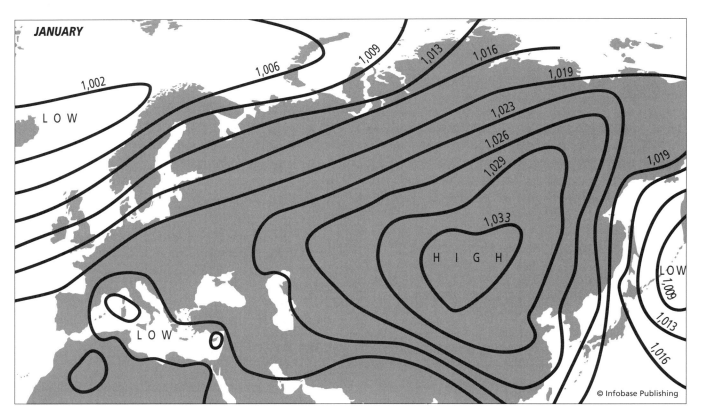

The usual distribution of pressure over Eurasia in January shows a large area of intensely high pressure centered on southern Siberia and Mongolia. This is the Siberian high.

a pressure of 1,084 millibars (32.01 inches, 81.31 cm) was recorded at the town of Agata, Russia (66.83°N 98.71°E). This is believed to be the highest pressure ever recorded.

Siderian The earliest period of the PROTEROZOIC era of the Earth's history, which began about 2,500 million years ago and ended 2,300 million years ago. Little is known about the geography or climates of the Earth during this time. *See* APPENDIX V: GEOLOGIC TIMESCALE.

Silurian The final period of the lower PALEOZOIC era of the Earth's history, which began 443.7 million years ago and ended 416 million years ago. Marine life flourished during the Silurian, and fossils from that time were first identified in the 1830s in South Wales by the English geologist Sir Roderick Impey Murchison (1792–1871). Murchison named the period Silurian after the Silures, a Celtic tribe living in Roman times along what is now the border between Wales and England.

The continents were clustered close to the equator during the Silurian, but with Gondwanaland, comprising the southern continents, drifting southward (*see* PLATE TECTONICS). Much of the low-lying land at the equator lay beneath shallow seas. Climates were generally warm, although there were GLACIERS in latitudes higher than about 65°. Some regions in latitudes within 40° of the equator had arid climates. *See* APPENDIX V: GEOLOGIC TIMESCALE.

silver iodide A yellow, solid, compound of silver, silver iodide (AgI) melts at 1,033°F (556°C) and boils at 2,743°F (1,506°C). It is used in CLOUD SEEDING because it can be made to form particles the size of FREEZING NUCLEI.

The material used for cloud seeding is a mixture of silver iodide and another compound, commonly sodium iodide (NaI) dissolved in acetone. The solution is then burned in a propane flame or a flare mounted on an airplane, releasing a smoke. As the smoke cools, the silver iodide solidifies as small crystals. Dropped into a cloud with a temperature of -4°F (-20°C), one ounce of silver iodide produces about 2.8×10^{17} of ice nuclei (10^{15} nuclei per gram of AgI).

singularities Certain types of weather occur fairly regularly at a particular time each year. These recurring events are known as singularities. The Indian summer (*see* WEATHER TERMS) and the sudden advent of ANTICYCLONES at the end of June are probably the best-known examples, but there are many more. The January thaw, affecting the northeastern United States around January 20–23, is a singularity. PRECIPITATION decreases sharply in California in March and April, due to an extension of the PACIFIC HIGH, while at the same time it increases in the Midwest due to an increase in cyclogenesis (*see* CYCLONE) in Colorado and Alberta, and an extension of maritime air (*see* AIR MASS) from the Gulf of Mexico. These are also singularities.

Singularities are major features of WEATHER LORE. Certain months are linked to particular kinds of weather, for example April showers and February fill-dyke (*see* LOCAL WEATHER). Many of these assumed singularities are imaginary, but others are genuine.

European singularities have been intensely studied, and a catalog of them has been compiled. Spring is often dry over much of northern Europe. This is due to a marked reduction in the frequency of weather systems arriving from the west. Westerly weather increases around the middle of June, bringing wet weather. Anticyclones often dominate the weather in the middle of September, broken by stormy weather in late September, and then more fine anticyclonic weather arrives at the end of September and in early October. FOGS and FROST are likely in the middle of November, caused by more anticyclones. DEPRESSIONS arriving from the west usually make early December a time of mild, wet weather.

Seasonal variations that last rather longer were identified by Hubert Lamb (*see* APPENDIX I: BIOGRAPHICAL ENTRIES). He called them NATURAL SEASONS.

November 3 is Culture Day in Japan and according to tradition the weather is usually fine. Naoki Sato and Masaaki Takahashi, who are scientists at the Center for Climate System Research at the University of Tokyo, have studied weather records for Tokyo over 38 years. They found that the November 3 singularity is real, and there are similar singularities in April and to a lesser degree in October.

Singularities also occur during the Asian summer MONSOON. These are associated with climatological intra-seasonal oscillations, and they produce times when the weather becomes suddenly wetter or drier. For example, the weather is usually dry from August

29 through September 2 in the western North Pacific monsoon region at longitudes 140°E and 15–20°E.

sink   An area that forms a receptacle for materials which are moving through a system. For example, some atmospheric CARBON DIOXIDE dissolves in seawater, where it is transported into the deep ocean. The ocean therefore acts as a sink for that carbon dioxide. Soil absorbs CARBON MONOXIDE and also solid particles that are removed from the air by rainout (*see* CLOUD CONDENSATION NUCLEI), snowout (*see* SNOW), and washout (*see* RAINDROPS), so the soil is a sink for these substances. A sink represents the end-point of a transport system that begins with the release of a substance into the environment. The place or process that releases the substance is called its source.

sky-view factor   The amount of sky that can be seen from a particular point on the surface, expressed as a proportion of the total sky hemisphere.

smog   A form of air pollution that used to occur frequently in winter in many large industrial cities but that was most closely associated with London, where pea soupers were a familiar feature of winter weather. Following the London smog incidents (*see* AIR POLLUTION INCIDENTS), legislation was enacted to prevent the burning of coal in open fires, and with the primary cause of smog removed this type of pollution ceased. Smog of this type is quite different from the PHOTOCHEMICAL SMOG that occurs in cities such as Los Angeles and Athens.

Smog is a contraction of *smoke* and *fog*. An atmospheric scientist named H. A. Des Voeux was the first person to use the term in 1905, but the phenomenon was far from new. It had been increasing in London since at least 1600, most rapidly in the 17th and 19th centuries.

Like many cities, London is low-lying, and a large river flows through its center. EVAPORATION from the river, as well as TRANSPIRATION from plants in the many parks and open spaces, ensures that the air is often moist, and DUST and other particles from the urban environment provide ample CLOUD CONDENSATION NUCLEI (CCN). Temperature INVERSIONS are also fairly common, and in winter, when the air is cool, the relative HUMIDITY beneath an inversion often exceeds 100 percent. WATER VAPOR condenses onto the CCN and the result is a FOG.

Coal and wood burn very inefficiently on domestic open fires. When heated, they emit combustible gases, of which only a proportion ignites to produce the fire. The remainder rises up the chimney. As the gases rise, they also cool and condense into solid particles. This is SMOKE, and in 1952 the domestic fires and coal-burning industrial furnaces of London emitted 157,000 tons (143,000 tonnes) of it. Between August 1944 and December 1946, in the London suburb of Greenwich there were an average of 20 days a month when the VISIBILITY was good at 0900 hours. In the city center there were fewer than 15 such days.

When smoke and fog were trapped together beneath an inversion, smoke particles adhered to and mixed with the water droplets. Sulfate particles, produced from sulfur dioxide emitted from the burning of coal with a high sulfur content, dissolved in the water to produce sulfuric acid (H_2SO_4). This made the smog acid, causing damage to buildings by ACID DEPOSITION. Sulfur dioxide came mainly from the coal burned in power plants and factories, rather than from domestic fires.

Burning coal also released CARBON DIOXIDE and increased the proportion of carbon dioxide to OXYGEN (the CO_2 partial pressure) beneath the inversion. This made some people breathe faster and more deeply, exacerbating the irritation to the respiratory passages caused by the acid smog itself.

Smog made window curtains and washing hanging outdoors filthy and deposited soot particles on the hands, faces, and clothes of people outdoors in it. It was visible as a haze even indoors.

A pea souper was a type of smog in which visibility was reduced to less than 30 feet (10 m)—and sometimes to very much less. The name refers to the yellowish color of the smog, reminiscent of the color of pea soup, when the mixture contained more smoke than fog. Traveling through a pea souper was difficult. Street lights remained lit throughout the day and road vehicles used their lights, but at times drivers were unable to see the edge of the road and became badly disoriented. Pedestrians fared only slightly better. In the thickest smogs it was difficult for them to see the ground beneath their feet. Factories, offices, and schools had to close early because of the long time it would take workers and students to travel home, and sometimes they would remain closed for two or more days, because travel was so difficult. Fortunately, pea soupers no longer occur.

smoke Smoke is an AEROSOL produced by the incomplete combustion of a carbon-based fuel. It consists of solid or liquid particles, most of which are smaller than 0.00004 inch (1 μm) in diameter.

When coal, wood, or some types of oil are heated, certain of their ingredients vaporize. Not all of the vapors ignite. Instead, they are carried up the chimney on the rising current of warm air. When these gases mix with the colder air higher up the chimney or outside, the vapors condense once more. These condensed particles, mixed with fine particles of ash, form soot (*see* AIR POLLUTION). Soot is readily ignited, producing more unburned vapors that also condense as they cool in the outside air. Smoke is the mixture of ash and recondensed volatile ingredients from the original fuel.

Smoke particles contain a large proportion of carbon. This gives them a dark color, because of which they absorb radiation and have a warming effect on the air. Smoke can also increase cloud formation and planetary ALBEDO, however, while at the same time reducing PRECIPITATION. Smoke particles are small enough to be active CLOUD CONDENSATION NUCLEI. WATER VAPOR condenses onto them (and becomes acid, *see* ACID DEPOSITION), but the very small size of the particles produces very small CLOUD DROPLETS. Observations over areas in Central and South America where surface vegetation was being burned during the dry season found that the average size of cloud droplets decreased from 0.0006 inch (14 μm) to 0.0004 inch (9 μm). Small droplets are less likely than large droplets to fall as precipitation. The cloud albedo increased slightly, reducing the amount of sunshine reaching the ground and therefore the surface temperature, but by an insignificant amount.

In colder climates, smoke can mix with FOG to produce SMOG. This has been responsible for many of the most serious incidents of AIR POLLUTION, including the London smog incidents (*see* AIR POLLUTION INCIDENTS).

Black smoke is produced when hydrocarbons are cracked (decomposed by heat) and then cooled suddenly. This releases particles of carbon, which are black.

Brown smoke contains particles of tarlike compounds. It is produced when coal is burned at a low temperature.

The top of a layer of smoke that is trapped beneath an INVERSION and that is seen from above and against the clear sky forms a smoke horizon. The smoke hides both the ground and the true horizon, so the boundary between the smoke and the clear air forms a false horizon.

smudging Using oil-burning heaters, called smudge pots, can protect a delicate farm crop against FROST damage. This is called smudging. In Florida citrus orchards as many as 70 burners are used to every acre (173 per hectare).

The smudge pots are lit on clear, cold nights when frost is likely. They may be used in conjunction with large propellers mounted on tall columns that are used to mix the air and prevent cold air from settling in frost hollows.

Protection can also be achieved by an alternative version of smudging in which materials are burned in order to produce voluminous amounts of black smoke. The smoke forms a layer above the ground, reducing the amount of heat that is lost by infrared radiation from the surface.

snow Snow is PRECIPITATION that falls as aggregations of ICE CRYSTALS. If large, these aggregations form SNOWFLAKES. An irregular crystal is a particle of snow that consists of a number of very small crystals that have formed erratically, so the resulting particle has an irregular shape. Irregular crystals are sometimes covered with a coating of RIME FROST. Snow that has an irregular crystalline structure is called amorphous snow. Snow grains, also called granular snow or graupel, are very small particles of white, opaque ice. The grains are flat and less than 0.04 inch (1 mm) in diameter.

Freshly fallen snow is usually white, but it may be brown. Brown snow is snow that is mixed with dust and by carrying dust or other particles to the ground the falling snow removes them from the air. This process is called snowout. The solid particles act as FREEZING NUCLEI or CLOUD CONDENSATION NUCLEI and the process is identical to rainout, but the precipitation falls as snow, not rain.

Snow formation begins in clouds with the freezing of WATER onto freezing nuclei. The resulting ice crystals continue to grow at the expense of supercooled water droplets (*see* SUPERCOOLING) until they are too heavy to remain airborne and fall from the base of the cloud. Unless the air TEMPERATURE between the CLOUD BASE and the ground is lower than 39°F (4°C), the snow will melt and reach the surface as rain.

Dry snow consists of ice crystals with no liquid water between them. The individual crystals are joined to each other directly or by necks of ice. Dry snow provides good thermal insulation. Small animals survive well beneath it and can move freely through tunnels they excavate. Wet snow includes 3–6 percent of liquid water held inside the crystal aggregations and in crevices between crystals.

The depth of snow that is lying on the ground at a particular time is known as the snow accumulation. Above the permanent SNOW LINE there is a net accumulation of snow. The term *snow cover* describes all the snow that is lying on the ground, or the proportion of an area that lies beneath snow. Ground is usually said to be snow-covered if more than 50 percent of its area lies beneath snow.

Freshly fallen snow that is very light and unstable is called wild snow. Once snow has fallen, the ice crystals begin to pack together under their collective weight and the layer of snow becomes denser. This happens more quickly with wet snow than with dry snow.

Fine, powdery snow that is suddenly thrown upward because of the disturbance caused when a thick layer of underlying snow settles abruptly is known as a snow geyser. Such settling is sometimes called a snow tremor or snowquake.

Snow melts when sunshine raises its surface temperature above freezing, but if the temperature subsequently falls below freezing again, the melted snow will refreeze as a crust of ice across the snow surface. Slush is a mixture of melting snow and/or ice and liquid water that is lying on the ground. It has a soft consistency and is very wet. Slush forms when snow and ice start to thaw, when rain mixes with them, or when they are treated, for example with salt, to lower the melting point (*see* FREEZING).

Blowing snow consists of snow grains that are lifted from the ground by the wind and transported through the air at a height of 6 feet (1.8 m) or more in amounts that are large enough to reduce VISIBILITY significantly. Snow that is lifted, but not to this height, will form snowdrifts, but it will not reduce visibility. Blowing snow can cause a WHITEOUT, and it forms one type of BLIZZARD. A snow banner, also called a snow plume or snow smoke, is the appearance, when seen from a distance, of snow that is being blown from a mountaintop or other exposed high ground. Very fine, powdery snow that is driven by the wind is called snow dust.

Driven snow is snow that has been transported by the wind and deposited in snowdrifts, which are accumulations of snow that is much deeper than the snow covering adjacent areas. A snowdrift that is shaped like a sand dune, having formed in the same way, is called a snow dune. A wind ripple, also called a snow ripple, is a wavelike pattern on the surface of snow that is produced by the wind and forms at right angles to the wind direction.

Snowflakes are carried by moving air, and the ability of air to transport them is proportional to its KINETIC ENERGY. If the air loses energy, it also loses its ability to keep snowflakes aloft, so they fall to the ground. Drifts occur where moving air has lost a significant amount of its energy.

Air loses energy by friction with the surface. When it encounters woodland or a belt of trees, moving air slows and some of its snow falls. The snow falls fairly evenly, however, so although the depth of snow is likely to be greater inside the wooded area than it is on open ground on the upwind side, the difference is not large. The overall effect is to collect snow from the passing air and reduce the amount falling on the LEE side. EDDIES will tend to accumulate small drifts on the lee sides of isolated plants.

Solid barriers, such as a wall or high banks on either side of a road, have a much bigger effect. Air approaching below the level of the top of a wall will strike it and be deflected in an eddy. The resulting loss of energy will cause the air to lose snow, but the eddy will also scour snow away from the side of the wall. There will be a space between the thin layer of snow lying against the base of the windward side of the wall and the deep drift. On the lee side, air crossing the top of the wall will eddy downward. This will have a similar scouring effect, and a space will separate the drift from the wall.

A road that is bordered by high banks will fill with snow if the air crosses the road approximately at right angles. The banks have the same effect as a wall, but in this case there are two barriers and the downwind drift caused by one overlaps the upwind drift caused by the other.

A snow fence produces a low drift on its upwind side and a much bigger drift on its downwind side. If the fence is erected to prevent snow from forming drifts on a road, the distance between the fence and the road should be equal to 10 times the height of the fence.

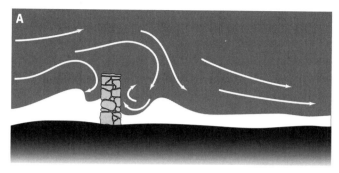

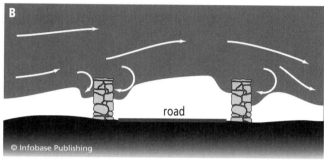

A wall (A) produces two snowdrifts, one on each side, with a space between the wall and the drift. High banks lining a road (B) also produce two drifts. In this case, the downwind drift behind one bank overlaps the upwind drift in front of the other bank, so the road quickly fills with snow.

A snowfield is an extensive, approximately level area that is covered uniformly with fairly smooth snow or ice. Snowfields are found in mountainous areas and high latitudes. A small GLACIER or an accumulation of snow and ice that is too small to be called a glacier is also known as a snowfield. It is snowfields that feed alpine glaciers, and when a glacier retreats it may be because of a reduction in the amount of snow falling over the snowfield feeding it.

The thickness of fallen snow does not provide a direct measure of the amount of precipitation, because snowflakes vary greatly in size, depending on the temperature of the air through which they fall, and their size determines the amount of air that they will trap between them as they accumulate. The colder the air, the smaller are the flakes and the more air they trap. A report of the thickness of a snowfall provides useful information for road users and anyone who needs to be outdoors, but it conveys very little meteorological information. Meteorologists need to know the amount of precipitation that has fallen. Consequently, snowfall amounts are always converted into their rainfall equivalents for meteorological purposes. The table below gives the snowfall equivalent of a standard amount of water at different surface air temperatures.

Suppose that a fall of snow covers the ground to a depth of 2 inches (5 cm) and that the air temperature is 15°F (-9.4°C). At this temperature, the ratio of snow to water is 20:1, which means that 20 inches (or centimeters) of snow are the equivalent of one inch (or cm) of water. Expressed mathematically using inches, this is:

$$20/1 = 2/x$$

where x is the amount of water that is equivalent to 2 inches of snow. Divide the equation by 20 and multiply by x and:

$$x = 2/20 = 0.1$$

At this temperature, therefore, 2 inches (5 cm) of snow are equivalent to 0.1 inch (0.25 cm) of water.

A light snow shower of brief duration is known as a snow flurry. A heavy fall of snow is called a snowstorm, although there is no precise definition of the term. If more than about 4 inches (10 cm) falls, snowplows are likely to be called out to clear roads, so this provides one possible way to define a fall of snow that is severe enough to be called a storm.

The weather conditions required to produce snow are no different from those that produce rain, and much of the precipitation that falls as rain begins as snow in the cloud. The heaviest snowfalls occur when the low-level air temperature is between 25°F and 39°F (-4°C to 4°C). This is warm enough for the air to hold a considerable amount of moisture and cold enough to ensure that precipitation falls as snow rather than rain.

Snow to Water Ratios

Temperature		
°F	°C	Ratio
35	1.7	7 : 1
29–34	-1.7–1.1	10 : 1
20–28	-6.7--2.2	15 : 1
10–19	-12.2--7.2	20 : 1
0–9	-17.8--12.8	30 : 1
less than 0	less than -17.8	40 : 1

That is why the heaviest snowstorms usually occur near the beginning and end of winter.

It is also possible that snow covering the ground chills the air in contact with it and that this causes air to subside, thereby sustaining low AIR PRESSURE high above the surface. These conditions can produce more snowstorms, so one snowstorm may cause another.

Four mechanisms account for most snowstorms. LAKE-EFFECT snow falls when cold air crosses relatively warm water and then encounters cold ground on the far side. Upslope snowfalls occur when air is cooled by OROGRAPHIC lifting. Snow is also produced at a FRONT when warm air is forced to rise over cold air, and by DEPRESSIONS. A particular snowstorm may result from one of these causes or a combination of two or more of them, but depressions cause more storms than any other cause.

In North America, lake-effect snow affects cities to the east of the Great Lakes and also Salt Lake City, to the east of the Great Salt Lake. Salt Lake City is a ski resort that receives abundant snow. One storm on October 18, 1984, delivered 27.2 inches (69 cm) and on February 2, 1989, the city received 20.9 inches (53 cm).

Upslope snow is especially common on the western side of the Rocky Mountains. In the month of January 1911, 390 inches (991 cm) of snow fell on Tamarack, California, and in the winter of 1998–99 Mount Baker, Washington, received 1,140 inches (28.96 meters).

snowball Earth Snowball Earth describes the Earth at times when, apart from the highest mountains, its entire surface is covered by ice. The term was coined in 1992 by the American geobiologist Joseph Kirschvink of the California Institute of Technology. This condition is believed to have occurred four times between 750 million and 580 million years ago. Two of the episodes have been identified. The Sturtian occurred about 710 million years ago and the Marinoan about 635 million years ago. Both ended abruptly, with the continental ICE SHEETS melting completely within about 2,000 years.

Mean temperatures during these glacial episodes were about -58°F (-50°C), and all of the oceans were frozen to a depth of more than 0.6 mile (1 km). Heat from the crust and core of the Earth prevented the oceans from freezing completely.

Dry land then comprised a number of small continents that were formed when a single large landmass broke apart. The interiors of these small continents were closer to the sea than they had been when the continents were joined together, and rainfall over land increased. This washed CARBON DIOXIDE from the air. The carbon dioxide reacted with minerals in the rocks and was carried to the sea, so the atmosphere came to contain less carbon dioxide. Temperatures fell and large ice packs formed in high latitudes. These increased the planetary ALBEDO causing temperatures to fall further (*see* SNOWBLITZ).

Once the oceans were completely sealed, no liquid water was exposed to the air, so EVAPORATION and PRECIPITATION ceased. VOLCANOES continued to erupt, however, releasing carbon dioxide into the air. With no precipitation to remove it, the carbon dioxide accumulated. After about 10 million years, its concentration was high enough to trigger a huge GREENHOUSE EFFECT. The ice melted in a matter of a few centuries, but this did not end the greenhouse warming. Surface air temperatures eventually rose to more than 120°F (50°C). More water evaporated and rainfall became intense. The rain washed carbon dioxide from the air and gradually the climate stabilized.

Further Reading

Allen, Philip A., and Paul F. Hoffman. "Extreme winds and waves in the aftermath of a Neoproterozoic glaciation." *Nature*, 433, 123–127, January 13, 2005.

Hoffman, Paul F. and Daniel P. Schrag. "Snowball Earth." *Scientific American,* January 2000.

———. "The Snowball Earth." Harvard University. Available online. URL: www-eps.harvard.edu/people/faculty/hoffman/snowball_paper.html. Accessed January 30, 2006.

snowblitz The snowblitz theory, popular in the 1970s, proposed that the Northern Hemisphere could be plunged into a full-scale GLACIAL PERIOD in a matter of a few decades. This is what is believed to have happened at the onset of the YOUNGER DRYAS and the proposed mechanism is the same.

The first stage requires a large release of freshwater into the northern part of the North Atlantic. Melting of the whole of the GREENLAND ICE SHEET might release sufficient water, but this is extremely unlikely, and there is no other large ice sheet in the Northern Hemisphere. The only source for the freshwater would therefore be greatly increased PRECIPITATION, perhaps as a consequence of a general rise in temperature.

The second stage involves shutting down the GREAT CONVEYOR. Freshwater, from the increased precipitation, is less dense than salt water and would float above it. At first, wave action would mix the two, but as more and more freshwater was added, mixing would become less effective until a layer of freshwater lay permanently above the salt water. The sinking of dense surface water drives the THERMOHALINE CIRCULATION, but if the surface water is less dense it cannot sink. This would shut down the Great Conveyor.

With the conveyor shut down, conditions would be set for the third stage in the process. Warm water would no longer flow northward, and the surface of the northern North Atlantic would cool rapidly by releasing its stored heat in the form of infrared radiation. In winter, it would freeze, partly because it would be colder than in previous years and partly because freshwater freezes at a higher temperature than salt water. The area of sea ice would expand, increasing the ALBEDO. This would cause further cooling.

The final stage could then follow. Polar continental AIR MASSES that form over North America are very cold in winter. Ordinarily, they warm as they cross the unfrozen Atlantic, but with the Atlantic frozen they would remain cold and would reach Europe essentially unaltered. In summer, when the sea ice melted, the ocean would still be cold, because the North Atlantic Drift (see APPENDIX VI: OCEAN CURRENTS) would have disappeared. Consequently, air would be cold and moist when it reached Europe, bringing reduced precipitation (because of reduced water-holding capacity due to the lower air temperature) but an increase in the proportion of precipitation falling as snow. In winter, increased albedo over both land and sea would bring very low temperatures to Europe. The general movement of air from west to east would spread this cooling to the whole of the Northern Hemisphere.

A year would come when not all the snow that fell on land during the winter melted in summer. The high winter albedo would continue through summer, so the temperature would remain low. More snow would accumulate the following winter and the snow-covered area would increase year by year. Temperatures would stabilize at below freezing, and as more snow accumulated its weight would compress the lower layers into glacial ice. ICE SHEETS would form in Europe and Canada, and each year their edges would advance farther south until a substantial part of both continents was in the grip of an ice age.

According to the snowblitz theory, the entire process, from the rise in temperature and increased rainfall to the formation and rapid advance of the ice sheets, might take no more than a few decades.

Few climate scientists now believe this scenario is credible. CLIMATE MODELS have estimated the consequences of a failure of the thermohaline circulation and the Great Conveyor, and have found that the resulting decrease in temperature is much less dramatic than the snowblitz theory suggests.

snow chill The effect of being covered in snow that then melts. In a BLIZZARD or snowstorm, snow itself affords protection from the wind, and it is possible to survive by digging an ice cave and sheltering inside it. If the snow in contact with a person's clothing begins to melt, however, the LATENT HEAT required to melt the snow is taken from the body, reducing its temperature. At the same time melted snow soaks the clothing, filling all the air spaces between the fibers. This water is close to freezing temperature, and water conducts heat much more efficiently than does air. The combined effects of using body warmth to melt snow and then of wearing clothes soaked in very cold water rapidly reduce the core body temperature.

snowflake A snowflake is an aggregation of ICE CRYSTALS that are grouped into a regular six-sided or six-pointed shape between about 0.04 inch and 0.8 inch (1–20 mm) across. Although all snowflakes have this basic hexagonal form, an extremely large number of variations is possible within it, and consequently every snowflake is unique.

Ice crystals form in clouds where the temperature is below freezing, and once they begin to form they grow at the expense of supercooled water droplets (see SUPERCOOLING). Crystals also fragment as protrusions extending from them are carried away by air currents, so the cloud soon contains splinters of ice that act as FREEZING NUCLEI for the formation of new crystals.

The individual crystals are composed of hexagonal units that are arranged in a number of possible shapes. The shapes depend on the temperature at which the crystals form. As they continue to grow beyond their unit shapes, they often do so by accumulating projecting arms or spikes. When crystals collide, their irregular shapes make it likely that they will lock together. The haphazard way this happens explains why no two snowflakes are identical. Snowflakes grow most readily when the temperature is about 32–23°F (between 0°C and -5°C). This is because at just below freezing a thin film of liquid water forms on the surface of each crystal and freezes solid where two crystals make contact.

Tens of individual crystals form a single snowflake, and the flake will continue to grow until its fall speed exceeds the speed of the air currents that carry it upward inside the cloud. Then it falls from the cloud. If it falls through air that is above freezing temperature, it will start to melt and if it remains in air at this temperature for long enough, it will reach the surface as rain rather than snow. It is unlikely to reach the surface as snow if the freezing level is higher than about 1,000 feet (330 m) and the surface air temperature is above 39°F (4°C). Rain is often melted snow and, as Wilson Bentley (*see* APPENDIX I: BIOGRAPHICAL ENTRIES) discovered, the size of RAINDROPS indicates the size of the snowflakes that melted to form them.

Bentley was the most famous student of snowflakes, but he was not the first. In a text called *Moral Discourses Illustrating the Han Text of the "Book of Songs,"* written between 140 and 131 B.C.E., Han Ying described the hexagonal shape of ice crystals and snowflakes. This is the earliest written reference to snowflakes, and their intricate patterns have continued to intrigue scientists ever since. In the 16th century Olaus Magnus (*see* APPENDIX I: BIOGRAPHICAL ENTRIES) referred to them, and in 1611 Johannes Kepler (1571–1630), the German physicist and astronomer, also described them, in a book called *A New Year's Gift, or On the Six-cornered Snowflake.* Robert Hooke (1635–1703, *see* APPENDIX I: BIOGRAPHICAL ENTRIES), the English physicist and one of the first microscopists, studied snowflakes under the microscope and drew what he saw. He published his drawings and descriptions in 1665, in a book called *Micrographia.* In 1936, Professor Ukichiro Nakaya, a Japanese physicist, was the first person to grow ice crystals in a laboratory, and in 1941 this led to the establishment of the Institute of Low Temperature Science at the University of Hokkaido. Professor Nakaya became a leading authority on snow. His book on the subject, called *Snow Crystals: Natural and Artificial,* was published in 1954 by Harvard University Press.

snow gauge A snow gauge is an instrument used to measure the amount of snow that has fallen over a stated period. It is a modified RAIN GAUGE.

In some snow gauge designs, the top edge of the collecting funnel is heated, so that snow falling on it melts and is collected as liquid water. This automatically converts the snowfall amount into its rainfall equivalent. An alternative design has a removable funnel and receiver, so that snow is collected and measured directly. Snowfall can also be measured by taking a core, using an open-ended cylinder pressed vertically all the way through the snow. The snow is then transferred to a measuring beaker. The beaker is calibrated to correct for any difference between its diameter and that of the coring cylinder. Transferring the snow to the beaker inevitably packs the grains together, so it must be melted and the depth read as the water equivalent. The simplest way to measure snowfall is to use a measuring stick, such as an ordinary ruler, and a THERMOMETER. The ruler will measure the thickness of snow, and the air temperature at the time it fell will indicate the water equivalent. Regardless of the method used, snowfall should always be measured in the open, well clear of trees and buildings that might deflect wind and alter the amount of snow reaching the ground.

The depth of a layer of snow after this has been converted to an equivalent fall of rain is called the water equivalent. Snow traps small pockets of air between its grains and flakes. This makes snow bulky, but to an extent that varies according to the type of snow. When snow is expected, weather forecasts predict the depth of fall people can expect. This is valuable information, but it cannot be used to compare the snowfall in different places and at different times. For comparisons, the depth of snow must be converted into an equivalent depth of water. This is done by pressing an open-ended cylinder vertically through the layer of snow, sliding a plate beneath the lower end to seal it, then removing the cylinder and melting the snow. The depth of liquid water is then recorded as the amount of PRECIPITATION. Liquid water cannot be compressed, so it provides a standard measure. As an approximate guide, the water equivalent is one-tenth the depth of snow. *See* SNOW for the table on converting snowfall to its water equivalent.

snow line The snow line is the boundary between the area covered by snow, for example on a mountainside, and the area that is free from snow. The edge of the snow that remains on a mountainside through the summer is also known as the snow line.

The location of the snow line depends on the TEMPERATURE, which in turn varies with latitude and with elevation. The location of the snow line on a particular mountain can be determined only by measurement, however. It is extremely difficult to calculate, because every mountain has crags shading the area behind them and deep gullies. These are places where snow lingers. The side of the mountain that faces the noonday Sun will be warmer than the side shaded from it and PRECIPITATION will be heavier on the side facing into the direction from which most weather systems arrive. These factors allow an imaginary mountain, which is perfectly conical and smooth, to be divided into four unequal sectors according to whether they are exposed to sunshine, shaded, sheltered from the weather, or both shaded and sheltered. It is possible then to estimate where snow is most likely to fall. The average snow line elevation at various latitudes is given in the table (top, right).

The climatic snow line is the altitude above which snow will accumulate over a long period on a level surface that is fully exposed to sunshine, wind, and

Mean Snow Line

Latitude	Northern Hemisphere		Southern Hemisphere	
	(feet)	(meters)	(feet)	(meters)
0–10	15,500	4,727	17,400	5,310
10–20	15,500	4,727	18,400	5,610
20–30	17,400	5,310	16,800	5,125
30–40	14,100	4,300	9,900	3,020
40–50	9,900	3,020	4,900	1,495
50–60	6,600	2,010	2,600	793
60–70	3,300	1,007	0	0
70–80	1,650	503	0	0

precipitation. Below this altitude, ABLATION between snowfalls will be sufficient to prevent snow from accumulating.

soil air   Soil air is the air held in the spaces between soil particles. In the upper layer of a cultivated soil, air accounts for about 25 percent of the volume. Soil air is similar in composition to atmospheric air and most of the constituents are present in similar proportions. The exception is CARBON DIOXIDE (CO_2). Atmospheric air contains about 0.04 percent CO_2, but soil air contains about 0.65 percent. The CO_2 in the soil comes from RESPIRATION by plant roots and soil organisms, including those engaged in decomposing organic matter. In some soils the proportion of OXYGEN decreases with increasing depth, and that of CO_2 increases by a similar amount. Some of the soil CO_2 dissolves in water and is removed as carbonic acid ($H_2O + CO_2 \rightarrow H_2CO_3$).

soil classification   Soil classification is any method for labeling different types of soil to distinguish their characteristics. Modern classifications are based on principles similar to those used to classify living organisms and clouds (see CLOUD CLASSIFICATION).

The earliest attempts to classify soils were made in classical times, but the first steps toward the modern system were made in the latter part of the 19th century by Russian scientists led by Vasily Vasilievich Dokuchaev (1840–1903). Dokuchaev based his classification on the climates in which soils form. His classification was widely adopted, and many of the original Russian names for soil types are still used, although now some of them are used only informally and others have been redefined,

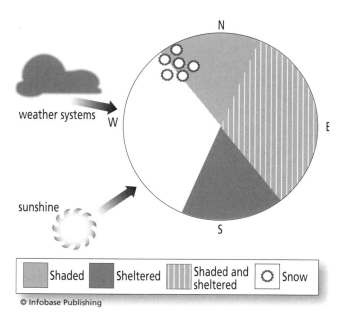

The direction of the noonday Sun and the direction from which most weather systems arrive make it possible to divide an idealized mountain into four sectors. The sectors that are sheltered and both shaded and sheltered receive little precipitation, because of the rain shadow effect. Snow is most likely to fall in the sector that is shaded from the Sun but exposed to the weather.

because the old system has been replaced. Soil names such as podzol, chernozem, and rendzina are taken from the early Russian work. Podzols are gray, ashlike, acid soils. Chernozems are black grassland soils sometimes called prairie soils, and rendzinas are brown soils found in humid or semi-arid grasslands.

American soil scientists were also working on the problem, and by the 1940s their work was more advanced than that of their Russian colleagues. By 1975, scientists at the Soil Survey of the United States Department of Agriculture had devised a classification they called Soil Taxonomy. It divides soils into 12 main groups, called orders. The orders are divided into 47 suborders, and the suborders are divided into groups, subgroups, families, and soil series, with six phases in each series. The classification is based on the physical and chemical properties of the various levels, or horizons, that make up a vertical cross section, or profile, through a soil. These were called diagnostic horizons. Some of the names seem strange on first acquaintance, and they become still stranger at levels below that of the order. The suborders include Psamments, Boralfs, and Usterts, the great groups include Haplargids, Haplorthods, and Pellusterts, and the subgroups include Aquic Paleudults, Typic Medisaprists, and Typic Torrox.

The 12 orders are listed below, with a brief description of each.

Gelisols are soils where there is PERMAFROST within 6.5 feet (2 m) of the surface.

Histosols are soils rich in organic matter that are found in bogs and marshes, where the climate is cool and wet.

Spodosols are sandy, strongly acid soils that are found in forests, especially coniferous forests.

Andisols are soils that form from volcanic ash; they are deep and light-textured.

Oxisols are deeply weathered (see WEATHERING), acid soils from which most of the plant nutrients have been washed away; they are found in the humid TROPICS and subtropics.

Vertisols are clay soils that swell when they are wet and develop deep cracks when they are dry; they are found in climates with marked wet and dry seasons.

Aridisols are desert soils that contain little organic matter; they often have salt layers.

Ultisols are strongly acid, deeply weathered, tropical soils.

Mollisols are very dark, grassland soils that are rich in organic matter and highly fertile.

Alfisols are soils found mainly in forests where the annual rainfall is 20–50 inches (510–1,270 mm); there is a layer of clay beneath the topsoil.

Inceptisols are soils found in cold, wet climates; they are at an early stage in their development.

Entisols are soils with little vertical development into layers; they are found on recent flood-plains, beneath recently fallen volcanic ash, and as wind-blown sand.

National classifications, such as the Soil Taxonomy, are often very effective in describing the soils within their boundaries, but there was a need for an international classification. In 1961, representatives from the Food and Agriculture Organization (FAO) of the United Nations, the United Nations Educational, Scientific and Cultural Organization (UNESCO), and the International Society of Soil Science (ISS) met to discuss preparing one. The project was completed in 1974. Like the Soil Taxonomy, it was based on diagnostic horizons. It divided soils into 26 major groups, subdivided into 106 soil units. The classification was updated in 1988 and has been amended several times since. It now comprises 30 reference soil groups and 170 possible subunits.

The FAO reference soil groups are listed below.

Histosols are soils with a peat layer more than 15.75 inches (40 cm) deep.

Cryosols are soils with a permanently frozen layer within 39 inches (100 cm) of the surface.

Anthrosols are soils that have been strongly affected by human activity.

Leptosols are soils with hard rock within 10 inches (25 cm) of the surface, or with more than 40 percent calcium carbonate ($CaCO_3$) within 10 inches (25 cm) of the surface, or less than 10 percent of fine earth to a depth of 30 inches (75 cm) or more.

Vertisols are soils with a layer more than 20 inches (50 cm) deep containing more than 30 percent clay within 39 inches (100 cm) of the surface.

Fluvisols are soils formed on river (alluvial) deposits with volcanic deposits within 10 inches (25 cm) of the surface and extending to a depth of more than 20 inches (50 cm).

Solonchaks are soils with a salt-rich layer more than 6 inches (15 cm) thick at or just below the surface.

Gleysols are soils with a sticky, bluish-gray layer (gley) within 20 inches (50 cm) of the surface.

Andosols are volcanic soils having a layer more than 12 inches (30 cm) deep containing more than 10 percent volcanic glass or other volcanic material, or weathered volcanic material within 10 inches (25 cm) of the surface.

Podzols are pale soils with a layer containing organic material and/or iron and aluminum that have washed down from above.

Plinthosols are soils with a layer more than 6 inches (15 cm) deep containing more than 25 percent iron and aluminum sesquioxides (oxides comprising two parts of the metal to three of oxygen) within 20 inches (50 cm) of the surface that hardens when exposed.

Ferralsols are soils with a subsurface layer more than 6 inches (15 cm) deep with red mottling due to iron and aluminum.

Solonetz are soils with a sodium- and clay-rich sub-surface layer more than 3 inches (7.5 cm) deep.

Planosols are soils that have had stagnant water within 40 inches (100 cm) of the surface for prolonged periods.

Chernozems are soils with a dark-colored, well-structured, basic surface layer at least 8 inches (20 cm) deep.

Kastanozems are soils resembling chernozems, but with concentrations of calcium compounds within 40 inches (100 cm) of the surface.

Phaeozems include all other soils with a dark-colored, well-structured, basic surface layer.

Gypsisols are soils with a layer rich in gypsum (calcium sulfate, $CaSO_4.2H_2O$) within 40 inches (100 cm) of the surface, or more than 15 percent gypsum in a layer more than 40 inches (100 cm) deep.

Durisols are soils with a layer of cemented silica (silicon dioxide, SiO_2) within 40 inches (100 cm) of the surface.

Calcisols are soils with concentrations of calcium carbonate within 50 inches (125 cm) of the surface.

Albeluvisols are soils with an irregular upper surface and a subsurface layer rich in clay.

Alisols are slightly acid soils containing high concentrations of aluminum and with a clay-rich layer within 40 inches (100 cm) of the surface.

Nitisols are soils with a layer containing more than 30 percent clay more than 12 inches (30 cm) deep and no evidence of clay particles moving to lower levels within 40 inches (100 cm) of the surface.

Acrisols are acid soils with a clay-rich subsurface layer.

Luvisols are soils with a clay-rich subsurface layer containing clay particles that have moved down from above.

Lixisols comprise all other soils with a clay-rich layer within 40–80 inches (100–200 cm) of the surface.

Umbrisols are soils with a thick, dark-colored, acid surface layer.

Cambisols are soils with an altered surface layer or one that is thick and dark-colored, above a sub-soil that is acid in the upper 40 inches (100 cm) and with a clay-rich or volcanic layer beginning 10–40 inches (25–100 cm) below the surface.

Arenosols are weakly developed soils with a coarse texture.

Regosols are all other soils.

An acid soil is one with a pH of less than 7.0. A basic soil is one with a pH greater than 7.0.

Further Reading
Ashman, M. R., and G. Puri. *Essential Soil Science.* Oxford: Blackwell Science Ltd., 2002.

soil moisture WATER that is present in the soil. Since all living organisms require water, the amount of soil moisture affects the rate of biological activity in the soil, including the rate at which organic matter decomposes.

Many organisms also require OXYGEN for RESPIRATION, and they obtain it from air that is trapped in spaces between soil particles. If the soil is completely saturated, these spaces are filled with water and the air is expelled, so too much water is as bad as too little. For most soil organisms, the optimum amount of soil moisture is 50–70 percent of field capacity.

Field capacity is the amount of water that a particular soil will retain under conditions that allow water to drain freely from it. It is measured by thoroughly soaking a measured weight of oven-dried soil, then leaving it to drain for a day or two before weighing it again. Field capacity is usually reported as a percentage of the oven-dried weight of the soil.

When water arrives at the soil surface, by falling as PRECIPITATION or flowing from adjacent land, it drains downward under the force of gravity. The antecedent precipitation index is a summary of the amount of precipitation that falls each day in a particular area, weighted so it can be used to estimate soil moisture. The speed with which water moves depends on the PERMEABILITY of the soil, but even when most of the water has drained away to join the GROUNDWATER, a film of water up to 15–20 molecules thick remains, covering all the soil particles.

The amount of water held in a soil is measured by weighing a sample of soil before and after drying it in an oven at 221°F (105°C) or by using an instrument such as a neutron probe to measure the electrical conductivity of the soil between two electrodes. The neutron probe is lowered into a pipe, about 2 inches (5 cm) in diameter and of an appropriate length for the conditions being studied. Electrodes are suspended in the pipe at the levels being monitored. The probe contains a radioactive source that emits fast (high-energy) neutrons. When the neutrons collide with HYDROGEN atoms, they change direction and slow down and some of the slow neutrons are deflected back into the probe, where they are counted. The greater the number of slow neutrons the probe detects, the greater is the water content of the soil, because water molecules are the major source of hydrogen atoms.

As water moves through the soil, the hydrogen atoms in its molecules, which carry positive charge, are attracted to oxygen atoms, bearing negative charge, at the surface of molecules of mineral particles. The water molecules attach themselves quite firmly to the mineral particles. This force is called adhesion.

Water molecules are also attached to one another by hydrogen bonds (see CHEMICAL BONDS). This force is called cohesion, and because of it other water molecules attach themselves to the molecules coating exposed mineral particles. The resulting film of water is held together by both adhesion and cohesion.

In addition, soil particles absorb moisture from the atmosphere. This is called hygroscopic moisture. It is held very tightly by the particles and cannot enter the root hairs of plants.

Adhesive and cohesive forces act on surfaces—of mineral particles and water, respectively. Consequently, the amount of water a soil can retain in this way depends on the amount of surface area its particles present. The bigger the soil particles, the smaller their total surface area is.

Soil particles are not spherical, of course, but the ratio of their radius to surface area obeys the same geometrical law. The surface area of a sphere is equal to $4\pi r^2$, where r is the radius. If $r = 2$, then the surface area is 50. If $r = 4$, then the surface area is 201. The volume of each of these particles ($4/3 \pi r^3$) is 33.5 and 268, respectively, so any volume of soil will contain many more of the smaller particles than of the bigger ones. For example, a soil volume of 10,000 will contain 298 of the small particles, but only 37 of the big ones, and the total particle surface area will be 14,900 in the soil with small particles and about 7,440 in the other soil. That is why SAND, with big particles, holds much less water than silt or clay, with very small particles.

Considerable force is needed to remove the film of water from soil particles. Adhesion water moves very little, but cohesion water is more mobile. Adhesion water is not available to plants, but molecules of cohesion water are constantly joining and leaving the water flowing through the soil. Water flowing downward through the soil is called gravitational water. It is held in the soil with a force that is less than about 30 kPa (300 mb, 4.5 lb/in^2). Water that can be absorbed by plant roots is called available water, and it is held with a force of about 30–1,500 kPa (0.3–15 bar, 4.5–218 lb/in^2). Adhesion water is held with a force equal to 1.5–100 MPa (15–1,000 bar, 218–14,500 lb/in^2).

Water that has drained to below the water table may then move upward again by CAPILLARITY.

Where the amount of water supplied by precipitation and available to plants (usually crop plants) is smaller than the amount needed to sustain healthy plant growth, the difference is called the water deficit. The water surplus is the difference between the amount of water that is needed to sustain the healthy growth of plants (usually crop plants) and the amount supplied by precipitation that is available to plants, where the amount available is greater than the amount required.

solar constant The solar constant is the amount of energy that the Earth receives from the Sun per unit area (usually per square meter) calculated at a point perpendicular to the Sun's rays and located at the outermost edge of the Earth's atmosphere. The value of the solar constant is not known precisely, but the best estimate is 127 watts per square foot (1,367 watts per square meter, 1.98 langleys, see UNITS OF MEASUREMENT).

The Sun generates energy by means of thermonuclear reactions in its core. The Sun is made from gas, about 75 percent HYDROGEN and 25 percent HELIUM, and its mass is 743 times that of the combined masses of all the planets in the solar system. It is 330,000 times more massive than Earth. This mass exerts a gravitational force on the Sun itself, pressing material inward. The inward gravitational pressure is so great at the core that the electrons are stripped away from atomic nuclei. These are predominantly hydrogen nuclei, each of which consists of a single proton.

Protons carry positive charge and repel one another, but the pressure at the solar core is sufficient to overcome this repulsion and to force protons together. This causes a reaction known as the proton–proton cycle in which four hydrogen nuclei combine to make one helium nucleus.

Later stages in the series of reactions that comprise the proton–proton cycle can follow several different paths, but they all begin with the combination of two hydrogen nuclei to make one deuterium (heavy hydrogen) nucleus with the release of a positron (positive electron, e^+) and a neutrino (ν).

$$^1H + {}^1H \rightarrow {}^2H + e^+ + \nu$$

The deuterium nucleus then captures a third proton. This produces a nucleus of helium-3 (^3He) with the release of a gamma ray (γ).

$$^2H + {}^1H \rightarrow {}^3He + \gamma$$

The final result of the cycle is the production of helium-4 (^4He) with the loss of 0.7 percent of the original mass of the protons. The lost mass is converted into energy according to the Einstein equation $E = mc^2$, where E is energy, m is mass, and c is the speed of light. A little of the energy is carried away by neutrinos, but most is converted into heat.

The TEMPERATURE inside the solar core is about 27 million degrees F (15×10^6 K = 15×10^6°C). The heat makes the material in the core expand, so it exerts an outward pressure. The outward pressure of expansion precisely balances the inward, gravitational pressure, so the Sun neither expands nor collapses.

The core heats the outer layers of the Sun. The outermost visible layer is called the photosphere. Its temperature is approximately 10,960°F (5,800 K = 6,073°C). Although the temperature is higher in the chromosphere and solar corona, neither of which is visible except during a solar ECLIPSE, these consist of matter that is so tenuous it has little effect on the emission of energy from the Sun. The effective temperature of the Sun is 10,960°F (5,800 K). The chromosphere and corona lie beyond the photosphere. The corona is the source of the SOLAR WIND.

The total amount of energy radiated by the Sun is given by the STEFAN–BOLTZMANN LAW. Earth receives only a small proportion of this energy, about 4.5×10^{-10} of the total. The radius of the Sun (R) is 109 times that of the Earth, but its average distance from the Earth, of 93 million miles (149.5 million km), is equal to $215R$ and it subtends an angle of only 0.5° in the sky. The amount of solar energy intercepted by Earth, the solar constant (S), is equal to $S_O/4\pi R_E^2$, where S_O is the solar output and R_E is the radius of the Earth. It is this calculation that yields the value of 127 watts per square foot (1,367 W/m^2). Although called a constant, the solar constant varies with changes in the solar output.

Clouds and the surface reflect a proportion of the solar radiation reaching the top of the atmosphere. This proportion represents the planetary ALBEDO (a) and it must be deducted from the solar constant ($S - a$) to give a value for the amount of solar energy reaching the surface.

Irradiance is the rate at which radiant solar energy flows through a unit area that is perpendicular to the radiation beam. Since the Sun emits radiation equally in all directions, illuminating a sphere around itself, irradiance decreases with distance from the Sun in accordance with the INVERSE SQUARE LAW. The irradiance at any point (I) is given by the equation: $I = I_s(R_s/R)^2$, where I_s is the amount of radiation being emitted by the Sun, R_s is the radius of the photosphere, and R is the distance of the point from the Sun.

solar energy Solar energy is energy that the Earth receives from the Sun and that can be used to perform useful tasks, such as providing space and water heating and generating electrical power. Although the concept of solar energy is sometimes extended to include WIND POWER, because wind is produced by weather systems driven by solar energy, the term strictly covers only the direct use of solar radiation in the form of heat or light.

Sometimes the generation of energy by burning biomass is also included. This is the growing of crops, such as fast-growing trees, that can be harvested for fuel, or that produce large amounts of sugar, such as corn (maize) and sugar beet, that can be fermented to produce alcohol (ethanol) for use as a liquid fuel. These technologies are included because the crops grow by means of PHOTOSYNTHESIS, which is powered by sunlight.

In principle, a huge amount of energy is available. The surface of the Earth receives about 9.5×10^{17} calories (4×10^{18} J) of energy a year, while the total annual energy consumption of the entire human population of the Earth amounts to only about 7.2×10^{13} cal (3×10^{14} J).

The simplest direct use of solar heat is also the most traditional: south-facing (in the Northern Hemisphere) windows. Add double-glazing to reduce heat loss, and the window allows solar energy to enter the building and warm its interior and contents, but prevents warm air and long-wave BLACKBODY radiation from leaving (see GREENHOUSE EFFECT).

A flat-plate solar collector heats water. It consists of a large, shallow box covered with glass and with a base that is painted matt black to absorb radiation. The box contains piping that winds back and forth across the base. Water is either pumped or flows by gravity through the piping. Solar radiation passes through the glass plate, is absorbed by the black base, and heat passes by conduction from the black base to the pipes and the water they contain. The piping carries the heated water inside the building to a heat exchanger in a water tank. Usually a number of solar collectors are mounted as an array on the roof and angled to face the noonday Sun.

Solar collectors are effective provided they are sited in a place where they receive an adequate amount of solar radiation. They are especially useful in latitudes between about 36°N and 36°S. Their disadvantage is that they require warm sunshine. That means they work best during daytime in summer and are much less useful at night, in winter, and when the sky is overcast, which are the time when heat is most needed. They must also be cleaned frequently, because dirt on the exterior glass shades the interior of the box.

Mirrors can be used to focus solar radiation onto a central point. Several devices, called solar furnaces, have been built to exploit this, and these produce temperatures high enough to heat water to drive generating turbines. Several of solar furnaces are located in the United States, the largest being at Albuquerque, New Mexico.

Solar cells, which are also known as photovoltaic cells, produce an electric current directly by the action of sunlight. When light strikes certain semiconductor materials, some of the light energy frees electrons in the material. An electric field in the material forces all the free electrons to move in the same direction. This constitutes an electric current. Solar cells were developed primarily for use in space, and they supply much of the energy for spacecraft and orbiting satellites. They are also starting to be installed for energy production on the surface. In sunny locations that are too remote to be reached by conventional power lines, solar cells may be less costly than alternative forms of generation.

A solar pond exploits the difference in DENSITY between freshwater and salt water. Typically, a pond several feet deep and with a large surface area is lined with black plastic to absorb radiant heat. A layer of water saturated with salt covers the lower part of the pond and a layer of freshwater is poured on top of it, taking care that the two layers do not mix. Freshwater then floats above the denser salt water. Sunshine heats the black plastic, which warms the salt water in contact with it. CONVECTION currents distribute the absorbed heat throughout the salt water layer, but without affecting the overlying freshwater layer. When the salt water reaches a satisfactory temperature—close to 212°F (100°C)—it is piped away to a heat exchanger, where the hot water heats a tank of freshwater. The salt water then passes back into the pond. The freshwater layer must be replenished from time to time to compensate for EVAPORATION, but covering the pond with clear plastic minimizes evaporation losses.

Further Reading
Allaby, Michael. *Deserts*. Rev. ed. Ecosystem. New York: Facts On File, 2007.

solar irradiance Solar irradiance is the total amount of energy that the Sun emits. The STEFAN-BOLTZMANN LAW relates this to the TEMPERATURE at the surface of the photosphere, which averages about 10,960°F (5,800K). At this temperature the total amount of energy radiated by the Sun (its EXITANCE) is about 70 MW/m², and over the entire surface of the photosphere it is about 4.2×10^{20} MW. Emitting energy at this rate, the Sun has so far converted about 0.1 percent of its mass into energy (see SOLAR CONSTANT). The temperature of the photosphere varies slightly, however, and therefore so does the amount of energy it radiates. In 1977, for example, the Kitt Peak National Observatory near Tucson, Arizona, measured a drop of 19.8°F (11K) in the temperature of the photosphere in a single year. This is very small (a drop of about 0.18 percent), but if it were sustained for a decade or more it would produce a slight but noticeable cooling of the Earth's climate.

During the 1990s, the solar output increased, peaking in 2000. The amount of cloud over the surface of the Earth, and especially of low cloud, is directly proportional to the intensity of COSMIC RADIATION reaching the atmosphere. Cosmic radiation consists of charged particles. When the charged particles enter the atmosphere, they ionize (*see* ION) AEROSOL particles, which then clump together to form CLOUD CONDENSATION NUCLEI. The charged particles are also deflected by the SOLAR WIND. Consequently, the intensity of cosmic radiation decreases at times of increased solar output and a stronger solar wind. This reduces cloud formation, and with fewer clouds the planetary ALBEDO decreases. More solar energy is then absorbed by the surface and the global climate grows very slightly warmer. Some scientists believe that the increased solar output during the 1990s is a significant cause of the slight increase in global temperatures that were recorded and that the effect of changing solar output on the intensity of cosmic radiation and cloud formation is the mechanism by which this occurred. Records since solar output peaked in 2000 have shown no change in the rate of GLOBAL WARMING, however, although this could be due to the time taken by the atmosphere to respond.

The amount of electromagnetic radiation emanating from a body such as the Sun or falling on a surface is called the radiant flux density. It is measured in watts per square meter (W/m^2), langleys (*see* UNITS OF MEASUREMENT), or calories per square centimeter (cal/cm^2). The total OPTICAL AIR MASS of the atmosphere that is penetrated by light from the Sun with the Sun at any given position in the sky is known as the solar air mass.

Transmissivity is a measure of the transparency of the atmosphere to incoming solar radiation. It is the fraction of the solar radiation incident on the top of the atmosphere that reaches the surface in a direct beam (not as diffuse radiation). Transmissivity varies with the state of the atmosphere—HAZE and AIR POLLUTION reduce it—and by the distance the radiation must travel through the atmosphere, or the PATH LENGTH. Zenith transmissivity is a measure of the fraction of solar radiation reaching the top of the atmosphere when the Sun is directly overhead (at zenith) that penetrates to the surface. The term is used to describe the transparency of the atmosphere and is calculated in relation to the zenith even for places outside the TROPICS, where the Sun is never directly overhead.

Several instruments are used to measure solar irradiance. A diffusograph measures diffuse radiation from the sky. It consists of a pyranometer surrounded by a circular strip set at such an angle that as the Sun crosses the sky the sensor on the pyranometer remains always in shade. Consequently, the sensor is exposed only to diffuse light from above.

A pyranometer, also called a solarimeter, is fairly robust and more suitable for field use than the more accurate, but more delicate pyrheliometer. There are several pyranometer designs, but all are built around a sensor that detects the difference in temperature

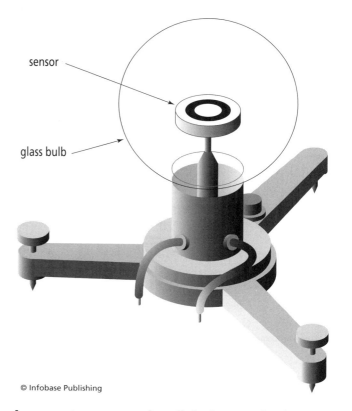

sensor

glass bulb

© Infobase Publishing

A pyranometer measures solar radiation by comparing the temperatures of the adjacent black and white surface of a sensor. The sensor is contained in a glass sphere. This is the Eppley pyranometer.

between two adjacent materials. This may be achieved with a sensor consisting of two concentric rings, the inner one painted black to give it a low albedo and the outer one painted white to give it a high albedo. Up to 50 electrical sensors, in good contact with the underside of the rings, provide readings of the temperature difference between the high- and low-albedo surfaces. The sensor is enclosed in a glass bulb filled with dry air and designed either to allow radiation of all wavelengths to enter or to filter out particular wavelengths, such as those shorter than visible light. A pyranometer is usually set on the ground with its sensor surface horizontal and facing upward. By comparing the readings from two pyranometers, one facing vertically upward and the other vertically downward, it is possible to measure surface albedo. A pyrgeometer is a pyranometer that measures infrared radiation.

A pyrheliometer is a very sensitive instrument. It contains a blackened surface positioned at right angles to the sunlight. In the Abbot silver disk pyrheliometer, which is the one most often used in the United States, the receiving surface is a disk of blackened silver supported on fine, steel wires at the bottom of a copper tube. A very accurate THERMOMETER is attached to the underside of the disk. Diaphragms arranged at intervals in the tube allow only direct sunlight and light from the sky immediately adjacent to the Sun to enter. The instrument can be used only when the sky is completely cloudless within a 20° radius of the Sun. As well as measuring solar intensity, the pyrheliometer measures the rate at which this changes. For this purpose there is a triple shutter above the tube. The shutter opens and closes to expose and shade the disk alternately, at very precise two-minute intervals. The final readings must then be corrected for a standard air temperature of 68°F (20°C) and a disk temperature of 86°F (30°C). The Abbot pyrheliometer remains calibrated for many years and can be used to calibrate other instruments. Pyrheliometers are so sensitive and require such constant maintenance that they are used only at research laboratories and some of the principal weather stations.

The Sun photometer measures the intensity of direct sunlight. It is held in the hand and pointed directly at the Sun. The Volz photometer is widely used to monitor air pollution. The Volz photometer was invented by Frederick E. Volz. It makes measurements that are defined by filters, that isolate particular wave-

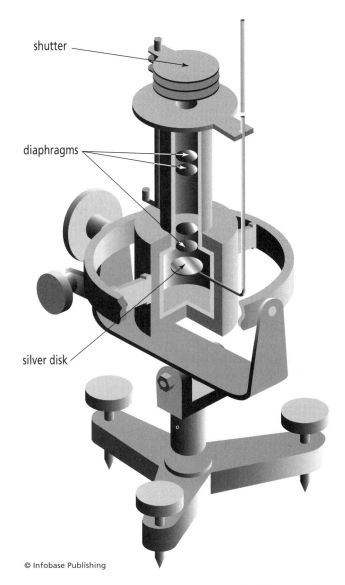

The Abbot silver disk pyrheliometer measures the intensity of solar radiation. Sunlight enters past a system of shutters that open and close at two-minute intervals to allow the rate of change of solar intensity to be measured. Beneath the shutter, light passes along a tube containing diaphragms that exclude light that does not come directly from the Sun and the immediately adjacent sky. At the bottom of the tube the light falls on a blackened silver disk attached to a thermometer.

bands. The measurements are not very precise, but relatively unskilled workers are able to take readings, and the sky does not need to be completely cloudless. The measurements indicate the amount of sunlight that is being scattered by haze and atmospheric particles.

solar spectrum The solar spectrum is the full range of electromagnetic radiation that emanates from the Sun. Solar radiation is emitted from the Sun's photosphere. The energy the radiation possesses is related to the TEMPERATURE of the photosphere by WIEN'S LAW. The photosphere is at about 10,960°F (6,070°C), and at this temperature the Sun radiates at every wavelength (*see* WAVE CHARACTERISTICS). The spectrum ranges from gamma rays with a wavelength of 10^{-5}–10^{-8} μm to radio waves with wavelengths of 1–10^9 m. The Sun does not radiate with equal intensity at all wavelengths, however. Solar radiation reaches its maximum intensity at about 0.5 μm, which is the wavelength of green visible light.

A wave band is a range of wavelengths within which all electromagnetic radiation is similar in character. The electromagnetic spectrum contains seven wavebands, listed in the table below.

Gamma rays or gamma radiation, often written using the Greek letter γ (gamma), have a wavelength of 10^{-8} μm to 10^{-4} μm. This is shorter than the wavelength of X-rays, so gamma rays possess more energy than X-rays. Less than 1 percent of the radiation emitted by the Sun is at gamma wavelengths, and all the solar gamma radiation reaching the Earth is absorbed in the upper atmosphere. None penetrates to the surface.

X-rays have wavelengths of 10^{-5} μm to 10^{-3} μm. Less than 1 percent of the radiation emitted by the Sun is at X-ray wavelengths, and all of the solar X-rays reaching the Earth are absorbed in the upper atmosphere. None reach the surface.

Solar Wave Bands

wavelength (m)	name	frequency (Hz)
10^{-11}–10^{-14}	gamma rays	300–30 EHz
10^{-9}–10^{-11}	X-rays	3 EHz–300 PHz
10^{-7}–10^{-9}	ultraviolet	30–3 PHz
4–7 × 10^{-7}	visible light	1 PHz–300 THz
10^{-3}–10^{-6}	infrared	300 THz–300 GHz
10^{-1}–10^{-3}	microwave	300 GHz–300 MHz
1–10^9	radio	300 MHz–3 Hz

(EHz (exahertz) = 10^{18} hertz; PHz (petahertz) = 10^{15} hertz; THz (terahertz) = 10^{12} hertz; GHz (gigahertz) = 10^9 hertz; MHz (megahertz) = 10^6 hertz.)

Ultraviolet radiation (UV) has wavelengths between about 4 nanometers (nm) and 400 nm. This waveband is approximate, and there is overlap at either end, with X-rays at the short-wave end and visible violet light at the long-wave end. UV radiation at 400–300 nm is sometimes called near UV, UV at 300–200 nm is far UV, and UV at less than 200 nm is extreme UV or vacuum UV. The UV waveband is also divided into UV-A (315–380 nm), UV-B or soft UV (280–315 nm), and UV-C or hard UV (shorter than 280 nm).

Visible radiation is electromagnetic radiation to which the human eye is sensitive. It is shortwave radiation at wavelengths between 0.4 μm and 0.7 μm. At wavelengths shorter than 0.4 μm there is ultraviolet radiation and at wavelengths higher than 0.7 μm there is infrared radiation. The atmosphere is completely transparent to radiation at wavelengths between 0.35 μm and 0.8 μm. This means radiation passes through the atmosphere without being absorbed, but it is scattered by air molecules (*see* SCATTERING).

When visible white light passes through a prism, it divides into its constituent wavelengths. Light of different wavelengths then appears as a band of colors on a screen placed in a suitable position. The separation of the component wavelengths as white light passes through a prism is called dispersion. Violet light, which has the shortest wavelength, appears on the left, and red, with the longest wavelength, on the right. This is the arrangement of colors in the rainbow (*see* OPTICAL PHENOMENA). The wavelengths of visible light are:

violet 0.4 μm
indigo 0.43 μm
blue 0.46 μm
green 0.5 μm
yellow 0.57 μm
orange 0.6 μm
red 0.7 μm

Infrared radiation has a wavelength from 0.7 μm to 1 mm. This is longer than visible red light and shorter than microwaves. Certain atoms and molecules vibrate at frequencies within this waveband, and because of this they absorb infrared radiation at characteristic wavelengths (*see* ABSORPTION OF RADIATION). Certain substances can be identified by the infrared wavelength at which they absorb and absorption by some atmospheric gases produces the GREENHOUSE EFFECT.

In a real greenhouse, the glass absorbs and is therefore opaque to infrared radiation with a wavelength greater than 2 μm. All warm bodies emit infrared radiation. Although infrared is invisible to the human eye, human skin glows at infrared wavelengths. Far-infrared radiation has a wavelength greater than 15 μm. Near-infrared radiation has a wavelength of about 1–3 μm. WATER VAPOR is a very efficient absorber of near-infrared radiation. Thermal-infrared radiation has a wavelength of about 3–15 μm.

Microwaves are electromagnetic radiation with a wavelength of 1 mm to 10 cm (0.04–4 in). Many SATELLITE INSTRUMENTS use microwave radiation to measure sea ice and various features of the atmosphere.

Radio waves, at the far end of the solar spectrum, have the longest waves of all. Their wavelengths range from 1 meter (3.3 feet) to 1 million kilometers (621,000 miles).

solar-topographical theory   A theory that explains past changes in climate in terms of variations in solar output and the formation and erosion of mountains. Mountains form when part of the Earth's crust is raised by volcanic activity (*see* VOLCANO) or as the result of a collision between two crustal plates (*see* PLATE TECTONICS). Their height then gradually decreases as material is lost by erosion.

The cycle of mountain-formation and erosion alters the elevation of the surface. It also affects the flow of air, tending to deflect air to the north or south of a mountain range, and forcing air to rise, and therefore cool and lose moisture, as it crosses high ground. Mountains have a clear climatic influence. So do variations in solar output, because they affect the amount of solar energy reaching the Earth. The theory considers variations of 10–20 percent to either side of the mean to be climatically significant.

solar wind   A stream of protons, electrons, and some nuclei of elements heavier than HYDROGEN that flows like a wind, outward from the Sun. The wind is generated in the outermost region of the Sun, called the corona, where the temperature is so high that particles acquire sufficient energy to escape from the Sun's gravitational field.

The mean distance between the Earth and Sun, known as one astronomical unit (AU), is about 93.2 million miles (150 million km). At this distance the solar wind carries 16–165 protons per cubic inch (1–10 per cm³) traveling at 220–440 miles per second (350–700 km/s). The solar wind deflects the tails of comets, so these always point away from the Sun, and it compresses the Earth's magnetic field. Interaction between the solar wind and the upper atmosphere produces auroras (*see* OPTICAL PHENOMENA). The intensity of the solar wind varies with the amount of SUNSPOT activity.

solstice   The solstice is one of the two dates in each year when the difference in length between the hours of daylight and darkness is most extreme. The Arctic and Antarctic Circles, at latitudes 66.5°N and S (*see* AXIAL TILT), are defined as the latitudes at which the Sun does not rise above the horizon at the winter solstice and does not sink below it at the summer solstice. In latitudes higher than 66.5°, the periods of continuous darkness and daylight are longer than the one day at each solstice.

The solstices are also called midsummer day and midwinter day. At noon on midsummer day the Sun is directly overhead at the tropic in the summer hemisphere and at noon on midwinter day it is directly overhead at the other tropic. The solstices fall on 21–22 June and 22–23 December. The length of the day varies because the rotational axis of the Earth is tilted with respect to the PLANE OF THE ECLIPTIC, and it is this tilt that produces the SEASONS.

sonic boom   The sound, like a clap of thunder, that is heard at the surface when an aircraft or other body flies at a speed greater than the SPEED OF SOUND. At subsonic speeds a moving object disturbs the air, sending out quite gentle waves of pressure that propagate in all directions, traveling at the speed of sound. These disturbances reach the surface continuously, and because they are slight and there is no sudden change in their intensity they make no sound. As the speed of the moving body approaches that of sound, and therefore of the pressure disturbances its motion produces, the pressure waves are confined in a decreasing volume. When the body exceeds the speed of sound and also the speed of its own pressure waves, the waves are behind the body and extend horizontally from it in a cone, called a Mach cone. The entire pressure field is contained within the Mach cone, where it produces a sudden, sharp increase in pressure, or SHOCK WAVE. An observer on the surface hears this as a loud bang when the edge of the Mach cone passes. Very large aircraft sometimes produce two sonic booms, one from the front of the aircraft and the other from the rear.

sounding Any measurement that is made through the column of air between the instrument and the level that is being monitored can be called a sounding. An instrument that makes a sounding is called a sounder.

Satellites usually carry sounders. These measure the amount of radiation being emitted at particular levels in the atmosphere. WEATHER BALLOONS also take soundings.

The word *sounding* was first used at sea to refer to the measurements of the depth of water that were made using a weighted rope. The word is derived from the French verb *sonder,* which comes from the Latin *subundare, sub-* meaning "under" and *unda* meaning "wave," so a sounding is literally a measurement taken from beneath the waves.

source The place at which a particular substance, usually a pollutant, is released into the environment, or the process by which it is released, is known as the source for that emission. For example, road transport is the source of CARBON DIOXIDE, NITROGEN OXIDES, and particulate matter (*see* AIR POLLUTION). The opposite of a source is a SINK.

Southern Oscillation

At intervals of 1–5 years a change occurs in the distribution of air pressure over the equatorial South Pacific. This is called the Southern Oscillation.

Ordinarily, air rises in the region of Indonesia, where the sea-surface temperature is high. The air flows from west to east at high level, and descends near the South American coast. This produces low pressure over Indonesia and high pressure over the eastern South Pacific. Low-level winds blow from east to west, balancing the flow. From time to time this situation is reversed. Pressure is high over Indonesia, low over the eastern South Pacific, and air flows from east to west at high level and from west to east at low level. The change in direction of the surface wind constitutes an El Niño, the change between the two patterns of pressure distribution constitutes a southern oscillation, and the two together constitute an ENSO event.

The distribution of pressure over the South Pacific is recorded as the Southern Oscillation Index (SOI).

This is a measure of the difference in sea-level atmospheric pressure between two monitoring stations, those most often used being at Darwin, Australia, and Tahiti, in the central South Pacific. A strongly negative SOI indicates a warming, and a strongly positive SOI indicates a cooling in the central and eastern equatorial South Pacific.

specific volume The specific volume is the volume that is occupied by a unit mass of a substance.

speed of sound Sound propagates as a SHOCK WAVE, so the speed with which a sound travels through a medium depends on the DENSITY and elastic modulus of the medium. The speed of sound through a medium is given by:

$$c = \sqrt{(E/\rho)}$$

where c is the speed of sound, E is the elastic modulus, and ρ is the density of the medium. The elastic modulus is the ratio of the stress applied to a body and the strain that stress produces.

For a gas, $E = \gamma p$, where γ is calculated from the HEAT CAPACITIES of the principal constituents and p is the pressure. This shows that the speed of sound in a gas is related to the temperature of the gas. This means that the speed of sound through a gas at a particular temperature can be expressed as:

$$c = c_0 \sqrt{(1 + t/273)}$$

where c_0 is the speed of sound in the gas at 0°C, t is the temperature in °C, and dividing by 273 relates the Celsius TEMPERATURE SCALE to the Kelvin scale.

In the case of a liquid, E is the bulk modulus, which is the ratio of the pressure applied to the medium and the extent to which its volume decreases.

For a solid, the elastic modulus is known as Young's modulus, after the British physicist Thomas Young (1773–1829), who proposed it, and it is the ratio between the stress applied to the solid and the change in its length.

Sound travels through air at 68°F (20°C) at 769.5 MPH (344 m/s). It travels through water at 68°F (20°C) at 3,268 MPH (1,461 m/s) and through steel at 68°F (20°C) at 11,185 MPH (5,000 m/s).

spore A spore is a microscopically small organic structure produced by bacteria, fungi, and some plants, such as algae and ferns. It is capable of giving rise to a new individual without first fusing with another cell, but it does not contain an embryo of the new individual and is therefore quite different from a seed. Some bacteria survive in hostile environments for long periods as spores. Bacterial, fungal, and plant spores can be carried into the air to form part of the aerial plankton (*see* AEROSOL).

Spörer minimum The German solar astronomer Friedrich Wilhelm Gustav Spörer (1822–95) identified a period from 1400 to 1510 during which very few SUNSPOTS were observed. Spörer described the phenomenon in an article published in 1889, which attracted the attention of the English astronomer Edward Walter Maunder (*see* APPENDIX I: BIOGRAPHICAL ENTRIES). The American solar astronomer John A. Eddy (born 1931) named the period the Spörer minimum.

Like the MAUNDER MINIMUM, the Spörer minimum was known as the LITTLE ICE AGE. Temperatures were abnormally low. Norse settlers were forced to leave Greenland because their crops failed and the SEA ICE failed to thaw, so they could not fish. Famine increased in many parts of the world and in the winter of 1422–23 ice covered the entire surface of the Baltic Sea.

Further Reading

Geerts, B., and E. Linacre. "Sunspots and Climate." Available online. URL: www-das.uwyo.edu/~geerts/cwx/notes/chap02/sunspots.html. Accessed February 1, 2006.

squall A squall is a sudden, brief STORM in which the WIND SPEED increases by up to 50 percent, then dies away more slowly. For a storm to be described as a squall in the United States, the wind must reach 16 knots (18.4 MPH, 30 km/h) or higher for at least two minutes. The wind speed may reach 30–60 MPH (50–100 km/h) in a severe squall and speeds of 100 MPH (160 km/h) have been recorded.

A squall that is accompanied by a short period of heavy rain is called a rain squall. The rain falls from cumuliform cloud, commonly cumulonimbus (*see* CLOUD TYPES). It is carried by wind that blows outward from the center of the storm, and it often precedes the rain associated with a THUNDERSTORM.

A series of squalls sometimes occurs along a line known as a squall line. This is a series of very vigorous cumulonimbus clouds that merge to form a continuous

A squall line in Pamlico Sound, North Carolina, with a sail boat heading toward it. *(Michael Halminski, Historic NWS Collection)*

line which is often up to 600 miles (965 km) long and that advances at right angles to the line itself. Cloud formation begins along a cold FRONT, where moist air in the warm sector ahead of the front is being undercut and lifted by the advancing cold air. This produces a number of severe local storms. The clouds grow very tall in the unstable air (*see* STABILITY OF AIR), often penetrating through the tropopause (*see* ATMOSPHERIC STRUCTURE). A weak INVERSION in the warm sector restricts the development of CONVECTION cells except at the front itself.

Wind speed increases with height, and the speed of the wind in the middle troposphere determines the speed with which the clouds move. Consequently, the upper part of each cloud overtakes its base, so the cloud overhangs its own base. Warm, moist air ahead of the cloud is then swept up into the convection cell inside it. The warm air is lifted to the free convection level and then rises rapidly to the top of the cloud, where the wind reaches its maximum speed and the cloud is drawn into an anvil shape. Where the base of the anvil meets the main body of the cloud, EDDIES produced by the vertical air currents produce a characteristic roll of cloud called a squall cloud, often with mammatus.

CONDENSATION in the updraft of air produces PRECIPITATION. This falls from the trailing edge of the cloud. Some of the water droplets evaporate in the drier air below the cloud. This chills the adjacent air, and falling RAINDROPS drag cold air down with them, causing a DOWNDRAFT. The downdraft produces a very local region of slightly raised surface atmospheric pressure. Part of the downdraft flows beneath the cloud and below the updraft. The updraft and downdraft meet at the leading edge of the cloud, where they produce strong GUSTS of wind along a line known as the gust front.

The gust front cuts beneath the warm air ahead of the cloud, lifting it into the updraft. Each individual cumulonimbus contains a single convection cell that lasts for only an hour or two before exhausting its supply of moisture and dissipating. Its gust front, however, scoops up moist air in front and to the right, which produces a new cloud. As soon as a cloud attains its maximum development it begins to dissipate, at the same time triggering the development of a new cloud on its right front.

The vigor with which the gust front is able to shovel up warm, moist air causes the line of clouds to move faster than the cold front which initiated their formation. The squall line then becomes detached from the cold front, advancing ahead of it into the warm sector and moving at 10°–20° to the right of the wind direction in the middle troposphere. Behind the local high-pressure band associated with the downdraft, there is a region of low pressure, called the wake low (*see* THUNDERSTORM).

In a cumulonimbus cloud containing a single convection cell, the updrafts and downdrafts conflict. This limits the development of the cloud and contributes to its dissolution. Along a squall line, however, the updrafts and downdrafts augment each other. This allows a squall line to last for several days, rather than the hour or two of an individual cloud, and to produce storms of much greater ferocity. TORNADOES often form along squall lines.

Squall lines are especially common in the central and eastern United States.

A squall in tropical or subtropical seas that occurs suddenly and without the prior appearance of a squall cloud is known as a white squall. The only indication of its approach is a line of white water caused by the wind.

stability index A stability index is one of a series of values that are used to summarize the STABILITY OF AIR and the severity of the THUNDERSTORMS, up to and including TORNADOES, that varying degrees of atmospheric instability are likely to generate. Stability indices must be used with caution, because they do not apply to every situation. The most commonly used indices are the K Index (K), Lifted Index (LI), Showalter

K Index

K Index	Probability of a thunderstorm
<15	0%
15–20	<20%
21–25	20–40%
26–30	40–60%
31–35	60–80%
36–40	80–90%
>40	nearly 100%

Stability Index (SSI), Total Totals (TT), SWEAT Index, and Deep Convective Index (DCI). Each index produces a single numerical value that indicates either the degree of instability or the probability of severe storms. The relationship between K values and the probability of severe thunderstorms is shown in the table above.

K measures the potential for thunderstorms from the LAPSE RATE and atmospheric moisture. LI and SSI both measure convective instability but to different heights. TT has two components, Vertical Totals (VT) and Cross Totals (CT), which measure static stability and DEW point temperature respectively at particular levels. SWEAT (Severe Weather Threat Index) takes a number of factors into account. DCI combines equivalent potential temperature (*see* PARCEL OF AIR) and instability.

The introduction of powerful computers has allowed meteorologists to calculate the more complicated convective available potential energy (CAPE). This has reduced their reliance on stability indices.

stability of air The tendency of a PARCEL OF AIR to possess neutral BUOYANCY, so that it remains at a constant height, is described as the stability of that air. Air with neutral or negative buoyancy is said to be stable and air with positive buoyancy is unstable. Whether air is stable or unstable depends on the difference between the environmental LAPSE RATE (ELR) and the adiabatic lapse rate (*see* ADIABAT). This difference is known as the POTENTIAL TEMPERATURE gradient, because it refers to a rate of change, or gradient, in the potential temperature of the air.

Neutral stability is the condition in which the atmosphere is stratified, and the potential temperature neither increases nor decreases with height. If a parcel of air moves vertically in either direction, it will enter a region where its DENSITY is similar to that of the surrounding air. Consequently, gravity will not restore it to its original level and it will remain where it is.

The amount by which the air temperature decreases between the surface and the tropopause (see ATMOSPHERIC STRUCTURE) determines the ELR. The ELR is assumed to remain constant throughout the troposphere. The temperature at the tropopause is always about -74°F (-59°C). If, then, the surface temperature is 59°F (15°C), and the height of the tropopause is 36,000 feet (11 km), the ELR will be about 3.7°F per 1,000 feet (6.7°C per km). This is an average value for the ELR, based on the fact that the mean surface temperature over the whole world is 59°F (15°C). The actual ELR varies with every change in the surface temperature. For example, if the surface temperature is 32°F (0°C), the ELR will be 2.9°F per 1,000 feet (5.4°C per km), but if the surface temperature is 80°F (27°C), the ELR will be 4.3°F per 1,000 feet (7.8°C per km).

Suppose unsaturated air at 59°F (15°C) is forced to rise. It will cool at the dry adiabatic lapse rate (DALR) of 5.38°F per 1,000 feet (9.8°C per km). This is greater than the ELR. Consequently, the temperature of the rising air decreases faster than the ELR. At any height the rising air will be cooler and denser than the surrounding air, so it will sink back to the level at which it is at the same temperature and density as the air around it. This air will be stable. The most stable condition of all occurs in a layer of air that lies beneath an INVERSION.

If the SALR is greater than the ELR, rising air will always be cooler and denser than the surrounding air, even if the air is saturated. This condition is called absolute stability. If a parcel of absolutely stable air is made to rise, at first it will cool at the dry adiabatic lapse rate (DALR). This is higher than the SALR and therefore it is also higher than the ELR. The rising air will quickly reach a level at which it is cooler than the surrounding air and so it will sink once more. Even if it is forced to rise high enough for its WATER VAPOR to start to condense, for example by being carried across a mountain, it will still tend to sink. CONDENSATION of its water vapor will release LATENT HEAT of condensation, warming the air and altering its lapse rate from the DALR to the SALR, but this is still greater than the ELR, so the rising air will always be cooler than the air around it.

When cold, stable air lies beneath a layer of potentially warmer air, a high föhn, also called a stable-air föhn, may develop. This situation occurs when cold, stable air moves against a mountain range. In stable air the upper layer is at a higher potential temperature than the lower layer. The upper layer of air spills over the mountains and descends, warming adiabatically as it does so. A high föhn can also develop when subsiding air in an ANTICYCLONE is chilled by a cold land or sea surface.

Static stability, also called convective stability, is the condition in which the atmosphere is stratified so that its density decreases with height and buoyancy increases. The potential temperature increases with height. If a parcel of air rises by CONVECTION, it enters a region where it is denser, and therefore heavier, than the surrounding air, so gravity will cause it to sink back to its original level. If the parcel should sink, it enters a region where it is less dense, and lighter, than the surrounding air, so it will rise to its original level. When the atmosphere is not statically stable, it may be neutrally stable or convectively unstable.

If the air is moist, it may reach a height at which it becomes saturated and water vapor starts to condense. Condensation will release latent heat, reducing the rate of cooling to the SALR. This is a slower rate of cooling than the ELR, so air cooling at the SALR will always be warmer and less dense than the air around it. It will therefore continue to rise. This air will be unstable. Instability is the tendency of air that has begun to rise to continue to rise.

Because the ELR is greater when the surface air is warm than it is when the surface air is cold, air is more likely to be unstable in warm weather than it is in cold weather. On really cold days in winter the air is usually very stable, because the ELR is low. When the surface temperature is -10°F (-23°C), the ELR is only 1.8°F per 1,000 feet (3.3°C per km), which is much lower than the SALR.

Any line that is not associated with a frontal system (see FRONT) and along which the air is subject to vigorous convection and is therefore unstable is called an instability line. If the instability leads to the formation of active cumulonimbus clouds (see CLOUD TYPES) and THUNDERSTORMS, the line is known as a SQUALL line.

The opposite of absolute stability is absolute instability, which is the condition of air when the ELR is greater than the DALR. As it rises, a parcel of absolutely unstable air will cool at the DALR, but because this is a slower rate of cooling than the ELR it will always be warmer than the surrounding air. If the air contains water vapor, it may reach a height at which this starts to condense to form cloud. The release of latent heat of condensation will warm the air, reducing the lapse rate from the DALR to the SALR. The difference between the ELR and the SALR is greater than that between the ELR and DALR, so the instability of the air will increase.

Being always warmer than the air immediately above it, the rising unstable air will retain its buoyancy and will continue to rise. Its rise will finally be checked when it reaches a height at which the ELR decreases to less than the SALR or to less than the DALR in the dry air above the cloud top or if no cloud has formed.

These conditions are likeliest to occur on very hot days. Air that is heated strongly from the ground then forms a layer that is at a much higher temperature than the air above it, producing a very steep ELR. Clouds that form in absolutely unstable air are seldom very deep, because the layer of very warm air that causes the instability is usually quite shallow, so the steep ELR does not extend very high. Consequently, absolutely unstable air does not usually produce STORMS or even SHOWERS.

Convective instability, also known as potential instability and static instability, is caused by convection. It happens when the atmosphere is stratified, but the potential temperature decreases with height, and as a layer of moist air rises, the lower part of the layer becomes saturated (see SATURATION) before the upper part. Rising air in the lowest layer, which is not yet saturated, is then cooling at the DALR, saturated air in the layer above it is cooling at the shallower SALR, and air in the uppermost layer is cooling at the DALR. If the lapse rate through the entire layer of rising air is greater than the SALR, saturated air will always be rising into air that is cooler. It will always be less dense, and therefore lighter, than the surrounding air. Consequently, its upward movement will accelerate. The entire layer will then be unstable and may overturn. As it continues to rise, however, the air will also cool and its density will increase, so eventually the atmosphere will become statically stable.

When the ELR is greater than the SALR but smaller than the DALR, the condition is known as conditional instability. Unsaturated air will not rise unless it is forced to do so, for example by crossing a mountain, because if it does it will cool at the DALR. This is greater than the ELR, and so the rising air will immediately be cooler than the surrounding air and will tend to sink again. While it remains unsaturated the air is stable.

Should it be forced to rise high enough for it to reach its lifting condensation level (see CONDENSATION), however, the water vapor it carries will start to condense into cloud droplets. This will release latent heat of condensation, warming the air and slowing its lapse rate from the DALR to the SALR. This is lower than the ELR, so the rising air will always be warmer than the surrounding air because it is cooling more slowly. Being warmer, it will remain buoyant and will continue to rise. The air is then unstable.

Some outside force must compel the air to rise before it will change from being stable to being unstable. This force is the condition needed to trigger instability, which is why the air is said to be "conditionally" unstable. This is the commonest type of instability.

The level of free convection is the height at which a parcel of air that is being forced to rise through a conditionally unstable atmosphere changes from being cooler than the surrounding air to being warmer than it. While it is near the surface, the parcel of air is moist, but not saturated. As it is forced to rise through air that is warmer than itself it cools at the DALR until it reaches the lifting condensation level, beyond which it cools at the SALR. Its temperature then decreases with height more slowly than that of the surrounding air, where the ELR is close to the DALR, until a level is reached at which it is warmer than the surrounding air. Beyond that point the parcel of air is unstable and continues to rise without the need of forcing until it reaches a second level, where its temperature is once more lower than that of the surrounding air and it becomes stable.

Latent instability is the condition in which a parcel of air that acquires sufficient KINETIC ENERGY to rise through a layer of stable air becomes unstable once it is above the level of free convection. It is a type of conditional instability that develops only if a parcel of air rises to the critical level.

There is a different type of conditional instability, known as conditional instability of the second kind (CISK). CISK leads to the formation of large, long-lived clusters of cumulonimbus cloud over tropical oceans.

Ordinarily, conditional instability produces cumuliform clouds that are isolated because each cloud is built by a convection cell that utilizes all the moisture and energy in its immediate vicinity, thus preventing the formation of another cloud nearby.

In the TROPICS it is possible for conditional instability to be intensified by horizontal air movements. This happens when there is an area of low pressure covering a surface area about 600 miles (1,000 km) across and the air is conditionally unstable. Air converges from all sides toward the low-pressure region (see STREAMLINE). Friction with the surface slows the air, reducing the magnitude of the CORIOLIS EFFECT so the PRESSURE GRADIENT FORCE becomes dominant. The converging air enters the low-pressure region and rises at the center. Alternatively, an ATMOSPHERIC WAVE may move across the area. Such waves often produce convergence in some places and divergence in others.

Convergence triggers convection. At the same time, the converging air is warm and moist, so it feeds moisture into the convection cells. It also pushes the developing clouds closer together. Latent heat is released as water vapor condenses in the rising air. This warms the air, fueling further convection. Divergence above the clouds reduces the amount of air beneath and therefore reduces the surface pressure. This intensifies and sustains the low-pressure system.

This alternative type of conditional instability is involved in the formation of TROPICAL CYCLONES. Meteorologists Jule Charney and Arnt Eliassen were the first scientists to recognize the difference between it and ordinary conditional instability in 1964. Katsuyuki Ooyama also described it and called it "conditional instability of the second kind," the name it has retained. Ordinary conditional instability is sometimes called conditional instability of the first kind to help distinguish the two types.

There are ways to test the stability of air. The parcel method calculates the consequences of displacing particular parcels of air. It is assumed that only those parcels are affected, but in fact their stability is similar to that of the larger body of air around them. The slice method takes account of air that is moving vertically in both directions through a slice of air.

stade (stadial) A period of cold that occurs during a GLACIAL PERIOD. It may be marked by a change in the vegetation of ice-free areas or by the advance of ICE SHEETS.

stage A series of sediments, fossils, or plant remains that are found in a particular place and that indicate the climatic conditions which obtained at the time they were deposited. The stage is named after the place where the evidence for it was first discovered. For example, the Hoxnian INTERGLACIAL was identified by deposits found at the village of Hoxne, in Suffolk, England, and is known as the Hoxnian stage.

standard artillery atmosphere A standard artillery atmosphere is a set of hypothetical values that are used in calculations of the trajectory of missiles through the air (ballistic calculations). These values assume there is no wind, the surface temperature is 59°F (15°C), surface pressure is 1000 mb (14.5 lb/in², 29.5 inches of mercury), surface relative HUMIDITY is 78 percent, and the LAPSE RATE directly relates air DENSITY to altitude.

A standard artillery zone is a layer of the standard artillery atmosphere that is of a specified thickness and at a specified altitude.

Statherian The most recent of the four periods of the PALEOPROTEROZOIC era of the Earth's history, the Statherian period began 1,800 million years ago and ended 1,600 million years ago. *See* APPENDIX V: GEOLOGIC TIMESCALE.

static electricity An electrical charge that is at rest, so it does not flow. A charge that does not flow is called an electrostatic charge.

One region, such as the upper part of a cumulonimbus cloud (see CLOUD TYPES), bears a positive charge and another, such as the lower part of the same cloud, bears a negative charge, but the two areas of charge are separated by an area that is electrically neutral. Consequently, no electrons move from the positive to negative areas. When the charge has accumulated sufficiently to overcome the insulating effect of the neutral region, the static charge is discharged in a spark, in this case a flash of LIGHTNING.

station model The formalized diagram that is used to report observations from a weather station is called a station model. The diagram uses standard symbols to represent cloud, PRECIPITATION, and wind direction and speed. Other information is given in numbers. Each item of information occupies a particular position around the station, which is represented by a circle at the center.

The central circle, called the station circle, is drawn in its geographic position on a weather chart, and the other information is positioned around it. Whether the station circle is open or partly or completely black indicates the amount of cloud cover.

A line from the station circle indicates the wind direction and lines or pennants at the end of this line indicate the WIND SPEED. The lines, or barbs, are drawn at an angle at the end of a longer line. Barbs are used only for wind speeds up to 47 knots (54 MPH, 87 km/h). The other end of the shaft to which the barb is attached joins the station circle, and the angle at which it projects indicates the wind direction. A pennant is a triangular symbol, resembling a pennant flag, that is used to indicate wind speeds greater than 48 knots

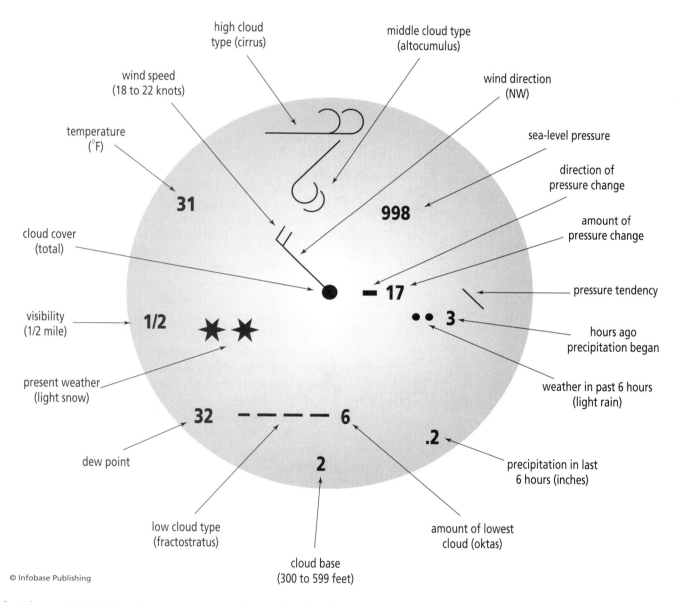

A station model is the formal arrangement of numbers and symbols that is plotted on a weather chart to indicate the conditions reported from a particular weather station.

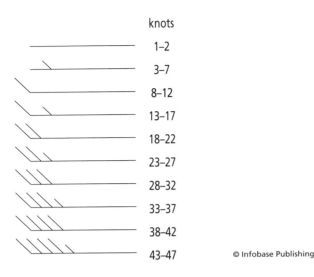

knots

	1–2
	3–7
	8–12
	13–17
	18–22
	23–27
	28–32
	33–37
	38–42
	43–47

© Infobase Publishing

Barbs are short lines drawn on the shaft indicating wind direction on a station model. The number of long and short barbs indicates the wind speed, which is given in knots.

(55 MPH, 89 km/h). It is drawn so that the shaft of the pennant indicates the wind direction. The barbs and pointed tips of the pennants point toward the region of low pressure.

Cloud symbols are a set of ideograms used to indicate cloud types. There are symbols for all 10 cloud genera (*see* CLOUD CLASSIFICATION) as well as two more, for fractostratus and towering cumulus (*see* CLOUD TYPES).

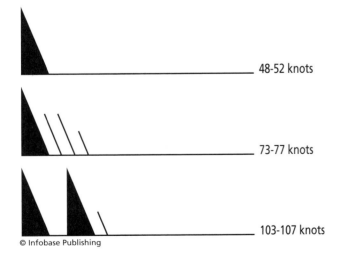

48-52 knots

73-77 knots

103-107 knots

© Infobase Publishing

Pennants are triangular symbols, resembling pennant flags, which are used on a station model to indicate wind speeds higher than those represented by barbs. These are three examples.

cirrus	cirrostratus	cirrocumulus
altostratus	altocumulus	nimbostratus
stratocumulus	stratus	fractostratus
cumulus	towering cumulus	cumulonimbus

© Infobase Publishing

There are 12 cloud symbols. These represent the 10 cloud genera, plus fractostratus and towering cumulus.

The CLOUD BASE is reported in a code that translates to a range of heights (in feet) and the amount of the lowest cloud is given in oktas (*see* CLOUD AMOUNTS.) Symbols are also used to indicate the present and past weather.

In addition, the model shows the sea-level AIR PRESSURE in millibars. The atmospheric pressure that is measured at the station elevation is called the station pressure or surface pressure. The station elevation is the vertical distance between a weather station and mean sea level. This is measured for a particular point at the station and is used as a reference DATUM for calculating atmospheric pressure. The station barometers show pressure at the station, but because pressure varies with elevation this is converted to sea-level pressure, and it is the sea-level pressure that is reported. Since all stations report sea-level pressure, their reports can be compared directly and isobars (*see* ISO-) can be plotted without a need for further corrections.

The lowest sea-level pressure ever recorded (in the eye of typhoon Tip; *see* APPENDIX II: TROPICAL CYCLONES AND TROPICAL STORMS) was 870 mb, and the highest (recorded on December 31, 1968, at Agata, Siberia, Russia) was 1,083.8 mb. In reporting sea-level pressure, therefore, the thousands and hundreds units (8, 9, or 10) can be assumed. They and also the decimal point can then be omitted for the sake of brevity, so the pressure is always represented by a three-digit number. For example, a pressure of 999.8 mb is reported as 998 (that is, (9)**99**(.)8), and a pressure of 1012.4 as 124 (that is, (10)**12**(.)4). The model also reports whether the pressure has increased (+) or has decreased (-) over the past six hours, the amount by which it has changed

rising then falling

rising then steady or rising more slowly

rising steadily or unsteadily

falling or steady then rising or rising then rising faster

steady

falling then rising but still same or lower than before

falling then steady or falling more slowly

falling steadily or unsteadily

steady or rising then falling or falling then falling faster

© Infobase Publishing

Barometric tendency is shown on a station model by symbols that indicate the way atmospheric pressure has changed over the reporting period.

(in tenths of a millibar), and the present barometric tendency. The barometric tendency, also called the air pressure tendency and pressure tendency, is the direction in which the atmospheric pressure has changed at a weather station since the last time it was reported.

TEMPERATURE and DEW point are reported (in degrees Fahrenheit or Celsius). Horizontal VISIBILITY is given in miles or kilometers. If there is precipitation, the number of hours since it began and the amount of precipitation in the last six hours are reported.

Station models are updated every six hours.

Stefan-Boltzmann law   The Stefan-Boltzmann law is the physical law that relates the amount of radiant energy a body emits to its temperature. The law applies to BLACKBODIES, including the Sun and Earth, and shows that the amount of radiation emitted is proportional to the fourth power of the temperature of the body.

The law is expressed as: $E = \sigma T^4$, where E is the amount of radiation emitted, T is the temperature, and σ is the Stefan-Boltzmann constant. The temperature is in kelvins (see UNITS OF MEASUREMENT) and the energy units are watts per square meter (W/m^2).

The Stefan-Boltzmann constant (σ) is the amount of radiant energy released by a blackbody. This is equal to 5.67×10^{-8} W/m^2/K^4 (watts per square meter per kelvin to the fourth power).

The law and constant were discovered in 1879 by the Austrian physicist Josef Stefan (1835–93; see APPENDIX I: BIOGRAPHICAL ENTRIES), and at first they were known as Stefan's law and constant. In 1884, the Austrian physicist Ludwig Eduard Boltzmann (1844–1906), Stefan's former student, showed that the law holds only for blackbodies, and his name was added to that of the law and constant. Stefan used the law to make the first fairly accurate estimate of the temperature of the photosphere (visible surface) of the Sun, as 10,800°F (6,000°C).

stem flow   PRECIPITATION that falls onto vegetation and reaches the ground by running down the stems of plants is called stem flow. The amount of water reaching the ground as a result of stem flow depends on the total amount of precipitation intercepted by the plants, the amount that evaporates from the leaves, and the amount that evaporates from the stem during stem flow.

If the vegetation is dry when precipitation commences, the plants will intercept a high proportion of the precipitation, especially if the precipitation is light. Water coating the leaves will immediately begin to evaporate, the rate of EVAPORATION depending on the temperature. In Brazilian forests, for example, about 20 percent of the rain falling onto the leaves evaporates, and almost all the rain falling on them evaporates from the leaves of some forests growing in Mediterranean climates, so there is no stem flow. The amount of stem flow and water dripping from leaves must be measured very precisely in studies of the amount of precipitation that reaches the ground in forests.

Stenian   The third and final period of the MESOPROTEROZOIC era of the Earth's history, the Stenian began 1,200 million years ago and ended 1,000 million years ago. During the Stenian most of the world's continents came together and joined, forming a supercontinent called Rodinia (see PLATE TECTONICS). Very little is known about conditions on Earth so long ago. See APPENDIX V: GEOLOGIC TIMESCALE.

Stevenson screen   The container that houses the THERMOMETERS and HYGROMETERS or HYGROGRAPHS used at a weather station is named after Thomas Stevenson (1818–87), who invented it. Stevenson was an amateur meteorologist, but nevertheless he had a professional interest in the weather. He was a civil engineer, and the Stevenson family firm specialized in building lighthouses throughout the world. Stevenson was born in Edinburgh, a city that he helped to make into a world center for the science of lighthouse lenses and lights. That is also where his son was born, the author Robert Louis Stevenson (1850–94).

The Stevenson screen is a box with louvered walls on all its four sides. The walls and top are of double thickness, with the louver strips forming inverted V shapes in cross section. The screen is painted white and stands on legs that raise it so that the bulbs of its thermometers are about 4 feet (1.25 m) above the ground. At this height the thermometers register air temperature, which is also called shelter temperature or surface temperature. This height is the lowest at which a thermometer will give a reliable reading for the temperature that is experienced by people and that is true for a large surrounding area. The temperature at this height

© Infobase Publishing

A Stevenson screen is a container of standard construction and dimensions that is used to house the thermometers at a weather station. The screen is painted white, has double-louvered sides, and is sited in the open and well clear of the ground.

is influenced, but not overwhelmed, by variations in temperature closer to the ground. Ground-level temperatures often change markedly over short distances, so simply moving the instrument to a different location nearby could alter thermometer readings significantly.

The purpose of the white paint and louvered construction is to shield the instruments from exposure to direct radiation without isolating them from the air they are to monitor. It is because of the shielding it provides that the container is called a screen.

Thermometers are sensitive to sunlight, which warms them directly, and they are also affected by radiation rising from the ground. The white color reflects most of the radiation that falls on it, and the thick, wooden walls and floor of the screen insulate the air inside. For added protection the screen is positioned so that access to the instruments is from a door on the side facing away from the equator (the northern side in the Northern Hemisphere). This prevents the Sun from shining into the screen while readings are being taken. The louvered construction of the sides allows air to circulate, so the temperature being measured is that of the air outside the screen.

Standardizing both the construction and the siting of Stevenson screens means that all weather stations are obtaining readings under similar conditions. This makes the readings directly comparable.

The disadvantage of the Stevenson screen is that although the double-louvered walls permit air to circulate, the ventilation inside the screen is poor. In hot weather the temperature inside the box can be a degree or more higher than the air temperature outside. Some screens are fitted with fans to improve ventilation.

A Stevenson screen usually contains four thermometers. These are the wet-bulb thermometer and dry-bulb thermometer used to calculate relative HUMIDITY and DEW point temperature, and the dry-bulb maximum and minimum thermometers. The wet-bulb and dry-bulb thermometers are mounted vertically and the maximum and minimum thermometers horizontally.

A Bilham screen is a small container with louvered sides like a Stevenson screen that contains a wet-bulb thermometer and dry-bulb thermometer mounted vertically and a maximum thermometer and a minimum thermometer mounted horizontally.

stochastic A stochastic system is one that obeys statistical laws. The future behavior of a stochastic system cannot be predicted precisely, but its probable future behavior can be calculated on the basis of its known past behavior.

Modern weather forecasters adopt an approach that is partly stochastic and partly deterministic (*see* DETERMINISM). They recognize that while natural laws certainly determine the behavior of the atmosphere, it is impossible to know the condition of the atmosphere in the detail required for an entirely deterministic calculation. Consequently, they make the best calculation they can, but qualify it by taking account of the way similar conditions developed in the past. This allows them to estimate the probability of a particular weather pattern emerging. A probabilistic forecast makes allowance for the chaotic behavior of weather systems, but because of CHAOS its reliability is inversely proportional to the forecast period.

stochastic resonance Stochastic resonance is the observable effect that results when a STOCHASTIC process acts in the same sense as a natural cycle that is too weak to produce any effect by itself. Stochastic processes ordinarily appear as noise, and scientists try to remove them in order to detect signals. If the signals are extremely weak, however, a stochastic process may reinforce them. The two are then said to resonate, and because it is the stochastic process that produces the resonance, it is called stochastic resonance.

OXYGEN isotope records from the GREENLAND ICECORE PROJECT indicate that stochastic resonance underlies a 1,500-year climate cycle, the most recent manifestation of which was the LITTLE ICE AGE. Stochastic resonance may also be responsible for the onset of GLACIAL PERIODS and INTERGLACIALS at intervals of approximately 100,000 years. These are driven by the MILANKOVITCH CYCLES, but changes in the ORBIT and rotation of the Earth are believed to be too small to account for ice ages. If stochastic warming and cooling of the climate coincide with the peaks and troughs of the Milankovitch cycles, their additional effect would be enough to trigger the observed effects.

Stochastic processes are unreliable and cannot be expected to coincide with every phase in a regular natural cycle. Sometimes resonance should occur and sometimes not. The result of this would be that the observed effect occurs most often at the peak or trough of one cycle, but sometimes skips one cycle and more rarely skips two. Very rarely, an event occurs at less than a complete number of cycles since the preceding one. This pattern is what emerges from the ICE CORE record.

In the case of the 1,500-year cycle, this means that most cool episodes resembling the Little Ice Age happen at intervals of 1,500 years, but some occur every 3,000 years and a few every 4,500 years. Similarly, ice ages and interglacials often begin every 100,000 years, but not always. Sometimes 200,000 years can pass without either.

Stokes's law Stokes's law describes the factors that determine the magnitude of the friction experienced by a spherical body (such as a RAINDROP or hailstone, *see* HAIL) falling by gravity through a viscous medium (such as air). Friction (*F*) is given by:

$$F = 6\pi r \eta v$$

where r is the radius of the body, v is its velocity, and π is the VISCOSITY of the medium.

From Stokes's law it is possible to calculate the TERMINAL VELOCITY (*V*) of the body as:

$$V = 2g \, (\rho_p - \rho_a) \, r^2 \div 9 \, \eta$$

where g is the acceleration due to gravity (32.18 ft/s^2 = 9.807 m/s^2), ρ_p is the density of the body, and ρ_a is the density of the medium through which it is falling.

A very small water droplet, with a diameter of about 0.0004 inch (10 μm), has a terminal velocity falling through air of about 0.03 feet per second (10^{-3} m/s). A drop 100 times larger, with a diameter of about 0.04 inch (1 mm), will fall at about 33 ft/s (10 m/s).

The relationship was discovered between 1845 and 1850 by the Irish physicist Sir George Gabriel Stokes (1819–1903; *see* APPENDIX I: BIOGRAPHICAL ENTRIES).

stomata (**sing. stoma**) Stomata are the pores in the surface of plant leaves, especially on the undersides, through which gases are exchanged. In most plants, stomata are open during the day and closed at night. Each stoma can be opened and closed by the expansion and contraction of two guard cells at its mouth.

When the stomata are open, air can enter the cell beneath to provide CARBON DIOXIDE for PHOTOSYNTHESIS and OXYGEN, a by-product of photosynthesis, can leave. WATER VAPOR also escapes while the stomata are open. The loss of moisture in this way is called TRANSPIRATION.

Plants adapted to dry climates usually minimize the time during which their stomata are open. Some desert plants (called CAM plants) open their stomata at night and store the carbon dioxide they absorb until daybreak, after which photosynthesis proceeds with the stomata remaining closed.

storm   In the BEAUFORT WIND SCALE, a storm is a wind of force 11, which blows at 64–75 MPH (103–121 km/h). In the original scale, devised for use at sea, a force 11 wind was defined as "or that which would reduce her to storm staysails." On land, a storm uproots trees and blows them some distance, and overturns cars.

Storm detection comprises the identification of conditions that are leading to the development of a storm, followed by its observation and tracking. As techniques have improved, the importance of storm detection has increased. Detection involves recognizing particular characteristics, especially wind strength and PRECIPITATION, that indicate the type of storm and measuring or calculating the area the storm covers. Balloon sondes (see WEATHER BALLOON), flights by aircraft equipped with meteorological instruments, RADAR including Doppler radar, and satellite images are used to detect and then study storms.

A windstorm is a storm in which the most significant characteristic is a very strong wind. In the TROPICS the areas of low pressure that produce such storms are not usually of frontal origin. Tropical windstorms can grow into TROPICAL CYCLONES. Windstorms in middle latitudes are often associated with deep frontal DEPRESSIONS, and they can be severe. A series of windstorms crossed France and Belgium for three days between Christmas 1999 and the New Year. Winds gusted to 105 MPH (169 km/h). The winds caused a major BLOWDOWN in which about 60,000 trees were damaged or destroyed in two forests on the outskirts of Paris, about 2,000 trees lining Paris streets were uprooted, and in France as a whole 160 square miles (259 km^2) of forest were destroyed. More than 120 people lost their lives.

A storm that affects only a small area is known as a local storm. THUNDERSTORMS and SQUALLS are considered to be local storms.

In a revolving storm the air moves cyclonically (see CYCLONE), so it rotates about a low-pressure center. Tropical cyclones are revolving storms. Revolving storms are contrasted to convectional storms that are produced by cumulonimbus clouds (see CLOUD TYPES) which are isolated or that form a SQUALL line.

A severe storm is any storm that damages property or endangers life. The U.S. NATIONAL WEATHER SERVICE defines a severe thunderstorm as one that produces HAIL with hailstones 0.75 inch (19 mm) or more across, or wind GUSTS of 58 MPH (93 km/h) or more, or a TORNADO, or more than one of these. A severe-storm observation is a report of a severe storm that has been positively identified. The report states the time of the observation and the location and direction of movement of the storm.

A storm beach is a linear pile of coarse material, such as pebbles, gravel, and seashells, that has been built on a beach by the action of sea storms. During the storm, waves throw the material into a heap, and in the course of many storms they form a distinct ridge or bank. The presence of such a pile of material indicates a beach that is exposed to storms.

A layer of sediment that is deposited over a surface by the action of a sea storm is called a storm bed. Shallow waves carry fine-grained material up the shore, where it is precipitated. A storm bed is sometimes called an event deposit because it is the result of a single physical event.

storm glass   A storm glass is an instrument that indicates a change in the weather. It is no longer used, but was popular in the 18th and 19th centuries. It consisted of a heavy glass tube, tightly sealed to prevent air entering from outside, that contained a supersaturated mixture of chemical compounds. The precise recipe for the contents of the glass varied from one instrument to another, but most recipes were based on camphor dissolved in alcohol, with other chemicals. Crystals would form and dissolve inside the glass, the changes apparently being linked to meteorological changes other than simple changes in temperature or air pressure. If the liquid was clear, it meant the weather would be fine. Cloudy liquid meant it would rain. If crystals formed at the bottom of the glass in winter there would be frost.

Admiral FitzRoy (see APPENDIX I: BIOGRAPHICAL ENTRIES) became very interested in storm glasses. He developed one based on his own chemical mixture that was attached to some versions of the FitzRoy barometer, and he explained how to interpret the patterns in it in *The Weather Book,* published in 1863.

storm surge A storm surge is a rise in sea level, accompanied by huge waves, that is produced by large storms at sea and especially by TROPICAL CYCLONES. Water sweeping inland, often for a considerable distance, causes severe flooding and in areas struck by a tropical cyclone the storm surge may cause more loss of life, injuries, and damage to property than the wind.

In the SAFFIR/SIMPSON HURRICANE SCALE, a category 1 storm produces a storm surge of 4–5 feet (1.2–1.5 m). Category 2 produces 6–8 feet (1.8–2.4 m). Category 3 produces 9–12 feet (2.7–3.7 m). Category 4 produces 13–18 feet (4.0–5.5 m). Category 5 produces a surge of more than 18 feet (5.5 m). Storm surges can be much larger than 18 feet (5.5 m), however. Hurricane Gilbert struck the Mexican coast in 1988 with a storm surge of 20 feet (6 m) that threw onto the shore a Cuban ship that had been several miles out at sea. In 1992, tropical storm Polly produced a 20-foot (6-m) surge at the port of Tianjin, China, and in December 1999, cyclone John produced one of similar size in Western Australia. Hurricane Katrina, which devastated New Orleans in 2006, generated a storm surge of 16–30 feet (5–9 m) at different points along the coast, although in this case the flooding was due to heavy rain. The rain raised the level of Lake Pontchartrain, breaching the levees protecting the city (*see* APPENDIX II: TROPICAL CYCLONES AND TROPICAL STORMS).

Three factors contribute to produce a storm surge. The first is the drop in surface air pressure at the center of the storm. A fall in pressure of 1 millibar (mb) below the mean sea-level pressure of 1,013 mb causes the sea level to rise by about 0.4 inch (1 cm). In a tropical cyclone, measuring on the Saffir/Simpson scale, the sea-level rise due to the low pressure in the eye of a category 1 storm will be about 14 inches (36 cm). In categories 2–4 it will be 14.6–20 inches (37–51cm), 20.5–28 inches (52–71 cm), 28.4–37.8 inches (72–96 cm) respectively. In category 5, the sea level will be raised by more than 37.8 inches (96 cm).

The second factor arises from waves driven by the winds. In a severe storm, spray whipped up from the sea surface turns the sea completely white and greatly reduces horizontal visibility. It looks as though the sea is boiling, and beneath the white water the waves are huge. Their size, speed, and wavelength (*see* WAVE CHARACTERISTICS) increase in proportion to the WIND SPEED, the distance over which the wind blows (the fetch), and the length of time the wind has been blowing. Waves reach their maximum size when the wind has been blowing for about 40 hours. A wind of 110 MPH (177 km/h) can raise waves 30 feet (9 m) high, and a tropical cyclone can raise waves with amplitudes up to 70 feet (21 m). This is close to the maximum size a sea wave can attain, because waves larger than this fall forward under the weight of water. While sailing through a Pacific typhoon on February 6–7, 1933, the USS *Ramapo* measured a wave that was 112 feet (34.14 m) high. This is believed to be the biggest storm wave ever recorded.

At sea, not all waves are the same size. This is because waves interfere with one another in complex ways, sometimes augmenting and sometimes diminishing them. Of 100 waves passing a fixed point, on average there will be one wave that is 6.5 times bigger than the others. If 1,000 waves pass, there is likely to be one that is eight times bigger.

Where the winds drive waves into a confined area, such as a partially landlocked sea, or into shallower water, the water level rises dramatically. Wave amplitude increases as wavelength decreases, so waves grow higher as they approach a shelving coast. Driving water into a confined area causes water throughout the entire basin to slop back and forth like water in a bathtub when a person moves a hand back and forth through it. This motion can become very large if the slopping resonates with the natural period of the sea basin, just as the waves a hand makes in a bathtub grow bigger if the waves resonate with the size of the tub.

The final component of a storm surge is the TIDE. If the raised sea level and storm-driven waves coincide with a high tide when they reach the coast, the effect of all three factors will be added. It is under these circumstances that a storm surge penetrates farthest inland and causes its greatest damage.

storm track Storms are carried by the prevailing WIND SYSTEMS in the regions where they occur. Generally, therefore, storms in middle latitudes, where the prevailing winds are from the west, tend to move from west to east. In the TROPICS the prevailing trade winds blow from the east, so storms move from east to west. Storms rarely move in straight lines, however. In North America, a major storm that develops in Alberta is likely to curve southward into Montana and North Dakota before passing to the north of Lake Superior and into Labrador. A storm originating in Colorado may travel northeast and cross the Great Lakes.

TROPICAL CYCLONES start by moving toward the west, but as they approach land they turn to the northwest in the Northern Hemisphere and southwest in the Southern Hemisphere. This track takes them toward the SUBTROPICAL HIGHS and they curve around the western boundaries of these ANTICYCLONES. They then enter the region of the midlatitude westerlies, and their track turns increasingly toward the northeast in the Northern Hemisphere and southeast in the Southern Hemisphere. As they move across cooler water, tropical cyclones weaken and finally dissipate.

Although these are the average tracks that tropical cyclones follow, individual cyclones are subject to local influences that cause them to deviate, even to the extent of traveling for a time in the opposite direction. This makes the task of forecasting their movements very difficult. The speed at which a tropical cyclone moves is determined mainly by the rate at which the warm air above the eye moves. Most travel at 10–18 MPH (16–29 km/h), but some move faster.

Strahler climate classification

A CLIMATE CLASSIFICATION that was proposed by Arthur Newell Strahler (1918–2002), professor of geomorphology at Columbia University. His classification is closely related to the KÖPPEN CLASSIFICATION and can be used in conjunction with it.

The Strahler classification is of the genetic type, which means it is based on the GENERAL CIRCULATION of the atmosphere and relates regional climates to the AIR MASSES and prevailing winds (see WIND SYSTEMS) that produce them. Strahler divided the climates of the world into three main types, or groups, according to the air masses that control them. These are subdivided further into 14 climatic regions, to which he later added highland climates.

His Group 1 comprises climates that are controlled by equatorial and tropical air masses. They occur in low latitudes and include: 1. wet equatorial climate; 2. trade wind littoral climate; 3. tropical desert and steppe climates; 4. west-coast desert climate; and 5. tropical wet–dry climate (see CLIMATE TYPES).

Group 2 comprises middle latitude climates controlled by both tropical and polar air masses. These are: 6. humid subtropical climate; 7. marine west-coast climate; 8. Mediterranean climate; 9. middle latitude desert and steppe climates; and 10. humid continental climate.

Group 3 comprises high latitude climates controlled by polar and arctic air masses. These are: 11. continental subarctic climate; 12. marine subarctic climate; 13. tundra climate; and 14. icecap climate. Group 3 also includes highland climates.

Further Reading
Strahler, A. N. *Introduction to Physical Geography.* New York: John Wiley & Sons, Inc., 1970.

strand line A strand line is a layer of material that was deposited at the edge of a former lake or sea and that marks the location of a shoreline that has since disappeared. Strand lines indicate higher sea levels or heavier PRECIPITATION at some time in the past, and they can therefore be used in the reconstruction of past climates.

streamline The track that is followed by moving air is called a streamline. A wind streamline is shown on a map or diagram as a straight or curved line that is parallel to the wind direction at every point along its path. The line may be drawn using information obtained from one or more synoptic charts (see WEATHER MAPS) showing the wind that was observed and measured at particular times.

Streamlines can also be used to show the way air (or water) flows over or around an obstruction. In addition, streamline is used as a verb to describe the inclusion of a design feature in which a body, such as a car, airplane, or boat, is so shaped as to offer the least possible resistance to the flow of air or water around it. Such a shape is said to be "streamlined," because streamlines show a smooth, laminar flow of fluid around the object.

A streamline is the path traced by moving air.

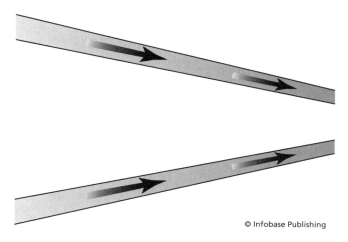

© Infobase Publishing

Confluence is the situation where two airstreams are approaching each other and accelerating.

The modification of a design to achieve this objective is called "streamlining."

A flow of air in which two or more streamlines approach one another is called a confluence. This accelerates the air, because a narrowing stream must carry the flow, but unlike convergence it does not lead to an accumulation of air or to any vertical movement.

Convergence is a flow of air in which streamlines approach an area from different directions. This happens when air flows into a region of low pressure. The effect is to accumulate air where the streamlines meet. The accumulation of air increases the quantity of air in that area and, therefore, increases the atmospheric pressure. The increased pressure produced by convergence near the surface of land or sea causes air to rise, so a region of low-level convergence is also a region of rising air. The rising air eventually reaches an INVERSION level that constitutes a ceiling beyond which it can rise no farther. If the vertical motion is strong enough, air may rise all the way to the tropopause (*see* ATMOSPHERIC STRUCTURE), where air spreads out, moving away from an upper-level area that corresponds to the area of low-level convergence. This produces a region of divergence and falling pressure above the area of low-level convergence and rising pressure. A horizontal line along which convergence is occurring is called a convergence line.

Air that rises in a region of convergence cools adiabatically (*see* ADIABAT) as it does so. This favors the formation of cumuliform clouds (*see* CLOUD TYPES) and PRECIPITATION.

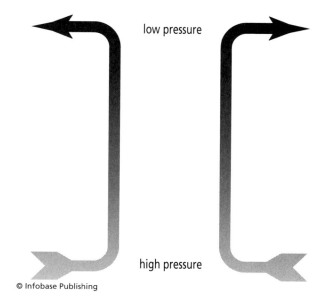

© Infobase Publishing

Convergence is the situation where air is approaching an area from several directions. This produces high pressure at the surface and low pressure above, where the air diverges.

A flow of air in which two or more streamlines move away from one another is called a diffluence. This slows the rate of flow, because there is a widening stream to carry the flow but, unlike divergence, diffluence does not result in an outflow of air from the area or to any vertical movement. A part of the atmosphere in which diffluence is occurring is called a delta region, because the diverging streamlines make a triangular shape, reminiscent of the Greek letter Δ (delta).

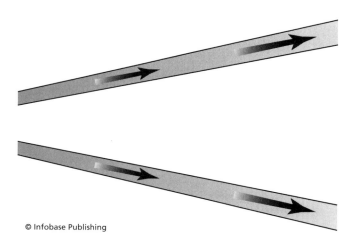

© Infobase Publishing

Diffluence occurs when airstreams flow away from each other and slow.

Divergence is a flow of air in which streamlines move outward from an area. This happens when air flows away from a region of high pressure. The effect is to disperse air from where the streamlines separate. Divergence decreases the quantity of air in that area and, therefore, decreases the atmospheric pressure. The decreased pressure produced by divergence near the surface of land or sea causes air at a higher level to sink and fill the low-pressure, so a region of low-level divergence is also a region of sinking air. At high level air flows inward to an area that corresponds to the area of low-level divergence. This produces a region of convergence and rising pressure above the area of low-level divergence and falling pressure. Air that sinks in a region of divergence warms adiabatically as it does so. This lowers the relative HUMIDITY of the air, favoring the dissipation of clouds and clear skies.

The horizontal movement of air when there is no FRICTION and the isobars (*see* ISO-) and streamlines coincide is called gradient flow. In this situation the tangential ACCELERATION is zero throughout the system.

A line drawn through the point of maximum curvature in the streamline of an easterly wave (*see* TROPICAL CYCLONE) is called the axis. The term is most often used in connection with equatorial waves (*see* INTERTROPICAL CONVERGENCE ZONE). The axis may be positive or negative. A positive axis indicates a TROUGH in

the Northern Hemisphere and a RIDGE in the Southern Hemisphere. A negative axis indicates a RIDGE in the Northern Hemisphere and a trough in the Southern Hemisphere.

sublimation Sublimation is the direct change of ICE into WATER VAPOR, without passing through a liquid phase. Sometimes the reverse process, in which water vapor changes directly into ice, is also called sublimation, but it is more correctly known as deposition.

The two processes can be seen happening in winter when patches of SNOW and ice disappear from the ground or when frost appears while the temperature remains well below freezing. They also occur in a freezer, when ice cubes that have been left there for too long dwindle in size and eventually disappear, and when the sides of the freezer become frosted.

The height at which ICE CRYSTALS entering dry air will change directly into water vapor by sublimation is called the ice evaporation level. Sublimation will occur if the temperature of the air is lower than -40°F (-40°C).

A snow eater is a warm, dry wind that removes snow by sublimation. A chinook wind (*see* LOCAL WINDS) often removes 6 inches (15 cm) of snow in a day and it can clear 20 inches (50 cm) in a day.

subpolar low A subpolar low is one of the two belts of low atmospheric pressure that lie between latitudes 60° and 70° in both hemispheres. They are where the polar easterly and midlatitude westerly winds converge.

The strong contrast in temperature between the tropical air arriving from one side and polar air (*see* AIR MASS) from the other gives rise to frequent DEPRESSIONS and storms. These are carried in a westerly direction by the prevailing winds (*see* WIND SYSTEMS) on the low-latitude side of the polar FRONT. The ALEUTIAN LOW and ICELANDIC LOW are the most prominent parts of the subpolar low.

subpolar region The subpolar region is the part of the world that lies between the low-latitude margin of land occupied by tundra and the high-latitude margin of lands with cool temperate or desert vegetation. Winters are long and cold, summers short, and the climate is fairly dry throughout the year. Coniferous forest is the vegetation most typical of subpolar regions.

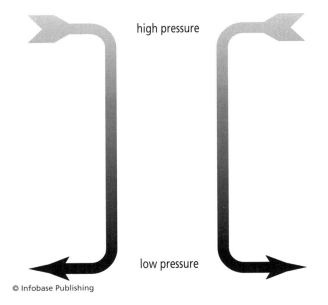

high pressure

low pressure

© Infobase Publishing

Divergence is the situation where air flows outward, producing low surface pressure.

subsidence In meteorology, subsidence is a general sinking of air over a large surface area. Subsidence brings high surface AIR PRESSURE and divergence (*see* STREAMLINE).

As air subsides, more air is drawn into the column at a high level, producing high-level convergence. Subsiding air is compressed by the increasing weight of air above it and it warms adiabatically (*see* ADIABAT).

Subsidence occurs on a global scale in the Hadley, Ferrel, and polar cells that are part of the GENERAL CIRCULATION of the atmosphere.

subtropical cyclone As the name suggests, a subtropical cyclone is a CYCLONE that occurs in the SUBTROPICS. It develops when the southern tip of a polar TROUGH in the upper atmosphere becomes cut off from the main part of the trough. The isolated pocket of low pressure, with cold air at its center, may then extend downward to the surface where it forms a very symmetrical cyclone. The center of the cyclone is up to 100 miles (160 km) in diameter, and the highest rainfall occurs about 300 miles (480 km) from the center. Subtropical cyclones are very persistent and are often absorbed into new polar troughs, rather than dissipating.

A subtropical cyclone can also develop from a TROPICAL CYCLONE that moves across land. Developing into a subtropical cyclone prolongs the rainfall from the decaying tropical cyclone. This transformation can also occur in the other direction, with a subtropical cyclone strengthening into a tropical storm. This happens if warm, moist air is drawn strongly toward the center of the subtropical cyclone. Rainfall then intensifies around the center and the temperature at the center rises until the cyclone has the warm core typical of a tropical cyclone.

Some subtropical cyclones, such as the kona cyclones of Hawaii (*see* LOCAL WEATHER), are an important source of rain.

subtropical front The subtropical front is a boundary that separates the cold water of the Southern Ocean around Antarctica from the warmer, subtropical waters farther north. The front extends from the coast of Antarctica to about 40°S. Within the front the temperature of the water rises to approximately 39°F (4°C) and the salinity increases by about 0.5 per mil, although in summer it can be as low as 33 per mil. The boundary was originally known as the subtropical con-

vergence zone, but was renamed the subtropical front in the 1980s.

subtropical high (subtropical high-pressure belt) A subtropical high, or high-pressure belt, is one of the semipermanent ANTICYCLONES that lie in the SUBTROPICS. There are several subtropical highs, centered at about latitudes 30°N and 30°S. They are located over the ocean, are most fully developed in summer, and they strongly influence the climates to the east of them by BLOCKING or diverting DEPRESSIONS traveling from west to east in middle latitudes. The AZORES HIGH is a subtropical high that affects the climates of western Europe. The intensity of the subtropical highs varies from time to time (*see* NORTH ATLANTIC OSCILLATION). The high pressure is caused by the SUBSIDENCE of air on the descending side of the Hadley cells (*see* GENERAL CIRCULATION).

The Saharan high is the subtropical high that lies permanently over the Sahara. It is produced by the subsidence of air on the high-latitude side of the Hadley cell. The subsiding air is warm and dry, and because of the anticyclone the PRESSURE GRADIENT drives air out of the DESERT. This prevents moist air from flowing into the region and therefore maintains the arid conditions. The highest temperature ever recorded on Earth was 136°F (57.8°C) at Azizia, Libya, on September 13, 1922. Azizia lies beneath the Saharan high.

subtropics The subtropics are the two belts surrounding the Earth in both hemispheres that lie between the TROPICS and the TEMPERATE BELT. The subtropics are not sharply defined, but they are bounded by the Tropics on the side nearest the equator and by approximately latitude 35–40° on the side nearest the pole.

The latitudinal boundary represents an average, however. The actual boundary is farther from the equator on the western sides of continents and closer to the equator on the eastern sides. This difference is due to the anticyclonic (*see* ANTICYCLONE) circulation in the subtropical high-pressure belt (*see* SUBTROPICAL HIGH), which carries cooler air toward the equator on the western sides of the continents and warmer air toward the equator on the eastern sides.

sulfur cycle The pathways by which the element sulfur moves between the rocks, air, and living organisms constitute what is known as the sulfur cycle. Sulfur is an essential ingredient of proteins and protein–carbohydrate complexes, so all living organisms require it.

Fairly small amounts of sulfur enter the air from volcanic eruptions (*see* VOLCANO), but the principal sources are the WEATHERING of rocks and the oceans, both of which contribute approximately equal amounts. Whether it is released through weathering, dissolves into surface waters from the air, or is incorporated in living tissue, sulfur eventually reaches rivers that carry it to the sea. Some sulfur is deposited as a variety of compounds in the airless muds of estuaries and bogs where BACTERIA obtain energy by reducing it and releasing gases as a by-product of the chemical reactions involved. Hydrogen sulfide (H_2S) is the gas released in the largest amounts by sulfur-reducing bacteria.

In the oceans several species of phytoplankton (very small, plantlike organisms) release dimethyl sulfide (*see* CLOUD CONDENSATION NUCLEI), some of which enters the air where it is oxidized eventually to sulfate (SO_4) AEROSOL. Sulfate aerosol particles act as cloud condensation nuclei, returning sulfur to the surface.

The burning of fuels now adds significantly to the natural cycle. Depending on its quality, coal contains an average of 1–5 percent sulfur and oil contains 2–3 percent. When these fuels are burned, the sulfur enters the air as sulfur dioxide (*see* AIR POLLUTION) unless it is removed from the waste gases. Averaged over the whole world, the amount of sulfur entering the air as a consequence of burning fuel is approximately equal to the amount entering the air in the course of the natural cycle. The difference is that the burning of fuel is concentrated in the industrialized regions of the world.

Sullivan winter storm scale A five-point scale for classifying winter storms was devised in 1998 by Joe Sullivan, a meteorologist employed by the NATIONAL WEATHER SERVICE in Cheyenne, Wyoming. The Sullivan scheme ranks storms as: 1. minor inconvenience; 2. inconvenience; 3. significant inconvenience; 4. potentially life-threatening; and 5. life-threatening.

Sun The Sun is the star around which the planets of the solar system revolve, and the source of all of the light and almost all of the heat (a small amount is released from the Earth's crust) on which life depends. Astronomically the Sun is classed as a G2 V star, where G2 indicates the second-hottest class of yellow G stars,

and V indicates a main-sequence or dwarf star; despite the name, the Sun is a medium-sized star, not a very small one. G stars are so designated because of the prominence of certain atoms and molecules in their spectral line; this was first observed by the German astronomer Joseph von Fraunhofer (1787–1826), who called them class G stars. V stars are typical G2 stars, with a surface temperature of about 10,960°F (5,800 K, 6,073°C).

The Sun comprises more than 99 percent of the total mass of the solar system. The mass of the Sun is 743 times that of all the planets of the solar system combined. It is 330,000 times more massive than Earth.

The visible surface of the Sun or any other star is called the photosphere. The Sun's photosphere is a layer of gas about 300 miles (500 km) thick that is opaque to radiation at the base but transparent at higher levels. It is the layer from which the solar radiation is emitted, and it is visible because it emits light. The temperature of the photosphere is about 10,960°F (5,800 K, 6,073°C) at the base and falls to about 7,725°F (4,000 K, 4,270°C) at the top. Above the photosphere there lies the chromosphere.

The chromosphere is the gaseous layer that lies above the photosphere. The temperature rises through the chromosphere from about 7,725°F (4,000 K, 4,270°C) at the base to about 18,525°F (10,000 K, 10,275°C) at the top. When the Sun is hidden by a solar ECLIPSE, the chromosphere appears as a pink glow, the color giving the layer its name of chromosphere, or "colored sphere."

A plage is a bright area on the chromosphere that is associated increased emission of radiation in the X-ray, extreme short-wave ULTRAVIOLET, and radio wavelengths (*see* SOLAR SPECTRUM).

sunshine recorder A sunshine recorder is an instrument that measures the intensity and duration of INSOLATION. There are several designs. The Campbell-Stokes sunshine recorder is one of the oldest and is still widely used.

The Campbell-Stokes sunshine recorder provides a daily record of the number of hours of sunshine. It comprises a spherical lens that acts as a burning glass, focusing the sunlight onto a card graduated with a timescale that partly encircles the lens. When the sky is clear, the sunshine makes a scorch mark on the card.

Two types of sunshine recorder that were replaced by the Campbell-Stokes sunshine recorder. The instrument on the left is a Jordan's sunshine recorder and the one on the right is a Marvin's sunshine recorder. The pictures are taken from "The Aims and Methods of Meteorological Work" by Cleveland Abbe, in volume I of *Maryland Weather Service*, published in 1899. *(Historic NWS Collection)*

The position of the mark on the card is determined by the position of the Sun in the sky and the graduation on the card interprets this as a time of day. The scorch mark ends whenever cloud obscures the Sun. The recorder will produce a reading for as long as the Sun is more than about 3° above the horizon.

Pyranometers (*see* SOLAR IRRADIANCE) measure solar radiation at all wavelengths (*see* WAVE CHARACTERISTICS) and over a complete hemisphere. A pair of pyranometers, one facing upward and the other downward, are used to measure surface ALBEDO. Pyrheliometers measure direct solar radiation perpendicular to a surface. Diffusographs, which are pyranometers modified by being surrounded by a shade ring, measure diffuse sunlight.

Possibly the simplest device is the actinometer, which consists of two THERMOMETERS, one with a blackened bulb. This thermometer absorbs more heat than does the thermometer with a clear-glass bulb, and the difference between the two temperatures they record can be used to calculate insolation.

sunspot A sunspot is a dark patch on the visible surface (photosphere) of the Sun, up to 31,070 miles (50,000 km) across, where the temperature is about 2,700°F (1,500°C) cooler than the surrounding area. Sunspots are caused by intense magnetic fields, which strongly affect the strength of the SOLAR WIND: the greater the number of sunspots, the more intense the solar wind.

The number of sunspots increases and decreases over an approximately 11-year cycle. This is associated with the approximately 22-year cycle, known as the Hale cycle, during which the magnetic polarity of the Sun reverses. Climate records for the last 300 years show that DROUGHTS in the western United States are most severe in the 2–5 years following one of the sunspot minima which occur every 11 years, and therefore twice in the course of the Hale cycle. The Hale cycle was discovered by the American astronomer George Ellery Hale (1868–1938). Professor Hale was also famous for developing large astronomical telescopes, including the 200-inch (508 cm) Palomar telescope and the 60- and 100-inch (152.4–225.4 cm) reflecting telescopes at the Mount Wilson Observatory, as well as for his research into solar physics.

There have been episodes in history when very few sunspots have formed. These episodes have coincided with climatic changes on Earth, the first such correlation to be recognized now being called the SPÖRER MINIMUM, although the MAUNDER MINIMUM is better known. The solar wind affects the intensity of COSMIC RADIATION reaching the Earth's atmosphere, which in turn influences cloud formation.

Sunspot activity is also linked to the intensity of ultraviolet radiation (*see* SOLAR SPECTRUM). This, through its absorption by OZONE, affects the TEMPERATURE in the upper atmosphere, which in turn affects the JET STREAM and the temperature in the lower atmosphere. Some scientists believe GLOBAL WARMING can be partly or even wholly attributed to variations in solar output indicated by changes in the number of sunspots. The regular increase and decrease in the amount of radiation that is emitted by the Sun is called the solar radiation cycle.

supercell A supercell is the type of CONVECTION cell that sometimes develops in a very massive cumulonimbus storm cloud (*see* CLOUD TYPES). A supercell storm is capable of producing TORNADOES.

Three conditions are needed for the growth of a supercell storm. There must be a warm, moist AIR MASS in which the air is conditionally unstable (*see* STABILITY OF AIR). There must also be an advancing cold FRONT to lift the warm air. Finally, there must be a strong WIND SHEAR at a high level to carry away the rising air.

Under these conditions, warm, moist air is first lifted by the cold air moving beneath it and then drawn upward by convection. The rising air cools adiabatically (*see* ADIABAT), and its WATER VAPOR condenses, releasing LATENT HEAT. The release of latent heat warms the air and increases its instability. Convergence (*see* STREAMLINE) at a low level is accompanied by divergence at high level, due to the wind shear. This accelerates the rising air. When the storm is fully developed, the upcurrents may travel at 100 MPH (160 km/h).

Supercell storms at Miami, Texas, on June 19, 1980 *(NOAA Photo Library, NOAA Central Library; OAR/ERL/National Severe Storms Laboratory [NSSL])*

PRECIPITATION falls as HAIL, SNOW, or RAIN from the upper part of the cloud. As it falls, the precipitation chills the air around it and drags cold air to lower levels. This produces cold downcurrents. In most storm clouds the precipitation and its cold air fall into the upcurrents. They chill the rising air, and within an hour or so they suppress the upcurrents altogether. Then the cloud dies.

A supercell is different, because its upcurrents rise at an angle to the vertical. Instead of falling directly into the rising air, the precipitation falls to the side of it. Consequently, the upcurrents are not suppressed. This allows the cell to continue growing. The biggest supercell clouds break through the tropopause (*see* ATMOSPHERIC STRUCTURE) and reach heights of up to 60,000 feet (18.3 km).

A supercell is also different for another reason. In ordinary storm clouds there are usually several convection cells. These share the energy of the storm, the size of each individual cell is limited, and the cold downcurrents of one cell suppress the warm upcurrents of the adjacent cell. In a supercell storm there is only one cell. It is huge and it occupies the whole of the interior of the cloud. The resulting storm releases very heavy rainfall and hail consisting of large hailstones.

Supercell storms last for much longer than ordinary storms do, but they seldom survive for longer than about three hours. This does not mean that the storm dies after a maximum of three hours, however, because the downdraft emerging at the base of the cloud can lift more conditionally unstable air and trigger the formation of a new cloud. This is what happens in a SQUALL line, where each storm produces a new storm as it dies.

supercooling Supercooling is the chilling of water to below its freezing temperature without triggering the formation of ICE. Ordinarily, pure water freezes when its temperature falls to 32°F (0°C). If the water contains dissolved impurities, such as salt, it freezes at a slightly lower temperature. In the air, however, a cloud droplet that falls slowly through very cold air can be chilled to well below freezing temperature without freezing. It is then said to be supercooled, and clouds often contain supercooled water droplets at temperatures as low as -20°F (-29°C).

Water droplets will freeze at temperatures between about -5°F and -13°F (between -15°C and -25°C), but only if FREEZING NUCLEI are present. In the absence of freezing nuclei, water droplets have been cooled under laboratory conditions to temperatures as low as -40°F (-40°C). Below this temperature ice starts to form by homogeneous nucleation.

An ice nucleus is any small particle onto which supercooled water will freeze. Both freezing nuclei and sublimation nuclei (*see* DEPOSITION) are classed as ice nuclei.

surface tension Surface tension is a property of a liquid that makes it seem as though a thin, flexible skin covers its surface. The phenomenon is caused by the mutual attraction of molecules in the liquid.

Below the surface of the liquid each molecule is attracted equally in all directions. The forces acting on it are balanced and it can move in any direction. Molecules at the surface are attracted by molecules to either side and beneath them, but there are no molecules beyond the surface to balance this attraction. Consequently, surface molecules are held firmly by molecular attraction from the sides and below. This attraction draws the surface molecules into a spherical shape, in which the forces are most evenly distributed. It is surface tension that draws small volumes of liquid into drops and the surface of liquids contained in tubes or vessels into a convex shape called a meniscus.

A RAINDROP falling through the air is surrounded by air on all sides, and the water molecules are subjected only to the attraction of other molecules within the drop. A raindrop is therefore able to assume a spherical shape that is then distorted into a teardrop shape by the force of gravity. Bubbles, in which air is contained by water, are spherical because of surface tension, but ordinarily the surface tension of water is so high that they survive for only a very short time before the water molecules move toward one another to form a drop, excluding the air. This bursts the bubble. Bubbles can be made to last longer by adding detergent to the water. Detergent reduces the surface tension that causes the bubble to collapse.

A drop of liquid that is lying on a solid surface also experiences attraction to molecules in the solid material. This produces adhesion tension. If the liquid is held in a container, adhesion tension will draw it up the sides, so the surface of the liquid assumes a concave shape. CAPILLARITY is caused by adhesion tension. If the liquid is in contact with another liquid, the surfaces of both liquids generate forces of attraction. This is called interfacial tension.

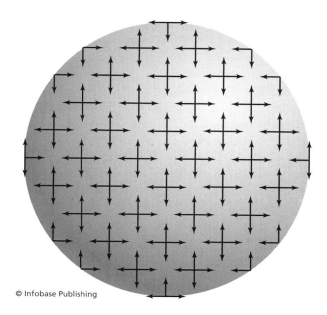

© Infobase Publishing

Surface tension is caused by the mutual attraction of molecules at the surface of a liquid. The arrows indicate the forces acting on the molecules. Inside the liquid mass, the forces act equally in every direction. At the surface, there are no molecules to exert an attraction from the outside, so the molecules are held by forces acting to the sides and toward the center.

Surface tension (symbolized by γ) is measured in thousandths of a newton per square meter (mN/m^2, *see* APPENDIX VIII: SI UNITS AND CONVERSIONS). It decreases with increasing temperature. For water in contact with air, γ = 74.2 mN/m^2 at 50°F (10°C), 72.0 mN/m^2 at 77°F (25°C), and 62.6 mN/m^2 at 176°F (80°C).

Water has a much higher surface tension than many other liquids, but that of mercury is higher (γ = 472 mN/m^2 at 68°F, 20°C and 456 mN/m^2 at 212°F, 100°C). The alcohol used in some THERMOMETERS has γ = 23.6 mN/m^2 at 50°F (10°C) and 21.9 mN/m^2 at 86°F (30°C).

surface weather observation A surface weather observation is an observation or measurement of weather conditions made at ground level or on the surface of the sea. Surface observations have been made and recorded for many years, so they provide the longest continuous series of data. Surface data from the past are difficult to interpret, however. At one time the methods of observing, measuring, and recording were not standardized, so it is unwise to draw any conclusions from comparisons between data from different stations.

It is not safe even to compare data from the same station at different times unless the full history of the station is available. Most WEATHER STATIONS have been moved at one time or another, and a change in location of only a few yards may alter the winds, TEMPERATURES, AIR PRESSURES, and HUMIDITY to which the instruments are exposed. A station may remain in the same place, but its surroundings may change. Urban developments may envelop it, but the change may be subtler than that. The closure of a factory upwind or the installation of equipment to remove pollutants from its chimney emissions may alter the temperature of the air or intensity of INSOLATION at the weather station. Even the removal of trees can affect wind patterns.

It is impossible to correct for all these influences and dangerous to assume that they will tend to cancel one another because for any factor an increase in one place is compensated by a decrease somewhere else. Surface weather observations made today under standard conditions are essential for the compilation of synoptic charts (*see* WEATHER MAPS) and the preparation of weather forecasts. They can also be used, with caution, to detect climatic changes over recent years, but they provide an unreliable base for estimates of climatic change since earlier times.

synoptic An adjective that is derived from the Greek word *sunoptikos,* from *sun,* meaning "with," and *optikos,* which means "seen." Synoptic refers to something that is based on a general view of conditions over a large area at a particular time. In synoptic METEOROLOGY, data gathered from many places is assembled to provide a picture of atmospheric conditions over a large area.

Conditions that are seen to cover an area that is large, but not so large as in the picture presented in a synoptic view, are said to be subsynoptic. Satellite images of cloud patterns over the ocean are subsynoptic in extent. They are able to show features the size of a TROPICAL CYCLONE in an area that is approximately 1,000 miles (1,600 km) square, but a synoptic chart (*see* WEATHER MAP) would cover a much bigger area.

T

Tay Bridge disaster The Tay Bridge disaster was a devastating catastrophe that occurred in the 19th century on the rail bridge that crosses the Firth of Tay, Scotland, linking Fife to the city of Dundee on the northern side of the river. The present bridge is the second (and a road bridge also spans the Firth).

The first Tay Bridge was opened on June 20, 1877. It was rather more than 1 mile (1.6 km) long, and the engineers who designed it believed it was strong enough to withstand any weather. On the evening of Sunday, December 28, 1879, a train departed as usual from Edinburgh with six carriages carrying passengers bound for Dundee. There were between 75 and 90 persons on the train as it began to cross the Tay Bridge. The weather was stormy, with gale-force winds blowing at up to 75 MPH (120 km/h). At about 7.15 P.M., when the train was about halfway across, the bridge collapsed and the entire train fell into the river below. There were no survivors. The subsequent investigation concluded that the bridge had not been properly built and maintained and that its design failed to allow adequately for wind loading. Some scientists now believe it was destroyed when two TORNADOES struck it simultaneously.

Further Reading

Allaby, Michael. *Tornadoes.* New York: Facts On File, rev. ed. 2004.

Martin, Tom. "The Tay Bridge Disaster." Available online. URL: www.tts1.demon.co.uk/tay.html. Accessed February 15, 2006.

teleconnections Linked atmospheric changes that occur in widely separated parts of the world are known as teleconnections. The SOUTHERN OSCILLATION index is a typical example. When the sea-level atmospheric pressure rises above normal at Darwin, Australia, it falls by an approximately similar amount at Tahiti, in the central South Pacific, thousands of miles away. El Niño (*see* ENSO) brings dry weather to northeastern Brazil and also to the western Mediterranean, the Sahel region of Africa, northeastern China, and Australia, but wet weather to much of the United States, Israel, and northwestern Europe.

There are many other examples. Changes in the water temperature of the tropical eastern North Pacific Ocean are linked to changes in AIR PRESSURE in the upper atmosphere over the Rocky Mountains. The water temperature in the center of the tropical North Pacific Ocean is linked to the temperature of the water in the Indian Ocean. The rainfall in northeastern Brazil is linked to the sea-surface temperature in the eastern tropical South Pacific Ocean.

telegraphy The transmission of information by means of electric pulses that travel along a wire cable from a sender to a receiver is called telegraphy. The word is derived from two Greek words, *tele,* which means "far," and *graphein,* which means "to write."

Until the invention of telegraphy, information could be communicated over a long distance no faster than a horse could gallop. Weather forecasting was impossible, because it took so long to assemble detailed information about conditions over a sufficiently large area that by the time the data had been analyzed the weather system the data described had disappeared.

Scientists knew by the middle of the 18th century that an electric current would travel a considerable distance along a metal wire if the wire were connected to the earth to complete the circuit. The first practical idea for a telegraph was suggested in 1753, in an article in the *Scots Magazine* by someone identified only as "C.M." C.M. proposed a separate insulated wire for each letter of the alphabet. At the receiving end, each wire was to be attached to a ball that hung above a piece of paper with a letter written on it. When the current reached the ball, the paper would jump up to it, and so the message could be spelled out letter by letter. Alternatively, each wire could end at a bell that would be struck by a ball when a current traveled along the wire. Various other inventors suggested similar systems and some of them were tried out.

These early methods were too slow and cumbersome to be practical, however. A breakthrough came in 1819, when the Danish physicist Hans Christian Oersted (1777–1851) discovered that an electric current produces a magnetic field. Other scientists, including André-Marie Ampère (1775–1836) and Pierre Laplace (1749–1827; *see* APPENDIX I: BIOGRAPHICAL ENTRIES), developed this idea and in 1825 the English inventor William Sturgeon (1783–1850) enclosed a needle within a coil of wire. The needle became a magnet when a current passed through the wire, and its polarity changed when the current changed direction. Sturgeon called his device an "electromagnet."

In 1831, Joseph Henry (1797–1878; *see* APPENDIX I: BIOGRAPHICAL ENTRIES) made a signaling apparatus that used an electromagnet. It consisted of a magnetized steel bar that was pivoted in a horizontal position and could be attracted by an electromagnet. When the bar was drawn toward the magnet, the end of it struck a bell. Messages could thus be conveyed by a sequence of sounds. In 1835, Henry invented the relay. This was a series of similar circuits in which each circuit activated the next. It overcame the diminution in the signal passing through a length of wire that is caused by resistance in the wire itself. Samuel Morse (1791–1872; *see* APPENDIX I: BIOGRAPHICAL ENTRIES) devised a code suitable for telegraphic use, and the MORSE CODE is still used today.

The first telegraph line in the world linked Baltimore and Washington and was opened in 1844. The first message, sent in Morse code, was "What hath God wrought?" Telegraph lines were soon being installed in many countries and also between them. The first submarine cable was laid in 1850 to link England and France. An attempt to lay a transatlantic cable was made in 1857, but the cable broke and the project had to be abandoned. The first successful transatlantic cable was laid in 1866.

Joseph Henry had been elected secretary of the Smithsonian Institution in 1846, and he used his position to establish a network of weather observers throughout the United States. The network became operational in 1849, and the observers used the telegraph to send data to the Smithsonian. This system formed the basis on which the United States Weather Bureau was formed in 1891. The first national network of meteorological stations linked to a central point by telegraph opened in France in 1863.

Further Reading
Allaby, Michael. *A Chronology of Weather.* Rev. ed. New York: Facts On File, 2004.

Television and Infrared Observation Satellite (TIROS) TIROS was the world's first weather satellite, launched by the United States on April 1, 1960. By 1965, nine more TIROS satellites had been launched, several of them into polar ORBIT rather than equatorial orbits.

TIROS-8 carried the first automatic picture transmission equipment. The first Advanced Very High Resolution Radiometer was carried on *TIROS-N*, launched in October 1978. *See* SATELLITE INSTRUMENTS.

TIROS satellites are now known as NOAA-class satellites. These satellites travel in polar orbits, scanning the entire surface of the Earth over a 24-hour period. Their instruments are sensitive to visible light and infrared radiation, and they scan to the sides of the orbital path, covering an area 1,864 miles (3,000 km) wide and 1.2 miles (2 km) high. At one time, for example, they would be able to observe the entire area from southern Florida to Hudson Bay and from the Atlantic to the Great Lakes. They transmit a constant stream of data by automatic picture transmission or high-resolution picture transmission. As well as monitoring the weather, the satellites carry search and rescue transponders that are used to help locate ships and aircraft in distress.

temperate belt The region that lies approximately between latitudes 25° and 50° in both hemispheres. The lower latitude is close to the TROPICS, and the higher latitude is about at the 50°F (10°C) isotherm (*see* ISO-) of mean SEA-SURFACE TEMPERATURE.

The temperate belt lies in the middle latitudes, and its climates are often described as middle latitude (or midlatitude). The climates of the temperate belt correspond to category C in the KÖPPEN CLIMATE CLASSIFICATION, with temperatures in the coldest month between 26.6°F (-3°C) and 64.4°F (18°C) and in the warmest month temperatures higher than 50°F (10°C).

temperature Temperature is a measure of the relative warmth of an object or substance that allows it to be compared to another object or substance (one is warmer or cooler than the other) or to a standard (so many degrees). Temperature and heat are not the same thing; heat is a form of energy, temperature the effect that energy produces.

All objects and substances, including the air and our own bodies, are made from atoms and molecules. Atoms and molecules move. If they are in the form of a gas, they move freely and rapidly. Molecules in a liquid move more slowly and have less freedom. In a solid the molecules are unable to move around, but they vibrate. How fast the atoms or molecules move or vibrate depends on the amount of energy they possess. Their KINETIC ENERGY (energy of motion) can increase or decrease.

Energy cannot be created or destroyed, but one form of energy can be converted into another (*see* THERMODYNAMICS, LAWS OF). When an atom or molecule absorbs heat (one form of energy), that heat is converted to kinetic energy (another form of energy).

The kinetic energy possessed by the atoms or molecules composing an object or substance is measured as motion in relation to the center of mass of the object or substance. When moving or vibrating atoms or molecules strike another object, a proportion of their kinetic energy is transferred to that object. An appropriate sensor can detect this transferred energy. Nerve endings in our skin are sensors that detect the impact of fast-moving or vibrating atoms and molecules. The message that the nerves send to the brain is interpreted as temperature. We feel that something is hot or cold, either in relation to our own skin temperature or in an absolute sense if the skin is exposed to a temperature so high it will burn or so cold it will freeze the tissues.

A THERMOMETER is an instrument that absorbs kinetic energy from impacting atoms and molecules and converts it to a reading against a scale. There are several TEMPERATURE SCALES, but only three are widely used. Scientists usually prefer the Kelvin scale, in which the temperature is written in the unit K (for kelvin), without a degree sign. The Celsius temperature scale is the most widely used everyday scale. Sometimes still called the centigrade scale (the Latin *centum* means "hundred"), because there are 100 of its degrees between the freezing and boiling points of water, Celsius temperatures are written as °C. Its name was officially changed from centigrade to Celsius in 1948, at the Ninth General Conference on Weights and Measures. The Fahrenheit temperature scale is more often used in the United States and Britain (where it is being replaced by the Celsius scale). Its temperatures are written as °F. A fourth scale, the Réaumur, is used in very few places today.

There are several ways to report the temperature. The mean temperature is the air temperature measured at a particular place over a specified period, such as a day, month, or year, and then converted to a MEAN. Mean temperatures can also be shown, and plotted as isotherms (*see* ISO-), for large regions, continents, and for the whole world. The rate of temperature change over a horizontal distance is known as the temperature gradient.

The temperature range is the difference between the highest and lowest mean temperatures that have been recorded for a particular place. If the annual range is required, only daytime temperatures should be used. The DIURNAL range compares the mean daytime and mean nighttime temperatures for a month, season, or year. The mean range uses only mean temperatures, but the absolute range takes account of the highest and lowest temperatures ever recorded by day or night. Chicago, for example, has a mean annual temperature range of 49°F (27°C), but an absolute temperature range (measured over 75 years) of 128°F (71°C). A temperature belt is the area that lies between two lines on a graph that show the daily maximum and minimum temperatures for a particular place. The belt indicates the temperature range for that place.

Obviously, the air temperature is not everywhere the same even within a small area. Consequently, where it is measured makes a difference. Temperature is often reported as the shade temperature. This is the air temperature that is measured inside a STEVENSON SCREEN or other shelter, or anywhere out of direct sunlight. Air is heated almost entirely by contact with the ground and not by directly absorbing solar radiation. The glass of a thermometer, on the other hand, will absorb solar radiation directly, and this heat will be transferred to the liquid, raising its temperature to a level higher than that of the surrounding air. The thermometer will then give a false reading. For this reason, reported temperatures are always shade temperatures unless it is stated otherwise.

The concrete minimum temperature is the lowest temperature that is registered by a minimum thermometer that remains in contact with a concrete surface for a specified period. It is used as an alternative to the grass minimum temperature, because it gives more uniform results and provides a better indication of the likelihood of ice forming on road surfaces.

The grass minimum temperature, also called the grass temperature, is the temperature that is registered by a minimum thermometer set in the open with its bulb at the level of the tops of the blades of grass in short turf. This is the temperature to which crop plants are exposed, and it is therefore of relevance to farmers and horticulturists.

The surface temperature is the temperature of the air or sea measured close to the surface of land or water (see SEA-SURFACE TEMPERATURE).

The dry-bulb temperature is the temperature that is registered by a thermometer with a bulb that is dry and directly exposed to the air. A dry-bulb thermometer is used to measure the air temperature. This reading is compared with the wet-bulb temperature to determine the relative HUMIDITY and DEW point temperature. A wet-bulb thermometer registers the wet-bulb temperature. In saturated air (relative humidity 100 percent) the wet-bulb temperature will be equal to the dry-bulb temperature, indicating that no evaporation is taking place. At any relative humidity below SATURATION the wet-bulb temperature is lower than the dry-bulb temperature.

temperature scales Measuring the temperature of the air is simple enough for anyone possessing a THERMOMETER, but as scientists of past centuries explored the causes of weather phenomena, devising a thermometer that would give an accurate reading proved extremely difficult. Many scientists worked at the problem. By itself, however, a reliable thermometer is not enough. All it will show is the level of a liquid rising or falling inside a tube or a needle moving on a dial. This will indicate whether the temperature at one time is higher or lower than it was at another time, but it will not allow the difference to be quantified. Quantification calls for a recognized scale that can be used to calibrate the thermometer. Over the centuries several temperature scales have been proposed.

The Römer temperature scale was possibly the first. It was devised in about 1701 by the Danish astronomer, physicist, and instrument maker Ole Christensen Römer (1644–1710; see APPENDIX I: BIOGRAPHICAL ENTRIES). Römer, also spelled Rømer, never published a description of the method he used, but in 1708 Daniel Fahrenheit visited him, watched him at work, and wrote his own account.

In 1701, Isaac Newton (1642–1727) pointed out that any temperature scale must be calibrated between two fixed, or fiducial, points. Newton suggested that average body temperature and the temperature of freezing water should be used as the two fiducial points, but Römer used the freezing and boiling temperatures of water.

According to Fahrenheit, Römer inserted his thermometer, filled with alcohol (in fact, wine), into freezing water and marked the point reached by the alcohol in the thermometer. He then placed the thermometer into tepid water, which Fahrenheit wrote was at blood heat (*blutwarm*). He then added half of the distance between these points below the lower fiducial point and marked this lowest point as 0. There is some confusion about the lower fiducial point, however. Some historians hold that Römer used a mixture of WATER, ICE, and ammonium chloride to determine the lower fiducial point and called that 0, others that he used melting snow only and called that point 7½. In either case, on the Römer scale water freezes at 7½°Rø, boils at 60°Rø, and average body temperature is 22½°Rø. The scale is no longer used, but it is important historically, because it is the one on which Fahrenheit based his scale.

Daniel Fahrenheit (see APPENDIX I: BIOGRAPHICAL ENTRIES) devised his temperature scale in about 1714. It remains in use in Britain, the United States, and other English-speaking countries, although the Celsius temperature scale is steadily taking its place. Scientific publications always use either the Celsius scale or the Kelvin scale.

Fahrenheit derived his scale from the Römer temperature scale, in which two fiducial points are used, the freezing and boiling points of water. Fahrenheit modified the Römer scale by using body temperature for his upper fiducial point, and for his lower fiducial point he used the freezing point of a mixture of ice and salt. This was then believed to be the lowest temperature that could be attained, so by calling it 0° all temperature values would be positive. He marked this point on his mercury-filled thermometer and then measured body temperature. In order to be able to measure small temperature differences, Fahrenheit divided the distance between the upper and lower fiducial points into 90 degrees. On this scale the freezing point of pure water was 30° and body temperature was 90°.

Later Fahrenheit adjusted the scale. He substituted the boiling point of pure water for the upper fiducial point and divided the distance between the new upper point and the freezing point of pure water into 180 degrees, while still retaining the lower fiducial point. On the revised scale, pure water freezes at 32° and boils at 212°, and average body temperature was 96°, which was later adjusted to 98.6°.

$1°F = 0.56°C$; $1°C = 1.8°F$. To convert Fahrenheit temperatures into Celsius, $°F = (°C × 1.8) + 32$; $°C = (°F - 32) × 5 ÷ 9$.

In 1730, the French physicist and naturalist René-Antoine Ferchault de Réaumur (1683–1757; *see* APPENDIX I: BIOGRAPHICAL ENTRIES) devised another scale. Réaumur measured the expansion of a mixture of water and alcohol as its temperature increased. The liquid was held in a bulb at the base of a tube, just like any thermometer. When it was at the freezing point, he marked the point it reached on the tube as zero. He then graduated the remainder of the tube into units, each of which was equal to one-thousandth of the volume of the liquid in the bulb and tube when it was at freezing. When the liquid reached boiling point, he found its length had increased to 1,080 units, so it had risen 80 units (or degrees). Consequently, his Réaumur temperature scale ran from 0°R at freezing to 80°R at boiling point.

In 1742, the Swedish astronomer Anders Celsius (*see* APPENDIX I: BIOGRAPHICAL ENTRIES) published a paper called "Observations on two persistent degrees on a thermometer" in the *Annals of the Royal Swedish Academy of Science*. Celsius proposed that the two fixed points should be the temperature of melting snow or ice and the temperature of boiling water, and he described his reasons for choosing these two points. Obviously, water freezes and thaws at the same temperature, but it is more difficult to measure the point at which liquid water begins to freeze than it is to measure the temperature at which snow and ice melt. Celsius reported that he had used one of the thermometers made by Réaumur to measure the temperature of melting snow. He repeated the measurement many times in the course of two winters and during all kinds of weather and at different atmospheric pressures. He even brought snow from outdoors and placed it in front of the fire in his room to measure its temperature as it melted. The temperature was invariably the same. Snow also melted at the same temperature in Paris and in Sweden at Uppsala (60°N) and Torneå (66°N). He was confident, therefore, in the first of his fixed points.

Measuring the temperature of boiling water was more complicated. Although the temperature of water will rise no further once it is boiling, Celsius thought the intensity of boiling might affect the thermometer, and he noticed that when he removed the thermometer from the boiling water the mercury rose before it began to fall. This, he suggested, was because the glass tube contracted before the mercury began to cool. Daniel Fahrenheit had observed that the boiling temperature of water varies according to the atmospheric pressure. Celsius confirmed this, but found a way to correlate the two, because the height of the mercury in the thermometer was always proportional to the height of the mercury in the nearby BAROMETER.

Celsius then proposed a standard method for calibrating a thermometer. First the bulb of the thermometer should be placed in thawing snow and the position of the mercury marked. Then the thermometer bulb should be placed into boiling water when the atmospheric pressure is approximately 1,006.58 millibars (29.75 inches or 755 mm of mercury). The position of the mercury should be marked.

The distance between the two points should then be divided into 100 equal parts, or degrees, so that 0 degree corresponds to the boiling temperature of water and 100 degrees corresponds to its freezing temperature.

This was the Celsius scale. It was later reversed, so that 0 degree represents the freezing temperature of water and 100 degrees the boiling point. It is uncertain who made the change. It may have been Martin Strömer, a pupil of Celsius. It has also been suggested that it was Carl von Linné (Linnaeus). The most likely person, however, was the leading Swedish instrument maker of the time, Daniel Ekström.

The Celsius temperature scale is the one that is used throughout most of the world. In English-speaking countries the Fahrenheit temperature scale is also used, although not for scientific measurements, where the Celsius scale is preferred.

Because there are 100 degrees, the scale is sometimes called the centigrade scale, from the Latin *centum*, meaning "hundred," and *gradus*, meaning "step." The name was officially changed from centigrade to Celsius in 1948, at the Ninth General Conference on Weights and Measures.

The Kelvin scale is the one most often used in scientific publications, and the kelvin is the unit of temperature in the SI scheme (*see* APPENDIX VIII: SI UNITS AND CONVERSIONS). The scale was devised by William Thomson (1824–1907), who later became Lord Kelvin. He called it the "absolute" scale. One kelvin, written as 1 K with no degree sign (not as 0°K), is equal to 1°C (1.8°F), and 0 K, or absolute zero, is equal to −273.15°C (-459.67°F).

The Rankine scale is sometimes used as an alternative to the Kelvin scale, mainly in the United States. It is used, for example, in calculating density altitude (*see* DENSITY), but its main use is in engineering. The scale was devised by the Scottish engineer and physicist William John Macquorn Rankine (1820–72). The Rankine degree (R) is equal to a Fahrenheit degree, but the Rankine scale extends to absolute zero. 0°R = −459.67°F. Water freezes at 492°R and boils at 672°R. The advantage of the Rankine scale is its compatibility with the Fahrenheit scale. Fahrenheit temperatures can be converted to Rankine temperatures without first having to convert the degrees themselves.

Absolute temperature is the temperature measured on the Kelvin scale, which has no negative values. To convert a Fahrenheit temperature to the equivalent absolute temperature, K = ((°F - 32) × 5 ÷ 9)) + 273.15. The absolute temperature at which water freezes (32°F) is 273.15 K and the temperature at which it boils (212°F) is 373.15 K. Absolute zero, or zero on the absolute (Kelvin) scale, is the temperature at which the KINETIC ENERGY of atoms and molecules is at a mini-

mum. It is the lowest temperature possible (and unattainable according to the third law of thermodynamics; *see* THERMODYNAMICS, LAWS OF). The existence of absolute zero was first implied in the work of Jacques Charles (*see* APPENDIX I: BIOGRAPHICAL ENTRIES and GAS LAWS). Absolute zero (0 K) is equal to -459.67°F (-273.15°C).

Further Reading

Poulsen, Erling. "Early Danish Thermometers: The Thermometers of Ole Rømer." Available online. URL: www.rundetaarn.dk/engelsk/observatorium/tempeng.htm. Accessed February 16, 2006.

tephigram (TΦgram)   A tephigram is a THERMODYNAMIC DIAGRAM on which the TEMPERATURE and HUMIDITY of the air are plotted against AIR PRESSURE. This reveals the entire structure of a column of air.

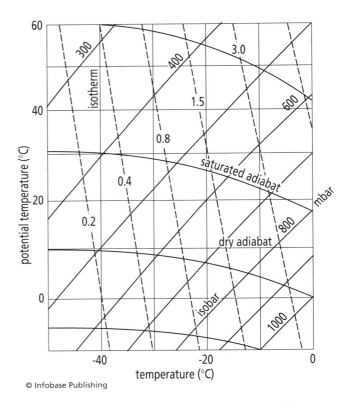

© Infobase Publishing

A tephigram is a graph on which temperature and humidity are plotted against pressure to illustrate the structure of an entire column of air. The actual temperature is plotted against the horizontal axis, the logarithm of the potential temperature against the vertical axis, pressure by the solid diagonals, and saturation specific humidity by the broken lines.

The actual temperature is plotted against one axis and the logarithm of the POTENTIAL TEMPERATURE against the other. Vertical lines then represent isotherms (*see* ISO-) and horizontal lines are isotherms of potential temperature. The horizontal lines are also dry ADIABATS and the distance between them decreases as the potential temperature increases.

Saturated adiabats appear as curved lines. In the lower troposphere (*see* ATMOSPHERIC STRUCTURE), the saturated adiabats cross the dry adiabats at about 45°, but the angle becomes smaller with increasing altitude. Isobars (*see* ISO-) are slightly curved lines running diagonally across the diagram from lower left to upper right and almost bisecting the right angles at which the isotherms intersect. Isopleths, showing the saturation specific humidity, are shown as dotted lines that make a small angle with the vertical isotherms.

When the tephigram has been constructed, it is sometimes rotated clockwise by about 45° so that the isobars lie horizontally. These then correspond to altitude and can be labeled as such, in addition to being labeled in units of pressure (usually millibars). The conventional symbol for potential temperature is the Greek letter phi (Φ), so the tephigram is a t (for temperature) phi (for potential temperature) -gram. The tephigram was devised by Sir Napier Shaw (*see* APPENDIX I: BIOGRAPHICAL ENTRIES).

terminal velocity Terminal velocity is the maximum speed that a falling body can attain. Once the body has accelerated to its terminal velocity, it continues its descent at that constant speed. The terminal velocity of a body is therefore proportional to its weight and to the drag exerted on it as a result of the resistance to its passage offered by the medium through which it is falling. That drag is proportional to the surface area of the body, because this is the surface against which resistance acts, and the weight is proportional to its volume.

If the body is very small, the flow of air around it is dominated by the VISCOSITY of the air. For larger bodies, the downward force acting on the body is equal to the weight of the body minus the weight of the air it displaces. In the lower troposphere (*see* ATMOSPHERIC STRUCTURE), the terminal velocity (V) of a falling body the size of a small raindrop is given by $V = 8 \times 10^3 r$, where r is the radius of the body. The airflow around

large raindrops and hailstones is much more turbulent, and their terminal velocity is given by $V = 250r^{1/2}$. *See also* STOKES'S LAW.

The speed with which a body, such as a RAINDROP, SNOWFLAKE, or hailstone, falls through the air is called its fall speed. This is equal to the terminal velocity of the body minus the velocity of any upward air current to which it is exposed.

Tertiary The sub-era of the CENOZOIC era of geologic time (*see* APPENDIX V: GEOLOGIC TIMESCALE) that began 65.5 million years ago and ended 1.81 million years ago. The Tertiary includes the PALEOGENE and NEOGENE periods.

The name Tertiary was introduced in 1758 by Giovanni Arduino (1714–95), a professor at the University of Padua, Italy. Arduino divided the rock strata of the Apennine Mountains of central Italy into three, calling them Primary, Secondary, and Tertiary. The names Primary and Secondary were abandoned long ago, and Tertiary has now been abandoned for formal use.

thermal belt A thermal belt is a fairly well defined area on many mountainsides in middle latitudes where nighttime temperatures are higher than the temperatures at higher and lower elevations. Below the thermal belt cold air subsides katabatically at night to produce low temperatures and often a FROST hollow in the valley bottom. Above the thermal belt the adiabatic (*see* ADIABAT) decrease in temperature with height also produces cold air.

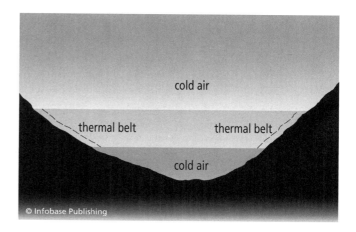

The thermal belt is a region on a mountainside where the nighttime temperature is higher than it is at higher or lower elevations.

thermal equator The belt around the Earth where the temperature is highest is called the thermal equator. Its location changes with the SEASONS between 23°N and between 10°S and 15°S. The mean location of the thermal equator is about 5°N.

The thermal equator is not the same as the meteorological equator, although both occupy the same position. The meteorological equator is the mean latitude of the equatorial trough (*see* INTERTROPICAL CONVERGENCE ZONE), which is also 5°N.

thermal wind A thermal wind is generated when the air temperature changes by a large amount over a short horizontal distance. The JET STREAM and the easterly jet are the most important examples.

Warm air is less dense than cool air. Where warm and cool air lie adjacent to each other, therefore, AIR PRESSURE decreases with height more rapidly in the cool air than in the warm air, because the cool air is the more compressed. Consequently, a CONSTANT-PRESSURE SURFACE (a surface across which the atmospheric pressure is the same everywhere) slopes upward from the cool, dense air to the warm, less dense air, and the THICKNESS of each atmospheric layer increases along a gradient from the dense air to the less dense air. This gradient becomes steeper with increasing height, because the thickness of a layer depends on the degree to which the air is compressed, and compression decreases with height more slowly in dense air than in air that is less dense.

The speed of the GEOSTROPHIC WIND is proportional to the PRESSURE GRADIENT or, to put it another way, to the slope of the constant-pressure (or isobaric) surface. This means that if the slope angle of the isobaric surface changes with height, so must the speed of the geostrophic wind. The thickness of a layer that is bounded by two isobaric surfaces is proportional to the mean temperature in the layer. It therefore follows that the change in the speed of the geostrophic wind across the layer is proportional to the temperature gradient across the layer. This means the layer must be BAROCLINIC. The relationship between the geostrophic WIND SHEAR and baroclinicity is known as the thermal wind relation.

Since the temperature gradient increases with height, so does the wind speed. It blows with the cool air to its left in the Northern Hemisphere and to its right in the Southern Hemisphere, which is why the

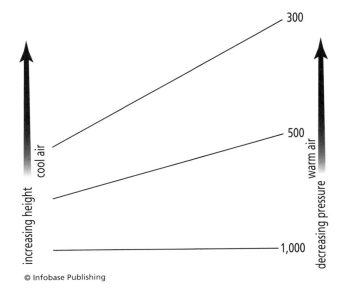

Pressure decreases with height more rapidly in cool, dense air than it does in warm air, which is less dense, and the thickness of each layer defined by pressure is proportional to the temperature. Where warm and cool air lie adjacent to each other, this produces a temperature gradient that increases with height. It is the gradient that generates the thermal wind.

polar FRONT and subtropical jet streams blow from west to east in both hemispheres.

The movement of an atmospheric disturbance in the direction the nearest thermal wind is blowing is called thermal steering. The thickness of the layer used to calculate the thermal wind is usually taken to extend from the surface to the middle troposphere (*see* ATMOSPHERIC STRUCTURE) and thermal steering is equivalent to movement along THICKNESS lines.

thermocline Literally, a thermocline is a change of temperature that occurs along a gradient between two places. More specifically, the thermocline is a layer in the ocean where the temperature decreases with depth much more rapidly than it does in the water above or below. The depth of the thermocline varies from place to place and with the seasons, but it may commence as little as 33 feet (10 m) or as much as 660 feet (200 m) below the surface and end at depths between 500 feet (150 m) and 5,000 feet (1,500 m). In Arctic and Antarctic waters there is usually no thermocline, because the sea surface is covered by ice during the winter and there is only slight warming by solar radiation in summer. The strongest thermocline is found in the TROPICS.

Water at the ocean surface is warmed by solar radiation. In the Tropics, the SEA-SURFACE TEMPERATURE commonly exceeds 68°F (20°C) and can reach 80°F (27°C). This is probably a maximum because the rate of evaporation increases with temperature to a point at which the LATENT HEAT of vaporization cools the surface layer sufficiently to prevent the temperature rising any higher. Radiant heat does not penetrate very deeply, and the ocean loses heat to the atmosphere by longwave radiation, but winds and currents mix the upper waters, and it is this mixing that carries warm water to a greater depth. Mixing also helps constrain the sea-surface temperature.

At about 13,000 feet (4,000 m), which is the average depth of all the oceans, the water temperature is between 34°F and 36°F (1°C–2°C). This temperature remains constant throughout the year, regardless of latitude. The deep ocean is the most unchanging environment on Earth.

Mixing produces an upper layer of water in which the temperature decreases only very slightly with depth. Below the mixed layer, water temperature begins to decrease sharply with depth, and by about 3,300 feet (1,000 m) it has fallen to approximately 40°F (4.4°C). From there it decreases much more gradually.

The thermocline is the layer in which temperature decreases rapidly, and it is most strongly marked in the Tropics because there the temperature must fall from about 68°F (20°C) to about 40°F (4.4°C). In midlatitudes, where the water in the mixed layer is cooler, the

temperature gradient is shallower, especially in winter when the surface temperature is about 54°F (12°C). In summer, when the sea is warmer, there is a more sharply defined summer thermocline very close to the surface. In latitudes higher than about 50°N and 50°S, there is no thermocline.

Solar radiation is absorbed by the oceans, but it warms only the water above the thermocline. When warmed, the oceans warm the air in contact with them. Cooler water from the thermocline that becomes incorporated into the mixed layer is immediately replaced by cold water from a higher latitude. It is partly through this coupling of oceans and atmosphere that heat is transferred by ADVECTION from low to high latitudes. A thermocline also develops in many lakes in summer.

thermodynamic diagram A thermodynamic diagram summarizes the factors affecting the temperature and pressure of a PARCEL OF AIR. A point on the diagram then refers to the thermodynamic (energy) state of the air in that location.

A simple thermodynamic diagram measures altitude along its vertical axis, and temperature along its horizontal axis. A line showing the change of air temperature with height corresponds to the LAPSE RATE and indicates the height of the lifting condensation level, marking the boundary between the dry and saturated adiabatic lapse rates (see ADIABAT) and the height of the CLOUD BASE. The most widely used types of thermodynamic diagram are the Stüve chart and TEPHIGRAM.

A line drawn on a thermodynamic diagram to show the lapse rate of air that is rising past the lifting condensation level is called a pseudoadiabat. The pseudoadiabatic lapse rate is almost the same as the saturated adiabatic lapse rate.

A Stüve chart, also called an adiabatic chart or pseudoadiabatic chart, is a thermodynamic diagram in which temperature is plotted along the horizontal axis and pressure along the vertical axis, with the highest pressure at the bottom. Pressure is calculated as the AIR PRESSURE raised to the power of 0.286. The chart was devised by G. Stüve and has been widely used, although meteorologists now find the tephigram more useful.

An aerological diagram is a thermodynamic diagram on which data from soundings of the upper atmosphere are plotted. The diagram usually shows isobars (see ISO-), isotherms (lines joining points at equal temperatures), and dry and saturated adiabats.

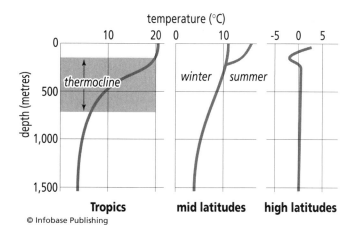
© Infobase Publishing

The thermocline is the ocean layer in which temperature decreases most rapidly with depth. The tropical thermocline is of much greater vertical extent than the high-latitude thermocline. In midlatitudes, the thermocline almost disappears in winter.

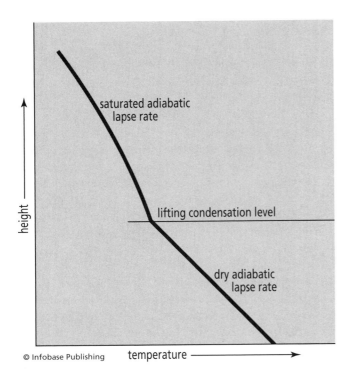

A thermodynamic diagram illustrates the amount of energy a parcel of air possesses. In this example, temperature is plotted against height to show the lapse rate. The lifting condensation level marks the height at which the dry and saturated adiabatic lapse rates meet.

thermodynamics, laws of Thermodynamics is the scientific study of the ways in which energy changes from one form into another, the way it is transmitted from one place to another, and its availability to do work. It is based on the idea that in any isolated system, anywhere in the universe, there is a measurable amount of internal energy. The internal energy is the sum of the KINETIC ENERGY and POTENTIAL ENERGY possessed by all the atoms and molecules within the system. This total amount of internal energy cannot change without intervention from outside the system, in which case the system ceases to be isolated. This principle can be described by four laws.

The first law of thermodynamics was suggested in the 1840s by the German physicist Julius Robert Mayer (1814–78, *see also* John Tyndall; APPENDIX I: BIOGRAPHICAL ENTRIES) and was verified in 1843 by the English physicist James Prescott Joule (1818–89). Lord Kelvin (1824–1907), the Scottish physicist, and the German physicist Hermann Ludwig Ferdinand von Helmholtz (1821–94) also made important contributions to the development of the law.

The first law states that energy can be neither created nor destroyed, but that it can be changed from one form into another. This is sometimes called the law of conservation of energy.

There was a difficulty with this law. Joule had measured the mechanical equivalent of heat, which is the change of energy from one form (heat) to another (kinetic energy). Heat engines, such as steam and internal combustion engines, exploit this transformation. The French theoretical physicist Nicolas-Léonard Sadi Carnot (1796–1832) had shown that the efficiency of such an engine depends only on the difference in temperature between the source of heat and the sink into which the heat is finally discharged. Some heat is lost, however, as energy flows through the engine. So where does the lost energy go, given that according to the first law energy cannot simply vanish? The loss was explained in 1850 by the German theoretical physicist Rudolf Clausius (1822–88). Clausius asserted that heat does not pass spontaneously from a colder to a hotter body. For example, if you leave a cup of hot coffee to stand in a cold room, the coffee will become colder by losing its heat, and not hotter by absorbing heat from its cold surroundings. In 1851, Lord Kelvin arrived at the same conclusion. The apparently lost energy is absorbed into the surrounding environment.

This is the second law of thermodynamics. There are several ways it can be expressed. Clausius summarized it in two ways: Heat cannot be transferred from one body to a second body at a higher temperature without producing some other effect; or to put it differently, the entropy of a closed system increases with time. Clausius coined the word *entropy*. The second law means that most physical processes are irreversible.

Entropy is a measure of the amount of disorder that is present in a system. As the amount of disorder increases, so does its entropy, and disorder increases all the time and has done so throughout the history of the universe. Entropy always increases because there is a much greater statistical probability that the random motion of atoms and molecules will lead them to form chance arrangements than that they will come together in highly organized structures. If a vase falls onto a stone floor, it will probably shatter into many pieces. This is a less ordered state than the one that existed when the fragments were joined seamlessly together and the vase was complete. The shattering of a vase is a common event and one that increases disorder. History records no instance of the fragments of a shattered vase spontaneously moving back together again to reconstruct the vase as it was before it fell.

Order can be described as a difference in energy level between an object and its surroundings. As entropy increases, this difference decreases. Consider what happens when a cup of hot coffee is left to stand on a table. The coffee cools until it is at the same temperature as the air around it. The energy difference between the coffee and the air has disappeared, and both are at the same energy level. It never happens that the hot coffee absorbs energy from the cooler air and its temperature increases.

In exactly the same way, when solar radiation warms the surface of the Earth that heat is dissipated. Some is lost immediately as infrared radiation (*see* BLACKBODY) into space and some warms the air in contact with the surface. This warming produces all our weather phenomena, but the energy driving the weather also dissipates, because the atmospheric gases also radiate energy into space. The weather continues because there is a constant supply of solar energy to drive it. Without that energy, the Earth would cool until it was at the same temperature as the space surrounding it.

Entropy means that most everyday events are irreversible. For this reason it is sometimes known as the "arrow of time."

Entropy increases, and so there must be a point at which it reaches a maximum and can increase no further. This point is described by the third law of thermodynamics, which was discovered in 1905–06 by the German physical chemist Hermann Walther Nernst (1864–1941).

The third law states that in a perfectly crystalline solid there is no further increase in entropy when the temperature falls to absolute zero (*see* TEMPERATURE SCALES). This also means that it is impossible to cool any substance to absolute zero (although substances have been cooled to a tiny fraction of a degree above absolute zero).

The discovery of the third law should have completed the list of laws of thermodynamics, but there is a further principle that is more fundamental than any of them. It had been well known for centuries and was taken for granted until the English physicist Sir Ralph Fowler (1889–1944) drew attention to it. It could not

be called the fourth law, because it is the principle that underlies the second law (and it can be derived from it). Nor could it be called the first law, because that would mean renumbering the existing laws, which would cause endless confusion. So Fowler proposed that it be called the zeroth law (law number 0).

The zeroth law states that if two isolated systems are each in thermal equilibrium with a third, then they are in thermal equilibrium with each other. Thermal equilibrium is the condition of two or more bodies that are at the same temperature and therefore possess the same amount of kinetic energy. Unless some outside process intervenes, energy will not be exchanged between bodies that are in thermal equilibrium.

This law makes it possible to use a THERMOMETER to measure the temperature of a substance or body. The thermometer is an isolated system that is brought into equilibrium with the system being measured, and the temperature scale marked on the thermometer is derived from a third system that was used to calibrate the instrument.

thermograph A thermograph is an instrument which provides a continuous record of TEMPERATURE. It consists of a component that changes shape with changes in temperature. This component is connected by a system of levers to an arm that terminates in a pen held against a calibrated chart mounted around a cylinder. The cylinder rotates at a constant speed, so the pen

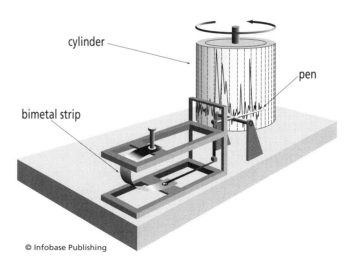

A thermograph provides a continuous record of temperature as a line drawn by a pen on a chart attached to a cylindrical drum.

traces around the chart a line that rises and falls with temperature changes.

The principal component may be a bimetal strip or a Bourdon tube. A bimetal strip consists of two pieces of different metals that expand and contract by widely different amounts when heated and cooled. The two strips are bonded together, so that their differential expansion and contraction causes the combined strip to bend. A Bourdon tube is a curved container made from phosphor bronze and filled with alcohol. Like a bimetal strip, it bends in response to changes in temperature.

thermohaline circulation   The thermohaline circulation is the exchange of surface and deep water that takes place in the oceans, but only in high latitudes, due to differences in temperature (*thermo-*) and salinity (*-haline*).

Over most of the oceans, the surface layer remains warmer than the deep water at all times, because it absorbs solar radiation. This means that the surface layer, being warmer and therefore less dense, mixes only slightly with the water below it. The warmer the surface layer, the greater is the difference in DENSITY between the surface and deep water. Consequently, very little mixing can occur between the surface and deep waters in the tropical oceans, but rather more in mid-latitude oceans.

In Arctic and Antarctic waters, however, the surface layer loses so much heat by radiating it toward the sky that it becomes colder than the water beneath it. As it freezes, salt is expelled from the ice crystals and into the water adjacent to the ice. Water near the ice is therefore colder than the subsurface water and also more saline. Both factors increase its density, so this water sinks and deep water rises to take its place, establishing a thermohaline circulation. This vertical circulation is also known as convective overturning.

In the North Atlantic, the cold, saline water sinks all the way to the ocean floor, where it forms the NORTH ATLANTIC DEEP WATER (NADW). The NADW flows southward as a slow-moving current all the way to the Southern Ocean, thus forming the movement of water that drives the GREAT CONVEYOR.

The thermohaline circulation in the North Pacific Ocean is very weak, and there is no convective overturning, because there the surface waters are too fresh to sink. Dense, saline water sinks near the edge of the ice in the Southern Ocean, but there the effect is greater than in the North Atlantic due to a high rate of evaporation and also the possibility that water becomes supercooled (*see* SUPERCOOLING) and sinks at the base of the ice shelves.

thermometer   A thermometer is an instrument that is used to measure the TEMPERATURE of a substance. Thermometers used in METEOROLOGY are used to measure the temperature of AIR and WATER. The measurement of air temperature by instruments that are mounted on aircraft is called aircraft thermometry.

There are several ways to measure changes in temperature. The most common method is based on the fact that many substances expand when they are warmed and contract when they are cooled. If this property is to be used to measure small changes in temperature, the substance chosen must expand and contract by the largest amount possible. Air was the first to be tried, in the air thermoscope that was invented by Galileo (*see* APPENDIX I: BIOGRAPHICAL ENTRIES).

Galileo's air thermoscope consisted of a glass bulb connected to a narrow glass tube that was mounted vertically with its lower end immersed in colored water contained in a sealed vessel. Air in the bulb expanded and contracted as the temperature rose and fell, pushing the water in the tube downward as it expanded and drawing it upward as it contracted. The thermoscope was very sensitive to changes in temperature, but it made no allowance for changes in AIR PRESSURE that also alter the volume of air, and so it was very inaccurate. Consequently, the air thermoscope was very inaccurate.

Liquids were found to be better than air. In 1641, Ferdinand II of Tuscany (*see* APPENDIX I: BIOGRAPHICAL ENTRIES), a contemporary of Galileo, used liquid to make the first reliable thermometer. Ferdinand used alcohol, and in 1714 Daniel Fahrenheit (*see* APPENDIX I: BIOGRAPHICAL ENTRIES) made a similar thermometer using mercury. His was the first thermometer to give readings that were both reliable and accurate, and Fahrenheit is credited with having invented the thermometer.

Until recently, both alcohol and mercury were used, but mercury is no longer permitted because it poses a risk to health. Alcohol expands and contracts more than mercury does, but an alcohol thermometer is less accurate than a mercury thermometer. This is because the rate at which substances expand with increasing temperature varies slightly across the temperature

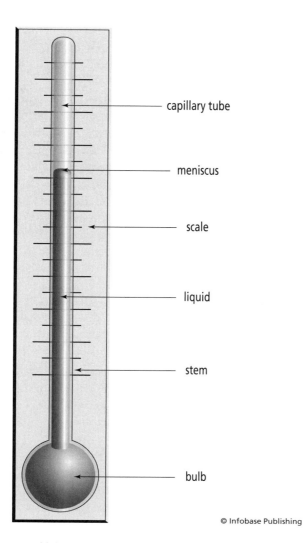

capillary tube

meniscus

scale

liquid

stem

bulb

© Infobase Publishing

The most widely used thermometer measures the expansion of a liquid held in a reservoir, called the bulb, opening into a capillary tube. The expansion and contraction of the liquid are read against a scale.

range, and the variation, although small, is greater for alcohol than for mercury. Both alcohol and mercury remain liquid at temperatures encountered in the lower atmosphere. A dye is added to alcohol to make it visible. Most alcohol thermometers use a red dye, but some use blue.

Alcohol and mercury thermometers consist of a narrow capillary tube (*see* CAPILLARITY), sealed at one end and blown into a bulb at the other end. The bulb acts as a reservoir for the liquid. As the liquid expands and contracts, the change is exaggerated by the narrowness of the tube in which it moves.

The thermoscope failed because Galileo did not know that the volume of a body of air changes with the atmospheric pressure as well as with the temperature. Account can now be taken of this factor, and modern gas thermometers are the most accurate of all thermometers that are based on the changing volume of a fluid. The gas used is not air, but either NITROGEN, HYDROGEN, or HELIUM. It is held in a vessel, so its volume remains constant. As the temperature changes, the pressure the gas exerts also changes according to the GAS LAWS, and this is the change that is registered. Gas thermometers are used in industry.

The amount by which the volume of a substance changes with changing temperature is called the coefficient of thermal expansion of that substance. This varies greatly from one substance to another. As the temperature changes, two metals with different coefficients of thermal expansion will increase and decrease in length by different amounts. If strips of these metals are securely bound together into a bimetal strip, as the length of one strip changes more than that of the other, the bimetal strip will curl by an amount proportional to the change in temperature. This property provides the basis for a thermometer using a bimetal strip. It is often used to operate thermostats. Its principal meteorological use is in THERMOGRAPHS. A bimetal strip is less accurate than an alcohol or mercury thermometer, however.

The properties of dissimilar metals can also be used to measure temperature electrically. A thermocouple consists of two wires or rods of different materials, each of which is made into a half loop. The two half loops are welded together at their ends to make a circuit. If one of the joints is at a different temperature from the other, an electric current flows through the circuit. This phenomenon was discovered in 1821 by the Estonian–German physicist Thomas Johann Seebeck (1770–1831) and is known as the Seebeck effect. The first thermometer to use it was made in 1887 by the French physical chemist Henri-Louis Le Châtelier (1850–1936), using platinum and rhodium as the two metals.

As the temperature of a metal rises, the electrical resistance of the metal increases. This property is exploited to make resistance thermometers, commonly using platinum, nickel, tungsten, copper, or alloy wires. Resistance thermometers are accurate to within 0.2°F (0.1°C).

Ceramic semiconductors have a similar property. They are used to make thermistors. The principal component in a thermistor is an electronic device that resists the flow of an electric current by an amount that varies with the temperature. As the temperature rises, the resistance increases, and the current flowing through the device decreases. As the temperature falls, the resistance decreases, and the current increases.

The Beckman thermometer is used for measuring very small changes in temperature. It is a mercury-in-glass thermometer with two bulbs. One bulb is located at the bottom of the thermometer tube, as in an ordinary thermometer. The top of the tube is shaped like an inverted U and the other, smaller bulb is at the end of one arm of the U. At the base of the upper bulb there is a second, upright, U-shaped tube. Mercury can be run from the upper bulb into the lower one. This alters the range of temperature the thermometer measures. The scale covers only about 9°F (5°C). The thermometer was invented by the German chemist Ernst Otto Beckman (1853–1923).

Once a thermometer has been made it must be calibrated. This is usually done by marking the position of the liquid at two reference temperatures, called fiducial points, commonly the freezing and boiling temperatures of pure water under standard sea-level atmospheric pressure. The distance between these two points is then divided into a convenient number of gradations, called degrees. Three calibration systems are in common use. The Fahrenheit TEMPERATURE SCALE is still popular in the United States and Britain, but it is being replaced by the Celsius temperature scale, which is the one used in the rest of the world. Scientists use the Kelvin scale.

It is sometimes useful to record not only the present temperature, but also the highest and lowest temperatures that have occurred during a given period, for example of 24 hours. A maximum–minimum thermometer makes this possible. It consists of two thermometers mounted side by side.

The maximum thermometer records the highest temperature reached during the time since it was last reset. The thermometer uses mercury in a tube that has a constriction. As the temperature rises, the force with which mercury expands is sufficient to push it past the constriction. As the temperature falls, however, the mercury is unable to pass the constriction and return to the bulb, so it continues to indicate the highest temperature attained. Shaking the thermometer to jerk the mercury through the constriction and back into the bulb resets it.

A minimum thermometer records the lowest temperature reached since it was last reset. The thermometer contains a fluid with a low density, most commonly colored alcohol. Inside the thermometer tube there is a small strip of metal, often in the shape of a dumbbell, called the "index." When the temperature falls, the liquid contracts toward the bulb. As the upper surface of the liquid reaches the top of the index, the index is drawn down the tube by SURFACE TENSION. When the temperature rises, the liquid flows past the index, leaving it in the position it reached when it was drawn toward the bulb. The tip of the index farthest from the thermometer bulb therefore registers the lowest temperature attained. To reset the thermometer, it is held vertically with the bulb uppermost. The index then sinks to the top of the liquid.

Alternatively, some minimum thermometers use an iron index that is repositioned by using a small magnet to drag it. Because the index can move along the tube by gravity, the most accurate minimum thermometers must be mounted horizontally.

Thermometers can also measure the rate at which water evaporates and the relative HUMIDITY and DEW point temperature can be calculated from the resulting reading. This also requires two thermometers, one with a dry bulb and the other with a wet bulb. The resulting instrument is known as a psychrometer (see PSYCHROMETRY).

A wet-bulb thermometer is fitted with a layer of wetted cloth around its bulb. The cloth is usually muslin, and it extends below the bulb, where it is immersed in a reservoir of distilled water, so it acts as a wick. Water is drawn into the wick by capillarity and evaporates from it, thus maintaining a constant amount of moisture around the bulb provided the reservoir does not run dry, and there is a free circulation of air around the cloth. The LATENT HEAT of vaporization is taken from the thermometer bulb. This depresses the temperature registered by the thermometer. The difference between the temperature that is registered by a dry-bulb thermometer and that registered by a wet-bulb thermometer adjacent to it is known as the wet-bulb depression. The extent of the wet-bulb depression indi-

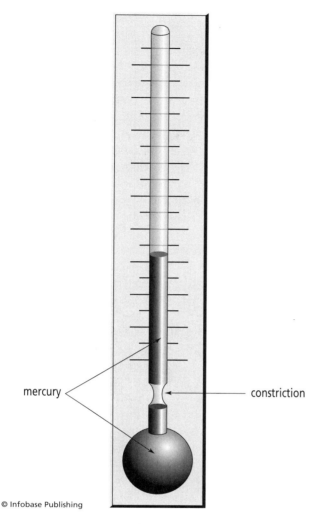

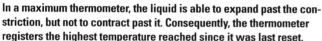

© Infobase Publishing

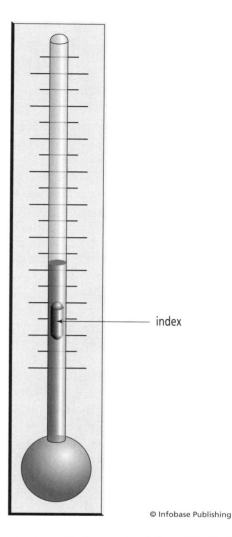

© Infobase Publishing

In a maximum thermometer, the liquid is able to expand past the constriction, but not to contract past it. Consequently, the thermometer registers the highest temperature reached since it was last reset.

In a minimum thermometer, as the temperature falls and the liquid contracts the index is drawn toward the bulb. As the temperature rises, the liquid flows past the index, leaving it in position.

cates the rate of EVAPORATION, which varies according to the atmospheric humidity.

thickness In METEOROLOGY, thickness refers to the difference in altitude between two CONSTANT-PRESSURE SURFACES. This difference varies with the temperature of the air, because the warmer air is the less dense it is, which means that a given mass of warm air occupies a greater volume than a similar mass of cold air. Consequently, a layer of warm air bounded by two constant-pressure surfaces will be thicker than a layer of cold air bounded by the same surfaces. The resulting gradient is responsible for generating the THERMAL WIND.

A line drawn on a map that joins places where the thickness of a given atmospheric layer is the same is called a thickness line, also known as a relative isohypse. The pattern that is made by the thickness lines on a thickness chart (*see* WEATHER MAP) is called the thickness pattern, or relative hypsography.

Thomson effect A water molecule at the surface is more tightly bound in a body of liquid that has a plane (level) surface than it is in a spherical droplet, and the smaller the droplet the easier it is for the molecule to escape. The effect was first described mathematically by the Scottish physicist William Thomson (1824–1907), who later became Lord Kelvin. (It should not be confused with the Thomson effect in thermodynamics, which was also discovered by Lord Kelvin.)

The equation describing the Thomson effect is:

$$\rho_r / \rho s = \exp (A/rT)$$

where ρ_r is the equilibrium vapor density (*see* BOUNDARY LAYER), ρ_s is the density at SATURATION of the layer adjacent to it, A is a constant for the liquid, r is the radius of curvature of the droplet, T is the absolute temperature (*see* TEMPERATURE SCALES), and exp indicates that the relationship is EXPONENTIAL.

The forces that bind water molecules are exerted in all directions. Molecules at a plane surface are attracted from the sides and from below, but there are no molecules to attract them from above. A molecule at a curved surface is also attracted from below and from the sides, but because of the curvature molecules to the sides are also a little below it, and so the lateral attraction is reduced. This means it is easier for a molecule to escape into the air from a curved surface than from a plane surface. The smaller the radius of curvature, the greater the reduction in the lateral attraction.

The ratio of ρ_r to ρ_s increases as the radius of curvature (r) decreases and the droplet becomes smaller. Consequently, the smaller the droplet, the more water molecules that will escape from the equilibrium vapor around it into the air beyond and droplets can survive only if ρ_s increases until it is greater than ρ_r. In other words, the air surrounding the droplet must be supersaturated.

Water molecules have a definite size, which limits the minimum size it is possible for a droplet to be. When this is taken into account, it is found that spontaneous nucleation of liquid droplets (*see* CONDENSATION) can occur only when the relative HUMIDITY (RH) reaches about 300 percent. Air is never this humid. The fact that water vapor is able to condense so readily

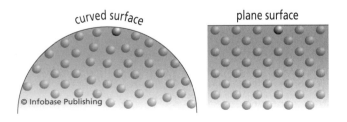

A molecule at the surface of a spherical droplet is bound to the liquid less strongly than a molecule at a plane surface.

demonstrates the importance of CLOUD CONDENSATION NUCLEI to the process of condensation.

The exponential nature of the relationship between ρ_r and ρ_s is critical. As soon as r begins to increase, by only the smallest amount, condensation accelerates rapidly. Once a droplet has a radius of 0.15 μm (0.000006 inch) an RH of 101 percent is sufficient for it to grow.

The Thomson effect can be modified by RAOULT'S LAW when the droplet is not of pure water, but is a solution. Where a mass (m) of a solute is dissolved in a droplet of water, the overall effect is modified by a constant (B) that is determined by the composition of the solute. The equation is then:

$$\rho_r / \rho_s = 1 + (A/rT) - (Bm/r^3)$$

Thornthwaite climate classification The Thornthwaite climate classification is a scheme for classifying climates that was devised by C. W. Thornthwaite (*see* APPENDIX I: BIOGRAPHICAL ENTRIES). The first version of the scheme, published in 1931, applied only to North America, but in subsequent years it was expanded to cover the entire world.

Like the KÖPPEN CLIMATE CLASSIFICATION, the Thornthwaite classification is generic, in that it uses quantitative criteria of TEMPERATURE and PRECIPITATION to define the boundaries of climatic types. It differs from the Köppen classification in its use of the concepts of precipitation efficiency and thermal efficiency, but its most important contribution came in the 1948 revision of the scheme, in which Thornthwaite introduced the concept of potential EVAPOTRANSPIRATION and a moisture index.

In the Thornthwaite scheme, precipitation efficiency (PE) is calculated as the sum of the ratios of precipitation (P) to evaporation (E)—the amount of water that evaporates from an exposed water surface in the course of one month—for each month through the year. Evaporation varies according to the temperature, so temperature is included in the calculation. Precipitation efficiency is equal to $115(r/t - 10)^{10/9}$, where r is the mean monthly rainfall in inches and t is the mean monthly temperature in °F. This calculation is made for each month and the sum of the indexes for 12 months gives the precipitation-efficiency index (P-E index), which is a value indicating the amount of water that is available for plant growth through the year.

The moisture index (Im) is a value that is calculated to show the monthly surplus or deficit of water in the soil. It is given by: $Im = 100(S - D)/PE$, where S is the monthly water surplus (see SOIL MOISTURE), D the monthly water deficit, and PE is the potential evapotranspiration. It can also be calculated from: $Im = 100(r/PE - 1)$, where r is the annual precipitation.

Thornthwaite also introduced the humidity index, which is a measure of the extent to which the amount of water available to plants exceeds the amount needed for healthy growth. It is calculated as $100W_s/PE$, where W_s is the water surplus and PE is the potential evapotranspiration.

A humidity province is one of the five categories into which climates are divided on the basis of their precipitation efficiency index value. The five provinces are labeled A, B, C, D, and E. Province A, with a P-E index greater than 127 is the rain forest climate. B, with a P-E index of 64–127, is the forest climate. C, with a P-E index of 32–63, is the grasslands climate. D, with a P-E index of 16–31, is the steppe climate. E, with a P-E index of less than 16, is the desert climate.

Thermal efficiency similarly relates temperature (T) to evaporation to yield a thermal-efficiency index (T-E index). Because temperature and evaporation are so closely linked, the T-E value is shown in the table of climate types as the potential evapotranspiration. The thermal-efficiency index indicates the amount of energy, as heat, that is available for plant growth in the course of a year. It is calculated from measurements of the amount by which the mean temperature in each month is above or below freezing. For each month the thermal efficiency is $(t - 32)/4$, and the thermal-efficiency index is the sum of the thermal efficiencies for each month through the year. A value of 0 indicates what is called a frost climate, and a value of more than 127 indicates a tropical climate.

From these calculations, Thornthwaite recognized nine moisture provinces based on the P-E index, and 9 temperature provinces based on the T-E index. These are given names and also designated by letters and numerals. The moisture provinces are related to vegetation and correspond to rain forest (A), four types of forest (B), two types of grassland (C), steppe (D), and desert (E).

In addition, the classification adds code letters that qualify these main categories by referring to the amount and distribution of precipitation associated with them. This brings to 32 the total number of climate types recognized in the scheme. These additional qualifications are based on an aridity index for moist climates and a humidity index for dry climates.

Moisture Provinces

Climate type		Moisture index
A	perhumid	100 or more
B_4	humid	80–100
B_3	humid	60–80
B_2	humid	40–60
B_1	humid	20–40
C_2	moist subhumid	0–20
C_1	dry subhumid	-20–0
D	semi-arid	-40–20
E	arid	-60–40

Temperature Provinces

Climate type		Potential evapotranspiration inches	centimeters
E′	frost	5.61	14.2
D′	tundra	11.22	28.5
C'_1	microthermal	16.83	42.7
C'_2	microthermal	22.44	57.0
B'_1	mesothermal	28.05	71.2
B'_2	mesothermal	33.66	85.5
B'_3	mesothermal	39.27	99.7
B'_4	mesothermal	44.88	114.0
A′	megathermal	>44.9	>114

Moist climates (A, B, C_2)	Aridity index
r little or no water deficiency	0–16.7
s moderate water deficiency in summer	16.7–33.3
w moderate water deficiency in winter	16.7–33.3
s_2 large water deficiency in summer	more than 33.3
w_2 large water deficiency in winter	more than 33.3

Dry climates (C_1, D, E)	Humidity index
d little or no water surplus	0–10
s moderate water surplus in winter	10–20
w moderate water surplus in summer	10–20
s_2 large water surplus in winter	more than 20
w_2 large water surplus in summer	more than 20

The moisture provinces, temperature provinces, and moist and dry types of climate are shown in the tables above.

THORPEX The Observing System Research and Predictability Experiment (THORPEX) is the successor to the Global Atmospheric Research Programme (GARP). It was established in May 2003 by the 14th World Meteorological Congress under the auspices of the World Meteorological Organization (WMO) Commission for Atmospheric Sciences. It is a component of the WMO World Weather Research Programme (WWRP).

THORPEX is planned to run for 10 years, until 2013. During this time it will conduct experiments aimed at improving short-range (up to 3 days) and medium-range (3–10 days) deterministic and probabilistic forecasts of extreme weather in the Northern Hemisphere. The THORPEX projects will study the ways in which satellite and surface observations acquire data and how those data are assimilated. The overall objective is to ensure that weather forecasts are timely and accurate, and that they are translated into specific information that supports practical decisions to reduce the loss of life, injuries, and property losses associated with extreme weather events such as tropical cyclones, blizzards, floods, and droughts.

Further Reading

World Meteorological Organization. "THORPEX, A World Weather Research Programme." WMO. Available online. URL: www.wmo.int/thorpex/about.html. Accessed February 17, 2006.

thunder Thunder is the sound caused by the discharge of energy during a lightning flash. As the flash moves along its lightning channel, it raises the temperature of the ionized (see ion) air inside the channel by up to 54,000°F (30,000°C) in less than one second. This causes the air to expand violently, increasing the pressure inside the channel to as much as 100 times its normal value. The expansion is so rapid as to be explosive, and it emits shock waves that immediately become the sound waves people hear as thunder.

Sound waves travel at about 670 MPH (1,080 km/h), which is the speed of sound in air. Light from the flash travels at the speed of light, which is approximately 1 million times faster. Consequently, a distant observer sees the lightning flash before hearing the clap of thunder associated with it. By counting the time that elapses between seeing the lightning and hearing the thunder, it is possible to calculate the approximate distance to the storm: every five seconds represents a distance of about one mile (three seconds per kilometer).

As they travel, the sound waves are damped by the air. Those with a short wavelength (see wave characteristics), carrying sounds of a high pitch, disappear before those with a long wavelength, carrying the low pitches. The greater the distance from the lightning, therefore, the deeper the note of the thunder. Very distant thunder is heard as a deep rumble.

The sound often continues for several seconds. That is because of the length of the lightning flash that causes it. A lightning flash can be more than a mile (1.6 km) long, and its forked shape means some parts of it are closer to the observer than others. Sound from the nearest part of the flash reaches the observer first, and sound from the more distant parts arrives later, extending the duration of the sound.

Thunder can seldom be heard from more than about 6 miles (10 km) away, because by then all the sound waves have been either absorbed by the air or refracted upward by the effect of the decrease in air temperature with height. Lightning can be seen over a much greater distance, especially at night. Silent flashes of sheet lightning are most often seen on warm summer nights and are sometimes called "heat lightning."

The cause of thunder puzzled people for centuries. Germanic peoples believed it was the sound of the god Thor, either beating his huge anvil or throwing his hammer at other gods. The word *thunder* is from the Old Norse *thórr*. According to a poem called *De Rerum Natura* ("On the Nature of Things") written in about 55 B.C.E. by the Latin poet Lucretius (his dates of birth and death are not known, but his full name was Titus Lucretius Carus) thunder is the sound of great clouds crashing together. Native Americans thought it was the sound of huge thunderbirds flapping their wings.

thunderbolt A thunderbolt is a rock, piece of metal, or dart that storm gods such as Zeus (Jupiter), Yahweh (Jehovah), and Thor were once believed to hurl at the Earth. As they traveled through the air, thunderbolts could be seen as lightning, and they were believed to cause the damage that occurs when lightning strikes an object or person.

When lightning strikes sand, the sudden discharge of energy is often sufficient to melt the grains, producing an irregular mass of glass coated in sand grains called a fulgurite. Fulgurites are usually about half an inch (1 cm) in diameter, with side branches. Most are less then 10 feet (3 m) long, but they can be more than 60 feet (18 m) long. A piece of such glass was assumed to be a thunderbolt.

In ancient Greece and Rome the area around a lightning strike was fenced off and considered sacred, and persons killed by lightning were buried where they died, rather than in the usual burial ground. It was not until 1752 that the true, electrical nature of lightning was proved by Benjamin Franklin (*see* APPENDIX I: BIOGRAPHICAL ENTRIES) and, independently, by the French scientist Thomas-François d'Alibard (1703–99) and the link between storm gods and so-called thunderbolts was shown to be entirely mythical.

thunderstorm   A thunderstorm is a violent storm that causes THUNDER and LIGHTNING as well as heavy PRECIPITATION and strong GUSTS of wind. Every day more than 16 million thunderstorms occur in the world as a whole, and about 2,000 are taking place at any moment. About 100,000 thunderstorms occur in the United States every year. A day on which a thunderstorm is observed at a WEATHER STATION is known as a thunderstorm day.

Development of a thunderstorm begins with the growth of cumulus clouds (*see* CLOUD TYPES). Inside such clouds warm air is rising by CONVECTION. As it rises, the air cools adiabatically (*see* ADIABAT) and some of its WATER VAPOR condenses. This releases LATENT HEAT of CONDENSATION, warming the surrounding air and causing it to continue rising.

The cloud builds rapidly, with updrafts traveling at up to 100 MPH (160 km/h). The very strong upcurrent found in a rapidly developing cumulus cloud is called an uprush or a vertical jet. A cloud with an uprush is likely to grow into a cumulonimbus and produce a thunderstorm.

As the upper part of the cloud grows past the FREEZING LEVEL, the Bergeron–Findeisen mechanism (*see* RAINDROPS) causes precipitation to commence. The precipitation may fall as HAIL, the size of the hailstones indicating the vertical extent of the cloud and the force of its upcurrents. There will also be RAIN or SNOW depending on the temperature in the lower part of the cloud and in the air between the cloud base and the surface. The precipitation, in whatever form, is carried in the downdraft through the base of the cloud.

An isolated severe thunderstorm in central Oklahoma. The core of the storm, containing the main updraft, is in the background. The cloud in the foreground is the incus, or storm anvil. *(NOAA Photo Library, NOAA Central Library; OAR/ERL/National Severe Storms Laboratory [NSSL])*

By this stage the cloud has become a cumulonimbus, and it extends vertically to a height of 50,000 feet (15 km) or more. Upper-level winds draw out the ICE CRYSTALS at its top into an anvil (incus) shape. The storm is then in its most active stage.

A small area of low AIR PRESSURE found to the rear of a fully developed thunderstorm is called a wake low. It develops because the cold downdrafts at the leading edge of the advancing thunderstorm produce a local area of high pressure, and air is drawn into the rear of the storm to compensate, producing low surface pressure where the air begins to rise. Wake lows form behind the main cloud mass and are associated with clearing skies and the end of the precipitation. They are common behind SQUALL line storms.

The storm derives its energy from the latent heat of condensation, so it requires a constant inflow of warm, moist air to sustain it. Once precipitation commences, the falling raindrops, ice crystals, and hailstones drag with them small envelopes of chilled air. These accumulate and form downdrafts, carrying cold precipitation and cold air out of the base of the cloud. Some of the precipitation evaporates as it falls, drawing the latent heat of vaporization from the surrounding air and cooling it further. The location in a thunderstorm where air that has been cooled by the EVAPORATION of moisture meets warm, moist air is called the outflow boundary.

As the precipitation intensifies, so do the downdrafts. These fall into the updrafts and eventually come to dominate the cloud and suppress the updrafts by cooling the rising air. The storm is then deprived of energy and the cloud is deprived of moist air. The precipitation becomes lighter and the cloud begins to dissipate.

It is while it is in its most active stage that the storm produces lightning. This occurs because electrical charge becomes separated inside the cloud. Positive charge accumulates in the upper regions and negative charge in the lower regions. The negative charge near the base of the cloud induces a positive charge on the ground beneath the cloud. Lightning consists of electrical discharges between the separated charges. This neutralizes them, but only temporarily in an active storm.

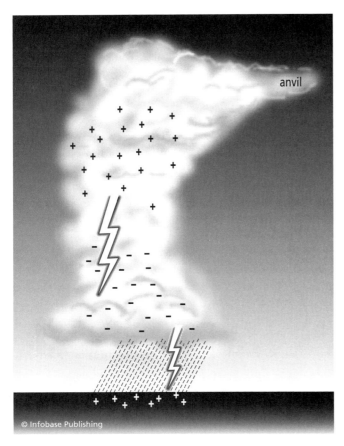

Electric charge has become separated inside the cumulonimbus cloud. Positive charge accumulates in the upper part of the cloud, negative charge in the lower part, and the negative charge induces a positive charge on the ground surface beneath the cloud. Lightning is the spark that partly neutralizes the charge distribution. Near the tropopause, the top of the cloud is swept into an anvil shape by the wind.

It takes no more than about 20 seconds for the charges to separate again.

Scientists are not certain what it is that causes charge separation, but it is known that the small particles or droplets acquire positive charge and the large ones acquire negative charge. It is gravity that separates them, therefore, as the heavy ones sink to the bottom and the light ones are carried to the top. When a droplet freezes, positive ions migrate toward the colder regions, leaving the less mobile negative ions in the warmer regions. Liquid droplets freeze from the outside in, so a shell of positive charge surrounds a core of negative charge. As the core starts to freeze, it expands, shattering the shell into minute fragments. The fragments carry their positive charge upward, and the heavier core carries its negative charge downward.

It may also happen that small ice splinters collide with hailstones the outsides of which have been warmed by the release of latent heat as supercooled water (*see* SUPERCOOLING) freezes onto them. At each collision positive ions move toward the colder end of the splinter. This increases the negative charge both at the warmer end of the splinter and on the hailstone. The charge changes by only a very small amount at each collision, but the very large number of collisions means there is a big cumulative effect. The splinters, with their positive charge, are carried upward and the hailstones, with their negative charge, drift downward.

There may also be an upward flow of ions carrying positive charge from tall objects such as trees and buildings. This is called a point discharge, and it is induced by the negative charge at the base of a cloud that is producing a thunderstorm. Point discharge is the more important of the two processes that produce a negative charge at the surface of the Earth (the other is lightning) and in this way replenish the charge lost as positive ions are conducted downward through the air from the ionosphere (*see* ATMOSPHERIC STRUCTURE). In the absence of point discharges and, to a lesser extent, lightning strokes, within about 15 minutes the surface would acquire a positive charge equal to that in the ionosphere and the Earth's electrical field would break down. Occasionally a point discharge is strong enough to be visible as a glow, known as Saint Elmo's fire, around the structure from which it flows. This is sometimes seen near the top of a ship's mast.

An AIR MASS thunderstorm results from convection in unstable air (*see* STABILITY OF AIR). The cumulonimbus cloud producing the storm grows vertically, and probably lacks an anvil because there is no change in wind speed with height. The storm is produced entirely by conditions pertaining to the air mass. This type of storm is frequent in the TROPICS. It is also the type of storm that occurs in middle latitudes late on summer afternoons when the air is humid and the ground has been strongly heated, making the air above it unstable. This type of thunderstorm is sometimes called a heat thunderstorm.

An advective thunderstorm is one triggered by the ADVECTION of warm air across a cold surface, or of cold air above a layer of warmer air at a high level. Warm air is cooled when it crosses a cold surface. This causes its water vapor to condense, releasing latent heat, which may make the air sufficiently unstable to produce cumulonimbus cloud. Cold air moving above warm air may also cause instability as it sinks beneath the warmer air, thus raising it and causing the warm air to cool adiabatically (*see* ADIABAT) and its water vapor to condense.

A frontal thunderstorm is one that develops in warm, moist air that has been made unstable by frontal lifting. Such storms can be violent and where the cold FRONT is advancing very fast and pushing vigorously beneath the warm front, the rapid lifting along a substantial frontal length can produce a squall line. A surge line is a line just ahead of a local group of thunderstorms where there is a sudden change in the wind speed and direction. A cold front thunderstorm is produced on a cold front. As cold air pushes beneath warm air, or warm air rises over cold air, moist air in the warm sector may become sufficiently unstable to generate cumulonimbus clouds that are vigorous enough to cause storms.

Tibetan high An ANTICYCLONE, called the Tibetan high, develops in summer over the Tibetan Plateau. In early summer the ground warms strongly. Air rises by CONVECTION, producing a shallow layer of low pressure near the ground and high pressure at heights above the 500-millibar level. The anticyclonic flow (*see* ANTICYCLONE) is from the east on the southern side of the anticyclone. This contributes to the breakdown of the westerly JET STREAM. The change from a midlatitude westerly flow to an easterly flow and the disappearance of the jet stream are linked to the onset of the MONSOON over southern Asia and the mai-u rains over China (*see* LOCAL CLIMATES).

tides Tides are the regular movements of surface waters, the atmosphere, and the solid Earth that are caused by the gravitational attraction of the Moon and, to a lesser extent, of the Sun. This attraction produces two bulges at the surface. The bulges move around the Earth in step with the ORBIT of the Moon, producing two tidal cycles in every 24-hour period.

Tidal forces affect the liquid core of the Earth, and the resulting Earth tides can be measured by the bulges they produce in surface rocks. These are small, however, the greatest tidal movement nowhere exceeding about 3 feet (1 m). ATMOSPHERIC TIDES are synchronized to the daily solar cycle, rather than the lunar cycle. It is in the oceans that the tides are most clearly seen.

All the parts of the Earth, including the atmosphere and oceans, are drawn toward the center of the Earth by the gravitational force. Because the Earth is rotating, the gravitational force balances the inertial tendency of a moving body to continue to move in a straight line (*see* CENTRIPETAL ACCELERATION).

Both the Moon and Sun also exert a gravitational force. Their effect is much smaller than that of the Earth's own gravitational force, because gravity is subject to the INVERSE SQUARE LAW, which means that magnitude decreases rapidly with increasing distance. Lunar and solar gravity act in the opposite direction to terrestrial gravity and therefore reduce its magnitude. This effect alters the balance between the terrestrial

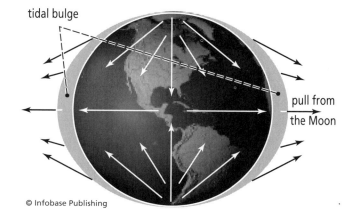

© Infobase Publishing

Lunar gravity reduces the magnitude of terrestrial gravity. Because the earth is rotating, this increases the tendency of every part of the Earth to continue moving in a straight line and fly away into space. The result is two bulges, produced where terrestrial and lunar gravity pull in opposite directions. One bulge is directly beneath the Moon, and the other is on the opposite side of the Earth.

gravitational force and the inertial tendency of each part of the Earth, producing two bulges in the oceans. One bulge lies on the surface of the Earth that is directly beneath the Moon. The lunar attraction acts in a straight line, reducing the magnitude of terrestrial gravity everywhere along that line. Consequently, the other bulge lies on the side of the Earth directly opposite the Moon. As the Moon orbits the Earth, the two bulges follow it.

The Moon takes 24 hours and 50 minutes to orbit the Earth, so high and low tides occur at intervals of 12 hours 25 minutes, which is why the times of the tides change from day to day. The declination (see PLANE OF THE ECLIPTIC) of the Moon changes during the month and is sometimes as much as 28.5°. This means that the lunar gravitational attraction acts at an angle to the equator, and therefore the tidal bulge is also at an angle to the equator. At any location on the surface of the Earth, therefore, one tide will rise higher than the other tide, and the two tides will have the same amplitude only when the Moon is directly above the equator.

The time of slack water, when the tide is about to turn, is called the stand of the tide. During this short period the height of the tide does not change, and tidal currents slow down and then cease before starting to flow again in the opposite direction.

The gravitational force exerted by the Sun is about 47 percent of the force exerted by the Moon. This is because, despite being much more massive than the Moon, the Sun is also much more distant. That is why the lunar influence on tides predominates.

The influence of the Sun is felt at times of spring and neap tides. Spring tides occur when the Earth, Sun, and Moon are aligned, so the gravitational influence of the Sun is added to that of the Moon. The approximate alignment of the Earth, Sun, and Moon, or of the Earth, Sun, and another planet is called syzygy. It is the Earth–Sun–Moon syzygy that causes spring tides.

Spring tides rise to a higher level than average and ebb to a lower level. The height of spring tides varies according to the accuracy of the Earth–Moon–Sun alignment; the more closely the three bodies are aligned, the higher the spring tides will be. Neap tides occur when lines drawn from the Sun and Moon meet in a right angle at the center of the Earth. The solar and lunar gravitational forces then act partly against each other, reducing the overall tidal effect. Neap tides are smaller than the average tides. As with the spring tides, their height varies according to the accuracy of the misalignment.

Although the two tidal bulges circle the Earth like waves, their magnitude and timing vary from place to place because of FRICTION between the waves and the ocean floor and because of the shape and orientation of coastlines. The amplitude of the waves (see WAVE CHARACTERISTICS) also varies with the volume of water through which it moves. Ocean tides are much larger than the tides in seas that are almost completely enclosed by land, such as the Mediterranean and Baltic Seas. The difference between the mean height of high and low water at a particular place is known as the tidal range.

The tidal range varies according to the phase of the Moon. Tidal range is greatest during spring tides and least during neap tides. Mean tidal ranges take account of these cyclical variations to provide a general value. Tidal range also varies according to the configuration of coastlines. Tides propagate as waves with a PERIOD similar to that of the forces generating the tides. When these sea waves arrive at coastlines or enter bays and estuaries, they may be reflected. The water may then form a standing wave, or seiche, with a period determined by the length and depth of the basin that contains it. If this period coincides with that of the tides, the amplitude of the tide can be increased greatly. That coincidence is the cause of the huge tidal range in the Bay of Fundy, in eastern Canada. The bay is about 168 miles (270 km) long, and its depth averages 230 feet (70 m). These dimensions mean the standing wave produced by the tide has a period of 12 hours, with the result that the tidal range at spring tides can exceed 50 feet (15 m). Elsewhere, tidal ranges are smaller. The mean tidal range at Boston, Massachusetts, is about 9 feet (2.7 m), for example. A tidal range of less than 6.5 ft (2 m) is called microtidal, one of 6.5–13 ft (2–4 m) is mesotidal, and one of more than 20 ft (6 m) is macrotidal.

Oceanic tidal movements play an important part in the transport of heat from the equator to polar regions. Tidal energy is dissipated in the deep oceans, and its dissipation causes some mixing of ocean waters. Some scientists now suspect that this mixing, combined with the wind-driven ocean currents, is more important than the THERMOHALINE CIRCULATION in the oceanic transport of heat.

Tides are also thought to be important in climate change. The declination of the Moon changes over a cycle that maximizes the tides every 1,800 years, and the amplitude of the cycle also varies over a cycle with a period of 5,000 years. Strong tides increase the amount of mixing in the oceans. Mixing with cold, deep water lowers the temperature of the surface water. Weak tides have the opposite effect. These cyclical variations coincide with, and may cause, abrupt fluctuations in climate that occur on much smaller timescales than the change between ice ages and INTERGLACIALS.

Very gradually, the tides are weakening. This is because the Moon is receding from the Earth by about 1.6 inches (4 cm) every year.

Titan atmosphere   Titan is one of the moons of the planet Saturn, also known as Saturn IV. Its radius is 1,600 miles (2,575 km), mass 1.48×10^{20} tons (1,345.5 $\times 10^{20}$ kg), and mean density 117 pounds per cubic foot (1,881 kg/m^3). Titan is larger than Earth's moon, but much less dense. Titan was discovered in 1655, by the Dutch astronomer and physicist Christaan Huygens (1629–1695).

In 2004, the Cassini–Huygens space mission reached Titan, and on January 14, 2005, the Huygens lander descended through the Titanian atmosphere and landed on the surface. Huygens returned data and images during its descent and from the surface itself.

Titan has a thick atmosphere. The atmospheric pressure at the surface is 1.5 times that at Earth's surface, but the climate is much colder, with a mean surface temperature of -300 °F (90 K; -183°C). The northern hemisphere was found to be colder than the

An artist's impression of Saturn as it might appear to an observer on its satellite Titan. Titan's surface is made from rock and ice, and it has a thick atmosphere of nitrogen and organic compounds, mainly methane and ethane. Its atmosphere gives Titan its orange color. Scientists are interested in the atmosphere of Titan because it may be similar to the primitive atmosphere on Earth and contain the compounds from which life originated. *(Chris Butler/Photo Researchers Ltd)*

southern hemisphere, but this may have been because Huygens landed during the northern-hemisphere winter.

The atmosphere consists mainly of NITROGEN, together with METHANE, ethane, and hydrogen cyanide. ULTRAVIOLET RADIATION and impacts by high-energy electrons dissociate atmospheric methane and drive chemical reactions that yield a range of other organic (carbon-based) compounds, principally acetylene, ethylene, ethane, and diacetylene. The methane must be constantly replenished, but its source is still unknown.

It is likely that methane condenses in the atmosphere and falls to the surface as rain. Every few centuries it is possible that Titan experiences a "methane monsoon" lasting for several months. During this time the PRECIPITATION would be intense.

Titan has a well-defined stratosphere and mesosphere. The tropopause is at about 30 miles (50 km), and the stratopause at about 186 miles (300 km).

Further Reading

Flasar, F. M. et al. "Titan's Atmospheric Temperatures, Winds, and Composition." *Science,* 308, 975–978. May 13, 2005.

Schilling, Govert. "Volcanoes, Monsoons Shape Titan's Surface." *Science,* 309, 1985. September 23, 2005.

Tonian The Tonian was the earliest period of the NEOPROTEROZOIC era of the Earth's history. It began 1,000 million years ago and ended 850 million years ago. During the Tonian, the continents were probably joined together in the supercontinent Rodinia. The equator passed through the center of Rodinia, with East Antarctica, Siberia, Australia, and India in the Northern Hemisphere, and Congo, Amazonia, West Africa, and northwestern Europe (called Baltica) in the Southern Hemisphere. North America and the rest of Eurasia, together known as Laurentia, formed the central part of the supercontinent. Rodinia began to break apart about 900 million years ago, toward the end of the Tonian.

Single-celled aquatic organisms were the principal forms of life during the Tonian. Multicellular organisms may have appeared toward the end of the period. *See* APPENDIX V; GEOLOGIC TIMESCALE.

Topex/Poseidon A joint mission by NASA and the Centre National d'Études Spatiales (CNES) in France that used an ALTIMETER mounted on a satellite to measure the height of the surface of the ocean with unprec-

edented accuracy. The satellite, Topex, was launched from Kourou, French Guiana, in August 1992 on a Poseidon rocket. Topex orbited at a height of 830 miles (1,336 km) on a track with an INCLINATION of 66° and an orbital period of 112 minutes. It carried two altimeters, one built by NASA and the other by CNES. Both instruments measured the height of the ocean at the same points at intervals of 10 days.

The resulting data was distributed to scientists in nine countries, who used them in connection with other research programs investigating the global climate. The measurements showed changes caused by meanders in strong ocean currents, local vortices, and large eddies associated with BOUNDARY CURRENTS and subtropical GYRES. The data helped in forecasting and tracking TROPICAL CYCLONES, forecasting ENSO events, and in other aspects of ocean and climate research of value to offshore industries, shipping, fisheries management, and marine mammal research.

In January 2006, after completing 62,000 orbits of Earth, the Topex spacecraft lost its ability to maneuver and its mission came to an end.

Further Reading

University of Colorado. "Topex/Poseidon home page." University of Colorado. Available online. URL: http://ccar.colorado.edu/research/topex/html/topex.html. Accessed February 20, 2006.

University of Texas. "Topex/Poseidon Educational Outreach." University of Texas. Available online. URL: www.tsgc.utexas.edu/topex/. Last modified January 29, 2001.

tornado (1) The most violent of all weather phenomena, a tornado develops inside a huge cumulonimbus storm cloud (*see* CLOUD TYPES), extends from its base, and produces winds that in extreme cases can reach 300 MPH (480 km/h). A storm that is capable of generating tornadoes is described as tornadic. Tornadic storms are sometimes isolated, but they are more likely to occur along a SQUALL line.

The storm cloud extends from below 1,000 feet (300 m) all the way to the tropopause (*see* ATMOSPHERIC STRUCTURE), at an average height of 50,000 feet (15.25 km), or even higher, penetrating the lower stratosphere. Inside the cloud the air is extremely unstable (*see* STABILITY OF AIR). As rising air cools, its WATER VAPOR starts to condense, releasing LATENT HEAT. The latent heat warms the air and makes it continue rising.

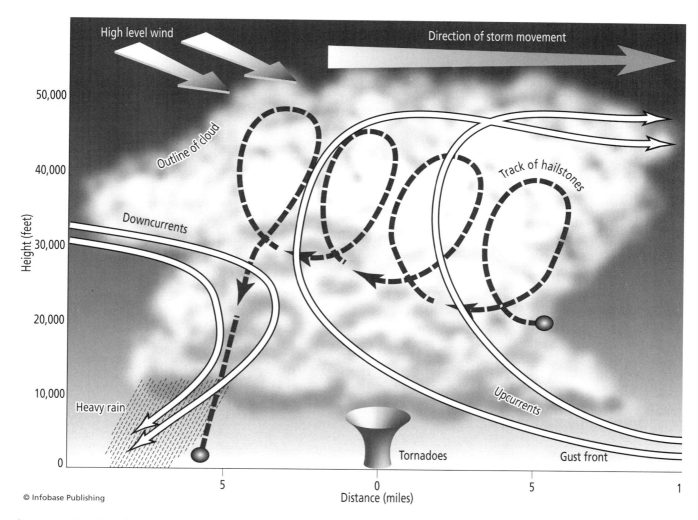

A cross section through a storm cloud that is producing a tornado. Ahead of the cloud there is a gust front, with strong winds, and at the rear of the cloud, behind the tornado, there are hail and heavy rain. The storm is driven by fierce air currents.

At the top of the cloud, rising air cannot penetrate far into the stable air of the stratosphere, and it spreads horizontally, forming an extension shaped like a blacksmith's anvil. The more vigorous the vertical air currents in the cloud, the bigger the anvil will be. The anvil extends at an angle of about 45° to the direction in which the storm is moving. In the Northern Hemisphere, the anvil extends to the left of the direction of motion as seen from above; in the Southern Hemisphere, it extends to the right.

Near the top of the cloud the air is very cold. The cold air sinks, warming adiabatically (*see* ADIABAT) and collecting moisture as it does so. This convective movement forms a number of cells (*see* CONVECTION). Ordinarily, the downcurrents in one cell interfere with the upcurrents in the neighboring cell, eventually suppressing them. When this happens, the storm dies. If the anvil is big enough, however, the upcurrents are swept clear of the downcurrents and a SUPERCELL forms, greatly extending the lifetime of the cloud.

Hemispherical protrusions called mammatus often form on the underside of the anvil of isolated tornadic storm clouds, but rarely on those along squall lines. At this stage of the storm's development, air is being drawn into the base of the cloud along its leading edge, so that strong GUSTS of wind blow toward the cloud as it approaches. This is the gust front. Air rises to the top and spreads into the anvil. Behind the anvil, equally strong downcurrents, producing first HAIL and then torrential rain as the storm passes, emerge as winds of up to hurricane force (more than 75 MPH, 120 km/h) from the rear of the cloud.

Small tornadoes sometimes form in the gust front. A tornado of this kind is called a gustnado. Because they spin in diverging air (*see* STREAMLINE), gustnadoes often rotate anticyclonically (clockwise in the Northern Hemisphere), unlike most tornadoes, which rotate cyclonically (counterclockwise in the Northern Hemisphere).

WIND SHEAR deflects the upcurrent at about the mid-height of the cloud. This causes the rising air to rotate about its own axis (*see* VORTICITY), creating a VORTEX with very low pressure at its center. The rotating air constitutes a MESOCYCLONE, and its rotation begins to extend downward through the cloud. As the mesocyclone grows downward, so its diameter decreases. The conservation of its angular MOMENTUM causes the wind speed to accelerate around the vortex.

Some tornadoes develop in relatively weak cumulonimbus clouds that contain no mesocyclone or supercell, and in which air is not rising vigorously. These are known as landspouts or nonsupercell tornadoes. Landspouts have been observed and the lack of a mesocyclone confirmed by Doppler RADAR. Wind shear inside the cloud sets the air rotating, but without developing into a mesocyclone.

Fragments at the base of the cloud start to rotate. Then part of the cloud base—in fact the bottom part of the mesocyclone where a mesocyclone exists—descends below the main cloud base, rotating slowly, to become what is known as the wall cloud. A funnel cloud, consisting of a rapidly spinning upcurrent, may emerge through the base of the wall cloud. If the funnel touches the ground, it will become a tornado, or twister, which is their popular name.

The funnel consists only of air, and it may be invisible. Most funnel clouds are visible, however, because moisture condenses in the relatively low pressure inside the vortex. The funnel looks as though it is an extension of the cloud; in fact it is a cloud in its own right, produced by condensation in the air being drawn into the updraft. When it touches the ground, the tornado funnel darkens because of the dust, debris, and other material that is swept into it and carried upward, and a cloud of dust and debris forms around its base.

A small column of spinning air may develop, rotating about its own axis and also moving in a circle around the main vortex of the tornado. This is called a suction vortex, and a major tornado may generate two or more suction vortices. Suction vortices are often hidden in the dust cloud that surrounds the base of the tornado, but they are responsible for many of the freakish effects a tornado sometimes produces. A suction vortex has been known to destroy half of a house, but leave the other half unscathed and vanish before it reached the house next door. Suction vortices may spin in either direction, but one that rotates in the same clockwise direction as the main tornado will generate winds up to 50 percent more powerful than those around the core of the main tornado. This is because the angular

Raining rats during a particularly violent storm; from *Der Wunderreiche Überzug unserer Nider-Welt* by Erasmus Francisci, published in 1680. *(Historic NWS Collection)*

Tornadoes can produce freak effects. On May 27, 1896, a tornado at St. Louis, Missouri, hurled this shovel with so much force it penetrated 6 inches (15 cm) into the tree. The picture is from "The New Air World," by Willis Luther Moore, published in 1922. *(Historic NWS Collection)*

tex. The vortex is so short-lived that hardly has it made its scar before it dies down. At one time, some people believed suction scars were the footprints of giants.

Tornadoes can happen anywhere in the world outside the TROPICS and at any time of year, but they are most common in the Great Plains of the United States. There their frequency is greatest between May and September, and they are most likely between 2 P.M. and 8 P.M.

When one develops, there may be more, occasionally many more. A chain of 148 tornadoes that occurred on April 3 and 4, 1974, became known as the Super Outbreak. Those tornadoes were produced by three separate squall lines that developed simultaneously and moved eastward. Together they extended from the southern shore of Lake Michigan to Alabama and at one point from Canada to the Gulf of Mexico. More than 300 people were killed. In 2003, a total of 395 tornadoes touched down in the United States between May 1 and May 10. One of these tornadoes lifted up a farmhouse in Iowa and set it down again more than 30 feet (9 m) away. Two of its three teenage occupants were unharmed, and the third boy suffered scratches from broken glass.

Most tornadoes die before they have traveled very far, but there are exceptions. The Tri-State outbreak in March 1925, so-called because it crossed Missouri, Illinois, and Indiana, included one tornado that traveled 219 miles (352 km) at an average speed of 60 MPH

VELOCITY of the suction vortex is added to the angular velocity of the main tornado and its mesocyclone, and also to the forward speed with which the tornado is moving over the ground. A tornado with winds of 200 MPH (322 km/h) may be surrounded with suction vortices spinning at 300 MPH (483 km/h).

Few suction vortices are more than 100 feet (30 m) in diameter, and some are no more than 10 feet (3 m) across. They are very short-lived. Few last longer than about three minutes, and few survive for long enough to complete even one full orbit of the main tornado.

A suction vortex may leave its mark on open ground in the form of an approximately circular, shallow depression a few feet across called a suction scar. As it passes, the intense updraft of the suction vortex draws in loose dirt that is scattered from the top of the vor-

Tornadoes are not always alone. Here two tornadoes are seen moving together through Elkhart, Indiana, on April 11 (Palm Sunday), 1965. *(Paul Huffman, Historic NWS Collection)*

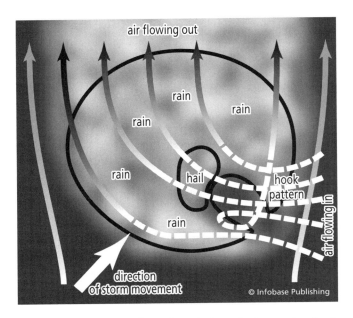

A hook pattern is visible from above in the radar image of a cloud, indicating the imminent risk of tornadoes. The broken lines indicate air entering the storm at low level and climbing. The solid lines indicate air at high level leaving the storm.

(96 km/h). In 1977, a tornado crossed Illinois and Indiana, covering a record 340 miles (547 km) in 7 hours 20 minutes.

Tornadoes are classified by the damage they cause according to the FUJITA TORNADO INTENSITY SCALE. WATERSPOUTS, WHIRLWINDS, DUST devils, and WATER DEVILS are related phenomena.

Tornadic storms can sometimes be identified by a distinctive shape called a hook pattern that is often visible in radar images, taken from directly above, of the clouds associated with a supercell storm. The hook pattern usually occurs at the edge of a mesocyclone, but it is not entirely reliable, since not all mesocyclones produce one and not all mesocyclones lead to tornadoes. Doppler RADAR provides more reliable information on conditions inside a storm cloud.

Further Reading

Allaby, Michael. *Tornadoes*. Rev. ed. Dangerous Weather. New York: Facts On File, 2004.

Bluestein, Howard B. *Tornado Alley*. New York: Oxford University Press, 1999.

tornado (2)   A particular type of violent but brief thunderstorm that occurs in West Africa is also known as a tornado. A West African tornado does not develop a MESOCYCLONE leading to a narrow column of rapidly rotating air that may extend below the storm cloud, and consequently it bears no relation to the "twister" type of storm.

West African tornadoes are associated with the southwesterly MONSOON and air from the harmattan (*see* LOCAL WINDS). Near the coast they develop between March and May and in October and November, and they occur inland between May and September.

The storms lie along a SQUALL line, from 10 miles (16 km) sometimes to 200 miles (320 km) long in Nigeria. They travel from east to west at about 30 MPH (50 km/h) and produce huge, dark clouds, frequent LIGHTNING, DUST and winds of up to 80 MPH (130 km/h) inland, but only half of that near the coast, and torrential rain once the squalls have passed. A tornado may last as little as 15 minutes and rarely for longer than two hours.

Tornado Alley   The area of the Great Plains, in the United States, where tornadoes occur more frequently than they do anywhere else in the world and where the most violent TORNADO outbreaks are experienced. Tornado Alley is centered on Texas, Oklahoma, and Nebraska, but the area also covers Kansas, Iowa, Arkansas, Missouri, Alabama, and Mississippi, and tornadoes are also fairly frequent in northern Florida. In all these states there is an average of five tornadoes every year.

This region suffers more than any other because of its geography and the AIR MASSES that affect it. When it reaches the North American coast, air that has crossed the North Pacific is cool and moist. This maritime air rises to cross the Rocky Mountains, losing much of its moisture as it does so. On the eastern side of the mountains it descends gently, warming slightly by compression, and advances slowly across the Great Plains behind a weak cold FRONT that is aligned approximately southwest to northeast.

At the same time, continental tropical air forms over Mexico, New Mexico, and Texas and moves northward. This air is warm and dry, and in spring and summer, when the land warms rapidly, it is heated further as it advances. It enters the plain as hot, dry air.

A third mass of air forms over the Gulf of Mexico. This air is warm and moist, and it moves in a northwesterly direction.

When the two air masses, from Mexico and the Gulf, meet over northern Mexico and the southern United States, the moist air from the Gulf is held beneath the less dense continental air. The two move northwestward, the lower air warming still more as it crosses the hot land surface. Small clouds form, but although the air is being warmed strongly from below, the warm air cannot rise very far by CONVECTION, because of the overlying layer of dry air, and the clouds bring very little rain.

On the western side of the Great Plains the two air masses meet the weak cold front advancing from the opposite direction. The cold air undercuts the warm air. There is now a "sandwich" of air, with dry air at the bottom and top and warm, moist air between them. The moist air is forced to rise as the cold air pushes beneath it. As it does so, it expands, cools, and its WATER VAPOR starts to condense. At the same time, its rate of cooling changes from the dry to the saturated LAPSE RATE. The air becomes very unstable (*see* STABILITY OF AIR) as it rises up the cold front. Eventually convection within it becomes so vigorous that upcurrents start to break through the overlying dry air. When that happens, the clouds rise all the way to the tropopause (*see* ATMOSPHERIC STRUCTURE) and often beyond it into the lower stratosphere.

It is these clouds that produce some of the most violent thunderstorms known, and because they form along the cold front, which is still moving in a southeasterly direction, they tend to link together in SQUALL lines. These are what produce the tornadoes that give Tornado Alley its name.

Tornado and Storm Research Organization

A British organization, founded in 1974, that exists to gather data and undertake research into TORNADOES and other severe weather phenomena in Europe. It has representatives in Austria, France, Germany, Ireland, and Switzerland.

Further Reading

Tornado and Storm Research Organisation. "TORRO: The Tornado and Storm Research Organisation." Available online. URL: www.torro.org.uk/TORRO/index.php. Accessed February 20, 2006.

torque (couple, moment of a force) Torque is a twisting force that is equal to the product of a force and its distance from a point about which it is causing rotation. It is measured in newton meters (Nm) or pound-force feet (lbf ft). Torque is the force that causes air to rotate about a vertical axis.

Totable Tornado Observatory (TOTO) TOTO is a package of instruments designed to survive the conditions inside a TORNADO with winds up to 200 MPH (322 km/h) and measure wind speed, atmospheric pressure, temperature, and electrical discharges. Its acronym, TOTO, refers to Dorothy's dog in *The Wonderful Wizard of Oz*.

TOTO was built in 1980, with limited funds and using spare parts, by Alfred L. Bedard and Carl Ramzy, scientists working at the NATIONAL OCEANIC AND ATMOSPHERIC ADMINISTRATION (NOAA) Environmental Research Laboratory in Boulder, Colorado. Weighing 400 pounds (182 kg), TOTO comprises a cylinder housed in a casing of half-inch aluminum set in a frame made from angle iron, with arms that hold the instruments extending from the casing. It is powered by batteries and records its measurements. Inside the cylinder there are strip-chart recorders connected to instruments that measure TEMPERATURE, AIR PRESSURE, WIND SPEED, and electrical discharges.

TOTO could be carried in the back of a pick-up truck and deployed within 30 seconds. It remained in use for many years, finally to be joined by the "Turtle," developed by University of Oklahoma meteorologist Fred Brock. The Turtle is smaller and lighter than TOTO, and it records data digitally. Its name refers to its appearance. Its instruments are housed inside a hemispherical metal shell.

Further Reading

Bluestein, Howard B. *Tornado Alley*. New York: Oxford University Press, 1999.

Tower of the Winds (horologion) The Tower of the Winds is believed to be the first device ever invented with the purpose of forecasting the weather. It was designed by the Greek astronomer Andronicus of Cyrrhus (who flourished around 100 B.C.E.) and was built in Athens at some time in the first century B.C.E. A substantial part of it is still standing.

The tower had eight sides. Figures representing the eight principal wind directions were carved at the top of each side. Boreas, the north wind, was portrayed as a man wearing a cloak and blowing through a twisted seashell. Kaikas, the northeast wind, was a man carrying a shield from which he poured small, round objects that were possibly hailstones. Apeliotes, the east wind, was a young man holding a cloak filled with grains and fruit. Euros, the southeast wind, was an old man wrapped in a cloak. Notos, the south wind, was a man emptying an urn to produce a shower of water. Lips, the southwest wind, was a boy pushing a ship. Zephyros, the west wind, was a young man carrying flowers. Skiron, the northwest wind, was a bearded man carrying a pot filled with charcoal and hot ashes.

On top of the tower there was originally the bronze figure of Triton with a rod in his hand. The Triton (in

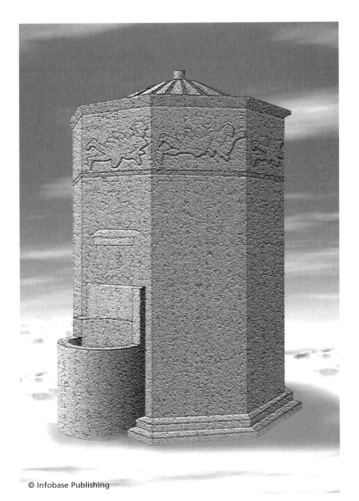

© Infobase Publishing

Most of the octagonal Tower of the Winds is still standing. It was built in the first century c.e. and helped Athenians predict the weather (and tell the time).

some stories there are several Tritons) was a mythological being with the head, trunk, and arms of a human and the lower body of a fish. He was the son of Poseidon, the god of the sea, and Amphitrite, daughter of Oceanos and Tethys. The Triton figure turned in the wind, indicating the wind direction. This statue gave rise to the custom of placing WIND VANES, often in the form of a weathercock or other figure, on the tops of church steeples.

Each side also had a sundial. As the Sun crossed the sky in the course of the day, at least one dial would always be showing the time. Thus, the tower was also a public timepiece. Even in Athens, however, the Sun does not always shine. To help people tell the time on cloudy days the tower contained a very elaborate clock driven by water (and called a clepsydra) that showed the hours on a dial.

In addition, the tower also had a disk that rotated, showing the movements of the constellations and the Sun's yearly course through them. Andronicus also built another tower with sundials around its sides, on the Greek island of Tenos.

The principle behind the weather forecast was simple. Traditionally, the Greeks had believed that their gods produced the weather. Make an appropriate offering to Zeus, father of the gods, and he would send good weather; please Poseidon, and he would send a storm to destroy your enemies. Aristotle (*see* APPENDIX I: BIOGRAPHICAL ENTRIES), on the other hand, had maintained that the weather has entirely natural causes and that in principle it can be understood and even predicted. The fifth-century B.C.E. dramatist Aristophanes (*ca.* 450–*c.* 388 B.C.E.) wrote a play about the weather, called *Clouds,* that includes a debate between an educated philosopher and an unsophisticated person from the country over whether thunder is made by Zeus or whether it results from the collision between clouds.

It was the more scientific, Aristotelian attitude that underlay the tower. The direction of the wind hinted at the weather that would follow. Athenians could look at the tower, see the direction in which the Triton was pointing, and deduce from that what the weather would be like over the next few hours or days. To help them, the gods of the winds, depicted on the faces of the tower, were also associated with particular types of weather. Everyone would have known that Boreas, the north wind, was a rude fellow who found it hard to breathe gently and was quite unable to sigh. Zephyrus, the west wind, was gentle. The sweetness of his breath brought forth flowers. Notus, the south wind, was wet, his forehead covered by dark clouds. Euros, the southeast wind, needed his warm cloak because he would bring cold, dry weather. The significance of the winds, of course, was true only for Athens. In other parts of the world, or even of Europe, wind directions might have quite different connotations.

While they thought about the kind of weather they might expect, the Athenians could also check on the time. The Greek word *hōra* means time, *logos* means account, so a horologion is a timepiece, or clock.

transparency Transparency is the capacity of a medium for permitting radiation to pass through it with no significant SCATTERING or ABSORPTION. The transparency of the atmosphere is usually measured by its transmissivity (*see* SOLAR IRRADIANCE) when the Sun is at its zenith (and the PATH LENGTH is 1).

The property of being transparent to radiant heat is called diathermancy (the adjective is diathermous).

Turbidity is a reduction in the transparency of the atmosphere that is caused by HAZE or AIR POLLUTION.

transpiration Plants need WATER for four reasons. Water is the source of HYDROGEN for PHOTOSYNTHESIS. Mineral nutrients dissolve in water present in the soil and enter through the root in solution. Nutrients are transported in solution to all parts of the plant, and the sugars produced by photosynthesis are also carried in solution through the phloem tissue. It is water that fills plant cells and keeps them rigid. In hot weather the EVAPORATION of water from leaf surfaces takes LATENT HEAT from the leaf, thereby lowering its TEMPERATURE. If a plant is deprived of water, it wilts, and unless the supply of water is restored, it may die.

Plants obtain their water from the soil. Water travels through the stems along channels that form tissue called xylem, and it evaporates through the tiny pores in the leaves called STOMATA (the singular is stoma). A much smaller amount evaporates from pores in the stem called lenticels. The evaporation of water from its surfaces helps keep the plant cool in very hot weather. This loss of water from the plant is called transpiration, and the amount involved is large. For example, in summer, a silver birch tree (*Betula pendula*), with about 250,000 leaves, may transpire 95 gallons (360

liters) of water a day, and a full-grown oak tree (*Quercus* species) transpires up to 185 gallons (700 l) a day. A maple (*Acer* species) transpires about 53 gallons (200 liters) an hour. This is water that the plant takes from the ground and returns to the air. Not surprisingly, this transport of water dries the ground and increases the HUMIDITY of the air. It is one of the ways in which plants influence the climate.

If a plant stem is cut, liquid will seep from it. This liquid, called sap, is mainly water, and it is flowing upward. A small herb may lift water a few inches above the ground, but a Sierra redwood tree (*Sequoiadendron giganteum*) can grow to a height of 300 feet (90 m) and has roots that extend many feet to either side. The tree raises water all the way from its roots to the leaves on its topmost branches.

Transpiration is driven from the leaves and not from the ground; the water is pulled from above, not pushed from below. Beneath the surface cells with their stomata, leaves have a layer of tissue called mesophyll. The inside of mesophyll cells is coated in a film of water, and when the stomata are open some of this water evaporates. Hydrogen bonds (*see* CHEMICAL BONDS) between water molecules draw in more water to replace the water that has been lost. Adhesion and cohesion (*see* SOIL MOISTURE), processes which function in living tissues in just the same way as they do in the soil, then transmit this attraction all the way through the plant to the tips of its roots. There the attraction exerts a force that draws water toward the roots. The magnitude of this force varies from one plant species to another. In some it is very weak, but in others it can amount to more than 200 pounds per square inch (1,380 kPa), which is about 13.8 times sea-level atmospheric pressure. The force can be so strong that on a hot day, when the evaporation rate from the leaves is very high, water moves through the plant at up to 30 inches (76 cm) a minute and the sides of the xylem vessels are pulled in, making a measurable difference in the diameter of the stem.

It is possible to measure the rate of transpiration from individual plants under laboratory conditions, but extremely difficult to do so reliably outdoors. For practical purposes it is impossible to distinguish the WATER VAPOR entering the air by evaporation from exposed surfaces and that entering by transpiration. The two are therefore considered together, and the combined process is called EVAPOTRANSPIRATION.

Further Reading

Allaby, Michael. *Temperate Forests.* New York: Facts On File, rev. ed., 2007.

Kent, Michael. *Advanced Biology.* Oxford: Oxford University Press, 2000.

tree line   The tree line is the climatic limit beyond which temperatures are too low to permit trees to grow. The tree line marks the boundary between tundra, with vegetation that includes small, stunted trees, and the bare rock, SNOW, and ICE of the high Arctic and Antarctic.

On mountains the height of the tree line varies with latitude and with the CONTINENTALITY or OCEANICITY of the climate. The variation with latitude on a mountainside occurs because the tree line is determined by TEMPERATURE, and the temperature on a mountainside is determined by the environmental LAPSE RATE (ELR), so the temperature at any given elevation measured from a sea-level DATUM depends on the temperature at sea level. In the central Alps of Europe, for example, the tree line is at 6,500–7,000 feet (2,000–2,100 m), in the Rocky Mountains it is at about 12,500 feet (3,800 m), and in the mountains of New Guinea it is at 12,200–12,600 feet (3,700–3,800 m).

Usually trees will not grow where the mean summer temperature is lower than 50°F (10°C), so it is possible to calculate the approximate height of the tree line from the temperature at the foot of the mountain. Assume that the temperature at the tropopause (*see* ATMOSPHERIC STRUCTURE) is -74°F (-59°C) and the height of the tropopause is 36,000 feet (11 km). Subtract the temperature at the foot of the mountain from -74°F (-59°C) and divide the result by 36 (or 11) to give the ELR per 1,000 feet (or kilometer). Then calculate the height at which the temperature falls below 50°F (10°C). For example, the mean summer temperature in Seattle, Washington, is 71°F (22°C). Therefore:

$$-74 - 71 = 145$$

$$145 \div 36 = 4.03 \text{ (ELR)}$$

$$71 - 50 = 21$$

$$21 \div 4.03 = 5.2$$

If there were a mountain in Seattle, the tree line on it would be at a little over 5,200 feet (1.59 km).

tree rings   The concentric rings that can be seen in a cross section of the trunk or large branch of a tree are called tree rings. The pattern of rings results from secondary growth.

All plants, including trees, grow by extending the length of the main stem and branches. This is called primary growth. Woody plants such as trees and shrubs also grow thicker stems and branches. This thickening is called secondary growth, and it is secondary growth that produces the rings in woody plants.

The outside of a tree trunk is protected by bark. This is not simply a rough layer of dead tissue, however, but three layers, two of which are living. The rough outer layer consists of dead cells with a waxy coating. These are called cork cells, and they form the outer skin that protects the living cells beneath from injury and also keeps out water. Immediately beneath the cork cells there is a layer of living cells, called the cork cambium. These cells divide, but after a few weeks the cells die and become cork cells. A while later, the outer skin splits and more cork cambium cells are produced to fill the gaps, and then these also die and become cork cells in their turn. That is how the trunk acquires its rough surface.

Beneath the cork cambium, the innermost layer of the bark consists of tubular arrangements of cells composing tissue called phloem. Sugars produced in the leaves by PHOTOSYNTHESIS, are transported through the phloem to every part of the plant. If the bark of a tree is cut all the way around the trunk, sugars can no longer reach the roots because the phloem has been severed, and the tree will die. Cutting the bark in this way is called girdling or ring-barking.

There is another layer of cambium beneath the phloem. This is called the vascular cambium. It consists of cells that divide to produce more cambium cells, phloem cells, and xylem cells. Xylem cells also form a tubular system for transport, in this case to carry water upward from the roots to every part of the plant. Xylem and phloem tissues are composed of dead cells. Each time a cell of the vascular cambium divides into two, one of the new cells remains as a cambium cell and the other cell dies to become either a xylem cell or a phloem cell. Xylem cells accumulate on the inside of the vascular cambium. The resulting accumulation of xylem cells shifts the cambium layer progressively farther from the center of the trunk. This is the secondary growth that increases the thickness of the trunk and branches.

On the inside of the xylem, old cells fill with metabolic waste products, including the lignin that toughens

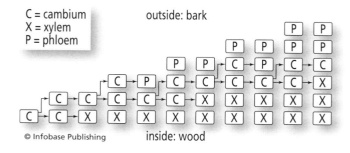

C = cambium
X = xylem
P = phloem

outside: bark

© Infobase Publishing

inside: wood

The history of a single cell in the vascular cambium. Each time the cell divides one of the resulting cells remains as a cambium cell and divided again, and the other dies to become either a phloem cell or a xylem cell.

cell walls. The inner part of the trunk or branch, consisting of dead tissue, makes up the heartwood, and the outer part, containing the active xylem and vascular cambium, is the sapwood.

In those parts of the world with a seasonal climate (*see* SEASONS), there is a period during which plant growth ceases. In temperate regions growth ceases in the winter, and in other parts of the world it ceases during the dry season, which may be the summer. The cessation is of both primary and secondary growth.

When the dormant period ends, the vascular cambium starts to produce new xylem tissue. The cells are large and have thin walls. They are a pale color. These cells form a cylinder surrounding the trunk or branch. Toward the end of the growing season—by late summer and fall in temperate regions—the vascular cambium produces xylem made from smaller, darker cells, with thicker walls. These form as another cylinder on the inside of the earlier xylem. When growth ceases for the year, the plant will have laid down two layers of xylem. In a cross section through the trunk or branch these will be visible as a pale, circular band and a narrower, dark band. Each pair of circles, or rings, marks one year's growth in the life of the plant. These are tree rings, and by counting them it is possible to determine the age of the tree.

Tree rings also reveal information about the conditions for plant growth during the year when they formed. If the weather was good, the rings will be broad, because growth was vigorous and many cells were produced. If the weather was poor, the rings will be narrow. If the weather was very bad, no growth at all may have occurred, and in this case one year's rings

will be missing. Missing years can be detected only by cross-dating the rings from that tree with a standard reference compiled from many trees. The dating of wood by means of tree rings is called dendrochronology.

The possibility of dating material in this way was discovered in 1901 by Andrew Ellicott Douglass, an astronomer working at the Lowell Observatory, in Flagstaff, Arizona, who was interested in SUNSPOTS and who thought it might be possible to correlate tree rings with climate. Later, Douglass used the technique to date buildings in the prehistoric settlement of Pueblo Bonito, New Mexico, by matching tree rings from timber in the buildings with samples taken from trees of a known age.

The technique involves obtaining several samples that are first compared with each other to make certain no rings have been missed and there are no inconsistencies. If trees are being dated, the samples are also checked against specimens taken from other trees in the immediate vicinity and in the region. Finally, the tree rings are matched with an accepted reference standard. Using bristlecone pine rings from living trees and cross-dating them with rings from nearby dead trees, dendrochronologists can date material that is more than 8,200 years old.

Bristlecone pines are often used. These trees grow in the arid regions of California, and they survive to a remarkably old age. Some living specimens are 4,600 years old. There are two species, the Great Basin bristlecone pine (*Pinus longaeva*) and the mountain bristlecone pine (*P. aristata*). A complete chronology covering 5,500 years has been developed for one group of bristlecone pines. This chronology is used as a standard reference against which other tree-ring sequences can be calibrated. It is also used to calibrate RADIOCARBON DATING methods, by measuring the ratio of $^{12}C:^{14}C$ in individual rings and compiling a record of fluctuations in the ratio.

The relative widths of tree rings are indicative of general growing conditions year by year, but closer examination of the rings yields more detailed information. When water evaporates, water molecules containing the lighter OXYGEN isotope (^{16}O) vaporize more easily, so rainwater is enriched in ^{16}O, compared with the heavier isotope ^{18}O, or, to phrase it another way, rainwater is depleted in ^{18}O. The separation of isotopes is called Rayleigh distillation. Plant roots absorb rainwater, and oxygen from that water becomes incorporated in plant cells, including the cells produced by secondary growth. Consequently, Rayleigh distillation may leave a trace in the tree rings. This signal is especially strong when the rainfall is extremely heavy, as it is during a tropical cyclone, because the torrential downpours associated with such storms supplies tree roots with water strongly depleted in ^{18}O.

A team of scientists led by Claudia Mora, a geochemist at the University of Tennessee at Knoxville, has studied the oxygen isotope ratios in tree rings from longleaf pine (*P. palustris*) trees that have been preserved in water or swamps around Lake Louise, Georgia. Mora and her colleagues have found tree-ring evidence for the Great Hurricane that struck Cuba in

A mountain bristlecone pine (*Pinus aristata*) in Pike National Forest. Bristlecone pines are very long-lived, and their annual growth rings—obtained by drilling a narrow core from the trunk, not by felling the tree—are used as a standard reference against which other tree-ring sequences can be checked. *(USDA Forest Service, Rocky Mountain Region Archives)*

1780 and for a period of 40 years in the 17th and 18th centuries when it seems that not a single hurricane struck Georgia.

Further Reading

Allaby, Michael. *Temperate Forests.* Rev. ed. Ecosystem. New York: Facts On File, 2007.
Morton, Oliver. "Storms Bow Out, But Boughs Remember." *Science,* 309, no. 1321 (August 26, 2005).
Sonic.net. "Dendrochronology." Available online. URL: www.sonic.net/bristlecone/dendro.html. Accessed February 21, 2006.

Triassic The Triassic is the first period of the MESO-ZOIC era of the Earth's history. It began 251 million years ago and ended 199.6 million years ago.

During the Triassic the supercontinent of Pangaea straddled the equator, surrounded by Panthalassa, the "world ocean." No sooner had Pangaea formed, however, than rifts appeared between what were to become North America and Africa, and the supercontinent began to break apart. The climate everywhere was warm and generally dry, and the interior of Pangaea had an extremely arid climate.

Starfish and sea urchins are among the marine invertebrate animals that first appeared during the Triassic. The seas also contained ichthyosaurs and other marine reptiles as well as lungfish. On land there were new species of insects and the ancestors of the dinosaurs, called archosauromorphs, evolved. (*See* APPENDIX V: GEOLOGIC TIMESCALE.

Tropical Atmosphere Ocean (TAO) A monitoring network that uses an array of moored buoys positioned across the Pacific Ocean to gather data from the sea surface and sea-level atmosphere. The data are used to improve the understanding, detection, and prediction of ENSO events. The array took 10 years to assemble and was completed in 1994.

Further Reading

McPhaden, Michael J. "The Tropical Atmosphere Ocean Project." NOAA. Available online. URL: www.pmel. noaa.gov/tao/. Updated daily. Accessed February 21, 2006.

tropical cyclone A tropical cyclone is an area of low surface AIR PRESSURE that generates fierce winds and rain to become the biggest and most violent type of atmospheric disturbance experienced on Earth. TORNA-DOES often have greater wind speeds, but they are very local. Tropical cyclones affect a much larger area than tornadoes and contain storm clouds that often produce tornadoes.

Tropical cyclone is their scientific name, but these storms have several common names. In some parts of the Greater Antilles a tropical cyclone is known as a taino. Those that occur in the North Atlantic are called hurricanes, and that name is often applied to all tropical cyclones regardless of where they occur. Atlantic hurricanes move in a westerly direction, along a track that may carry them across several inhabited islands in the Caribbean. Then they turn onto a more northerly track that may carry them toward the United States, where they may make landfall along the Gulf coast, Florida, or the Carolinas. Many die while they are over land, but those that survive continue to turn until they are on an easterly track that carries them back over the Atlantic, occasionally as far as northwestern Europe, where they may still have enough strength to cause considerable damage.

Only one tropical cyclone has ever been recorded in the South Atlantic, because ordinarily the equatorial trough (see INTERTROPICAL CONVERGENCE ZONE) does not move far enough south of the equator over the Atlantic to provide the conditions they need. The solitary South Atlantic hurricane, called Catarina, occurred on March 22–28, 2005, near the coast of Brazil and caused considerable damage when it moved onshore. It attained only category 1 on the SAFFIR/SIMPSON HURRICANE SCALE. Meteorologists have not yet discovered how Catarina developed.

Tropical cyclones that occur in the North and South Pacific and the East China Sea and South China Sea are called typhoons. In the vicinity of Indonesia a tropical cyclone is known as a baguio, which is the name of a town in Luzon, Philippines, and near Australia it sometimes used to be called a willy-nilly or willy-willy. These storms develop in the Timor Sea, move southwestward and then swing to an easterly direction, a track that carries them toward the Australian coast. They are often very severe, but they die away as they move inland.

The coronazo de San Francisco is a tropical cyclone that forms over the eastern Pacific Ocean, off the coast of Central America between Costa Rica and Point Eugenio in Baja California, Mexico. The tracks of these storms usually carry them northward or northwestward

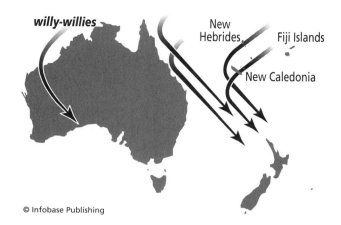

Tropical cyclones that strike northeastern Australia form over the Timor Sea. The map shows the direction in which they travel. They used to be called willy-willies or willy-nillies, but this name is no longer in use.

and many strike the coast. They are less violent than hurricanes that form over the Atlantic and Caribbean and they cover a smaller area. Those that form over the Bay of Bengal are called cyclones. Of all the tropical cyclones that occur, 90 percent are either cyclones or typhoons, and they are often extremely severe.

Tropical cyclones develop in or close to the equatorial trough around an area where the pressure is slightly lower than that of the surrounding air. This is called a tropical disturbance, and in the Atlantic it is often associated with an easterly wave.

An easterly wave, also called an African wave or tropical wave, is a long, weak, low-pressure TROUGH that moves from east to west across the tropical North Atlantic. It deflects the easterly trade winds, producing a wave pattern in the surface STREAMLINE. The wavelength (see WAVE CHARACTERISTICS) is usually 1,200–2,500 miles (1,900–4,020 km), and the waves travel about 6°–7° of longitude per day. Easterly waves last for 1–2 weeks before disappearing, and they are especially marked in the Caribbean region. In vertical profile the troughs producing easterly waves usually slope toward the east, so the weather associated with them occurs behind the line at which the trough lies on the surface. Ahead of the trough there is a RIDGE of high pressure, with generally fine weather, scattered cumulus cloud (see CLOUD TYPES), and some HAZE. Close to the line of the trough, the cumulus clouds are bigger, giving some showers, and improved VISIBILITY as the rain washes away the haze. Behind the trough, the wind veers, the temperature is lower, and the cloud thickens, with some cumulonimbus. SHOWERS are heavy, with some thunder. Easterly waves sometimes intensify to become tropical disturbances.

A tropical disturbance is an incipient tropical storm that is not associated with a frontal system. It is caused by the convergence of air at a low level. A disturbance may produce nothing more than single cumulus clouds that survive for only a few hours. The clouds are often aligned to form cloud streets (see CLOUD TYPES). More intense disturbances can produce much bigger clouds and SQUALL lines. If its central pressure falls, a tropical disturbance may develop into a tropical DEPRESSION.

Winds around a tropical depression blow at less than 38 MPH (61 km/h). If the depression deepens and these speeds increase, the depression is reclassified as a tropical storm.

A tropical storm is a tropical depression that has deepened until the winds around it are blowing at speeds of 38–74 MPH (61–119 km/h). When the mean wind speed exceeds 74 MPH (119 km/h), the storm is reclassified as a tropical cyclone. For purposes of identification tropical storms are given names, and these names are retained if they subsequently strengthen to become cyclones. Between 80 and 100 tropical storms develop in most years. Up to about half of those that survive long enough to cross an ocean develop into full cyclones.

Air flows toward the low-pressure area. As it does so, the CORIOLIS EFFECT swings it to the right, and as it spirals into the low-pressure center it is accelerated by the conservation of its angular MOMENTUM. The air then spirals upward. The magnitude of the Coriolis effect is zero at the equator. A tropical cyclone cannot develop closer than latitude 5° north or south of the equator, because at that distance the Coriolis effect is too weak to set the air turning.

If the temperature of the sea surface is at least 80°F (27°C) over a large area, the rate of EVAPORATION will be high and the rising air will carry a large amount of WATER VAPOR. The rising air cools adiabatically (see ADIABAT), and as it does so its water vapor condenses, releasing LATENT HEAT. This warms the air, causing it to rise further, and towering cumulonimbus clouds form.

If the air rises vigorously enough, it will pierce the trade wind INVERSION. If the rising air is then able to flow into a high-level TROUGH, the upper-level low pressure will draw air upward, intensifying the low-level convergence of air (see STREAMLINE). A trough may be present in the upper troposphere (see ATMO-

SPHERIC STRUCTURE) as the remains of a weather system that has almost dissipated or as an easterly wave.

Three conditions must be met in order for a tropical cyclone to develop. The sea-surface temperature must be at least 80°F (27°C) over a large area. There must be an area of low pressure no closer to the equator than 5° north or south. There must be vertical WIND SHEAR at high level to accelerate the rising air. The second and third of these conditions may be met at any time of year, but the surface water reaches a high enough temperature only after it has warmed through the summer. It is then almost as warm as it is possible for the sea to become, because at this temperature the rate of evaporation is such that the latent heat of vaporization absorbed from the sea prevents the temperature rising higher. At the same time, wind across the surface mixes the warm surface water with cooler water below the surface. Consequently, there is a season for tropical cyclones. It begins in late summer and ends in late autumn, when the equatorial trough moves toward the equator and the sea starts to cool.

When fully developed, a tropical cyclone consists of a central eye surrounded by a solid bank of cumulonimbus clouds forming an eyewall, and then by several

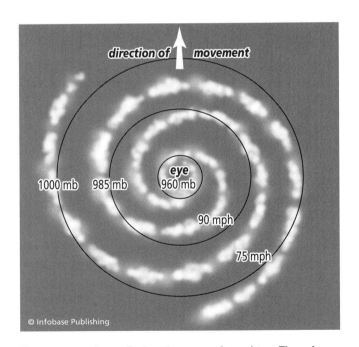

The structure of a tropical cyclone, seen from above. The pale patches are clouds. Around the eye, the eyewall is the area of densest cloud, strongest wind, and heaviest rain. Pressures and wind speeds at various distances from the eye are indicated.

In the eye of Hurricane Debbie, August 20, 1969. Air is calm and warm in the eye, but the storm's fiercest winds occur near the surface in the wall of cloud surrounding the eye—the eyewall. *(Edward E. Hindman, NOAA/AOML/Hurricane Research Division)*

concentric bands of cloud. The air is cool inside the eye of a tropical storm. There are clouds, some of which give precipitation. As the storm intensifies into a tropical cyclone, subsiding air in the eye is warmed adiabatically and becomes warmer than the air surrounding the eye. Meteorologically, it is the warm air in the eye that distinguishes a tropical cyclone from a tropical storm, rather than the increase in wind speed. In the eye the sky clears and conditions are calm, with a wind of no more than 10 MPH (16 km/h).

Convection in the eyewall is more intense than it is anywhere else in the storm, and it is in the eyewall that the strongest winds are generated and PRECIPITATION is heaviest. The eyewall consists of cumulonimbus cloud towering sometimes to 59,000 feet (18 km). Beyond this bank of cloud, in which air is rising vigorously, there is a region of clear skies and subsiding air, which is surrounded by a further bank of cumuliform cloud. There are usually several bands of cloud, the individual clouds becoming smaller with each band. The cyclone may extend to a diameter of up to 600 miles (965 km). Tropical cyclones are classified on the SAFFIR/SIMPSON SCALE of hurricane intensity according to the surface pressure in the eye and the wind speeds in the eyewall.

Low pressure in the eye causes the sea to rise. A fall in pressure of 1 millibar from the sea-level average of 1,013 mb produces a sea-level rise of 0.4 inch (1 cm). In the eye of a category 1 hurricane the sea level rises by about 14 inches (35.6 cm), and in a category 5 hurricane by about 40 inches (102 cm). The elevated sea

level contributes to the severity of the storm surge that is produced as a tropical cyclone crosses a coast.

Tropical cyclones move in a westerly direction in both hemispheres, then turn away from the equator. In the case of hurricanes, this carries them into the Caribbean and, depending on where they turn northward, they may cross the coast of the United States, or miss it and remain over the sea. Those that remain over the sea for most of the time may travel all the way to Canada and even reach northwestern Europe, weakening all the time. Typically, tropical cyclones intensify as they start to move away from the equator, but they weaken rapidly once they leave the warm surface waters of the TROPICS.

Tropical cyclones travel initially at 10–15 MPH (16–24 km/h), but as they move away from the equator they accelerate, sometimes to double that speed. Because the entire system is moving, the wind speed on one side of the eye is greater than that on the opposite side. If the storm is moving at 30 MPH (50 km/h) and the wind speed in the eyewall is 100 MPH (160 km/h), then on one side the two speeds combine to produce winds of 130 MPH (210 km/h) and on the other side one speed must be subtracted from the other, producing winds of 70 MPH (110 km/h).

The side of a tropical cyclone where the winds are strongest and where they tend to push ships into the path of the approaching storm is called the dangerous semicircle. Because the circulation around the storm is cyclonic and the storms move in a generally easterly direction, driven by the trade winds (*see*

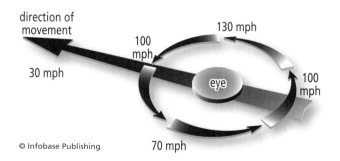

Wind speeds around the eye of a tropical cyclone vary as a result of the movement of the cyclone itself. On one side of the eye, the speed of the cyclone must be added to the wind speed, and on the other side it must be subtracted from it. In this example, the cyclone is traveling at 30 MPH (50 km/h) and the wind speed is 100 MPH (160 km/h). On one side of the eye, the actual wind speed is therefore 130 MPH (210 km/h), and on the other side it is 70 MPH (110 km/h).

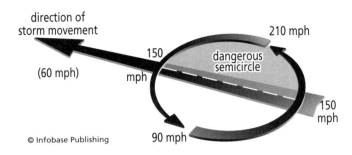

direction of storm movement

210 mph

150 mph

(60 mph)

dangerous semicircle

150 mph

90 mph

© Infobase Publishing

The dangerous semicircle is on the northern side of a hurricane or typhoon in the Northern Hemisphere. If the storm is moving westward at 60 MPH (96 km/h) and generating winds of 150 MPH (241 km/h), in the dangerous semicircle the wind speed is 210 MPH (338 km/h) and blowing toward the storm track, pushing ships into the path of the storm.

WIND SYSTEMS), then turn away from the equator, the dangerous semicircle is on the side of the storm farthest from the equator. That is the northern side in the Northern Hemisphere and the southern side in the Southern Hemisphere. The other side, where the winds are lightest and where they tend to push ships out of the path of the approaching storm, is called the navigable semicircle. It is on the side of the storm closest to the equator. This is the southern side in the Northern Hemisphere and the northern side in the Southern Hemisphere.

Pacific typhoons tend to be larger and more intense than Atlantic hurricanes, because they have a larger area of warm ocean over which to develop. Occasionally, though, they can grow much bigger. One that covers an area very much larger than the area covered by most is called a supertyphoon. A supertyphoon can be nearly 2,000 miles (3,200 km) across, with an area of 3 million square miles (8 million km²). For comparison, the area of the United States is about 3.7 million square miles (9.5 million km²). Fortunately, supertyphoons are very rare.

The Fujiwara effect is a phenomenon that occurs on average once every year and that was first described in 1921 by the Japanese meteorologist Sakuhei Fujiwara. If two typhoons of approximately similar size approach to within about 900 miles (1,450 km) of each other, they begin to interact. They start to turn about a point that lies about halfway between them. If one storm is much bigger than the other, they turn about a point that is closer to the larger storm. The big storm then absorbs the smaller one.

A tropical cyclone derives its energy from the condensation of water vapor, and therefore it retains its strength only for as long as it has an ample supply of warm water. As soon as it crosses land or sea water cooler than 80°F (27°C), the cyclone begins to weaken.

El Niño episodes (see ENSO) influence the number and intensity of tropical cyclones. During an El Niño there are fewer hurricanes in the Atlantic and Caribbean. The number of tropical cyclones remains unchanged on both sides of the equator in the western Pacific Ocean, but the storms tend to move farther east and into higher latitudes. There are more typhoons in the eastern Pacific, but fewer around Australia.

The number of tropical cyclones in a season also varies in a cycle of 20–30 years according to some scientists and 25–50 years according to others. From the 1970s until 1994, tropical cyclones were fairly uncommon. Their frequency increased in 1995. Between 1950 and 1990, there were an average of 9.3 tropical storms, 5.8 hurricanes, and 2.2 major hurricanes a year in the Atlantic. Between 1995 and 1999, there were an average 13 tropical storms, 8.2 hurricanes, and 4.0 major hurricanes a year. The number increased in the

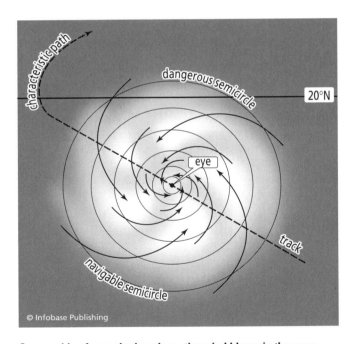

characteristic path

dangerous semicircle

20°N

eye

track

navigable semicircle

© Infobase Publishing

On one side of a tropical cyclone, the wind blows in the opposite direction to the direction in which the storm is moving. This reduces the effective wind speed, and the winds tend to blow ships behind the center of the storm, rather than into its path. This is the navigable semicircle.

early years of the 21st century. In 2005, there were 26 named storms in the Atlantic and Caribbean, of which 14 were classed as hurricanes and 7 as intense hurricanes. Forecasters expect the increase to be maintained into the early decades of the 21st century. Overall, however, there has been no increase in either the frequency or severity of tropical cyclones since 1940, and between 1940 and 1990 the mean sustained wind speed decreased in hurricanes developing in the Atlantic Ocean and Caribbean Sea.

Some climate scientists believe that global warming could produce conditions that would lead to an increase in the frequency and intensity of tropical cyclones. Others accept that this is possible, but consider it unlikely.

The effect of hurricanes making landfall over the United States is related to the energy of each individual storm. Meteorologists at the NATIONAL OCEANIC AND ATMOSPHERIC ADMINISTRATION calculate this from the hourly maximum sustained wind speeds over land from storms producing winds of at least the strength of a tropical storm. The resulting data, accumulated for a hurricane season, is then adjusted to make it compatible with similar data from storms at sea to produce an Accumulated Cyclone Energy (ACE) index. The ACE index allows hurricane seasons to be compared.

Meteorologists monitor the development and movement of tropical cyclones closely, using satellite images and data, data from ocean buoys, and reports from ships and aircraft. The buoys include specialized hurricane monitoring buoys. These are free-floating instrument packages that detect the approach of a tropical cyclone and are designed to be expendable. The hurricane rainband and intensity change experiment (RAINEX) is a three-year program funded by the U.S. National Science Foundation to study the dynamics of tropical cyclones. Using experimental models combined with data from aircraft carrying Doppler RADAR to examine turbulence and its effects, the aim is to improve the accuracy of forecasts of storm intensity. The WORLD METEOROLOGICAL ORGANIZATION sponsors research into the effects of tropical cyclones that cross coasts, through the International Tropical Cyclone Landfall Programme.

In the early 1960s, the U.S. government appointed a panel of scientific advisers to explore the possibility of bringing tropical cyclones under control. Project Stormfury comprised experiments to determine whether CLOUD SEEDING techniques would modify the condensation process sufficiently to rob a developing hurricane of its power by increasing rainfall in the first cloud band outside the eyewall and thereby slowing the development of the storm. Stormfury produced ambiguous results. Early in the 21st century another team of scientists, led by Ross N. Hoffman of Atmospheric and Environmental Research, at Lexington, Massachusetts, began using computer models to try other ways of modifying tropical cyclones. The model results suggested that in years to come it may become possible to deflect approaching storms, preventing them from making landfall.

Further Reading

Allaby, Michael. *Hurricanes.* Rev. ed. Dangerous Weather. New York: Facts On File, 2003.

Hoffman, Ross N. "Controlling Hurricanes." *Scientific American,* October 2004.

Landsea, Chris. "What may happen with tropical cyclone activity due to global warming?" NOAA. Available online. URL: www.aoml.noaa.gov/hrd/tcfaq/G3.html. Accessed February 21, 2006.

Saunders, Mark A. and Adam S. Lea. "Seasonal prediction of hurricane activity reaching the coast of the United States." *Nature,* 434, 1005–1007, April 21, 2005.

Tropical Ocean Global Atmosphere (TOGA) A program that ran from 1985 until 1994. It implemented an observational system for oceanic and atmospheric measurements with the aim of improving the understanding of ENSO events and, from that, their prediction. TOGA comprised satellite and surface measurements.

Tropical Prediction Center (TPC) The TPC is the part of the NATIONAL WEATHER SERVICE that issues watches and warnings about dangerous weather conditions in the TROPICS. The TPC is based at the campus of Florida International University, in Miami, and it is a component of the National Centers for Environmental Prediction. Under international agreement through the WORLD METEOROLOGICAL ORGANIZATION, the TPC has responsibility for generating and coordinating tropical cyclone forecasts for 24 countries in the Americas, Caribbean, and for the waters of the North Atlantic Ocean, Caribbean Sea, Gulf of Mexico, and the eastern North Pacific Ocean.

The TPC has three branches. The NATIONAL HURRICANE CENTER (NHC) maintains a watch on TROPICAL

CYCLONES. The Tropical Analysis and Forecast Branch (TAFB) concentrates on forecasting, especially for ships and aircraft, and it interprets satellite data and provides satellite rainfall estimates. The TAFB also supports the NHC. The Technical Support Branch (TSB) provides assistance with the computer and communications systems.

Further Reading

National Hurricane Center. "About the Tropical Prediction Center." NOAA. Available online. URL: www.nhc.noaa.gov/aboutintro.shtml. Last modified December 13, 2005.

Tropics   The Tropics are the two lines of latitude at which the Sun is directly overhead at noon on one of the two SOLSTICES. There is one tropic to each side of the equator, at latitudes 23.5°N and 23.5°S. The northern tropic is known as the tropic of Cancer and the southern is the tropic of Capricorn. Tropical regions (the Tropics) are those that lie between the two tropics.

With respect to the PLANE OF THE ECLIPTIC, the rotational axis of the Earth is tilted 23.45° from the vertical (*see* AXIAL TILT). Because of this, in the course of its yearly orbit about the Sun, first one hemisphere and then the other is tilted toward the Sun. To an observer at the equator, the noonday Sun would be directly overhead at each EQUINOX, but at the solstices, in mid December and mid June, it would be approximately 23.5° to the south and north, respectively.

The belt around the Earth within which the mean annual temperature exceeds 68°F (20°C) is known as the hot belt.

tropophyte   A plant that is adapted to a climate with pronounced wet and dry seasons.

trough   A trough is a long, tonguelike protrusion of low AIR PRESSURE into an area of higher pressure. The waves in the polar front JET STREAM associated with the index cycle (*see* ZONAL INDEX) that extend toward the equator are also called troughs.

Trowal is a Canadian term for a trough of warm air that is held high above the surface by an OCCLUSION. It often produces layered clouds similar to those associated with a warm FRONT, and PRECIPITATION. As the trowal passes and cold air replaces it, the sky clears. Trowals are often shown on Canadian weather maps.

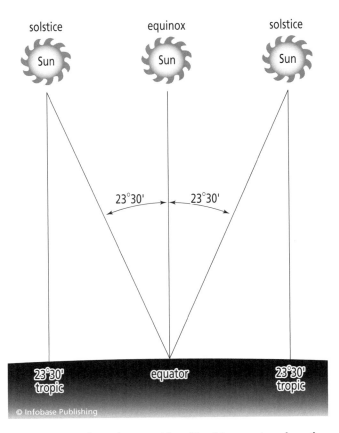

The Tropics are the regions on either side of the equator where the Sun is directly overhead at noon on at least one day every year. The Tropics are bounded by the tropics of Cancer and Capricorn, located at 23.5°N and 23.5°S, respectively. This latitude is also the angle between a point on the equator and the Sun when the Sun is overhead at one or other tropic, and the angle by which the rotational axis of the Earth is tilted with respect to the plane of the ecliptic.

A polar trough is a trough in the upper troposphere (*see* ATMOSPHERIC STRUCTURE) that extends toward the equator from the Arctic or Antarctic far enough to reach the TROPICS. The part of this high-level cold air that is closest to the equator sometimes becomes separated from the main part of the trough, to form a cutoff low (*see* CYCLONE) with a cold center. Air flows cyclonically (counterclockwise in the Northern Hemisphere) around the center, and if this flow extends to the surface, it will trigger a SUBTROPICAL CYCLONE.

A trough that forms at some distance from a major trough and that is related to it is called a resonance trough. The distance between a resonance trough and a dominant trough is measured in wavelengths (*see* WAVE CHARACTERISTICS) of ROSSBY WAVES. The trough that forms in winter over the Mediterranean Sea may be a resonance trough between the major troughs that lie above the eastern coasts of North America and Asia.

tsunami A tsunami is an ocean wave that is sometimes very large when it reaches the coast and that can cause great devastation. Unlike other waves, it is caused neither by wind nor by tidal movement—despite its old common name of tidal wave. The modern name, *tsunami*, is Japanese for "harbor wave," which is a much more accurate description.

Tsunamis are caused by major disturbances on the ocean floor. These include earth quakes, the eruption of submarine volcanoes, and large submarine mudslides that sometimes occur when sediments become unstable and slide down the outer edges of a continental shelf.

The tsunami that occurred on December 26, 2004, known in Asia as the Asian Tsunami and in other parts of the world as the Boxing Day Tsunami, resulted from a movement along the subduction zone where the Indian Plate is moving beneath the Burma Plate (*see* PLATE TECTONICS). Over a period of several minutes, approximately 750 miles (1,200 km) of the ocean floor aligned approximately north–south, rose by about 50 feet (15 m), causing an earthquake of about Richter magnitude 9.15 (the U.S. Geological Survey calculated 9.0 and other seismologists calculated 9.3). The earthquake hypocenter (the place where it happened) was about 100 miles (160 km) west of Sumatra, Indonesia, and 18.6 miles (30 km) below the seabed. The earthquake triggered tsunamis that sped across the Indian Ocean producing waves that in some places were 100 feet (30 m) high when they struck the coasts of Indonesia, Sri Lanka, India, and Thailand. In total the tsunami killed more than 280,000 persons.

A seabed disturbance of this kind sends a shock wave through the entire depth of ocean around it. This can be seen in war movies, where an explosion at depth causes a momentary shudder at the surface before the water displaced by the explosion is thrown into the air. A tsunami, therefore, is a shock wave transmitted through the ocean. Its wavelength and period are very long (*see* WAVE CHARACTERISTICS). Typically, a tsunami has a wavelength of about 160 miles (200 km) and a wave period of 15–20 minutes. Across the open sea tsunamis travel at a speed given by $\sqrt{(gd)}$, where g is the acceleration due to gravity (32 feet per second per second, or 9.81 m/s^2) and d is the depth of water. Because the average depth of the oceans is about 13,000 feet (4,000 m), a tsunami wave travels through the open ocean at an average speed of about 440 MPH (708 km/h). The wave is very small, however; its amplitude is rarely more than about 20 in. (50 cm). Two hours after the earthquake that triggered the Boxing Day Tsunami, RADAR instruments on satellites measured the wave crossing the ocean at a maximum of 2 feet (60 cm) high. The wave is so small, in fact, and it travels so fast, that out at sea sailors on ships often fail to notice it.

When the wave enters shallow coastal waters it slows down, because the wave speed is proportional to the depth of the water. Its wavelength shortens and its height increases, because water is still advancing toward the shore at the original wave speed, so water accumulates.

The size and form of tsunamis vary greatly, depending on the type, magnitude, and location of the disturbance that causes them and the configuration of the coasts they reach. Some appear as a breaker or a series of breakers. Others do not break as surf does, but resemble a rapidly rising TIDE that continues to rise until it has traveled far beyond the ordinary tidal limit. Often there is a warning of the impending approach of a tsunami. Water that is rising and falling against the shore with the normal movement of the waves retreats much farther than usual and remains for a few moments at a very low level. Then, when it advances, its advance takes it very much higher than usual. Anyone observing this phenomenon is well advised to seek safety immediately, by moving as quickly as possible to the highest ground within reach. They should not attempt to return to their homes until the emergency services tell them it is safe to do so.

Sensors maintain a constant watch for tsunamis in the Pacific Ocean, feeding data to centers in Alaska and Hawaii. No such warning system existed in the Indian Ocean at the time of the Boxing Day Tsunami, but one is being installed. Tsunami centers broadcast warnings to people living in a coastal area when a tsunami is approaching.

Further Reading

González, Frank I. "Tsunami!" *Scientific American*, May 1999.

NASA. "Tsunami: The Big Wave." TRW Inc. Available online. URL: http://observe.arc.nasa.gov/nasa/exhibits/tsunami/tsun_bay.html. Accessed February 21, 2006.

Wikipedia. "The 2004 Indian Ocean earthquake." Available online. URL: http://en.wikipedia.org/wiki/2004_Indian_Ocean_earthquake. Last modified February 21, 2006.

turbulent flow (**turbulence**) Turbulent flow occurs when elements of a moving fluid follow STREAMLINES that cross one another, so the flow passing any particular point changes speed and direction in an irregular and unpredictable fashion. Except in the laminar boundary layer (*see* LAMINAR FLOW) air movement is almost always turbulent.

The Richardson number is a mathematical value, devised by the English mathematician and meteorologist Lewis Fry Richardson (1881–1953; *see* APPENDIX I: BIOGRAPHICAL ENTRIES), that makes it possible to predict whether atmospheric turbulence is likely to increase or decrease. The Richardson number (Ri) is calculated from the strength of the WIND SHEAR, BUOYANCY of the air, and the POTENTIAL TEMPERATURE. It represents the ratio of the rate at which the KINETIC ENERGY of the turbulent motion is being dissipated by buoyancy due to natural or free CONVECTION to the rate at which kinetic energy is being produced by mechanical or forced convection. If Ri is greater than 0.25, turbulence will decrease and disappear. If Ri is less than 0.25, turbulence will increase.

Turbulent flow is the cause of the gustiness (*see* GUST) and the continually changing speed and direction of the wind felt at ground level, but turbulent flow acts vertically as well as horizontally. Where two belts of air are moving at different speeds, air at the bound-

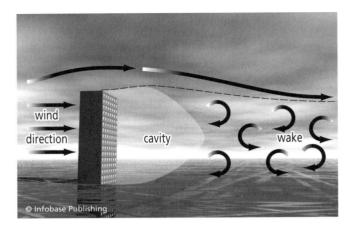

Immediately downwind of a building, there is a region called the cavity, and beyond that is the wake, a region of turbulence with erratic, gusty wind.

ary between them can be set rotating in a vertical plane. Turbulent flow perpetuates itself, because erratic movement in one place jostles the adjacent air and sets it moving, and FRICTION, convective movement, and pressure differences are continually introducing new disturbances to the flow.

The range of frequencies of the oscillations that make up turbulent flow is called the spectrum of turbulence. In turbulent flow, air is moving at different local VELOCITIES that vary over different lengths and times. Variation in an oscillation over distance (wavelength) and time is a variation in frequency. The spectrum at a given point and in a given direction is the ROOT-MEAN-SQUARE velocity of the frequencies contributing to the motion for each bandwidth.

Turbulent flow produced in moving air that encounters physical obstacles, such as buildings or trees, is called mechanical turbulence. Mechanical turbulence causes EDDIES, with the result that close to the ground the wind may blow from almost any direction, especially in cities. It is why the wind direction must be measured in the open. A region of turbulent flow that lies downwind of a surface obstruction or behind a body that is moving through a fluid is known as a wake. The moving fluid passes over the surface of the object and becomes detached from the surface on the LEE side. The region where it becomes detached is called the cavity. The wake forms on the downwind side of the cavity. As the wind detaches, eddies form within the flow. These produce gusts of wind and rap-

A turbulent flow is a flow of air or other fluid that is irregular, with many eddies.

idly changing wind directions in the wake of a building or other obstruction. If there is no further obstruction, the wake extends for a distance equal to about 10 times the height of the obstruction. The turbulent flow in the wake of a boat is clearly visible. The wake behind an aircraft is invisible, but it represents a serious hazard to any aircraft that enters it.

Local variations in TEMPERATURE and HUMIDITY that are produced by turbulent flow and that persist in the air after the movement that caused them has ceased and the DENSITY of the air has become uniform are known as fossil turbulence. Fossil turbulence scatters radio waves and can cause small clouds to form where air is made to rise.

U

ultraviolet index (UVI) The UV index is a guide to the intensity of ultraviolet radiation (*see* SOLAR SPECTRUM), reported as an index value that is related to the duration of exposure that will cause sunburn in the most susceptible people, which are those with pale skin. The index was developed by the U.S. Environmental Protection Agency and since June 1994 predicted UVI values for certain cities have been issued regularly by the NATIONAL WEATHER SERVICE. The index runs from 0 to 15, and the reported values are usually accompanied by recommended precautionary measures, which include the appropriate sun protection factor (SPF) for sunscreens.

The index is set out in the following table.

United Nations Environment Programme (UNEP) A program established by the General Assembly of the United Nations in accordance with a resolution from the United Nations Conference on the Human Environment, which was held in June 1972 in Stockholm, Sweden. UNEP is charged with coordinating intergovernmental measures for the protection of the environment. Its headquarters are in Nairobi, Kenya.

In UN terminology, a program has a lower status than an agency. *See* EARTHWATCH PROGRAM, GLOBAL ENVIRONMENTAL MONITORING SYSTEM, GLOBAL ENVIRONMENT FACILITY, GLOBAL RESOURCE INFORMATION DATABASE, INTERGOVERNMENTAL PANEL ON CLIMATE

UV Index

UV category	UVI value	Time to burn (minutes)	Precautions
Minimal	0–2	30–60	Wear a hat
Low	3–4	15–20	Wear a hat; use sunscreen SPF 15+
Moderate	5–6	10–12	Wear a hat; use sunscreen SPF 15+; keep in shade
High	7–9	7–8.5	Wear a hat; use sunscreen SPF 15+; keep in shade; stay indoors between 10 am and 4 pm
Very high	10–15	4–6	Stay indoors as much as possible; outdoors wear a hat and use sunscreen SPF 15+

CHANGE, APPENDIX V: LAWS, REGULATIONS, AND INTERNATIONAL AGREEMENTS (Montreal Protocol on Substances that Deplete the Ozone Layer and Vienna Convention on the Protection of the Ozone Layer).

United States Weather Bureau

The U.S. Weather Bureau is the federal agency that was instituted to gather meteorological data and to prepare and issue weather forecasts and warnings of severe weather. The inspiration to form the bureau was initially that of Thomas Jefferson (1743–1826), who had an abiding interest in meteorology and kept weather records over a number of years. In 1849 Joseph Henry (*see* APPENDIX I: BIOGRAPHICAL ENTRIES) inaugurated the collection of meteorological data at a central point, and on February 9, 1870, President Ulysses S. Grant signed a joint resolution of Congress that authorized the secretary of war to establish a weather service within the army. The Army Signal Corps operated the service until July 1, 1891, when the Weather Bureau was created as a civilian service within the Department of Agriculture. The bureau was transferred to the Department of Commerce on June 30, 1940; became part of the Environmental Science Services Administration on July 13, 1965; and in 1967 was renamed the NATIONAL WEATHER SERVICE.

Instruments for measuring wind deployed on the roof of the Headquarters Building of the Meteorological Service of the United States Signal Service. The picture is from a supplement to *Frank Leslie's Illustrated Newspaper,* published in New York on May 1, 1880. *(Historic NWS Collection)*

In 1898, the bureau started regular kite observations, which continued until 1933, and it began regular balloon soundings in 1909. From 1925, data were also collected by aircraft, and in 1926 the Air Commerce Act made the bureau responsible for providing weather services for aviation. Experiments with radio communication began in 1901, and in 1939 the bureau introduced the first telephone weather service, in New York City.

When the WORLD METEOROLOGICAL ORGANIZATION was formed in 1951, the chief of the Weather Bureau, Francis W. Reichelderfer, was elected its first president.

Further Reading

National Weather Service. "NOAA History: A Science Odyssey." NOAA. Available online. URL: www.history.noaa.gov/legacy/nwshistory.html. Last updated April 21, 2004.

units of measurement   Many units of measurement appear in books and articles on the atmospheric sciences. Most of these are SI units (*see* APPENDIX VIII: SI UNITS AND CONVERSIONS), but not all of them are. The units in the following list are arranged in alphabetical order. Each unit is defined, and where appropriate the definition also explains how the unit acquired its name. Most of the units are metric and defined in relation to SI units ("SI units" is an abbreviation for Système International d'Unités).

Ampere (A) is the SI unit of electric current. It is equal to the constant current that, if maintained in two straight, perfectly cylindrical, parallel conductors of infinite length and negligible cross-section placed 1 meter apart in a vacuum would produce between the two conductors a force of 2×10^{-7} newton per meter of their length. The unit is named in honor of the French physicist and mathematician André-Marie Ampère (1775–1836).

Ångström (Å) is a unit of length that was formerly used to measure very small distances, such as those between molecules and the wavelengths of electromagnetic radiation. It was devised by the Swedish spectroscopist Anders Jonas Ångström (1814–74) and is equal to 10^{-10} m. It has been replaced by the SI unit the nanometer (1 Å = 0.1 nm).

Arcsecond is the unit in which very small angles are measured. It is equal to one-sixtieth of an arcminute, and 1/3,600th of a degree.

Atmosphere is a measurement of AIR PRESSURE. One standard atmosphere is equal to 0.101325 megapascals (MPa), 1.01325×10^5 newtons per square meter (N/m²), or 1.013.25 bars.

Bar is a unit of pressure that is equal to 10^5 newtons per square meter (= 10^6 dynes/cm²). The unit was introduced by Vilhelm Bjerknes (see APPENDIX I: BIOGRAPHICAL ENTRIES) in *Dynamic Meteorology and Hydrography*, published in Washington in 1911. Meteorologists and climatologists now measure atmospheric pressure in pascals (1 Pa = 1 N/m²), which is a much smaller unit (1 bar = 0.1 MPa), but weather reports and forecasts published in newspapers and broadcast on radio and TV still use the millibar (1 bar = 1,000 mb).

Becquerel (Bq) is the SI unit of radiation. It is equal to an average of one transition of a radionuclide (one decay) per second. The unit is named in honor of the French physicist who discovered radioactivity, Antoine-Henri Becquerel (1852–1908).

British thermal unit (Btu) is a unit of work, energy, or heat. The unit was first used by the English physicist James Prescott Joule (1818–89) in a paper on the relationship between heat and mechanical energy that he presented to a meeting of the British Association for the Advancement of Science in 1843. The unit was given its name in 1876. One British thermal unit is the energy that is required to raise the temperature of one pound of water through one degree Fahrenheit. This varies according to the starting temperature, and so this is sometimes specified. A mean value for the Btu is given by dividing by 180 the energy needed to raise the temperature of one pound of water from 32°F to 212°F (from freezing to boiling). This Btu is equal to 1055.79 joules. The accepted international value for the Btu is 1055.06 joules.

Bubnoff unit (B) is used to measure rates of EROSION. The unit is equal to the erosion of 1 micrometer (μm) of material per year. It was named for the German geologist Serge von Bubnoff (1888–1957).

Calorie (cal) is a C.G.S. unit of heat. It is equal to the amount of heat that is needed to raise the temperature of 1 gram of water by 1°C (= 1K). This value varies with the temperature of the water, however, so this had to be stated, with the result that there were eventually four separate calories in use. These were the International steam calorie (= 4.1868 J), the 15°C calorie (= 4.1855 J) which measured the heat needed to raise the temperature from 14.5°C to 15.5°C, the 4°C calorie (= 4.2045 J), from 3.5°C to 4.5°C, and the mean (0–100°C) calorie (= 4.1897 J). The unit was introduced in 1880. Except for the kilocalorie, or Calorie, equal to 1,000 calories, that is sometimes still used in reporting food-energy values, in 1950 the calorie was replaced by the SI unit the joule (J); 1 cal = 4.1868 J (based on the value of the International steam calorie).

Candela (cd) is the SI unit of luminous intensity, which is defined as the luminous intensity, measured perpendicularly, of a surface 1/600,000 m² in area of a BLACKBODY at the temperature of freezing platinum under a pressure of 101,325 pascals.

The c.g.s. system of units was introduced for scientific use before the SI system was developed. The c.g.s. system was based on the centimeter, gram, and second, but it proved unsatisfactory and confusing when applied to electrical quantities and heat measurements. Its use persisted for some time, but it has now been replaced by the SI system.

Coulomb (C) is the derived SI unit of quantity of electricity, or electric charge, which is defined as the charge that is transferred by a current of 1 ampere per second. The unit is named in honor of the French physicist Charles-Augustin de Coulomb (1736–1806).

Dobson unit (DU) is used to report the concentration of a gas which is present in the atmosphere or in a particular part of the atmosphere. It refers to the thickness of the layer that gas would form if all the other atmospheric gases were removed and the gas in question were brought to sea level and subjected to standard sea-level pressure. The amount of OZONE present in the stratospheric OZONE LAYER is usually reported in Dobson units. In the case of ozone, 1 Dobson unit corresponds to a thickness of 0.01 mm (0.0004 inch), and the amount of ozone in the ozone layer is typically 220–460 DU, corresponding to a layer 2.2–4.6 mm (0.09–0.18 inch) thick. The unit is named for Gordon Miller Bourne Dobson (1889–1976; see APPENDIX I: BIOGRAPHICAL ENTRIES), the British physicist who studied stratospheric ozone in the 1920s and who invented the DOBSON SPECTROPHOTOMETER.

Farad (F) is the derived SI unit of CAPACITANCE, which is defined as the capacitance of a capacitor that has a potential difference of 1 volt between its plates when it is charged with 1 coulomb. This is a very large unit and so the unit most commonly used is the microfarad (10^{-6} F). The unit is named in honor of the English chemist and physicist Michael Faraday (1791–1867).

Henry (H) is the derived SI unit of INDUCTANCE, which is equal to the inductance of a closed circuit that varies uniformly at one ampere per second and produces an electromotive force of one volt. The name of the unit was adopted in 1893 in honor of the American physicist Joseph Henry (1797–1878; *see* APPENDIX I; BIOGRAPHICAL ENTRIES).

Hertz (Hz) is the derived SI unit of frequency (*see* WAVE CHARACTERISTICS), which is equal to one cycle per second. The name of the unit was adopted in 1933, in honor of the German physicist Heinrich Rudolph Hertz (1857–94).

Joule (J) is the derived SI unit of energy, work, or quantity of heat. It is equal to the work that is done when a force of 1 newton moves a distance of 1 meter, or the energy that is expended by 1 watt in 1 second. The unit was adopted in 1889, and in 1948 it was adopted as the unit of heat. It is named in honor of the English physicist James Prescott Joule (1818–89).

Kelvin (K) is the SI unit of thermodynamic temperature, which is defined as being 1/273.16 of the triple point of water (*see* BOILING). $1 K = 1°C = 1.8°F$. The unit is named in honor of the Scottish physicist and electrical engineer William Thomson, the first Baron Kelvin of Largs (1824–1907).

Kilogram (kg) is the SI unit of mass. It is equal to the mass of a prototype that is kept at the International Bureau of Weights and Measures at Sèvres, near Paris, France.

Knot is a unit of speed that was devised originally for use at sea, but that is also widely used by aircraft. It is a speed of one nautical mile per hour and, by international agreement, equal to 1.852 km/h. The United States adopted the international knot in 1954, but British ships and aircraft continue to use a knot equal to 1.00064 international knots. The knot came into use as a unit in the late 16th century. Then, the speed of a ship was measured by dropping over the side a float attached by a knotted rope. The knots were spaced 7 fathoms apart (a fathom is 6 feet, or 1.8 m) and the sailor measuring the speed counted the number of knots that passed in 30 seconds. This gave the speed of the ship in nautical miles per hour. Subsequently, the length of the nautical mile was changed, as was the distance between knots and the period of time used for counting, until the unit became standardized.

Langley (ly) is a unit of solar radiation that was suggested in 1942 by the German physicist F. Linke (in *Handbuch der Geophysik*, 8, 30) and named for S. P Langley (1834–1906; *see* APPENDIX I: BIOGRAPHICAL ENTRIES). It is defined in the c.g.s. system as one calorie per square centimeter (using the 15°C calorie) per minute. The SOLAR CONSTANT is equal to 1.98 langleys.

Lumen (lm) is the derived SI unit of the rate of flow of light, known as luminous flux. It is equal to the rate at which light flows from a point emitting a uniform intensity of 1 candela in a solid angle of 1 steradian.

Lux (lx) is the derived SI unit of illuminance, which is the amount of light energy that reaches a unit area of surface in a unit of time. The lux is equal to the illuminance produced by a lumous flux of 1 lumen distributed uniformly over an area of 1 square meter.

Meter (m) is the SI unit of length. It is defined as a length equal to 1,650,763.73 wavelengths in vacuum corresponding to the transition of an atom of krypton-86 between levels $2p_{10}$ and $5d_5$.

Milli atmospheres centimeter (milli atm cm) is a unit used to measure volcanic emissions. It is identical in concept to the Dobson unit, in that it measures the quantity of a substance present in the air as the thickness of the layer that substance would form if it were the only constituent of the air and subjected to sea-level pressure. The atmospheric concentration of sulfur dioxide, for example, is usually about 15 milli atm cm. This means that if all the sulfur dioxide contained in the column of air from the measuring station to the top of the atmosphere were compressed into a layer of gas at sea-level pressure, that layer would usually be about 0.015 cm (0.006 inch) thick.

Millibar (mb) is the unit in which atmospheric pressure is commonly reported in weather forecasts. It is equal to one-thousandth of a bar.

Mole (mol) is the SI unit of amount of a substance. It is equal to the amount of any substance that contains as many elementary units as there are atoms in 0.012 kilogram of carbon-12. The elementary units may be atoms, molecules, IONS, electrons, or other particles or groups of particles, but they must be specified.

Nautical mile is a unit of length that was introduced early in the 17th century, based on a suggestion by the English mathematician and inventor Edmund Gunter (1581–1626). Gunter thought navigation at sea might be simplified if distances were related directly to degrees of latitude. To achieve this, he proposed establishing a unit of length that is equal to the distance subtended by one minute of arc. Consequently, one nautical mile is the average meridian length of one minute of latitude. That is, the length of one line of latitude divided by 21,600, which is the number of minutes of arc in 360°. It is necessary to take an average length, because the distance subtended by one arc minute of latitude varies slightly with latitude owing to the fact that the Earth is not perfectly spherical. The International Hydrographic Conference of 1929 recommended that this distance be measured at latitude 45°, to produce a nautical mile of 6,076 feet (1,852 m). This is the length that is most widely used. In Britain, however, the nautical mile is measured at latitude 48°, which gives a length of 6,080 feet (1,854 m), so that 1 English nautical mile = 1.00064 international nautical miles. A speed of 1 nautical mile per hour is known as 1 knot.

Newton (N) is the derived SI unit of force, which is equal to the force needed to accelerate a mass of 1 kilogram at 1 meter per second per second (1 N = 1 kg/m/s^2). The name of the unit was adopted in 1938 in honor of the English physicist and mathematician Sir Isaac Newton (1642–1727).

Ohm (Ω) is the derived SI unit of electrical resistance. It is defined as the resistance between two points in an electric conductor when applying a constant potential difference of 1 volt between them produces a current of 1 ampere in the conductor. It is the oldest of all the electrical units and was adopted in 1838. The unit is named in honor of the German physicist Georg Simon Ohm (1789–1854), and its symbol is the Greek letter omega (Ω).

Okta (octa) is a unit that is used to report the extent of cloud cover in eighths of the total sky; 1 okta = 1/8 of the sky covered by cloud. The CLOUD AMOUNT is measured by examining a reflection of the sky on a mirror marked out in a grid of 16 squares and counting the number of squares filled by cloud.

Pascal (Pa) is the derived SI unit of pressure, which is equal to 1 newton per square meter. The unit is named in honor of the French physicist, mathematician, and philosopher Blaise Pascal (1623–62; *see* APPENDIX I: BIOGRAPHICAL ENTRIES).

Radian (rad) is the supplementary SI unit of plane angle, which is the angle subtended at the center of a circle by an arc equal to the radius of that circle. If the radius is *r*, an arc of length *r* will subtend an angle of 1 rad and the circumference of the circle, $2\pi r$ will sub-

tend an angle of $2\pi r \div r$ rad = 2π rad. The circumference subtends an angle of 360°, therefore 360° = 2π rad and 1 rad = 57.296°.

Second (s) is the SI unit of time. It is defined as the duration of 9,192,631,770 periods of the radiation that corresponds to the transition of an atom of caesium-133 between two hyperfine levels. The second is also a unit of angle (sometimes called an arcsecond), equal to 1/3,600 of a degree, and symbolized by ″.

Standard atmosphere (standard pressure) is a unit of pressure that is defined as a pressure of 1.013250 × 10^5 newtons per square meter (101.325 kPa, 1013.25 mb, 760 mm mercury, 29.9 inches of mercury, 14.7 lb/in^2). This is the average atmospheric pressure at mean sea level (sea level varies slightly from place to place). The definition assumes that the atmosphere consists of a perfect gas (a gas that obeys all the GAS LAWS) at a temperature of 59°F (15°C, or 188.16 K) and that the acceleration due to gravity is 9.80655 m/s^2. Sea-level pressure can be measured directly, but it is usually calculated from station pressures measured at known elevations above sea level. Unless it is stated otherwise, all reported atmospheric pressures are reduced to sea-level values. When reduced to a sea-level value, the pressure is called the reduced pressure (*see* BAROMETER).

Standard temperature and pressure (s.t.p.) are the conditions applied when measuring quantities that vary with temperature and pressure and especially when comparing the properties of gases. The conditions are a temperature of 273.15 K and pressure of 101,326 Pa.

Steradian (sr) is the supplementary SI unit of solid angle. It is defined as the angle, measured from the center of a sphere, that cuts off an area on the surface of the sphere equal that of a square with sides equal to the radius of the sphere.

Tesla (T) is the derived SI unit of magnetic flux density, which is the amount of magnetism per unit area of a magnetic field measured at right angles to the magnetic force. The tesla is equal to one weber per square meter (1 T = 1 Wb/m^2). The unit was adopted in 1954 and was named in honor of the Croatian-born American physicist and electrical engineer Nikola Tesla (1856–1943).

Torr is a unit of pressure equal to one millimeter of mercury. It is named after Evangelista Torricelli (1608–47; *see* APPENDIX I: BIOGRAPHICAL ENTRIES), and it was used for a time in some European countries, but is now little used. The tor, another unit named after Torricelli, and equal to one pascal, or one-hundredth of a millibar, was also proposed in 1913, but it, too, is rarely used.

Volt (V) is the derived SI unit of electromotive force, electric potential, or potential difference. It is defined as the potential difference between two points on an electric conductor that is carrying a constant current of 1 ampere and the power being dissipated between the two points is 1 watt. The unit was adopted internationally in 1881 and is named in honor of the Italian physicist, and inventor of the battery, Alessandro Giuseppe Anastasio, Count Volta (1745–1827).

Watt (W) is the derived SI unit of power, which is equal to 1 joule per second. The unit was adopted in 1889 and is named in honor of the Scottish engineer and instrument maker James Watt (1736–1819).

Weber (Wb) is the derived SI unit of magnetic flux, which is the amount of magnetism in a magnetic field calculated from the strength and extent of the field. The weber is equal to the magnetic flux that will produce an electromotive force of 1 volt in a conducting coil of one turn, as the flux is reduced to zero at a uniform rate in one second. The unit was adopted internationally in 1948 and is named in honor of the German physicist Wilhelm Eduard Weber (1804–1891).

Universal Time (UT)

Universal time is a name for Greenwich Mean Time that was introduced in 1928, on the recommendation of the International Astronomical Union to avoid confusion. Universal Time is counted from midnight, but at the time of its introduction Greenwich Mean Time was counted from noon.

Greenwich Mean Time (GMT, Z) is the mean time, calculated from the position of the Sun, at the 0° meridian. The 0° meridian passes through Greenwich, England, which was formerly the site of the Royal Observatory. The 0° meridian was agreed at an international conference held in Washington, D.C., in 1884. At first, Greenwich Mean Time was calculated from noon. This was altered to midnight in 1925, but because of the possibility of confusion the name was changed to Universal Time in 1928.

Most time services that broadcast the time against which users may check their own clocks use coordinated Universal Time (UTC). UTC is based on the uniform atomic timescale.

Upper Atmosphere Research Satellite (UARS)

The UARS is a NASA satellite that was launched on September 12, 1991. It carries six instruments that are designed to measure the TEMPERATURE, chemical composition, and WINDS in and above the stratosphere. These are the cryogenic limb array etalon spectrometer, improved stratospheric and mesospheric sounder, microwave limb sounder, HALOGEN OCCULTATION EXPERIMENT, high-resolution Doppler interferometer, and wind imaging interferometer (*see* SATELLITE INSTRUMENTS). Together, these instruments generate the most complete set of observational data for the upper atmosphere that have ever been produced.

Further Reading

NASA Goddard Space Flight Center. "Upper Atmosphere Research Satellite (UARS)." NASA Facts On Line. Available online. URL: www.gsfc.nasa.gov/gsfc/service/gallery/fact_sheets/earthsci/uars.htm. Posted January 1994, accessed February 23, 2006.

upwelling Upwelling is the name given to the rise of water all the way from near the ocean floor to the surface that occurs in particular regions of the ocean. Deep water is cold and, by bringing it to the surface, upwellings affect the temperature of the air in contact with the surface.

Upwellings are caused by the EKMAN SPIRAL. Ocean currents are driven by the wind, and they are subject to the CORIOLIS EFFECT (CorF). The balance between CorF and FRICTION between the wind and ocean surface produces two component forces of equal strength, one acting in the direction of the wind and the other acting at right angles to the wind direction. This results in the current flowing at an angle of 45° to the wind direction—to the right of the wind in the Northern Hemisphere and to the left in the Southern Hemisphere.

Below the surface the influence of the wind decreases, so the CorF becomes progressively more dominant with increasing depth. This has the effect of deflecting the current, and the overall result is that water in the surface boundary layer, down to about 82 feet (25 m), is slowly transported in a direction at right angles to the wind. Deeper water then rises to the surface to replace it. It is a slow movement, often at less than 3.3 feet (1 m) per day. The surface boundary layer that is subject to the Ekman spiral is known as the Ekman layer.

Where winds blow parallel to a north–south coastline and drive a current that flows near the coast and approximately parallel to it, the Ekman spiral pushes the water away from the coast, allowing deep water to rise to the surface. This is known as coastal upwelling, and it happens along the coasts forming the eastern margins of oceans, such as the western coast of North, Central, and South America. There, the ocean currents flow toward the equator—the California Current in the North Pacific and the Peru (or Humboldt) Current in the South Pacific. These carry cold water, but upwelling increases their cold influence. The difference is very marked. In August, the temperature of surface water in the Atlantic off the coast of North Carolina is often 70°F (21°C) and it can be warmer. The August temperature in the Pacific, in the same latitude off the coast of California, is about 59°F (15°C).

On either side of the equator the trade winds (*see* WIND SYSTEMS) drive currents from east to west in both hemispheres. Although it is very weak close to the equator, the CorF nevertheless acts on the Equatorial Currents, pushing them away from the equator. The result is that these currents are displaced a little way from the equator and there is a region of upwelling between them, the rising water dividing at the surface and flowing north and south. This is known as equatorial upwelling.

Upwelling, called open ocean upwelling, also occurs in the open ocean far from land. Atmospheric CYCLONES produce winds that drive currents. Like other currents, these are deflected by the CorF, and water rises to the surface layer. As a cyclone crosses the ocean, it leaves behind it a wake of relatively cool water. The deeper the cyclone the more pronounced this wake is, and it is very evident to the rear of a TROPICAL CYCLONE.

Ocean water is constantly turning over, but the process is extremely slow. The deep water that wells up to the surface was last at the surface centuries earlier (*see* GREAT CONVEYOR and NORTH ATLANTIC DEEP WATER). It sank in high latitudes and was very cold. Since the solubility of OXYGEN in water is inversely proportional to the temperature, the sinking water was rich in dissolved oxygen. It carried its oxygen to the ocean depths, where it sustained marine organisms. Inorganic nutrients, particularly nitrate (NO_3) and phosphate (PO_4), sink from the upper layers of the ocean and accumulate on the ocean floor. As it moves across the deep ocean floor, the deep water becomes enriched with these nutrients, and when it wells up to the surface, it brings them with it. This greatly enrich-

es the ordinarily nutrient-poor surface waters, and upwellings support large marine populations. If the prevailing winds should fail, or change direction, the wind-driven current may also cease to flow, and without its movement the upwellings also cease. This is what happens during an El Niño (*see* ENSO). Its social as well as its meteorological effects can be serious. Fisheries that are sustained by the upwelling nutrients fail, and populations that are economically dependent on them suffer badly.

upwind Upwind is the direction from which the WIND is blowing. Wind direction always refers to the direction from which the wind blows, not the direction toward which it is blowing, so if the wind is a westerly, for example, the upwind direction is to the west. The opposite of upwind is downwind, meaning in the direction the wind is blowing.

urban climate Climatic conditions in a large city are markedly different from those in rural areas adjacent to that city. The difference between the two climates prevails throughout the year, so city dwellers experience a climate that is generally warmer, wetter, and less windy than the climate in which country dwellers live. The urban climate is a genuine phenomenon.

Various factors combine to make cities warmer than rural areas. A city is therefore an area of relative warmth surrounded by the cooler countryside, a phenomenon that is called the HEAT ISLAND effect. In winter, the city is an average 1.8–3.6°F (1–2°C) warmer than the surrounding countryside, and over the year it is 0.9–2.7°F (0.5–1.5°C) warmer. Because the city is warmer, there are 10 percent fewer heating DEGREE-DAYS in urban than in rural areas.

City air is also dustier, however. There are 10 times more small particles in urban air than there are in country air. High buildings also shade much of the ground surface. Between them these two factors reduce the amount of sunshine at street level by 15–30 percent. They also reduce the intensity of ultraviolet radiation (*see* SOLAR SPECTRUM) by 5 percent in summer and 30 percent in winter.

Warmer temperatures mean that the relative HUMIDITY is about 6 percent lower in the city. There is also less EVAPORATION, because surface water from PRECIPITATION is removed rapidly through storm drains, and less TRANSPIRATION, because there are fewer plants.

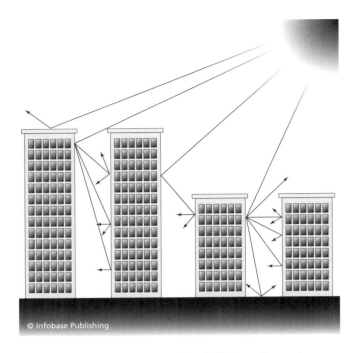

The pathways by which tall and short buildings reflect solar radiation. Little of the radiation reaches the ground. Almost all of it is absorbed by the fabric of the buildings.

The amount of precipitation over the city is 5–15 percent higher than it is in rural areas, however. This is probably due to the larger number of particles that are available as CLOUD CONDENSATION NUCLEI, although scientists are not yet sure. The city is 5–10 percent cloudier than the countryside. The warmer air increases instability (*see* STABILITY OF AIR) and the formation of cumuliform clouds (*see* CLOUD TYPES). Some of these develop into cumulonimbus, and cities experience 5 percent more THUNDERSTORMS than rural areas in winter and 29 percent more in summer.

Urban WIND SPEEDS are 25 percent lower than those in the countryside because of the effect of buildings, and there are 5–20 percent more days when the air is calm. At night there is often a country breeze (*see* WIND SYSTEMS).

A temperature INVERSION often forms above the city, especially in winter. This traps gaseous pollutants and particles, increasing the frequency of FOG. Fog occurs in winter twice as frequently in the city than in the countryside and 30 percent more often in summer.

The temperature inversion associated with an urban heat island encloses air inside what is called an

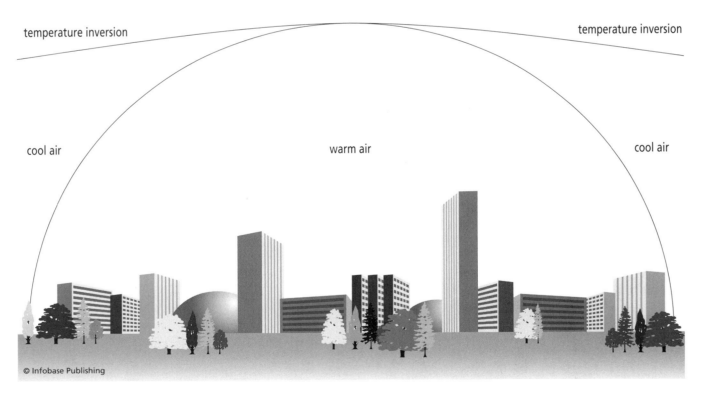

An urban dome forms when warm air rising over the city is trapped beneath an inversion. Air flows to the sides, cools, and returns to the ground, forming a domed shape

urban dome. Warm air rises over the city, encounters the inversion, and spreads to the sides. As it moves, it radiates away some of its own heat. This radiation cools the air, increasing its DENSITY, so the air subsides over the countryside just beyond the city boundary. From there it flows back into the city, toward the low-pressure region at the center. There is thus convergence in the inner part of the city and divergence above the city (*see* STREAMLINES), and the warm air beneath the inversion has an approximately domed shape. The urban dome is most pronounced on calm nights when the sky is clear, because that is when the heat island is most strongly developed.

Beneath the urban dome, the level of the rooftops resembles the canopy of a closed forest, and the layer of air below this canopy is called the urban canopy layer. The climate in this layer is strongly modified by the many microclimates (*see* CLIMATE CLASSIFICATION) produced by the streets and buildings. Winds blowing in from the surrounding countryside are slowed and deflected by the buildings and other obstructions they encounter, but they are also funneled (*see* WIND SPEED) along urban canyons. The burning of fossil fuels releases WATER VAPOR into the air, and some of the water piped into the city from outside evaporates. Together these modify the humidity of the air in the canopy layer. Its temperature is also modified by heat released from vehicle and other engines, industrial processes, and the air conditioning and heating of buildings. The magnitude of these effects changes in the course of the day, but it does so as a reflection of human activity rather than being wholly due to the daily cycle of sunshine and darkness.

A city street that is lined on both sides by tall buildings physically resembles a canyon. It is called an urban canyon. Wind tends to be funneled along the street, making it windier, especially if the street is aligned with the prevailing wind. The canyon also affects the way solar radiation is received and absorbed. Depending on the orientation of the canyon, the faces of buildings on one side may receive different amounts of radiation than those on the opposite side, or similar amounts but at different times of day. If the street runs north to south, for example, buildings on the western side will face the Sun in the morning and sunshine will reach buildings on the eastern side in the afternoon. If the street runs east and west, both sides will receive the same amount of sunshine through the course of the day, but protrusions from the buildings will cast deep shadows.

There is a rural boundary layer of air in the rural area adjacent to a large city. The boundary layer lies between the top of tall vegetation and the uppermost limit of the region within which the climatic properties of the air are modified by the surface below. The rural boundary layer is markedly thinner than the nearby urban boundary layer. Air in it is stable and capped by an inversion. Air moving outward from the city is carried above it, so the urban and rural bodies of air do not mix.

The urban boundary layer extends from the top of the urban canopy to the uppermost limit of the region in which the climatic properties of the air are modified by the surface below. The urban surface is usually rougher, warmer, and often drier than that of the surrounding countryside. The roughness reduces wind speed and generates EDDIES, the resulting TURBULENT FLOW mixing the air. The slowing of the wind causes air to accumulate and expand upward, and the upward expansion is increased by CONVECTION due to the heat island effect. Vertical expansion of the air produces and maintains the shape of the urban dome, and during the day it raises the upper margin of the urban boundary layer until it approximately coincides with the top of the PLANETARY BOUNDARY LAYER. At night the urban boundary layer contracts to about one-fifth of its maximum daytime depth. This is because air in the planetary boundary layer is stable at night, which restricts vertical movement.

vapor A vapor is any gas, although in METEOROLOGY the word is often used to mean WATER VAPOR. A gas, or vapor, is a substance that fills any container in which it is confined, regardless of its quantity.

The GAS LAWS describe the relationships between the TEMPERATURE and volume of an ideal gas, and the pressure that the gas exerts on the walls of its container. Gas molecules move freely, and bounce when they collide with one another or with a solid surface.

When a vapor is cooled, its molecules lose energy and move more slowly. If it is cooled sufficiently, the molecules will join in temporary groups and occupy a volume that reflects the number of molecules present, at which point the vapor has condensed and become a liquid.

Any substance that vaporizes readily at temperatures ordinarily found near the surface is said to be volatile.

varves Varves are sequences of light and dark bands that are visible in vertical sections taken from the sediments on the beds of some glacial lakes. The word *varve* is derived from *varv*, which is the Swedish word for layer. The study of varves, called varve analysis, helps scientists to measure how long the lake has existed and to determine the rate at which climate changed in the past.

One varve comprises one light band and one dark band. Within each varve, the pale layer is thick and has a coarse texture and the dark layer is thin and fine-grained.

In spring and summer, WATER that has melted from a nearby GLACIER flows rapidly into the lake. The meltwater carries small pebbles, sand grains, silt particles, and particles of clay, discharging all of them into the lake water. The bigger particles quickly settle to the bottom, forming a thick, pale layer of pebbles, sand, and some bigger silt particles. In winter, the edge of the glacier and the surface of the lake both freeze and the supply of meltwater ceases. All of the large particles now lie on the lakebed, but much smaller particles, of silt and especially of clay, sink much more slowly. They continue to settle through the winter, forming the thin, fine-grained, dark layer.

Since one layer is formed in summer and the other in winter, it is possible to measure the age of the lake by counting the varves in the same way that TREE RINGS can be counted to determine the age of a tree. The technique for doing this was introduced in 1878 by the Swedish geologist Gerhard Jacob de Geer (1858–1943). By counting the varves in Scandinavian lakes, de Geer concluded that southern Sweden was still covered by ice 13,500 years ago. As temperatures rose, the glaciers retreated to the north, and then, 8,700 years ago, the glaciers separated into two small ice caps.

Varves also form in milder climates that allow aquatic algae (single-celled, plantlike organisms) to grow for part of the year. These varves comprise one layer that is rich in organic matter and one that contains little organic matter.

In summer, there is a bloom of algae, and as they die the remains of dead algae settle to the lakebed to form the organic-rich layer. At the end of the summer, the blooms die down, so the winter layer contains much less organic matter. That is how the Green River Formation developed over an area of 48,250 square miles (125,000 km²) in southwestern Wyoming, northwestern Colorado, and northeastern Utah during the EOCENE epoch (55.8–33.9 million years ago).

Where the surface of the lake freezes in winter the process is a little different. The ice insulates the water from cold winds and also traps the heat that the water would otherwise lose by BLACKBODY radiation. Sunlight can penetrate the ice, however, and the algae continue to grow through the winter, albeit slowly. Their remains settle on the bottom, forming a thin, organic-rich layer. In spring, when the ice melts, water rushes into the lake carrying pebbles, silt, and sand. This mixes with the dead algae as it settles to the lakebed, so although the algae are growing more vigorously, the summer layer contains a smaller proportion of organic matter.

The thickness of varves provides an indication of the warmth of the weather during each year.

vector quantity   A vector quantity is a physical amount that acts in a direction, so that its description must include both the magnitude of the amount and its direction of action. This is contrasted with a SCALAR QUANTITY.

VELOCITY is a vector quantity. Wind velocity is the speed of the wind and the direction from which it is blowing. The speed of the wind, which is a scalar quantity, can also be stated, omitting the direction.

velocity   The velocity of a body is the speed at which that body is traveling in a specified direction. Speed is a scalar quantity, which means that only its amount is relevant and not the direction in which it acts. TEMPERATURE is also a scalar quantity. Velocity is a VECTOR QUANTITY, which means its direction of action must be specified.

It is correct, for example, to report the wind speed as, say, 25 MPH (40 km/h), but it would be incorrect to describe this as the wind velocity. The wind velocity might be 25 MPH from 240°, often abbreviated as 25/240, 240° being the compass direction from which the wind is blowing.

The speed of a body that is moving along a curved path is known as its angular velocity. Angular velocity is usually expressed in radians (*see* UNITS OF MEASURE-

MENT) per second (rad/s) and described by the symbol Ω. The circumference of a circle is equal to 2π radians. It follows, therefore, that $\Omega = 2\pi \div T$, where T is the time taken to complete one revolution. The tangential velocity (V)—the velocity in a straight line that can be measured in miles or kilometers per hour—is given by $V = \Omega r$, where r is the radius of the circle.

ventifact   A ventifact is a desert pebble that has been worn away by the abrasive effect of wind-blown sand in such a way that it has clearly defined faces. The name is derived from two Latin words: *Ventus* means wind, and *facere* means make. Provided the pebble is not moved, the positions of the faces indicate the direction of the prevailing wind (*see* WIND SYSTEMS).

An einkanter is a ventifact that has only one edge. This indicates it was formed by the action of wind-blown sand arriving from predominantly one direction.

A dreikanter has three edges. The abrasive action of wind-blown sand is believed to loosen and remove mineral grains from the surface directly exposed to the wind until the pebble falls over, exposing a second side. This is abraded in turn until the pebble falls again, exposing a third side and finally producing a dreikanter.

Venturi effect   When a flow of air is constricted, for example by tall buildings or hills, the wind accelerates. This is known as the Venturi effect, after the Italian physicist Giovanni Battista Venturi (1746–1822), who discovered it in 1791.

Air leaves the constricted area at the same rate as that at which it enters. In order to do so, the air must travel faster past the constriction. Constriction has the effect of bringing STREAMLINES closer together.

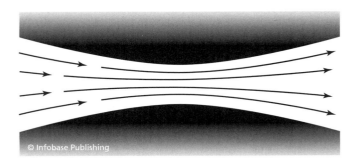

The constriction, which might be the walls of buildings or the sides of hills, draws streamlines closer together, accelerating the air through the constricted region.

The surface of Venus. This is a view of Cunitz crater. *(NASA/Science Source)*

Venus atmosphere The planet Venus is of approximately similar size to Earth. Its mean diameter is 7,521 miles (12,104 km); that of Earth is 7,918 miles (12,742 km). It is closer than Earth to the Sun, orbiting at 0.72 astronomical units (AU); 1 astronomical unit is the average distance between the Earth and Sun. Venus orbits at an average distance of 66,932,000 miles (107,712,000 km) from the Sun, and Earth orbits at 92,961,000 miles (149,600,000 km). Gravity at the surface of Venus is 8.87 m/s^2, and at the surface of Earth it is 9.8 m/s^2.

Earth and Venus are very similar in these respects and are sometimes described as twin planets. The atmosphere and climate of Venus are very different from those of Earth, however.

The atmosphere of Venus consists of 96 percent CARBON DIOXIDE and 3.5 percent NITROGEN with traces of about 150 parts per million (ppm) of SULFUR DIOX-IDE, 70 ppm of ARGON, 20 ppm of WATER VAPOR, 17 ppm of CARBON MONOXIDE, 12 ppm of HELIUM, and 7 ppm of NEON. The atmospheric pressure at the surface is 92 bars, compared with 1 bar on Earth.

The dense Venusian atmosphere, consisting almost entirely of carbon dioxide, produces a strong GREEN-HOUSE EFFECT, and combined with its closer proximity to the Sun this gives Venus an extremely hot, dry climate. The global average surface temperature is 867°F (737 K, 464°C). On Earth, lead melts at 621.5°F (327.5°C); the melting point would be higher under the greater atmospheric pressure on Venus, but probably lead would flow as a liquid on the surface of Venus, and this gives an indication of just how hot the climate is. It is hot enough to make rocks glow.

There is little variation in temperature between the equator and poles or between day and night, because of the strong and efficient transport of heat by the atmospheric circulation in both hemispheres. Surface winds blow at about 2 MPH (3 km/h) or less, but there are high-level winds blowing parallel to the equator at about 224 MPH (360 km/h). Such winds could not blow on Earth because they would be balanced by the CORIOLIS EFFECT (CorF). The magnitude of the CorF is proportional to the rotational speed of the planet, and Venus turns much more slowly than Earth, so one day on Venus is equal to 243 days on Earth. Consequently, the CorF is much weaker on Venus than it is on Earth. Venus also rotates in the opposite direction (the rotation is retrograde), so the Sun rises in the west and sets in the east.

Viewed from outside, the surface of Venus is entirely obscured by cloud. At one time the clouds were believed to be of WATER. Today they are known to consist of sulfuric acid droplets. Spacecraft have visited Venus on several occasions, and the surface has been comprehensively mapped, but the high pressure and temperature and strongly acidic atmosphere mean instruments on landers last for only a very short time on the surface.

Further Reading

Kordic, Ruby. "Venus." Available online. URL: http://ruby.kordic.re.kr/~vr/CyberAstronomy/Venus/HTML/index.html. Accessed Fenruary 23, 2006.

Soper, Davison E. "Atmosphere of Venus." University of Oregon. Available online. URL: http://zebu.uoregon.edu/~soper/Venus/atmosphere.html. Accessed February 23, 2006.

Viroqua Viroqua is a city in Wisconsin that was struck by one of the most severe tornadoes on record on the afternoon of June 28, 1865. On the basis of the damage it caused, the tornado has been judged F4 on the FUJITA TORNADO INTENSITY SCALE.

The tornado produced multiple vortices that sometimes merged into a single vortex, and it moved at an estimated 60 MPH (96 km/h) along a path that was 300 yards (275 m) wide and 30 miles (48 km) long. The tornado began to the southwest of Viroqua, passed through the southern part of the town, passed to the south of Rockton, and then dissipated. It lifted a schoolhouse from the ground, complete with the teacher and 24 students inside, then dropped it, killing the teacher and eight students. A total of 29 people were killed and at least 100 injured.

Further Reading

NOAA, National Weather Service Forecast Office, La Crosse, WI. "The Tornadoes of June 28, 1865." NOAA. Available online. URL: www.crh.noaa.gov/arx/events/tors_jun1865.php. Last modified November 8, 2005.

Viroqua. "The Viroqua Tornado June 28th, 1865." Available online. URL: www.wx-fx.com/viroqua.html. Accessed February 23, 2006.

virtual temperature The TEMPERATURE dry air would have if it were at the same DENSITY and pressure as moist air is called its virtual temperature. The virtual temperature of moist air is a little higher than its actual temperature.

By correcting for the effect due to the density of WATER VAPOR, the use of the virtual temperature makes it possible to apply the equation of state to moist air using a single gas constant (see GAS LAWS), rather than calculating constants separately for dry and moist air.

A PARCEL OF AIR that rises above the level of free convection (see STABILITY OF AIR) will reach a height at which its virtual temperature is equal to the temperature of the surrounding air. At this height the air will be neutrally buoyant and will rise no higher. It is said to have reached its equilibrium level, also known as the level of zero buoyancy.

viscosity Viscosity is the resistance a fluid presents to SHEAR forces, and therefore to flow. It is caused by the random intermingling of molecules at the boundary between two fluid bodies, one of which is moving in relation to the other. Such molecular mingling also transfers energy from one fluid body to the other. This transfer of energy is important in the transport of energy by liquids, but EDDY viscosity is by far the most important mechanism for energy transport in air.

The coefficient of viscosity is the force per unit area, applied at a tangent, that is needed to maintain a unit relative velocity between two parallel planes set a unit distance apart in a fluid. It is measured in newtons per square meter per second ($N/m^2/s$) in the SI system and dynes per square centimeter per second ($dyn/cm^2/s$) in the c.g.s. system.

The kinematic viscosity is the coefficient of viscosity of a fluid divided by its density. It is measured in square meters per second (m^2/s) in the SI system and square centimeters per second (cm²/s) in the c.g.s. system (*see* UNITS OF MEASUREMENT).

visibility Visibility is the distance from which an observer is able to distinguish an object such as a tree or building with the naked eye. Expressed another way, it is the transparency of the air to visible light.

Visibility is measured at a WEATHER STATION with reference to a number of familiar objects at known distances from an observation point. It is necessary to make a number of measurements in different directions to determine the all-round visibility. The prevailing visibility is the greatest horizontal visibility that extends over at least one-half of the horizon around an observation point. It is the visibility that is reported on a STATION MODEL.

Surface visibility is the horizontal visibility measured by an observer on the ground, rather than in an airfield control tower. Unrestricted visibility is horizontal visibility that is not obstructed or obscured for at least 7 miles (11 km).

Variable visibility is the condition in which visibility increases and decreases rapidly while it is being measured. The visibility must then be given as the average of the measured values. Variable visibility is reported only if it is less than 3 miles (4.8 km).

An object is visible if the eye can detect a contrast between it and the sky. For most people this requires the object to be at least about 5 percent darker than the sky. The minimum detectable contrast is symbolized by E (the Greek letter epsilon).

Visibility is reduced by the presence of water droplets and solid particles in the air between the object and the observer and, to a very much smaller extent, by the air itself. Objects are visible because of light reflected from them to the observer. Between the object and the observer some of the reflected light is scattered (*see* SCATTERING) and some is absorbed, so only a proportion of the reflected light reaches the observer. This light mixes with airlight coming from other directions, which "dilutes" the light from the object, making it harder to see.

The loss of reflected light between the object and the observer is known as extinction, and the fraction of the light lost per unit of distance under given conditions is known as the extinction coefficient, abbreviated as b_{ext}. The extinction coefficient varies according to the size and DENSITY of droplets or particles.

If the extinction coefficient is known, the visual range (r_v) is given by:

$$r_v = \log_e \varepsilon / b_{ext}$$

Obscuration is the situation when the sky is completely hidden by a weather feature at ground level, such as FOG. Any atmospheric feature, other than clouds, that obscures a portion of the sky as seen from a weather station is known as an obscuring phenomenon. An atmospheric feature that reduces horizontal visibility at ground level, such as fog, SMOKE, or blowing SNOW is called an obstruction to vision. These terms are used in United States meteorological practice. Blowing spray is water that is blown from the surface of the sea to form spray in an amount large enough to reduce visibility significantly.

The precipitation ceiling is the vertical visibility that is measured looking upward into PRECIPITATION. This measure is used when precipitation obscures the CLOUD BASE. An indefinite ceiling is the condition in which the vertical visibility cannot be measured precisely, because it is determined not by the cloud base, but by fog, HAZE, blowing snow, sand, or DUST. Precipitation does not produce an indefinite ceiling.

volcanic eruptions When a VOLCANO erupts, the cloud of gas it ejects into the atmosphere may affect the climate. Part of the ejected material, called ejecta, often consists of WATER VAPOR and steam, as well as SULFUR DIOXIDE (SO_2) that is quickly oxidized to sulfate (SO_3) particles, which then react with moisture (H_2O) to form sulfuric acid (H_2SO_4). The ejecta cloud also contains rock that was vaporized by the high temperature in the magma chamber. The vaporized rock condenses as it mixes with the cold air, forming small particles of rock. This is the volcanic ash that falls to the ground, blanketing the surface.

Volcanic ash is quite unlike the ash from a wood or coal fire. It is highly abrasive, and when it mixes with water it forms a slurry that sets hard as it dries. Meteorologists track the movement of ash clouds and warn aircraft to keep well clear of them. Volcanic ash could wreck an aircraft engine. It is also highly dangerous if inhaled.

Ash that is ejected into the lower atmosphere remains airborne for a matter of hours or days before it settles or rain washes it to the surface. If the ash enters the stratosphere (*see* ATMOSPHERIC STRUCTURE), however, it may remain there for months. Stratospheric air movements then spread the plume, sometimes into a belt of material that encircles the Earth. Sulfate, sulfuric acid, and fine ash particles all reflect sunlight, thus increasing the planetary ALBEDO, and the cloud also absorbs solar radiation, which warms the stratosphere. Both effects combine to reduce the amount of solar radiation reaching the surface and lower temperatures in the lower atmosphere.

Not all eruptions eject sufficient material to affect the climate, and not all eruptions are violent enough to hurl material all the way into the stratosphere. From time to time, however, an eruption triggers a marked fall in surface temperatures. Some of the most important of these are listed here.

El Chichón is a volcano in Mexico that became active in 1982, after having remained dormant for several centuries. It began to erupt on March 26 and continued erupting until the middle of May. The final death toll from the eruption was estimated at about 2,000. There was an especially violent eruption on March 29 that killed 100 people living in nearby villages, and on April 4 a huge explosion released a cloud of dust and gas. The gas included up to 3.6 million tons (3.3 million tonnes) of sulfur dioxide, all of which was converted into sulfuric acid. By May 1, this cloud had encircled the Earth. In late June, the densest part of the cloud was detected at a height of 17.4 miles (28 km), and by the end of July the top of the cloud reached about 22 miles (36 km). A year later, the cloud had spread to cover almost the whole of the Northern Hemisphere and a large part of the Southern Hemisphere. It was the first cloud of volcanic material to be tracked by instruments on satellites, and it produced a marked warming of the lower stratosphere. In June, the cloud reduced the average global temperature by about 0.4°F (0.2°C).

Krakatau, formerly known as Krakatoa, is a volcanic island that lies in the Sunda Strait between Java and Sumatra, Indonesia. On May 20, 1883, the volcano became active, and on August 26 and 27 it erupted in a series of increasingly violent explosions. These were heard in Australia, 2,200 miles (3,540 km) to the southeast, and at Rodriguez Island, 3,000 miles (4,800 km) to the southwest. The eruption threw about 5 cubic miles (21 km³) of solid particles to a height of more than 19 miles (30 km). Stratospheric winds distributed the DUST over most of the Earth, producing spectacular sunsets for the following three years, which

was the time it took for the dust to settle. During those three years, the Montpelier Observatory, in France, recorded a 10 percent decrease in the intensity of solar radiation, and there was a small but significant fall in average temperatures.

Mount Agung (Gunung Agung) is a volcano on the island of Bali, Indonesia, that erupted violently in March 1963, ejecting large amounts of particulate material, together with sulfur dioxide and sulfate AEROSOL. Some of the sulfate entered the lower stratosphere, and the wind patterns at the time carried it into the Southern Hemisphere. Within a very short time aerosol droplets were detected between Bali and Australia. The absorption of solar energy by the stratospheric aerosols warmed the lower stratosphere by 11–12°F (6–7°C) and surface TEMPERATURES fell by about 1°F (0.5°C). After about six months, the aerosols had spread around the world, and they remained in the air for several years, producing spectacular sunsets, but no measurable climatic effect.

Mount Aso-San is a volcano on the Japanese island of Kyushu. The volcano is 5,223 feet (1,593 m) high and has one of the biggest calderas (craters) in the world, measuring 17 miles (27 km) from north to south and 10 miles (16 km) from east to west. Mount Asosan erupted violently in 1783, injecting a large amount of dust and aerosol into the stratosphere. The eruption was followed by unusually cold weather from 1784 to 1786.

Mount Katmai, a volcano in Alaska, erupted violently in 1912, ejecting an estimated 5 cubic miles (21 km³) of dust high into the atmosphere. It was the most violent volcanic eruption of the 20th century. During the months that followed, observatories in California and Algeria recorded a 20 percent drop in solar radiation and the weather was unusually cool, although temperatures had been somewhat lower than normal before the eruption. A bishop's ring (see OPTICAL PHENOMENA) was seen following the eruption.

Mount Pinatubo is a volcano on the island of Luzon, Philippines, which erupted in 1991. It was the second most violent volcanic eruption of the 20th century (after Mount Katmai). The volcano had not erupted for 600 years. Activity began in April, and on June 14–16 the mountain split into pieces in a series of explosions. The eruption ejected dust and sulfate aerosol to a height of 25 miles (40 km). This spread over most of the world. The amount of material in the stratosphere reached a peak in September 1991, after which the amount began to decrease over the TROPICS. It did not peak until the spring of 1992 over latitudes 40–60°N and remained fairly constant until the end of 1992 over latitudes 40–60°S. It could still be detected over Hawaii and Cuba in January 1994. Absorption of solar radiation by material in the stratosphere caused a cooling of about 0.7–1.25°F (0.4–0.7°C) in the troposphere that lasted through 1992 and 1993.

Mount Spurr, a volcano in Alaska, erupted on June 27 and August 18, 1992, injecting dust and sulfur dioxide into the atmosphere. There is no record of any climatic effect.

Mount Tambora is a volcano in Indonesia (then the Dutch East Indies) that erupted in April 1815. The eruption was one of the most violent in the last few thousand years, and scientists calculated that it released at least 3.6 cubic miles (15 km³) of dust and sulfuric acid aerosols that spread to form a veil over much of the Northern Hemisphere. The particles may have added to some that were still present from earlier volcanic eruptions at St. Vincent in the West Indies and Awu in Sulawesi. Temperatures fell by up to 2°F (1°C) in many areas, and wind patterns were distorted. The following year, 1816, came to be known as "the year with no summer." Snow, driven by a northeasterly wind, fell over a wide area of eastern North America in June 1816 and from June 6 to 11 the ground inland was covered in snow as far south as Pittsburgh. Connecticut had frosts at some time in every month of 1816, and there were some June days when the temperature did not rise above freezing in Québec City. Harvests failed in many parts of North America and Europe, but the altered wind patterns brought a fine summer to the north of Scotland, and there was a heat wave in Ukraine. The bad weather of 1816 may have been partly responsible for the worst outbreak of typhus Europe had ever experienced, lasting from 1816 until 1819.

volcanic explosivity index (VEI) The VEI is a classification of VOLCANIC ERUPTIONS that includes the estimated amount of material each category injects into the atmosphere. This is relevant to climate, because the more violent the explosion the more likely it is that some of the fine particles it throws into the air will penetrate the stratosphere (*see* ATMOSPHERIC STRUCTURE) where they may remain for months or even years. Particles in the stratosphere reflect incoming sunlight and may therefore have a climatic cooling effect experienced over a wide area.

The VEI also includes the frequency with which eruptions of each type occur, based on historical records. The full VEI is given in the table below.

volcano VOLCANIC ERUPTIONS inject ash, DUST, rocks, and a variety of gases into the air. They play an important part in the cycles of elements by returning chemical elements to the atmosphere from below the Earth's crust.

Carbon, sulfur, and other elements that spend part of their time in the air eventually find their way into

Volcanic Explosivity Index

Index	Type	Plume height feet	Volume cu.feet	Eruption type	Frequency	Example
0	nonexplosive	<350	3,500+	Hawaiian	daily	Kilauea
1	gentle	350–3,500	35,000+	Hawaiian/Strombolian	daily	Stromboli
2	explosive	3,500–16,000	35 million+	Strombolian/Vulcanian	weekly	Galeras 1992
3	severe	10,000–50,000	35 million+	Vulcanian	yearly	Ruiz 1985
4	cataclysmic	33,000–82,000	350 million+	Vulcanian/Plinian	decades	Galunggung 1982
5	paroxysmal	>82,000	3.5 billion	Plinian	centuries	Saint Helens 1981
6	colossal	>82,000	35 billion+	Plinian/Ultra-Plinian	centuries	Krakatau 1883
7	supercolossal	>82,000	350 billion+	Ultra-Plinian	millennia	Tambora 1815
8	megacolossal	>82,000	3,500 billion+	Ultra-Plinian	tens of millennia	Yellowstone 2 Ma

(Ma means millions of years ago.)

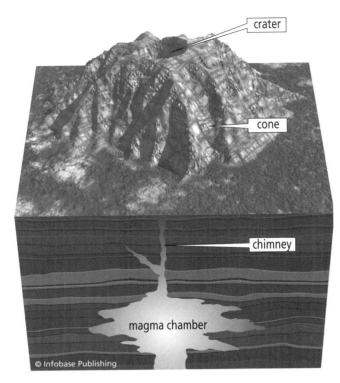

A cross section through a typical volcano of the Strombolian type

the composition of the magma. This may seep over the ground surface, be thrown into the air, or it may explode violently. Magma that pours out at the surface is called lava.

There are different types of volcanic eruption. A Hawaiian eruption produces fountains of fire and very fluid lava that flows down the side of the volcano. Peléean eruptions, named after Pelé, the Hawaiian goddess of volcanoes, are violent and explosive. Plinian eruptions are also explosive and eject large amounts of material into the air. They are named after Pliny the Elder, who died in 79 C.E., when Vesuvius erupted in this way. Strombolian eruptions, named after the Italian volcano Stromboli, throw out thick lava that falls back to build a steep-sided cone. Surtseyan eruptions are named after the island of Surtsey, near Iceland, which was formed by an eruption of this type in 1963. Surtseyan eruptions are very violent. They happen when water pours into the vent leading to the magma chamber, causing a huge explosion. Vesuvian eruptions, named after Vesuvius, the volcano beside the Bay of Naples, Italy, are explosive, but occur after long periods of dormancy. Vulcanian eruptions happen when the pressure from trapped gases blows away the overlying crust of solidified lava. Vulcan was the Roman god of fire—and the Romans sensibly built their temples to him outside the city.

Voluntary Observing Ship (VOS) A VOS is a merchant ship that is equipped to act as a WEATHER STATION. Port meteorological officers (PMOs) recruit ships into the VOS scheme. PMOs also supervise the provision and installation of the necessary instruments.

The VOS scheme began in 1853, at a conference in Brussels that was convened by Matthew F. Maury (*see* APPENDIX I: BIOGRAPHICAL ENTRIES) and attended by representatives from 10 maritime countries. In 1984–85 there were about 7,700 VOS in the world as a whole. Numbers have declined since then (probably because of the decline in the size of the merchant fleets of several major industrial nations), and today the VOS fleet comprises more than 6,000 vessels from 49 countries. The United States has the largest number of VOS, with more than 1,600. The scheme operates under the auspices of the WORLD METEOROLOGICAL ORGANIZATION (WMO).

A port meteorological officer (PMO) is an official belonging to a national meteorological service who is

the sea, and a proportion forms insoluble compounds that accumulate in the sediment on the seabed. Compression beneath the weight of overlying sediment and heat from the lower crust slowly turn this sediment into sedimentary rock. Seafloor spreading carries the layers of sedimentary rock lying on top of the rocks of the oceanic crust to a subduction zone, where one of the Earth's crustal plates is sinking beneath another (*see* PLATE TECTONICS). Subduction returns the sediment to the Earth's mantle, beneath the crust, and with it the elements that were once part of the air. Eventually, the mantle rock containing those elements may return to the surface in a volcanic eruption and the elements may be hurled high into the air.

A volcano begins as a space among the rocks of the crust into which mantle rock rises. Once it comes this close to the surface, the hot, semi-molten rock is called magma, and the space where it accumulates is a magma chamber. Magma continues to accumulate until its pressure forces it upward. Finding its way through weaknesses in the overlying rock, the rising magma makes one or more chimneys that eventually reach the surface. What happens then depends on appointed to supervise the Voluntary Observing Ships (VOS) scheme. PMOs are based at ports and spend much of their time visiting participating ships. They are responsible for enrolling vessels into the VOS scheme, supervising the supply, installation, and correct maintenance of the necessary instruments and other equipment, collecting the meteorological logbooks from ships returning to port, and generally ensuring that the weather reports from ships at sea meet the required standard.

Voluntary Observing Ships use SHIP code to send their reports. This is a version of the international synoptic code that is approved by the WMO for VOS use. Data contained in VOS weather reports are compressed into SHIP code before being transmitted by radio to a shore receiving station, from where they are passed to the WMO Global Telecommunications System.

A ship report is a weather report compiled on board a VOS at sea and transmitted to a shore receiving station. Regularly at midnight, 6 A.M., noon, and 6 P.M. UNIVERSAL TIME, officers on the VOS take observations of air TEMPERATURE, SEA-SURFACE TEMPERATURE, WIND SPEED, AIR PRESSURE, PAST WEATHER, and PRESENT WEATHER. They may also note the relative HUMIDITY, cloud cover, and state of the sea. The recorded data are compressed using SHIP code, stored in the onboard meteorological logbook, and transmitted to a receiving station on shore. The receiving station then passes the report on to the Global Telecommunications System of the World Meteorological Organization. When the ship reaches port, its meteorological logbook is handed over to the port meteorological officer. The contents of the logbook are used to augment the transmitted observations and help to build a long-term picture of the climate over the oceans.

Further Reading

JCOMM Voluntary Observing Ships' Scheme. "Port Meteorological Officers." Available online. URL: www.bom. gov.au/jcomm/vos/pmo.html. Accessed February 24, 2006.

NOAA. "The United States Voluntary Observing Ships Scheme." NOAA. Available online. URL: www.vos. noaa.gov/vos_scheme.shtml. Last modified September 29, 2003.

vortex A vortex is a spiraling movement in a fluid that affects only a local area. The fluid at the center of a vortex is usually stationary or slow-moving, but this calm center may be surrounded by gas or liquid that is moving very rapidly. In most vortices the fluid spirals inward toward a region of low pressure at the center. A circular vortex is a vortex in which the STREAMLINES are parallel to one another around a common axis.

A TORNADO is a vortex, and water flowing from a bathtub usually forms one. On a much larger scale, the polar vortex that forms each winter over Antarctica and in some winters over the Arctic encloses a large mass of very cold air. The tendency of a moving fluid to form a vortex is known as VORTICITY.

A bath plug vortex is a way of describing an atmospheric vortex by means of a familiar metaphor. Water leaving a bathtub usually forms a vortex that can be used to illustrate several features of atmospheric systems. Angular MOMENTUM is conserved. This can be seen by the acceleration of the water as it nears the center. The shape of the vortex resembles that of a tornado seen from above. The pressure surface is drawn down the center of the vortex, just as the tropopause (see ATMOSPHERIC STRUCTURE) is drawn down to the surface in the eye of a TROPICAL CYCLONE.

The polar vortex is the large-scale circulation that dominates the middle and upper troposphere in high latitudes and that is centered over the polar regions of both hemispheres. Air circulates cyclonically around the vortex. In the Northern Hemisphere, the vortex has two centers, one near Baffin Island and the other over northeastern Siberia.

A vortex also forms in winter in the stratosphere, with a strongly GEOSTROPIC WIND circulating around it. This is sometimes called the polar night vortex, and it is within this vortex that POLAR STRATOSPHERIC CLOUDS form. These are the sites of the chemical reactions by which OZONE is destroyed in the OZONE LAYER.

Vortices known as equatorial vortices develop at the center of equatorial waves over the Pacific Ocean. These travel westward toward the Philippines, but seldom intensify to become tropical cyclones.

A series of vortices that develops when a fast-moving flow of air passes an obstruction, such as a building, is known as a Karman vortex street. As the moving air approaches the obstruction, it divides into two streams. These pass to either side of the obstruction, and both of them accelerate due to the VENTURI EFFECT. If the air is moving gently and fairly steadily, the two streams decelerate and rejoin a short distance downstream from the obstruction. If the air is moving

© Infobase Publishing

A Karman vortex street is a series of vortices that develop downwind of an obstruction when the wind is strong.

rapidly, however, the two streams remain separated, and a series of vortices develop between them. These rotate alternately clockwise and counterclockwise at a slower speed than the mean speed of the main airflow. The vortices are constantly forming and disappearing and they greatly increase the TURBULENT FLOW of air. The Hungarian-born American aerodynamicist Theodore von Kármán (1881–1963) was the first person to describe this phenomenon.

The word *distrail* is an abbreviation of dissipation trail. This is a line of clear air that appears in the thin cloud behind an aircraft. It is caused by vortices in the wake of the aircraft. These draw dry air down into the cloud, causing the cloud particles to evaporate. The distrail may form a straight line or a line of holes in the cloud. What appears to be a distrail is sometimes a shadow cast onto the cloud by a CONTRAIL above it.

vorticity Vorticity is the measure of the rate at which a moving mass of fluid turns about an axis. If a moving fluid is pictured as an immense number of minute solid particles, and if each of these particles is rotating as well as moving forward with the general flow, then vorticity is present in the moving fluid. The rotation of the imaginary particles will impart curvature to the general flow.

The concept of vorticity is of great importance in fluid mechanics, which describes the movement of fluids of all kinds and under all circumstances. In METEOROLOGY, vorticity affects the large-scale movement of air around CYCLONES and ANTICYCLONES.

Vorticity is a VECTOR QUANTITY as is VELOCITY, but its coordinates have three components, each compo-

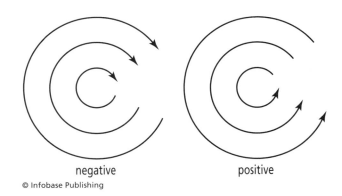

© Infobase Publishing

Vorticity is the tendency of a moving fluid to rotate about an axis. Vorticity is conventionally described as negative if the direction of flow is clockwise and positive if the flow is counterclockwise. In the case of air or water movement across the surface of the Earth, the axis is vertical and the direction of motion is seen from above.

nent on a line at right angles to the other two. Two coordinates are sufficient to specify the position and movement of the body on a flat (two-dimensional) plane. If the body is also rotating, a third coordinate in the direction of the axis of rotation specifies the rotation. This third coordinate is the vorticity coordinate, or vorticity scalar.

Vorticity develops because the mass of fluid is moving in relation to the fluid adjacent to it. This generates SHEAR forces arising from the difference in velocities of the two fluids, so that the faster tends to curve around the slower.

Near the ground, friction causes moving air to turn about a horizontal axis that is at right angles to the wind direction. This generates EDDIES, but has no large-scale effect on weather. In the more important type of vorticity, air turns about an approximately vertical axis, so its movement is horizontal and the vorticity scalar is vertical.

Convergence and divergence (see STREAMLINES) generate vorticity in respect of a fixed frame of reference, in this case the surface of the Earth. The vorticity of air that turns about a local vertical axis is equal to twice the angular velocity. It is known as the relative vorticity, and in meteorology relative vorticity is always shown by the symbol ζ (the lower case Greek letter zeta).

Vorticity is also caused by the rotation of the Earth. This is necessarily so because the fixed reference frame of the Earth's surface is also rotating about a vertical axis with an angular velocity that is proportional to the angular velocity of the Earth and the latitude of the rotational axis of the mass of fluid. This aspect is known as the planetary vorticity, and its magnitude is always equal to that of the CORIOLIS EFFECT, known as the Coriolis parameter. Both are designated by the symbol f. The magnitude of f is zero at the equator and at its maximum at the North and South Poles.

By meteorological convention, vorticity in a counterclockwise direction is said to be positive, and vorticity in a clockwise direction is said to be negative. Seen from a position directly above the North Pole, the Earth rotates in a counterclockwise direction, and it rotates clockwise as seen from above the South Pole. Consequently, planetary vorticity is positive in the Northern Hemisphere and negative in the Southern Hemisphere.

The sum of the planetary vorticity and the relative vorticity ($\zeta + f$) is known as the absolute vorticity.

Because of the conservation of angular MOMENTUM, the absolute vorticity remains constant ($d/dt (\zeta + f) = 0$). If minor components are omitted, such as those arising from the forces of BUOYANCY, FRICTION, and TORQUE, the value of absolute vorticity is given by the vorticity equation:

$$d/dt (\zeta + f) = - (\zeta + f)D$$

where d/dt is the rate of change and D is the rate of convergence (or divergence in the case of negative convergence, $-D$).

Because planetary vorticity is equal to the Coriolis parameter and absolute vorticity is constant, a change in the latitude of moving air causes a change in f that must be compensated by a change in ζ. If air in the Northern Hemisphere is diverted northward, for example, f increases, ζ decreases to compensate, and the vorticity becomes more negative, or anticyclonic. This turns the air in a southerly direction, decreasing f and increasing ζ, and vorticity then becomes more positive (cyclonic), turning the air northward again. This is how waves with a long wavelength develop in air that is flowing zonally (parallel to the equator). ROSSBY WAVES and LEE waves are examples of this effect.

Vostok, Lake

Vostok Lake lies beneath the Russian VOSTOK STATION in Antarctica. The presence of WATER beneath the ICE SHEET was recognized in the 1970s, and the size of Lake Vostok was revealed by satellite RADAR altimetry in 1996.

The largest of about 70 subglacial lakes, Lake Vostok measures approximately 139 × 30 miles (224 × 48 km) in area and is about 1,588 feet (484 m) deep—about the same volume as Lake Ontario. Drilling of the Vostok ICE CORE was stopped about 330 feet (100 m) above the surface of the lake while ways were devised to sample its waters without contaminating them. It is possible the lake is populated by microorganisms and molecules of biological origin in an environment that has been isolated from the rest of the world for hundreds of thousands or even millions of years.

Further Reading

West, Peter. "Lake Vostok." National Science Foundation Office of Legislative and Public Affairs. Available online. URL: www.nsf.gov/od/lpa/news/02/fslakevostok.htm. Posted May 2002. Accessed February 24, 2006.

Vostok Station   Vostok is a Russian research station in Antarctica, located at 78.46°S 106.87°E, and at an elevation of 11,401 feet (3,475 m). Its name means "east," and it was opened on December 16, 1957.

Vostok is sited at the geomagnetic South Pole and at the center of the East Antarctic ICE SHEET. It is also at the southern COLD POLE. Scientists from many other countries work there. The primary project at Vostok has been the drilling of ICE CORES. Drilling commenced in 1980 and in 1985 reached a depth of 7,225 feet (2,202 m), beyond which it was impossible to continue with that core. A second core was started in 1984. In 1989, it became a joint Russian–French–U.S. project and in 1990 reached a final depth of 8,353 feet (2,546 m). A third core was started in 1990. It reached 8,202 feet (2,500 m) in 1992, and in 1998 it reached 11,887 feet (3,623 m). Ice from this depth is about 420,000 years old. In 2005, Chinese workers began planning to drill a core through Dome A, which is the highest point on the ice sheet, 13,124 feet (4,000 m) above sea level. Ice at the base of the dome is more than 900,000 years old.

Further Reading

Antarctic Connection. "Vostok Station." Available online. URL: http://www.antarcticconnection.com/antarctic/stations/vostok.shtml. Accessed February 24, 2006.

Walker circulation The Walker circulation is a movement of tropical air that was proposed in 1923 by Sir Gilbert Thomas Walker (1868–1958) and that has since been found to be correct. The Walker circulation is a slight but continuous latitudinal movement that is superimposed on the Hadley cells (*see* GENERAL CIRCULATION).

The movement occurs between the equator and about latitude 30° in both hemispheres as a series of cells, called Walker cells. Air rises over the tropical western Pacific Ocean and over the eastern Indian Ocean, both near Indonesia. CONDENSATION in the rising air produces towering clouds and heavy rain. At high level each of the rising air currents separates into two streams, one flowing eastward and the other westward. The high-level streams from neighboring cells converge and subside over the eastern Pacific, near the South American coast, and over the western Indian Ocean, near the coast of Africa. They diverge at low level (*see* STREAMLINES).

This circulation produces regions of low surface AIR PRESSURE over land in tropical South America, Africa, and Indonesia, where air is converging and rising. Areas of high pressure over the oceans, where air is subsiding and diverging at the surface, separate these low-pressure areas. This distribution of surface pressure produces a prevailing easterly flow (east-to-west) near the surface, which strengthens the easterly trade winds (*see* WIND SYSTEMS) of the Hadley cells.

The Walker circulation produces a very wet climate in Indonesia and a dry climate over western South America. Every few years the pattern changes, and the Walker circulation over the Pacific weakens or reverses. This change is known as the SOUTHERN OSCILLATION.

Sir Gilbert Walker was a British meteorologist and professor of meteorology at Imperial College, London. He had been appointed head of the Indian Meteorological Service, and in 1904 he was asked to look for a pattern in the occurrence of the Indian MONSOONS. A failure of the monsoon means that crops will wilt and the harvest will fail, and excessively heavy monsoon rains can destroy crops. In either case famine is the likely result. The monsoon had caused famines in 1877 and again in 1899, and the 1899 famine prompted the British authorities (who then ruled India) to see if the variations in monsoons could be predicted sufficiently far in advance to take steps to minimize the suffering they caused.

Walker approached the task by studying climate records from all over the world. In those days weather was believed to be a fairly local phenomenon, but Walker noticed that events in one place were sometimes accompanied by different events a long way away. When there was low surface pressure over Tahiti, pressure was often high over Darwin, Australia. When the Indian monsoon failed, the winter in Canada was mild. Relationships such as these, between events that occur great distances apart, are now known as TELECONNECTIONS, and Walker first discovered them.

From these studies Walker discovered a relationship between oscillations in the AIR PRESSURE over the eastern and western Pacific Ocean and the Indian monsoon

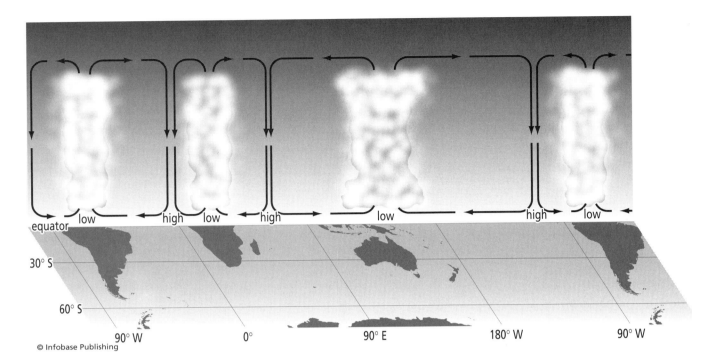

The Walker circulation is a pattern of latitudinal air movements over the Tropics that produce an easterly flow of air near the surface of the Pacific Ocean.

and rainfall in Africa. He called this periodic change in pressure distribution the Southern Oscillation, but he did not link it to the El Niño effect (*see* ENSO). It was not until 1960 that that connection was made, by Jacob Bjerknes (*see* APPENDIX I: BIOGRAPHICAL ENTRIES), who called this circulation the Walker circulation.

water   Water, or dihydrogen oxide (H_2O) is able to exist in all three PHASES (gas, liquid, and solid) at temperatures that are common at the surface of the Earth. A pond in winter, when its surface is partly frozen, contains liquid water and solid ice, and the air immediately above the surface contains WATER VAPOR, which is an invisible gas. In this example, the gas, liquid, and solid are at different temperatures, but at 32.018°F (273.16K, 0.01°C) and at a pressure of 6.112 mb (0.089 lb/in², 0.18 inch of mercury) pure water exists in all three phases simultaneously. This is called the triple point of water (*see* BOILING).

Water is the only common substance that can exist in all three phases under the conditions found at the surface of the Earth. Ammonia (NH_3) is also a common substance, but at ordinary sea-level pressure (*see* UNITS OF MEASUREMENT) it freezes at -107.9°F (-77.7°C)

and boils at -29.83°F (-34.35°C). Hydrogen chloride (HCl), also called hydrochloric acid, freezes at -173.6°F (-114.22°C), and boils at -121.27°F (-85.15°C). HCl is stored as a liquid in glass bottles by dissolving it in water, so the acid in the bottle in fact is a solution.

When water freezes, its molecules form an open pattern. This causes the water to expand as it freezes, and it also means that ice is less dense, and therefore weighs less, than liquid water and floats on top of it. This is another unusual property. Most substances are denser as solids than they are as liquids. If ice were denser than water, ponds, lakes, and the sea would freeze from the bottom up, rather than from the top down, and life would be impossible for many of the plants and animals that inhabit the bottom sediments.

Water with a molecular structure that differs from the structure of the main body of water of which it forms part is known as distorted water. This water forms a layer, several molecules thick, at the surface of a body of water and adjacent to the BOUNDARY LAYER.

Water also has a much larger HEAT CAPACITY than any other common substance. The very high heat capacity of the oceans means they absorb large amounts of heat with very little change in TEMPERATURE and then release their stored heat very slowly. This has a powerfully moderating effect on air temperatures. Without the heat capacity of the oceans, summer temperatures would be a great deal higher than they are and winter temperatures a great deal lower.

The water molecule is polar (*see* POLAR MOLECULE). This property makes it an excellent solvent, and the hydrogen bonds (*see* CHEMICAL BONDS) that link liquid water molecules also contribute to the capacity of water to form solutions from a wide variety of substances. This works in three ways. Some compounds, such as ethanol (C_2H_5OH), form hydrogen bonds with water molecules, allowing the ethanol and water molecules to mix freely with each other. Other polar compounds that are held together by covalent bonds become ionized (*see* ION) in water. Hydrochloric acid (HCl), for example, separates into H^+ and Cl^- ions. The H^+ then joins a water molecule, changing it into a hydronium ion (H_3O^+), and the Cl^- ions attach themselves to the H^+ ends of water molecules. Compounds linked by ionic bonds, such as common salt (NaCl), are pulled apart by the strong forces of attraction exerted by the positive and negative ends of water molecules. The Na^+ and Cl^- ions then attach themselves to water molecules and move with them.

Water is almost a universal solvent—a solvent in which anything will dissolve. Because of this it never occurs naturally in its pure form, but always as a solution. Its capacity as a solvent also means that living organisms can use it to transport nutrients and metabolic waste products in solution and that the chemical reactions by which organisms grow and maintain their tissues take place in solution. Without water, life on Earth would be impossible.

water balance (**moisture balance**)   The water balance is the difference between the amount of water that reaches an area as PRECIPITATION and the amount that is lost by EVAPOTRANSPIRATION and RUNOFF. If the amount of precipitation is greater than the sum of evapotranspiration and runoff, there is water available for plant growth and to support animals and people. The water balance is therefore of great importance agriculturally, and it is also a strong influence on the type of natural vegetation an area will support.

Water balance is calculated for a specified area and for a specified time (usually one year) by:

$$p = E + f + \Delta r$$

where p is precipitation, E is evapotranspiration, f is filtration, which is the amount of water absorbed and retained by the soil, and Δr is net runoff.

water devil   A water devil is a phenomenon that resembles an aquatic DUST devil, but that is smaller, less violent, and of shorter duration. A water devil can result from vigorous CONVECTION over a water surface that is warmed unevenly. A column of air rises vigorously above a local hot spot, and surrounding air is drawn in at the base of the column. Convergence (*see* STREAMLINES) may then cause the air to rotate, forming a spiraling funnel of rising air. AIR PRESSURE is low at the center of the vortex. Air approaching it expands rapidly, cools adiabatically (*see* ADIABAT), and its WATER VAPOR condenses. This makes the funnel visible and, like a WATERSPOUT, it has a ring of spray around its base.

There is a second way a water devil can develop. Where a lake is bounded by low cliffs on one side and high cliffs on the opposite side, suitable wind conditions can produce EDDIES that generate a VORTEX over the water.

If the wind blows over the low cliffs, across the lake, and into the high cliffs, air striking the high cliffs will be deflected down the face of the high cliffs and back across the lake. When that happens, somewhere over the lake air moving away from the high cliffs will meet air moving toward the high cliffs. The resulting WIND SHEAR may set the air rotating. VORTICITY and

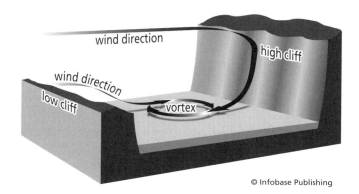

© Infobase Publishing

Wind blows over the low cliffs, blows across the lake, and rebounds from the face of the high cliffs, so two streams of air, moving in opposite directions, meet over the lake and produce a vortex.

A very large waterspout seen from an aircraft accompanying a North Atlantic convoy during World War II, from *Wenn die Elemente wüten* (When the elements rage) by Frank W. Lane *(Royal Air Force photograph, Historic NWS Collection)*

the conservation of angular MOMENTUM may then be sufficient to sustain a vortex, with air spiraling upward around it. Water devils of this type are freak occurrences. They last no more than a few minutes, and are seldom large, although they have been known to rise to a height of about 10 feet (3 m) and to have the strength to lift a small rowboat clear of the water and then drop it.

waterspout   A waterspout is a TORNADO or column of spiraling air that occurs over water. It resembles a tornado and is larger than a WATER DEVIL. Some waterspouts are true tornadoes. They form in a cumulonimbus cloud (*see* CLOUD TYPES) that contains a MESOCYCLONE and may originate either over water or over land and then drift over water. A waterspout of this type is as powerful as any other tornado, and if it moves from water to land, it is just as dangerous.

Tornado funnels are usually dark in color because of the DUST and debris that is drawn into them and then spirals upward. Waterspouts are white, and if a tornado crosses from land to water, its color quickly changes because while it remains above the surface of the sea or a lake there is no dust and debris to darken it. The funnel consists entirely of air and water.

Around the base of a waterspout, air accelerating into the VORTEX whips up spray that forms a white cloud called a spray ring. Some of this water is carried up into the funnel, but most of the water in the funnel forms there, by the CONDENSATION of water vapor.

Air above the water surface is moist. As it is accelerated toward the center of the vortex, the moist air moves down a very steep PRESSURE GRADIENT to a region where the atmospheric pressure is much lower than it is outside the vortex. The change in pressure causes the incoming air to expand rapidly. As it expands, it also cools because of the conversion of heat to the KINETIC ENERGY required for expansion. Its TEMPERATURE falls below the dew point, and water vapor condenses. Condensation releases LATENT HEAT, adding to the instability of the rising air (*see* STABILITY OF AIR).

Waterspouts can also form in the absence of a mesocyclone. They are most likely to do so over shallow water when the weather is very hot, especially in sheltered places such as bays. Because most waterspouts occur close to the shore on fine afternoons in summer, they are often visible to people relaxing on the beach.

Near-shore waterspouts are caused by CONVECTION. Summer sunshine warms the water. Where the water is deep, it is only the surface layer that is warmed, and there is a limit to the temperature it can reach because the warm surface water mixes with the colder water beneath it. Over the open sea or a large lake, winds ripple the surface, increasing the mixing. Sheltered, shallow water, however, can be warmed all the way to the bottom, and its temperature can rise much higher.

Air is warmed by contact with the water surface and rises by convection, carrying a large amount of water vapor with it. As it rises, the air cools adiabatically (*see* ADIABAT) and its water vapor starts to condense, releasing latent heat that increases its instability. Cumulus congestus cloud (*see* CLOUD TYPES) starts to form. This does not extend to a height where the tem-

Waterspouts off the Bahamas Islands *(Joseph Golden/NOAA)*

perature is below freezing, and the cloud may produce no PRECIPITATION, but if the air is rising rapidly enough, the air being drawn in below the cloud may start to rotate about a vertical axis. The resulting vortex may then extend below the cloud as a spiraling column of air and water.

Waterspouts of this type are weaker than tornadoes. Most have a funnel about 150 feet (45 m) in diameter, although some can reach 300 feet (90 m), and they produce wind speeds of about 50 MPH (80 km/h). They occur along tropical coastlines, in the Gulf of Mexico, and in the Mediterranean, and are most frequent in the southern Florida Keys, where up to 100 form every month during the summer.

water vapor Water vapor is the gaseous PHASE of WATER (H_2O) in which the molecules of H_2O are no longer attached to one another by hydrogen bonds (*see* CHEMICAL BONDS) but can move freely and independently. Energy must be applied in order to break the hydrogen bonds and change liquid water into a gas. This energy is called the LATENT HEAT of vaporization and is 600 calories per gram (2,501 J/g) of liquid water at 32°F (0°C). A similar amount of latent heat is released when water vapor condenses into a liquid (*see* CONDENSATION).

When ice is exposed to very dry air, some of its molecules enter the air, turning directly from the solid to the gaseous phase without passing through a liquid phase. This change is called SUBLIMATION. The change in the opposite direction, from gas to solid, is called DEPOSITION. Sublimation requires the absorption of an amount of latent heat that is equal to the sum of the latent heat of melting and the latent heat of vaporization. At 32°F (0°C) this is 680 cal/g (2,835 J/g). Deposition releases exactly the same amount of latent heat.

Water vapor absorbs infrared radiation (*see* SOLAR SPECTRUM) at wavelengths (*see* WAVE CHARACTERISTICS) of 5.3–7.7 μm and beyond 20 μm. It is the principal greenhouse gas (*see* GREENHOUSE EFFECT) in terms of the amount of BLACKBODY radiation it absorbs. However, water vapor is not usually counted as a greenhouse gas because its atmospheric concentration is widely variable and impossible to control.

The partial pressure (*see* AIR PRESSURE) exerted on a surface by water vapor present in the air is known as the vapor pressure. Over an open surface of water or ice, water molecules are constantly escaping into the

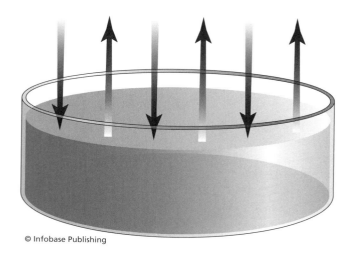

© Infobase Publishing

Water molecules are constantly escaping from an exposed water surface and being absorbed into it. This process is indicated by the broad arrows. Water molecules present in the air exert pressure on surfaces. This is the vapor pressure.

air by EVAPORATION or sublimation. These molecules add to those already present in the air, thus increasing the vapor pressure. Vapor pressure also drives water molecules to merge with the exposed surface. Consequently, there is a two-way motion of molecules leaving and entering the air. If the rates at which molecules are leaving and entering the air are equal, so there is no net gain or loss of water vapor, the water vapor is saturated, although it is usually the air that is said to be saturated.

wave characteristics A wave traveling through WATER (or AIR) possesses certain features that can be used to describe it.

Its wavelength is the horizontal distance between the crest or trough of one wave and the crest or trough of the next. The wavelength (λ) of a system of waves is related to their frequency (f) by $\lambda = c/f$, where c is the speed at which the waves advance (their celerity).

The wave amplitude is the vertical distance between the mean water level and the bottom of a wave trough or top of a wave crest. It is the distance by which the wave moves up and down. The amplitude is equal to half the wave height.

The height of the wave is the vertical distance between crests and troughs and is equal to twice the amplitude. Most sea waves are generated by the action of the wind, and the height of a wind wave is determined by the strength of the wind and the distance across which it blows (known as the fetch).

The steepness of the wave is given by the hypotenuse of a right-angled triangle, the other two sides of which are the wavelength and the wave height.

The number of crests (or troughs) that pass a fixed point in a unit of time is the wave frequency. Frequency is the rate at which any regularly repeating event recurs. In the case of waves, the frequency is the number of vibrations or oscillations that occur in a given time, usually one second. The frequency of a wave (f) is calculated by $f = c/\lambda$, where c is the speed at which the wave is moving and λ is the wavelength. The unit of frequency is the hertz (see UNITS OF MEASUREMENT).

Generally, the word *period* describes the amount of time that elapses between two events, and in particular the time that elapses between two repetitions of the same event. The wave period is the time that elapses between one crest or trough and the next passing a fixed point. The frequency of a wave is given by $1/T$, where T is the period, and the speed (c) of a wave is given by $c = \lambda T$, where λ is the wavelength. When air or water move within a confined space, such as a lake or coastal bay, the resulting waves may oscillate with a period that is determined by the configuration of the

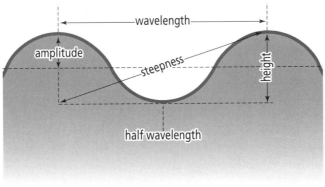

Waves are described in terms of wavelength—the distance between one crest or trough and the next; height—the vertical distance between crests and troughs; amplitude—half the height; and steepness—the angle between the horizontal and a line drawn from a point level with a trough and directly beneath a crest to the top of the adjacent crest. Wave motion ceases at a depth beneath the troughs equal to half the wavelength.

boundaries containing them. This is known as the natural period for that place, and it is equal to the reciprocal of the natural frequency.

The speed at which waves travel is known as their celerity. Celerity (c) is proportional to the wavelength (λ) and frequency (f) of the wave, such that $c = \lambda f$. This applies to waves in either air or water. In deep water, $c = (g\lambda/2\pi)^{1/2}$, where g is the acceleration due to gravity of 32.2 feet per second per second (9.81 m/s^2). Taking this into account, $c = 1.25\sqrt{\lambda}$. In shallow water, $c = (gd)^{1/2}$, where d is the depth of water, and therefore $c = 3.13\sqrt{d}$.

The wave equation is a partial differential equation that represents the velocity (v) and vertical displacement (ψ) produced by a wave as a function of space and time (t), where space is described by three coordinates x, y, and z. The equation can then be written as:

$$v^2\psi = \delta^2\psi/\delta x^2 + \delta^2\psi/\delta y^2 + \delta^2\psi/\delta z^2 = (1/v^2)\delta^2\psi/\delta t^2.$$

Waves travel in groups. Those with the longest wavelength are at the front of the group, and those with the shortest wavelength are at the rear. It follows (by $c = \lambda/T$) that the waves at the rear of the group are traveling fastest. They overtake the waves ahead of them, advancing through the group, but as they do so they lose height, and when they reach the front of the group, they disappear. The group advances as a whole at half the speed of the individual waves that compose it.

Waves travel through water or air, but they do not carry the water or air with them. A molecule of water or air, or a cork bobbing about on the water, describes a small circular motion. That is why a boat is rocked by waves, but not transported by them.

Molecules are affected by wave motion to a depth equal to half the wavelength, and at that depth the amount of molecular movement is negligible. Land slopes into the sea rather than meeting it abruptly. Consequently, as a sea wave approaches a coast, the sea becomes shallower. When the depth of water is less than half the length of the waves, the motion of particles near the bottom becomes flattened and the wave slows. This cannot alter the wave period, however, because waves continue to arrive at the same frequency. Instead, the wavelength decreases—the same number of wave crests pass in a given period of time, but the horizontal distance between them is reduced. The seafloor does not slope everywhere by the same amount, however, and the reduction in wavelength due to reduc-

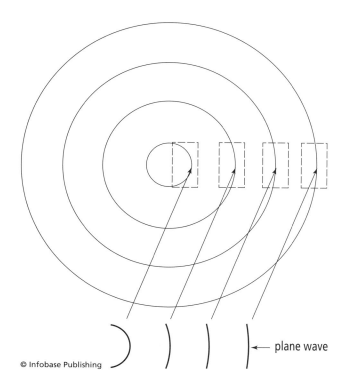

© Infobase Publishing

A short arc taken from the outermost circle appears almost straight because of the length of the circumference. This is a plane wave.

ing water depth has the effect of aligning the waves with the submarine contours. It is why waves usually approach a shore at right angles.

Waves transmit energy imparted to them by the wind, tidal forces, or, in the case of TSUNAMIS, disturbances of the ocean floor. As they enter shallow water and slow, they continue to convey energy at the same rate. This causes the height of the waves to increase, and as their height increases so does their steepness. The circular motion of molecules at the crest is accelerated, and when it exceeds that of the wave itself, the wave becomes unstable. Its crest curls forward and spills. It is then a breaker.

Objects floating at the surface will move in the same way as the water molecules. In deep water they will move only vertically, but close inshore, where the waves are breaking and molecules at the wave crests are traveling faster than the waves themselves, they will be carried toward the shore. That is what makes surfing possible.

A capillary wave is a very small wave. It is the "puckering" on the surface of very still water that is produced by the slightest breeze. The capillary wave

has a wavelength of less than 0.7 inch (1.7 cm) and the SURFACE TENSION of the water quickly restores the smooth surface.

An oscillatory wave is a wave that causes air or water to move about a point, but without advancing in the direction the wave is moving. Particles of air or water describe an approximately circular orbit, moving up and forward, then down and to the rear.

A plane wave is a wave front that is not curved owing to its distance from the source. As waves spread outward, like ripples on a pond, their circumferences increase in size until a point is reached at which any short section of the wave front is effectively straight. It is then a plane wave.

The wave front is the line joining all the points that are at the same phase along the path of an advancing wave.

A wave or group of waves that moves in relation to the surface of the Earth is called a progressive wave. Waves that remain stationary in relation to the surface are called standing waves.

A traveling wave is a wave that moves through a medium, although the particles from which the medium is composed oscillate about a fixed point. A wave is a regular pattern of vertical or horizontal displacements. If the wave is traveling, it is the pattern of displacements that progresses, but not the particles that are displaced. Sound waves (see SPEED OF SOUND) and most ATMOSPHERIC WAVES are traveling waves.

A sine wave is a curve on a graph that corresponds to an equation in which one variable is proportional to

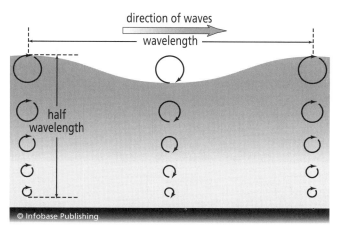

© Infobase Publishing

Waves cause the water to move in a circular motion. The waves move through the water, but the water itself does not advance.

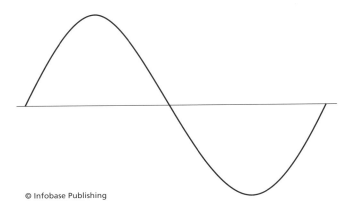

© Infobase Publishing

A sine wave oscillates in a very regular manner.

the sine of the other. A point moving along the curve oscillates about a central point so that crests and troughs are of equal amplitude and wavelength.

weather   Weather is the state of the atmosphere as it is in a particular place at a particular time or over a fairly brief period, with special emphasis on short-term changes. A description of the weather includes references to the current or expected TEMPERATURE, AIR PRESSURE, HUMIDITY, VISIBILITY, CLOUD AMOUNT and type (see CLOUD CLASSIFICATION), and PRECIPITATION. Weather is contrasted with CLIMATE, which is a much broader concept.

Any small-scale variation that occurs in the general state of the atmosphere at a particular time and place is called a disturbance.

A disturbance line is a weather system that occurs in spring and fall in West Africa. Moist air flowing from the southwest as part of the developing or fading MONSOON circulation is overrun by dry air from the Sahara. This produces a SQUALL line several hundred miles long that travels westward at about 30 MPH (50 km/h). The disturbance line produces squalls and THUNDERSTORMS, but dissipates once it has crossed the coast and encounters the cold water of the Atlantic. Disturbance lines are a major cause of rainfall in the periods immediately preceding and following the main monsoon season, between April and June and September and November.

weather balloon   Surface WEATHER STATIONS gather data on conditions at ground level, and for many years meteorologists had no way of routinely monitoring the

atmosphere anywhere higher than the roof of the tallest building in the area. Scientists used surface measurements and cloud observations to calculate pressures and winds at different altitudes. This allowed them to predict the track and speed of storm centers, but only approximately.

In 1893, Charles F. Marvin (1858–1943; see APPENDIX I: BIOGRAPHICAL ENTRIES), professor of METEOROLOGY at the UNITED STATES WEATHER BUREAU, invented the Marvin-Hargrave kite, together with its tethering mechanism and instrument package. The kite was a box kite, comprising a wooden frame covered with muslin. Typically there were 68 square feet (6.3 m^2) of fabric, although different weather conditions required different sizes of kite. Attached to the frame, the kite carried the Marvin Kite meteorograph, which was a set of instruments housed in an aluminum cylinder that moved pens linked to a chart on a rotating drum. The meteorograph weighed only 2 pounds (0.9 kg). The Marvin kite reel paid out the line of piano wire securing the kite. The reel could be operated manually or by a small gasoline engine, and as the kite rose the reel automatically registered the length of line it had paid out, the direction the line sloped, and the strength with which the kite was pulling on it. At intervals during its ascent, the kite paused for up to 10 minutes to allow its instruments to settle. The instruments recorded TEMPERATURE, AIR PRESSURE, HUMIDITY, and WIND SPEED.

Kites were deployed in 1898. Eventually there were 17 weather stations using them.

Kites were expensive, difficult to operate, and the scientists had to wait until the kite had been reeled in and they had removed the meteorograph before they had access to the data. By the late 1920s, they were being replaced by balloon sondes.

A balloon sonde is a package of instruments carried by a free-flying balloon that measure temperature, pressure, and humidity in the air through which they move. When the balloon bursts, the instruments fall to the ground by parachute. A balloon sounding is a measurement, or set of measurements, of atmospheric conditions. The word *sounding* is a nautical term, from the French verb *sonder,* which is derived from Latin *sub-undare, sub-* meaning under and *unda* meaning wave. To take a sounding originally meant to measure the depth of water.

Most sondes pass their data to a radio transmitter carried with them. A sonde from which data are transmitted by radio to a ground receiver is known as a radiosonde.

A radiosonde is a package of instruments, carried aloft beneath a balloon, that measure atmospheric conditions and transmits the resulting data by radio to a surface receiving station. The first weather balloon to be equipped with a radio transmitter flew in 1927. Modern radiosondes came into service about 10 years later. Each of the upper-air weather stations that form part of the network monitoring the upper air releases one radiosonde at midnight and one at noon UNIVERSAL TIME (Z) every day. By releasing all the balloons at the same time from stations all over the world, data from them can be compiled into a picture of conditions throughout the world at that time.

Midway between these launch times, at 6 A.M. and 6 P.M. Z, each station releases a balloon that carries only a RADAR reflector. This is tracked by radar to provide a profile of the wind speed and direction and the way this changes with height. A balloon of this type is called a wind sonde. A radiosonde that also carries a radar reflector, so it provides information on the wind at the same time as other meteorological data, is known as a rawinsonde. Wind direction is measured by noting the position of the sonde at intervals, usually of one minute. Because of the wind, the balloon does not rise vertically, but the data it records are taken to represent a vertical profile.

There are about 90 upper air stations in the United States, seven in Great Britain, and two in Ireland—approximately one upper air station for every 41,000 square miles (106,190 km^2) in the U.S. and one for every 13,000 square miles (33,670 km^2) in Great Britain and Ireland. In the world as a whole there are about 500 upper air stations.

The balloon is about 5 feet (1.5 m) in diameter when fully inflated, and it is filled with HELIUM. The instrument package is carried beneath the balloon at the end of a cable 98 feet (30 m) long. This length of cable prevents the contamination of instrument readings by effects from the balloon itself. Once released, the balloon rises at about 16 feet per second (5 m/s) to a height of 66,000–98,000 feet (20–30 km). As it climbs, air pressure around the balloon decreases. This allows the balloon to expand until it bursts. This happens about 1–1.5 hours after launch. The instrument package returns to the surface by means of a parachute. About one-quarter of the packages released are recovered and can be used again.

The instrument package consists of a main body that contains an aneroid pressure capsule (aneroid BAROMETER) for measuring atmospheric pressure, a battery to supply power, electronic devices that convert data into a form suitable for radio transmission, and the radio transmitter. Above the main body there is an open structure containing a skin HYGROMETER, and above that a plastic ring. The ring holds the fine wire of the electrical-resistance THERMOMETER. Humidity readings from the hygrometer are ignored at heights above 33,000 feet (10 km) because the instrument is unreliable at the very low temperatures prevailing at this altitude.

Radiosonde data are augmented by more than 2,000 reports every day from aircraft.

Much bigger balloons are used for upper-atmosphere research. These are also filled with helium, but they are only partly filled before launch. As they ascend, the helium expands to fully inflate the balloon. These research balloons are designed to return data from the middle and upper stratosphere (see ATMOSPHERIC STRUCTURE).

A balloon drag is a small balloon that is used to retard the first part of the ascent of a larger balloon in order to allow more time for making measurements. The drag balloon contains ballast to weight it and is inflated in such a way that it will burst at a predetermined height.

A pilot balloon, known as a pibal for short, is a weather balloon filled with a measured amount of HYDROGEN to ensure that it ascends at a predetermined rate. As it rises, the balloon is tracked using a theodolite. At intervals, the altitude of the balloon is calculated from its known rate of ascent, and its AZIMUTH angle is read from the theodolite. From this information the wind velocity is calculated for each height. If the sky is obscured by cloud, the height of the CLOUD BASE can be measured.

A theodolite is a surveying instrument that consists of a telescope with crosshairs in the eyepiece that is used for sighting and focusing on the balloon. The telescope is mounted on a tripod fitted with a spirit level to indicate when the instrument is horizontal, and it can be rotated in both the horizontal and vertical planes. The instrument is tilted until the telescope is focused accurately on the balloon. The horizontal and vertical angles between the instrument and the balloon are then read from graduated circles that are seen through a second eyepiece. These reveal the elevation and azimuth of the balloon. The method used to measure the speed and direction of high-level winds is called rabal.

A constant-level balloon is designed to rise to a predetermined altitude and then remain there. The balloon is contained within an inelastic cover, so its volume cannot exceed a certain value. This prevents the gas in the balloon from expanding until it becomes less dense than the air at the desired height. The altitude of the balloon cannot be controlled precisely, however, because it is affected by vertical air movements and by changes of temperature as it moves into and out of direct sunshine. These alter its volume and gas density.

Further Reading

Monmonier, Mark. *Air Apparent: How Meteorologists Learned to Map, Predict, and Dramatize the Weather.* Chicago: University of Chicago Press, 1999.

weather forecasting Meteorologists employed by their national meteorological and hydrological services work at central locations where they receive surface observations from WEATHER STATIONS and upper-air data from upper-air stations (*see* WEATHER BALLOON). They also receive data and images from orbiting WEATHER SATELLITES. It is their job to interpret all of this information and from it to predict how the weather will behave in the hours and days to come.

The scientists use a range of techniques in preparing their forecasts. Air mass analysis involves relating the characteristics of the AIR MASSES over a large area to the surface conditions illustrated on a synoptic chart (*see* WEATHER MAP). The surface chart is studied in conjunction with a number of other charts and graphs, including vertical cross sections of the troposphere (*see* ATMOSPHERIC STRUCTURE), charts showing the winds at different heights, constant-pressure charts, and Stüve charts (*see* THERMODYNAMIC DIAGRAM). The aim is to build up as complete a picture of the air masses as possible in order to improve the reliability of predictions of their future behavior.

The study of a SYNOPTIC chart of surface conditions is called surface analysis. A synoptic chart is one that is plotted from data contained in a synoptic report, which is any weather report that is based on synoptic WEATHER OBSERVATIONS, encoded using an authorized code, and transmitted to a weather center. Meteorologists abstract the information they need to identify air

masses, frontal systems, and other features and to plot their locations.

A facsimile chart is a weather chart that is distributed as a fax from a central meteorological office. The device that transmits the chart is known as a facsimile recorder. In the United States, facsimile charts are sent daily from a center in Washington, D.C., to stations throughout the country.

A meteorogram is a diagram showing the way weather conditions have changed. Variable meteorological phenomena, such as TEMPERATURE, HUMIDITY, and AIR PRESSURE, are plotted against time.

A statistical forecast is one compiled from studies of past weather patterns. These are compared with present conditions and used to assess the statistical probability of the pattern repeating. This method can be used fairly reliably to predict a particular climatic feature, such as temperature or PRECIPITATION, over a short period. Very general forecasts over longer periods can also be compiled from a detailed knowledge of past patterns, using analog models.

A thermotropic model of the atmosphere is used in numerical forecasting. The model aims to forecast the height of one CONSTANT-PRESSURE SURFACE, usually that at 500 mb, and the height of one temperature, usually the mean temperature between 1000 mb and 500 mb. The THERMAL WIND is assumed to remain constant with height. Given these two parameters, it is possible to construct a forecast surface chart.

Numerical forecasting is one of the most important and widely used of modern techniques. Weather folklore is based on the belief that weather patterns repeat themselves reliably. If this is true, it means that a particular indicator, such as the color of the sky or shape of the clouds, can be depended on to be followed by weather of a particular kind. Folklore applies this crudely, but this element of modern weather forecasting is based on a similar idea. It assumes that weather in the future will closely resemble the kind of past weather that followed conditions similar to those obtaining at present. Unfortunately, the method is difficult to apply because the state of the atmosphere is rarely if ever identical on two occasions and, because the atmosphere behaves chaotically (*see* CHAOS), quite small variations quickly develop into major differences.

Numerical forecasting aims to replace empirical methods with one that is firmly based on known physical laws. The forecaster begins with detailed measurements of the state of the atmosphere at many different places at regular intervals. These reveal the way conditions are changing. These changes are interpreted mathematically, by applying certain equations to them. These include the EQUATIONS OF MOTION, the laws of thermodynamics (see THERMODYNAMICS, LAWS OF), the equation of mass conservation, the equation of state (see GAS LAWS), and equations of continuity. The equation of mass conservation relates changes in the DENSITY of air to the transport of its mass. The equations of continuity relate changes in the concentrations of the various constituents of the air, such as WATER VAPOR, CLOUD DROPLETS, and CARBON DIOXIDE, to their transport and to their sources and SINKS. Together these comprise the hydrodynamical equations. When used in conjunction with the temperature, pressure, humidity, and wind velocity at a particular time, they make it possible to predict the state of the atmosphere at a future time.

The difficulty of applying this method is obvious: It calls for a truly prodigious number of separate calculations. Vilhelm Bjerknes (1862–1951; see APPENDIX I: BIOGRAPHICAL ENTRIES) saw the possibility of developing such a method as long ago as 1904. He was unable to proceed very far with it, however, because the detailed observations needed to supply the initial data were not available at the time. Lewis Fry Richardson (1881–1953; see APPENDIX I: BIOGRAPHICAL ENTRIES) made another attempt, which he published in 1922. This might have worked, but mathematicians have estimated that about 26,000 people would need to work full-time to perform by hand the calculations needed to predict the weather faster than it was occurring.

What Richardson needed was a computer, but it was not until 1953 that computers were sufficiently fast, powerful, and reliable to be used in weather forecasting. That was the year that the Joint Numerical Weather Prediction Unit (JNWP) was established in the United States. It began issuing forecasts in 1955.

Numerical methods are now used by most national weather services. Forecasters are supplied with prognostic charts that are generated by numerical models. Although these charts picture the way a weather system may develop, they do not in themselves constitute the forecast. Experienced meteorologists use them as tools to help them identify emerging patterns and to recognize the significance of what is happening. The forecast that is finally produced results from the combination of the numerical forecast and the interpretive skill of the meteorologist.

An important change in forecasting technique occurred in the early 1990s. Until that time the computational effort needed to produce a weather forecast was so great that from one set of input data forecasters were able to produce only one prediction. This was the weather they considered most likely, but the increase in computing power, and especially the availability of supercomputers for weather forecasting, allowed them greater flexibility. They began to use ensemble methods. To produce an ensemble forecast the scientist runs a numerical forecasting model repeatedly—typically between five and 100 times—with a slight change in either the initial condition or the numerical representation of the atmosphere for each run. This yields a number of forecasts, and the forecaster then evaluates the likelihood of each of them. Instead of a single forecast, the forecaster is able to predict the probability of a range of weather events. This is called a probability forecast.

A probability forecast states the expected likelihood of a particular type of weather, usually precipitation. The forecast might say there is a "60 percent chance of rain" or "a 0.6 chance of rain." Both mean the same thing, because probabilities can be expressed as a decimal between 0 (no chance at all) and 1 (absolute certainty). The statement means that during the forecast period there is a 60 percent chance that it will rain in a particular place and a 40 percent chance that it will not. It does not mean there is a 60 percent chance that rain will fall somewhere in the possibly large forecast area and a 40 percent chance that the entire area will remain dry. Precipitation is interpreted to mean 0.01 inch (0.25 mm) of rain or its equivalent at any particular place covered by the forecast at any time during the forecast period.

Prior to the introduction of ensemble methods, calculating the probability began by determining whether or not precipitation-bearing clouds will enter the area during the specified period. If it is fairly certain that they will (for example, because they are close to the edge of the area and advancing toward it), then there might be a 90 percent (0.9) probability of precipitation. The clouds may not pass over the entire area, however. Their size and predicted track might mean they will cross about 70 percent (0.7) of the area. The chance of precipitation in any particular place within the area is therefore 0.9 × 0.7 = 0.63, which is approximately 60 percent. This means there is a 60 percent chance that it will rain in any particular place, but there is a 90 percent chance that it will rain somewhere in the forecast area.

Forecasts must be tailored to the requirements of those who will use them. An area forecast is prepared for a specified geographic area. A local forecast covers a small area and is intended for the use of farmers, horticulturists, vacationers, and other people who need to know what conditions to expect in the next few hours or for up to about two days ahead. The local forecast is derived from the short-range forecast. The short-range forecast is compiled for a large area, and local conditions may be strongly affected by topography, distance from a coast, or the amount of exposed water surface in the area. The local forecaster modifies the general forecast in the light of these influences.

An aviation weather forecast is prepared for aircrews. It includes information relevant to the operation of aircraft, such as CLOUD BASE, CLOUD TYPE, WIND SPEED and direction at various heights throughout the troposphere, the risk of icing (*see* FLYING CONDITIONS), and CLEAR AIR TURBULENCE. A flight forecast is one prepared for a specific air journey. A route forecast is prepared for pilots. It provides them with details of weather conditions at various altitudes along the route they are planning to fly. Data from the route forecast are fed into a computer. This calculates the heading on which the aircraft should fly in order to follow the desired track over the ground or sea surface and the speed at which the aircraft will travel in relation to the surface (the ground speed) when it flies at its designated cruising airspeed.

A marine forecast is one prepared for the crews of ships at sea. Updated marine forecasts are broadcast at regular intervals. Forecasts issued by the British Meteorological Office, which cover the sea areas around the British Isles, including the Republic of Ireland, begin with a summary of conditions at 13 coastal weather stations, the positions of which are known to mariners. This is followed by a general summary of the way weather systems are expected to develop over the 24-hour forecast period. The summary states whether an area of high or low pressure will be centered in the area and, if so, the location of the center and the pressure at the center. More detailed forecasts, of pressure, change in pressure, wind direction and speed, precipita-

tion, and visibility are then broadcast for each coastal sea area in turn. Forecasts for more distant sea areas, as far as the central Atlantic and north into the Arctic, are relayed to ships from weather satellites. In addition, warnings are issued of severe weather such as storms and gales.

Forecasts are issued for different lengths of time. The forecast period may range from less than 12 hours to several months or a season. Common sense would

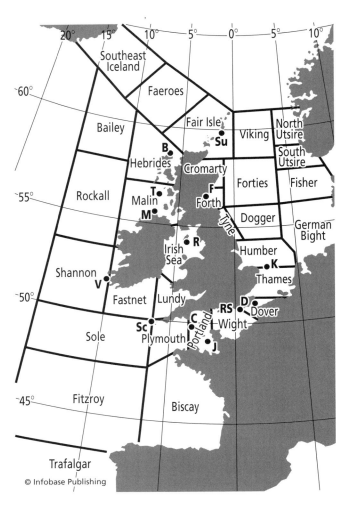

The map shows the sea areas around the British Isles, which are used in marine forecasts. Weather forecasts for coastal shipping cover each of these areas individually. The letters show the location of coastal weather stations. Each forecast begins with reports from each of these stations in turn. They are: Tiree (T); Butt of Lewis (B); Sumburgh (Su); Fife Ness (F); Smith's Knoll Automatic (K); Dover (D); Royal Sovereign (RS); Jersey (J); Channel Light-Vessel Automatic (C); Scilly (Sc); Valentia (V); Ronaldsway (R); and Malin Head (M).

suggest that the shorter the forecast period the more accurate the forecast is likely to be, but this is not necessarily so, because more detail is usually expected in forecasts for very short periods than in those for longer periods. Individual SHOWERS and storms are short-lived and affect small areas, for example. Their likelihood can be reliably predicted over a particular area, but they cannot be predicted to strike a specific neighborhood more than one hour ahead. A forecast for a season, on the other hand, would be expected to state no more than whether it would be warmer, colder, wetter, or drier than usual over an entire region or even continent. Most forecasts cover the period up to 24 hours ahead and add a summary of the outlook for two or three days beyond that. These forecasts achieve a fair degree of accuracy in predicting the track and behavior of middle latitude weather systems and in anticipating large-scale events such as the approach of CYCLONES and ANTICYCLONES over the outlook period.

Nowcasting is the issuing of local weather forecasts for the immediate future, up to two hours ahead. These forecasts give warning of approaching severe STORMS and TORNADOES.

A daily forecast is one issued for the period from 12 to 48 hours ahead.

A short-range forecast covers a period of up to two days. The forecast is based partly on synoptic methods, but nowadays more often on numerical forecasting. The synoptic method aims to predict future patterns of high-level pressure distribution and the THICKNESS of the layer between 1,000 mb and 500 mb, and then to judge the surface conditions that are likely to result. Short-range forecasts are generally fairly accurate, but they are limited. Weather systems may change the speed at which they move, and when the air is moist and unstable (see STABILITY OF AIR), it is impossible to predict where and when showers will occur. Consequently, these are forecast somewhat vaguely, as "scattered showers" or "showers with bright periods."

A medium-range forecast covers a period of 5–7 days. It is compiled in the same way as a short-range forecast.

A long-range forecast covers a period up to two weeks ahead and sometimes up to one month ahead. Forecasters compiling a long-range forecast cannot use the methods appropriate to short-range forecasts, because these describe atmospheric conditions with a lifetime of no more than about seven days. Instead,

Lighthouse in a storm. Reliable weather forecasts save lives at sea. *(Mariners Weather Log/Historic NWS Collection)*

forecasters rely on statistical methods in which current tendencies, such as BLOCKING, are projected into the future. This requires making allowance for the behavior of the JET STREAM and the stage that has been reached in the index cycle (see ZONAL INDEX). The influences of surface features, such as lying SNOW, are taken into account, and in coastal areas so is the SEA-SURFACE TEMPERATURE. Analog CLIMATE MODELS provide an alternative approach. Although large amounts of data are available to forecasters and they have access to powerful models, long-range forecasts are inherently unreliable. This is because weather systems behave in a chaotic fashion (see CHAOS), so variations in the initial conditions that are too small to detect can produce widely divergent outcomes.

The initial condition is one of the values that are used as the base from which later values are calculated. For example, a weather forecast represents a set of calculations that aim to predict the way the weather will change from one state to another. The change is calculated from a set of measured or estimated values for a range of factors including temperature, pressure, humidity, and wind at various heights. These values are the initial conditions. Very small errors in the initial conditions tend to become increasingly exaggerated as the weather develops and the calculated values diverge from them. Because such discrepancies are too small to be noticed, the reliability of a forecast decreases with time. The acute sensitivity of a developing system to its initial conditions makes the development chaotic.

Forecasters use a variety of mathematical models of the atmosphere to assist them. A BAROTROPIC model is one used in numerical forecasting. At each level this model assumes the atmosphere to be barotropic. In a barotropic atmosphere the winds are GEOSTROPHIC, and there is no convergence or divergence (*see* STREAM-LINES). In an equivalent-barotropic model it is assumed that air movements are not affected by FRICTION and are adiabatic (*see* ADIABAT) (that is, warmed and cooled as a consequence of the vertical movement of the air), and the vertical WIND SHEAR of the horizontal wind is proportion to the horizontal wind itself. The atmosphere is hydrostatic equilibrium (*see* HYDROSTAT-IC EQUATION) and quasi-geostrophic balance, which means air movement is controlled primarily by the PRESSURE-GRADIENT FORCE and the CORIOLIS EFFECT. In this atmosphere the wind direction does not change with height, all the contours on any CONSTANT-PRES-SURE SURFACE are parallel, and vertical movements are presumed to be equivalent to those at an intermediate level, known as the equivalent-barotropic level.

Envelope orography is a technique that is sometimes used in the mathematical models used in weather fore-casting. It assumes the valleys and passes in mountain ranges are filled mainly with stagnant air. This allows them to be ignored, effectively increasing the average height of the mountains. The disadvantage of this sim-plification is that increasing the average height of the mountains also increases the extent to which they block the passage of air to a value that exceeds the real value.

The accuracy of a forecast is known as its skill, and this must be checked. A skill score can be calculated by comparing the forecast with a reference standard. This standard may be a description of the weather situation prevailing when the forecast was compiled, and there-fore assuming no change in the weather, or a forecast made by selecting one feature at random. Many fore-cast methods are tested against a type of random fore-cast in which it is assumed that the weather will not change at all.

Applying any technique to measure the accuracy of a weather forecast is called forecast verification. Verifi-cation is based on comparisons that are made between the conditions that were predicted and those which actually occurred. The accuracy of the forecast is then given a numerical score. Forecast skill is measured on a scale that ranges from 0 (completely wrong) to 1 (com-pletely correct). The skill measures the predictive power of the forecast or of a forecasting method. For example, a forecast that it will not rain tomorrow in Death Valley is very likely to be correct, but making it demands little of the forecaster beyond a knowledge of the climate in Death Valley. Consequently, the forecast has very little predictive power. Forecast skill is usually measured by comparing the accuracy of a forecast with that of a cli-matological forecast or a persistence forecast.

A climatological forecast for a region is one that is based on its CLIMATE, rather than on a projection of the current synoptic situation, but with allowance made for such important features as FRONTS, pressure systems, and the location and strength of the jet stream.

A persistence forecast predicts a continuation of present conditions for several hours ahead. Such a forecast might state that the rain that is falling now will continue to do so, or that the present fine weather will remain unchanged. Such a forecast cannot predict changes in the direction or speed with which weather systems move or the formation or dissipation of fron-tal systems. Consequently, a persistence forecast usu-ally remains valid for no more than 12 hours, seldom for as long as a full day, and it often fails in as little as six hours.

A forecast-reversal test is used to measure the use-fulness of a method for forecast verification. The same verification method is applied simultaneously to two weather forecasts. One is an actual forecast, and the other is a fabricated forecast that predicts the opposite conditions. If the real forecast predicts rain, for example, the fabricated forecast predicts dry weather, and if the real forecast predicts wind, the fabricated one predicts calm. Each forecast is given an accuracy score on the basis of the test, and the two scores are compared. The comparison amounts to an evaluation of the verification test, because the real forecast should achieve a markedly higher score for accuracy than the fabricated one.

weathering Weathering is the general name for all of the processes by which solid rock is broken into ever smaller fragments and finally into particles ranging in size from those of clay, which are less than 0.00004 inch (4 μm) across, to sand grains up to 0.08 inch (2 mm) across. Not all weathering is due to the physical processes associated with weather. Chemical solutions that originate deep below the surface and rise through fissures in crustal rocks react with particular minerals in the rock. This makes some minerals soluble, so they are removed in solution. PRECIPITATION delivers WATER to the surface. Precipitation is naturally acid, because of the CARBON DIOXIDE and other gases that have dissolved in it. As the water filters downward through the soil, it dissolves some minerals, leaving rocks pitted. If air enters the cracks, other minerals will be oxidized. These changes, known as chemical weathering, weaken and fragment rock. They also tend to smooth it, because sharp corners and protrusions present large surface areas on which the chemical reactions can take place, so they are attacked more rapidly than flat rock faces.

Rock at the surface is directly exposed to physical weathering processes. In middle and high latitudes, water seeps across rock surfaces to places where water is freezing, forming ice lenses, and the freezing process alters the crystal structure of minerals, detaching fragments. In spring, the ice melts and the rock fragments are washed from the rock by the melting water or blown away by the wind. In warmer, drier climates, the rain reacts with minerals to form chemical salts that crystallize when the rain ceases and the rock dries, expanding as they do so. Crystallization, especially of common salt (sodium chloride, NaCl), causes rocks to crack. A violent STORM may cause rocks that have been weathered from below to fall down hillsides. As they fall, the rocks accelerate and detach other rocks, to produce a rockfall.

Weathered rock is subject to EROSION. On sloping ground, rocks and rock particles that have been broken from the main rock mass may suddenly slide downhill. Landslides occur when heavy rain turns soil to mud, which lubricates the ground beneath the rocks. Heavy rain can also shift large masses of soil, which descend as mudslides.

Rockfalls, landslides, and mudslides are known collectively as mass wasting. If they occur in populated areas, they can cause appalling devastation.

weather lore People have always tried to predict the weather. They have needed to know when to sow and harvest their crops, when to shelter their animals from an impending storm, and when it was safe to set sail on fishing expeditions or sea journeys. Until the invention of TELEGRAPHY there was no way weather observations made at points scattered over a large area could be brought to a central point quickly enough to produce a SYNOPTIC picture of weather conditions that might make accurate forecasting feasible. Instead, people had to rely on experience and local knowledge to interpret the signs they could see around them. These signs and their meanings became incorporated in sayings and short verses that made them easier to remember.

Predictions were also associated with gods and in Christian cultures with saints. This link derives from the time when the weather was believed to result from the direct intervention of supernatural beings. STORMS, HAIL, gales, and warm sunshine were all produced at the whim of these beings. Many of these old associations survive.

Lore, used in this sense (the word has other meanings), is a body of tradition or knowledge on a particular subject. Weather lore is the accumulated traditions and observations with which our ancestors attempted to interpret weather signs and forecast the weather. Possibly it was the Greek philosopher Theophrastus (371 or 370–288 or 287 B.C.E.) who compiled the earliest written collection of sayings about the weather. Theophrastus was a student of Aristotle (384–322 B.C.E.; *see* APPENDIX I: BIOGRAPHICAL ENTRIES). Evidently he talked well, because it was Aristotle who gave him the nickname Theophrastus, which means "divine speech." His real name was Tyrtamus, and after Aristotle retired he took over the Lyceum (school) Aristotle had founded. Theophrastus is best known as the founder of the science of botany, but he also wrote *On Weather Signs* and *On Winds*, two short books that describe natural signs indicating rain, wind, storms, and fair weather. The Greek poet Aratus (*ca.* 315–*ca.* 245 B.C.E.) also collected some weather sayings. About half of his only surviving complete work, *Phaenomena*, is devoted to them. The collections made by Theophrastus and Aratus were passed down from generation to generation, translated into Latin and repeated by Roman authors such as Virgil (70–19 B.C.E.), and were absorbed into many European cultures.

Religious festivals, mostly held to celebrate particular saints, mark the progress of the Christian year, and the weather on those days is often believed to set the pattern for the period that follows. Days that are traditionally held to be significant in this way are called CONTROL DAYS, and they include Candlemas, Easter Day, and

Christmas Day. The saints whose days predict the weather to come include Saints Bartholomew, Hilary, Luke, Martin, Mary, Michael and Gallus, Paul, Simon and Jude, Swithin, and Vitus. The dog days (*see* WEATHER TERMS) are inherited from Roman belief, and Groundhog Day is also a day when the weather is foretold.

Other beliefs are based on direct observation. Red sky in the evening indicates a fine day tomorrow, for example; dew in the night, rain before seven, and a gray mist at dawn also mean a fine day to come.

The appearance and behavior of familiar plants and animals is also held to foretell the weather. Cows lie down when rain approaches, but that is only one of the ways they can be used as forecasters.

Some of the traditional beliefs are accurate, but most are not. This may be because the original ideas behind them have become corrupted over the generations. There is not the slightest doubt that people whose lives sometimes depend on the weather, such as sailors and shepherds, are able to read signs of approaching wind, storms, and fine weather.

Many familiar animals are supposed to predict the weather through changes in their behavior. Cows can predict a range of conditions in addition to the supposed link between rain and whether the animals are standing or lying. If a cow tries to scratch its ear, there will soon be a SHOWER. If it beats its flanks with its tail, there will be a THUNDERSTORM. If the cows gather at the top of a hill, the weather will be fine, but if they move to lower ground it will be wet or stormy. If they stampede with their tails held high, there will be rain and thunder. In cold weather, cows will lie down at the approach of rain or huddle together in a sheltered place with their tails to the wind.

Some of these observations may be reliable, in particular the stampeding reaction to an approaching storm. Cows behave like this because parasitic flies that lay their eggs on the skin of cows are at their most active when the air is warm and humid—the conditions that precede a storm. They make a high-pitched buzzing sound as they fly, and this sound will make cows raise their tails and run.

If a goat grazes with its head facing into the wind, the weather will be fine. If it grazes with its tail to the wind, the weather will be wet.

An old English country belief holds that pigs can see the wind. When gales are imminent, they become very restless, running around their sties and scattering their bedding.

Barn swallows (*Hirundo rustica*) are migratory birds that spend the winter in low latitudes and migrate northward for the summer. Their appearance shows that summer has arrived, but the swallows do not arrive all together. It can happen that a few arrive first, probably carried by a favorable wind the others missed. Hence the saying "One swallow does not make a summer."

Even the barnyard rooster can predict the weather, though it is difficult to see how this saying works:

If the cock goes crowing to bed,
He'll certainly rise with a watery head.

There are plants that are believed to predict the weather. Scarlet pimpernel (*Anagallis arvensis*) is also known as poor man's weather-glass and shepherd's weather-glass. It is a small, herbaceous plant that grows on sand dunes and open grassland throughout most of Europe. It is quite common. Its tiny scarlet flowers, which appear from June through August, are reputed in weather lore to predict rain reliably. When the weather is to be sunny, the flowers open, and they close when rain threatens.

Months of the year are associated with particular types of weather, and sometimes months have descriptive nicknames. January is still winter and fine, mild days are deceptive. "A January spring is worth nothing" is one country saying. Another warns that "March in January, January in March."

"February fill dyke, black or white" is an English country expression that refers to the fact that in Britain February is usually a cold, wet month. In parts of England a dyke is a ditch (in Scotland and other parts of England it is a stone wall), and the saying means that in February the ditches will be full either of water (black) or of snow (white).

February 2 is Groundhog Day, the day on which an old tradition holds that the spring weather can be predicted. In parts of North America, this is the day on which the groundhog is said to emerge from hibernation to look for his shadow. If he sees it, he anticipates bad weather and returns to his hole for a further six weeks. If he cannot see his shadow, because the day is cloudy, he takes this as a good omen and remains above ground, anticipating fine weather.

February 2 is also Candlemas Day (so-named because it used to be celebrated by candlelight), and there is an ancient rhyme from which the Groundhog Day tradition may be derived:

If Candlemas be fair and bright,
Winter'll have another flight.
But if Candlemas Day be clouds and rain,
Winter is gone and will not come again.

March many weathers is an English expression that encapsulates the variability of the weather as winter is giving way to spring. This is one of several folk sayings about March weather. The saying "If March comes in like a lion it goes out like a lamb; if it comes in like a lamb it goes out like a lion" refers to windiness. Mists cannot form in strong winds. This is reflected in "As many mists in March as there are frosts in May." "March windy, April rainy, clear and fair May will be" is an English saying that also occurs in French and German.

The spring (in the Northern Hemisphere) EQUI-NOX occurs in March and gives rise to a folk belief in both the United States and Britain that gales are more common at the equinoxes. These are called equinoctial gales, but there is no basis for the belief.

May blossom is a country name for the flowers of the hawthorn (*Crataegus monogyna*), a common hedge-row shrub or small tree that flowers in the month of May. In temperate regions, May is a transitional month between spring and summer and consequently it can bring wide variations in temperature. In Britain, May frosts are fairly common, but so are pleasantly warm days. This variability gives rise to the old country say-ing: "Ne'er cast a clout till may is out." A clout (cloth) is any warm winter garment, but opinions differ as to whether "may" refers to the month or to may blossom.

Mild days that occur during autumn and early winter are sometimes associated with saints' days that fall around the same time. According to a British folk belief, a period of mild weather often occurs around St. Luke's Day (October 18) and ends at Saint Simon and Saint Jude (*see* CONTROL DAYS). Another period of mild weather occurs around St. Martin's Day (November 11).

Other predictions are based on observations of meteorological phenomena. Dew in the night is a piece of weather folklore that predicts a fine day in summer.

Dew in the night,
Next day will be bright.

This is often true.

According to weather folklore, gray mist at dawn is a sign that the summer day to follow will be fine.

Gray mists at dawn,
The day will be warm.

It is often true that:

Rain before seven,
Fine before eleven.

This folk saying is fairly reliable, because bad weather that occurs very early in the morning has ample time to clear.

Probably the most famous example of weather lore is the observation that a red sky often indicates the weather that is likely to occur in the next few hours.

Red sky at night, shepherd's delight;
Red sky in the morning, shepherd's warning.

The saying is not entirely reliable, because its accu-racy depends on the rate at which weather systems are traveling, but it is correct more often than not. There are several versions of the rhyme (one substitutes "sail-or" for "shepherd"), and the weather lore of many European cultures makes the link between a red sky and the approaching weather.

When the Sun is low in the sky, at dawn and sun-set, its radiation travels obliquely through the atmo-sphere, and so its path through the air is much longer than it is when the Sun is high in the sky. Blue and green light are scattered repeatedly by collision with air molecules (*see* SCATTERING), allowing orange and red light to pass. If the air contains DUST particles, these will scatter light at the longer wavelengths and, because the particles are much larger than gas mole-cules, they will scatter it predominantly in a forward direction. Then the sky in the direction of the Sun will appear orange or red.

The red sky color indicates the presence of dust particles. Dust particles are soon washed to the surface by rain, so their presence means the air is dry, and if the air is dry, the weather must be fine. The Sun sets in the west. In mid-latitudes most weather systems travel from west to east, so a red sky seen at sunset means fine weather is approaching and will probably arrive the following day. A red sky seen in the morning is in the east, where the Sun is rising. This means the fine weather has already passed and wet weather may be approaching from the west and, if so, it will arrive within a few hours.

Further Reading

Page, Robin. *Weather Forecasting The Country Way*. Lon-don: Penguin Books, 1981.

weather map   A weather map is a map that shows the distribution of AIR PRESSURE, winds, and PRECIPITATION over an area of the Earth's surface at a particular time. Weather maps that are shown on television and printed in newspapers are based on the more detailed SYNOPTIC charts used by meteorologists. There are several types, each of which is used for a special purpose.

A line on a weather map that joins points where the value of some atmospheric feature is the same is known as a contour line. Such lines have names beginning with ISO-, from the Greek work *isos,* which means equal, and they resemble the contour lines joining places at the same elevation on a topographic map. For example, isobars join points where the air pressure is the same, isohyets join points where precipitation is equal, isotherms join places where the TEMPERATURE is the same, and there are several more.

The angle shown on a weather map between the wind direction and the isobars is known as the inclination of the wind. This is an indication of the amount of FRICTION that is affecting the wind and also of the rate at which a CYCLONE is filling or an ANTICYCLONE weakening (*see* PRESSURE GRADIENT).

A normal chart, also called a normal map, is a chart on which the distribution of NORMAL values of the weather features is plotted. The chart then illustrates the average weather conditions in the area covered.

A mean chart is a weather map on which the average values for particular features are drawn as isopleths. Isopleths are lines joining points where values are the same for particular features, such as temperature, pressure, or rainfall.

A cluster map shows the weather situation over a large area, such as the coterminous United States, for a particular day. The map is prepared by a statistical technique that clusters data from many WEATHER STATIONS. This means that the reports from several stations of particular variables, such as temperature, air pressure, wind, and cloud cover, tend to be similar. These results form a cluster around a mean value. The more stations

An early weather map prepared by the Signal Service on September 1, 1872. It is reproduced from "Daily Bulletin of Weather-Reports." (Historic NWS Collection)

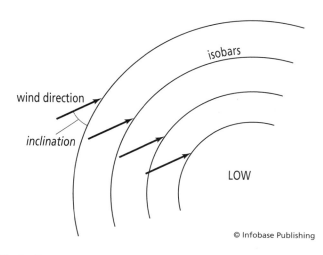

The inclination of the wind is the angle that is made between the direction of the wind and the isobars.

with data that fit into a cluster the more confidence meteorologists have in the cluster. On the final map, the generalized areas are indicated by different colors or shading to indicate the conditions within them.

A synoptic chart shows a general picture of the weather conditions over a large area at a particular time (*see* SYNOPTIC). Synoptic charts are produced at regular intervals, usually of six or 12 hours, and are based on reports from weather stations. These are plotted as STATION MODELS, and isobars are drawn to link stations where the air pressure is the same. Small differences are smoothed out, and the isobars then indicate the distribution of pressure and wind. The isobars show the reduced pressure (sea-level pressure; *see* BAROMETER) or the contours at several constant-pressure surfaces. The chart shows the CLOUD AMOUNT, surface air temperature, pressure, BAROMETRIC TENDENCY, wind direction, and wind strength. The chart also shows the surface position of cold FRONTS, warm fronts, and OCCLUSIONS.

A synoptic chart that shows the patterns of pressure, the height of pressure surfaces, temperature, wind speed and direction, or other features of the weather as these are expected to appear at some specified time in the future is called a prognostic chart or forecast chart. The position of fronts may also be drawn. If the forecast is for more than about two days ahead, the prognostic chart will show the average conditions expected, which are calculated from the range of predicted possibilities.

The center of an area of low or high pressure as these appear on a synoptic chart or other type of weather map is called the pressure center. The pressure pattern is the distribution of air pressure as it is shown by the isobars on a synoptic chart. The patterns made by the isobars indicate the location and intensity of cyclones, anticyclones, RIDGES, and TROUGHS, and the surface area that is affected by them. A pressure-change chart, also called a pressure-tendency chart, shows the barometric tendency. This is the change in air pressure that has occurred over a specified period across a surface at a constant height.

The precipitation area is the area on a synoptic chart over which precipitation is falling. Shading is often employed on TV weather maps to indicate the precipitation area. A snow-cover chart is a synoptic chart on which the areas covered by snow are marked and there are contour lines showing the depth of snow.

The synoptic scale, also called the cyclonic scale, is the scale of weather phenomena that can be shown on a synoptic chart. These events extend horizontally for about 600 miles (1,000 km), vertically for about 6 miles (10 km), and they last for about one day. These are the approximate dimensions of a cyclone. In the classification of meteorological scales that scientists use, events of these dimensions are said to occur on a meso-α scale, both spatially and temporally.

A constant-level chart is a synoptic chart on which meteorological conditions at a particular level are plotted. The level may be defined as the altitude above sea level, in which case the chart is a constant-height chart, or the level at which the atmospheric pressure remains constant, in which case the chart is a constant-pressure chart.

A constant-height chart is a synoptic chart on which the meteorological conditions at a particular altitude are plotted, based on radiosonde data (*see* WEATHER BALLOON). The heights most often used are 5,000, 10,000, and 20,000 feet (1,525, 3,050, and 6,100 m). Constant-height charts help in the identification of AIR MASSES and the boundaries between them.

A constant-pressure chart shows the distribution of atmospheric pressure, but not by means of isobars. Instead, the map assumes the existence of a level surface where the pressure is constant throughout. It then shows pressure contours that indicate heights above or below this imaginary surface. The resulting chart resembles an isobaric map and can be interpreted in the

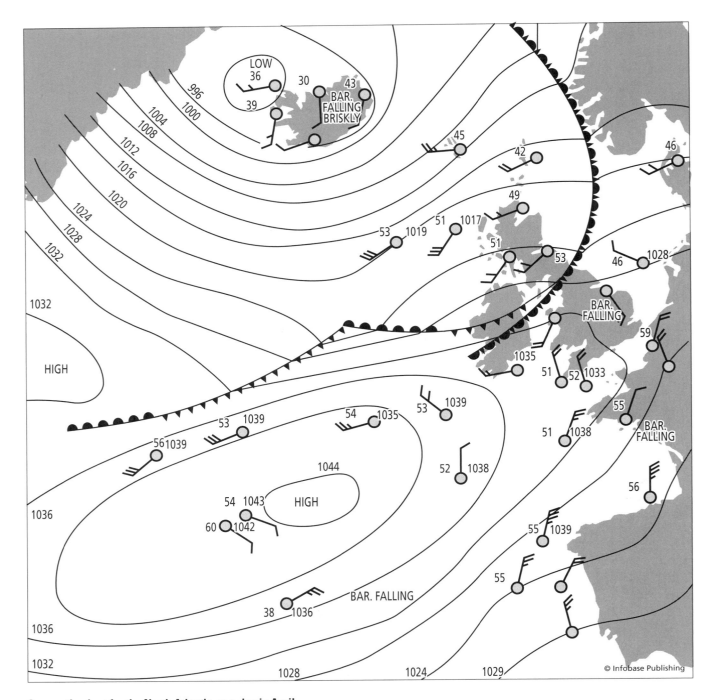

A synoptic chart for the North Atlantic on a day in April

same way, but its lines represent heights, not pressures. It is often a more convenient way to present the data.

A center of high or low air pressure that is shown by the pattern of isobars on a constant-height chart and as an elevated or depressed region on a constant-pres-

sure chart and that reappears on succeeding charts is called a singular corresponding point.

A surface pressure chart shows the distribution of station pressure. This is the atmospheric pressure that is measured at the surface rather than the reduced pressure, and it is indicated by isobars. An isobaric map shows the distribution of atmospheric pressure at any given height above sea level.

A tropopause chart shows the vertical distribution of pressure through the troposphere by height, isotherms, the height of the tropopause, and tropopause breaks (*see* ATMOSPHERIC STRUCTURE). An upper air chart, also called an upper-level chart, is a constant-pressure chart that depicts the condition of the atmosphere at a pressure level in the upper troposphere. Upper-air charts are usually prepared for the pressure levels at 925, 850, 400, 300, 250, 200, 150, and 100 mb. Charts are sometimes prepared for higher levels, but conditions there have little immediate effect on the weather experienced at the surface. The charts are prepared using a station model similar to those used for surface charts, but omitting information about cloud cover, precipitation, visibility, and present weather. The contour lines on a constant-pressure chart link points that are the same height above sea level.

An isentropic chart is a synoptic chart on which the elements of the weather, such as pressure, temperature, HUMIDITY, and wind, are plotted on a surface of equal POTENTIAL TEMPERATURE (an isentropic surface).

An isentropic thickness chart shows the THICKNESS of an atmospheric layer that is bounded above and below by isentropic surfaces. The thickness of such a layer is directly proportional to the convective instability (*see* STABILITY OF AIR) of the air within it. A chart that shows the difference in pressure between two isentropic surfaces is called an isentropic weight chart.

A freezing-level chart is a synoptic chart that uses contour lines to show the height of the constant-temperature surface of the FREEZING LEVEL. It shows the fronts between masses of warm and cold air and gives an indication of the likely availability of ICE CRYSTALS that may accelerate CONDENSATION, leading to precipitation.

A stability chart shows the distribution of values given by a particular STABILITY INDEX.

A vertical differential chart is a diagram that shows values for a particular atmospheric feature, such as temperature or pressure, at two different heights. A thickness chart is a vertical differential chart. A thickness chart shows the changing thickness of a particular atmospheric layer.

A föhn nose is the characteristic shape of the isobars that indicates a fully developed FÖHN WIND on a synoptic chart. There is a RIDGE on the windward side of the mountains and a föhn trough on the LEE side, producing a pattern of isobars that is reminiscent of a nose.

weather observation A weather observation is a record of weather conditions based on measurements that were made in a standardized fashion and written down according to a strict formula. This allows observations made by many people in many places to be compiled into an overall picture of weather over a large area at a certain time.

In most countries, thousands of volunteers make regular observations and communicate them to a central point. In the United States, these volunteers make up the Co-Op Network. Its members are supervised by the NATIONAL WEATHER SERVICE, which forwards the data they submit to the National Climatic Data Center. Some of the volunteers have been collecting data for more than 70 years.

A set of observations and measurements of surface weather conditions that is made at a WEATHER STATION at one of the times specified by the WORLD METEOROLOGICAL ORGANIZATION, and using standard instruments, calibrations, and methods is called a synoptic weather observation. The observations should include reduced pressure (*see* BAROMETER), TEMPERATURE, DEW point temperature, PRECIPITATION, CLOUD AMOUNT, VISIBILITY, WIND SPEED, and wind direction (*see* STATION MODEL), as well as any other details that may be relevant.

Further Reading
National Weather Service. "Cooperative Observer Program." NOAA. Available online. URL: www.nws.noaa.gov/om/coop/. Last updated February 8, 2006.

weather radar Weather RADAR is used to study processes inside clouds, especially the density of water droplets, where the water is most concentrated, and the level at which rising water droplets freeze and falling ice melts. This information is used to determine the likelihood of PRECIPITATION and the intensity of STORMS.

The technique is possible because water droplets strongly reflect electromagnetic radiation with a wavelength of 2–4 inches (5–10 cm). This was discovered early in the 1940s, but it was not until the 1960s that meteorologists first began using radar extensively. Today radar is used to monitor almost all severe storms in the United States and in many other countries.

A pattern of echoes that appears on the screen of a weather radar is known as a radar meteorological observation. The pattern reveals such features as clouds and precipitation, with their distance, density, and direction of movement. It also shows severe storms, TROPICAL CYCLONES, and TORNADOES.

Weather Radio   Weather Radio comprises a network of more than 480 stations that broadcast continuous weather information 24 hours a day over the whole of the United States, U.S. coastal waters, Puerto Rico, the U.S. Virgin Islands, and the U.S. Pacific Territories. As well as ordinary information about weather conditions, the network also broadcasts warnings and watches of hazards. It also provides warnings of other types of hazard, such as volcanic activity (see VOLCANO), earthquakes, and chemical and oil spills.

A special radio receiver is required to pick up the signal. It is a public service provided by the NATIONAL OCEANIC AND ATMOSPHERIC ADMINISTRATION, and it broadcasts at 162.400 MHz, 162.425 MHz, 162.450 MHz, 162.475 MHz, 162.500 MHz, 162.525 MHz, and 162.550 MHz.

Further Reading
National Weather Service. "NOAA Weather Radio All Hazards." NOAA. Available online. URL: www.weather.gov/nwr/. Last updated January 31, 2006; www.nws.noaa.gov/nwr/nwrbro.htm. Accessed March 2, 2006.

weather satellite   A weather satellite is a satellite that flies in Earth ORBIT and carries SATELLITE INSTRUMENTS that produce images of the Earth from which meteorological and climatological information can be obtained. Satellite images are now of vital importance in WEATHER FORECASTING and in monitoring climatic change. The first satellite dedicated to weather observation was the TELEVISION AND INFRARED OBSERVATIONAL SATELLITE (TIROS). There are now many, between them providing a complete coverage of the surface of the Earth through 24 hours every day of the year.

weather station   A weather station is a place equipped with the instruments needed to make standardized measurements and observations of weather conditions and the technical and communications facilities to transmit weather reports to a central point. The station may be manned, but many modern weather stations are fully automated. An automatic weather station transmits its instrument readings to a receiving center at predetermined times without assistance. No personnel are required to operate it.

A first-order station is any weather station in the United States that is staffed partly or wholly by personnel employed by the UNITED STATES NATIONAL WEATHER SERVICE.

In the United States, a voluntary weather observer who maintains a weather station and supplies data to the U.S. National Weather Service without remuneration is called a cooperative observer. The work of such volunteers is supervised and coordinated by the Cooperative Observer Program of the National Weather Service.

A precipitation station is a weather station where only the amount and type of PRECIPITATION is measured and recorded.

A high-altitude station is one located at a sufficiently high elevation for the conditions it records to be significantly different from those at sea level. High-altitude stations are sited no lower than about 6,500 feet (2,000 m) above sea level.

A polar automatic weather station is one designed to operate in extremely cold climates. The instruments are mounted on a sled with pontoons on either side to provide additional support.

A weather ship, also called an ocean weather station, is a weather station mounted on a ship dedicated for the purpose. The ship is anchored permanently in one location (except when it needs to return to port for repairs or maintenance), and it is staffed by observers equipped with instruments to measure both atmospheric and sea conditions, which are reported to a shore station at regular intervals. Weather ships are sited away from shipping lanes, in sea areas that are not monitored by VOLUNTARY OBSERVING SHIPS. Many are in remote parts of the North Atlantic and North Pacific Oceans and the seas off Scandinavia. The requirements for an ocean weather station are laid down by the WORLD METEOROLOGICAL ORGANIZATION.

weather terms   In addition to the technical terms used in METEOROLOGY, there are many informal descriptions of weather phenomena. The following list explains the meaning of a selection of popular words and expressions, arranged alphabetically for convenience.

A break is a sudden change in the weather. The term is usually applied to the ending of a prolonged period of settled dry, cold, or warm weather.

Burn-off is the clearance of FOG, MIST, or low cloud during the course of the morning, as the sunshine intensifies and the air TEMPERATURE rises. As the air grows warmer the DEW point temperature rises and suspended water droplets evaporate.

Weather conditions are sometimes said to be close, oppressive, muggy, sticky, or stuffy—all of which terms carry the same meaning. They describe a subjective feeling of discomfort that people sometimes experience when the air is still and warm (*see* COMFORT ZONE). The feeling can be experienced indoors or outdoors. It is caused by a combination of high temperature and a relative HUMIDITY that is high enough to inhibit the evaporation of sweat from the skin. The inability to cool the body by the evaporation of sweat produces an uncomfortably hot feeling and sweat that fails to evaporate soaks into clothing, which then tends to stick to the skin.

The dog days fall in July and the first half of August, which is the hottest part of the summer in the Northern Hemisphere. At this time Sirius, the brightest star in the sky, also known as the dog star, rises in conjunction with the Sun. The expression is from the Latin *caniculares dies* and arises from the Roman belief that the hot weather is due to heat from Sirius that adds to the heat from the Sun. This is not so. The amount of energy Earth receives from Sirius is infinitesimal and has no effect on the weather.

The doldrums are a sea area in which the winds are light and variable. The extent of the doldrums varies considerably with the seasons, but they are located in the INTERTROPICAL CONVERGENCE ZONE, on the side nearest the equator of the region in which the trade winds (*see* WIND SYSTEMS) originate. The calm weather of the doldrums is interrupted at intervals by fierce storms.

There are three principal doldrum zones. In the Pacific, the doldrums are located in the east, but from July to September they extend westward as a tongue reaching to about longitude 110°W. A second zone is located in the western Pacific, north of Australia and in the vicinity of Indonesia, and in the Indian Ocean. Its area increases from October to December, but it reaches its maximum extent in March and April, when it reaches from the coast of East Africa to longitude

180°E, a distance of about 10,000 miles (16,000 km). The third doldrum zone is in the eastern Atlantic. For most of the year it extends only a short distance from the African coast, but from July to September it reaches all the way across the ocean to Brazil.

Sailing ships could be becalmed in the doldrums. The variability in the location of these regions and the fact that at times they extended from one side of the Atlantic to the other made it difficult for captains to avoid them. Lack of sufficient wind to shift the vessel was a very real hazard because stores of food and drinking water could run low. Until modern times, sailors had no means of making seawater drinkable. The English poet Samuel Taylor Coleridge (1772–1834) described the plight of sailors in this condition in "The Rime of the Ancient Mariner:"

> *All in a hot and copper sky,*
> *The bloody Sun, at noon,*
> *Right up above the mast did stand,*
> *No bigger than the Moon.*
>
> *Day after day, day after day,*
> *We stuck, nor breath nor motion;*
> *As idle as a painted ship*
> *Upon a painted ocean.*
>
> *Water, water, everywhere,*
> *And all the boards did shrink;*
> *Water, water, everywhere,*
> *Nor any drop to drink.*

The origin of the word *doldrums* is obscure, but it probably comes from the Old English word *dol*, which meant dull or stupid. By early in the 19th century, a doldrum was a dull or stupid person, from which came the use of the doldrums to mean low spirits, and in the doldrums came to mean down in the dumps. Coleridge never used *doldrums* to describe the weather his Ancient Mariner experienced, because it was not until the middle of the century, after his death, that *in the doldrums* came to be associated with a geographical locality. Despite its association with the days of sailing ships, giving an impression of antiquity, it is a fairly recent word.

A dry spell is a period during which no rain falls, but that is of shorter duration than a DROUGHT. In the United States, a dry spell is said to occur if no measurable precipitation falls during a period of not less than two weeks.

The erosion of thermals is the mechanism by which a rising thermal (*see* CONVECTION) dissipates. As the warm air rises, cooler air from its surroundings is incorporated around its edges by ENTRAINMENT. The air at the edges then reaches its own equilibrium level (*see* VIRTUAL TEMPERATURE), leaving a smaller mass of air that is still rising. Entrainment continues, steadily eroding the warm air until all of it is neutrally buoyant (*see* BUOYANCY), at which stage the thermal has ceased to exist.

Exposure is the extent to which a site experiences the full effect of such meteorological events as wind, sunshine, frost, and precipitation. It is a measure of the lack of protection against the weather.

The eye of wind is the direction from which the wind is blowing or the point on the horizon from which it appears to blow. A person facing that point is said to be facing into the eye of the wind.

Fair is an adjective used to describe weather that is pleasant for a particular place at a particular time of year. The term is subjective and has no precise meaning, but it generally implies light winds, no precipitation, and less than half the sky covered by cloud.

A flurry is a sudden, brief SHOWER of SNOW that is accompanied by a GUST of wind. A mild wind SQUALL is sometimes called a flurry, even if it brings no snow.

Gloom is the condition in which daylight is markedly reduced by thick cloud or dense smoke, but horizontal VISIBILITY remains good. Gloom is not the same as anticyclonic gloom (*see* ANTICYCLONE).

Halcyon days are a period of calm, peaceful weather, especially in winter. In Greek mythology, the halcyon was a bird that laid and incubated its eggs around the time of the winter SOLSTICE in a nest that floated upon the sea. The halcyon charmed the wind and waves to make the sea calm. The mythical bird is sometimes identified with the kingfisher, and halcyon is sometimes used as a poetic name for this bird. The scientific name of the white-breasted kingfisher of North America is *Halcyon smyrnensis*.

A heat wave is a period of at least one day, but more usually lasting several days or weeks, during which the weather is unusually hot for the time of year. BLOCKING is often the cause of heat waves in middle latitudes. In North America, summer heat waves occur when the belt of prevailing westerly winds is shifted to the north. The SUBTROPICAL HIGH then expands and continental tropical air replaces maritime tropical air (*see* AIR MASS).

During a heat wave affecting much of the central United States in 1936 the temperature over parts of the Great Plains exceeded 120°F (49°C), and they reached 109°F (43°C) in several eastern states. Nearly 15,000 people died as a result of that heat wave. A heat wave in 1980 affecting Missouri, Georgia, Tennessee, and Texas brought temperatures so high that asphalt roads became plastic, concrete road surfaces expanded until they cracked and buckled, and sometimes they exploded violently. More than 1,200 people died. The highest temperatures were recorded in Memphis (108°F, 42.2°C), Augusta (107°F, 42°C), and Atlanta (105°F, 41°C). A heat wave is especially dangerous if the humidity rises as well as the temperature. Advice is available to help people protect themselves from the risks of high temperature and humidity.

Europe suffered a heat wave in August 2003, during which temperatures in France reached 104°F (40°C) and on August 10 the temperature reached 100°F (38°C) in London. A total of approximately 35,000 persons died in Europe.

The horse latitudes lie beneath the subtropical highs, centered at approximately 30°N and 30°S, where air that has risen over the equator is subsiding and diverging on the poleward side of the Hadley cells (*see* GENERAL CIRCULATION). These cells are not continuous, but where air from them is sinking to the surface the winds are light and variable and often the air is calm. Sailing ships were sometimes becalmed in these latitudes. The term *horse latitudes* refers to the fact that ships often carried cargoes of horses. When the ships were becalmed, supplies of water sometimes ran low and horses died and were thrown overboard.

The ice period is the time that elapses between the first fall of snow in winter and the melting of the last patches of snow in spring.

Indian summer is a period of warm weather with clear skies that occurs in late September and October in the northeastern United States and especially in New England. Frontal systems usually dominate the weather in early September, but late in the month these move southward and anticyclonic conditions become established. As the anticyclone extends southward, its airflow brings fine, cold weather that is followed by warm, dry air from the southwest. It is this air that produces the fine weather of the Indian summer. It is late in the year, however, and the clear skies allow the surface to cool rapidly at night by radiation, so nights are cold. There is not an Indian summer every year, and sometimes more than one occurs in a single year. A period of cool weather, with a killing FROST, must precede the warm weather for the change to be sufficiently marked to qualify as an Indian summer.

Other parts of the world also experience Indian summers. Anticyclones often become stationary over Britain in October and November, for example. They bring warm sunshine, but do not heat the ground strongly enough to produce vigorous convection resulting in cumuliform clouds (see CLOUD TYPES) and rain.

The origin of the term *Indian* is uncertain. It was first used in America in the 1790s and may reflect the idea that the fine weather comes from a part of the country that was then inhabited by Native Americans. The use of the term in other English-speaking countries is derived from the American usage.

The January thaw, also called the January spring, is a period of mild weather that sometimes occurs in late January in parts of the northeastern United States and in Britain. There is an English saying that "a January spring is worth nothing."

A lull is a temporary fall in the speed of the wind or cessation of PRECIPITATION.

A march is a variation over a specified period, such as the changes in weather associated with the seasons of the year, which are sometimes described as the march of the seasons. The daily march of temperature is the rhythmic cycle of temperature change in the course of 24 hours.

Persistence is the length of time during which a particular feature of the weather remains unchanged. In the case of the wind, persistence is calculated as the ratio of the mean wind vector (see VECTOR QUANTITY) to the average wind speed (ignoring the direction).

A rainy spell is a period during which more rain falls than is usual for the place and time of year. In Britain, the term is used more precisely to describe a period of 15 or more consecutive days during which the daily rainfall has been 0.008 inch (0.2 mm) or more.

Raw describes weather that is cold, damp, and sometimes windy.

Settled is an adjective that describes fine weather conditions that remain unchanged for a minimum of several days and more commonly for a week or more. Settled weather is usually associated with an anticyclone and is often caused by blocking.

A snap is a short period of unusually cold weather that commences suddenly.

A spell of weather is a period, usually of 5–10 days, during which particular weather conditions persist. The length of the period must be sufficient to make the spell a notable event, so it must take account of the effect of the weather on the lives of the people who experience it. A spell of fog might last only two or three days, but a spell of warm weather would need to last much longer.

A sprinkling is a very light shower of rain or snow.

Sultry is an adjective that informally describes the uncomfortable conditions that result when a high air temperature coincides with high relative humidity and still air.

The teeth of the wind is an old nautical term that means the direction from which the wind is blowing; "in the teeth of" means "in face of opposition." Sailing into the teeth of the wind meant sailing directly to windward.

Tendency is the rate of change of a vector quantity at a specified place and time.

The tendency interval is the period of time that elapses between the measurements that are used to determine the tendency of a meteorological factor. The interval is usually three hours.

A thaw is a warm spell of weather in winter or early spring during which snow and ice melt.

Unsettled is an adjective that describes weather conditions that are fine, but may change at any time in the near future with the development of cloud and possibly precipitation.

A wet spell is a prolonged period of rain. In Britain it is a spell of weather lasting for at least 15 days during which at least 0.04 inch (1 mm) of rain has fallen every day.

weather warnings National meteorological and hydrological services issue warnings of approaching severe weather. The warnings are directed to people living in the regions likely to be affected, and they are issued as far in advance as is possible. For their own safety, people hearing a warning should respond to it immediately.

Weather warnings are graded. By international agreement, warnings take a similar form throughout the world. They are graded as advisories, watches, and warnings.

An advisory means that conditions are giving cause for concern, but they are not yet severe enough to move to the next stage of alert. People should take note of an advisory and listen for any further warnings.

A watch means that conditions are such that severe weather or a hazard from water is likely to develop. The announcement specifies the type of hazard and provides as much information as is available on its intensity and direction of movement. People hearing the watch should prepare for the arrival of extreme conditions. This may include preparing to evacuate.

A warning means that the extreme conditions already exist nearby and their arrival in the designated area is imminent. It is time to take appropriate action.

In the United States the U.S. NATIONAL WEATHER SERVICE is the agency that issues weather warnings. These relate to specific threats. Advisories for each type of threat alert people to the possible risk and urge them to remain vigilant. Everyone hearing an advisory should listen to the local radio, television station, or WEATHER RADIO for further information.

A BLIZZARD watch or warning alerts people in a particular area to the imminent arrival of strong wind and heavy SNOW. This is likely to produce deep snowdrifts. VISIBILITY will be poor, possibly close to zero, and the wind will generate dangerously low WINDCHILL temperatures.

A FLASH FLOOD watch or warning alerts people to the risk of serious flash flooding. Persons receiving a flash flood watch should check their emergency supplies of food, drinking water, first aid equipment, and gasoline. On receipt of a flash flood warning they should move immediately to a place of safety. The warning may be accompanied by instructions to evacuate. Such an instruction should be obeyed immediately.

A flood watch or warning is similar to the alerts issued for flash floods, but warns of more gradual flooding than that experienced in a flash flood.

A frost-freeze warning is a little different, because freezing temperatures pose no immediate threat to human life. A frost-freeze warning to an area where cold weather is not expected informs people that the temperature is expected to fall below freezing. Some plants may be at risk and should be protected. People whose homes lack central heating should check their heaters are working and that they have adequate supplies of blankets and warm clothes.

A gale warning is a notification to shipping that winds of gale force (see WIND SPEED) are expected imminently in designated sea areas. The warnings are broadcast from coastal radio stations and are attached to routine weather bulletins for shipping. A gale warning is issued when a fresh gale or wind gusts of 43–51 knots (49.5–58.6 MPH; 80–94 km/h) are expected in part of a sea area, but not necessarily throughout the whole of it. A typical gale warning for the seas around Great Britain might be: "Gale warning issued by the Met Office at 0150 hours on Wednesday 1st March. Rockall, Malin, Hebrides, Bailey. Southwesterly gale force 8 imminent." The time of 0150 hours is 1.50 A.M. UNIVERSAL TIME. Rockall, Malin, Hebrides, and Bailey are the names of areas of sea around the British Isles. Force 8 is a measure on the BEAUFORT WIND SCALE and refers to a fresh gale with a wind speed of 39–46 MPH (62.7–74 km/h).

A hurricane watch or warning refers to the approach of a hurricane (see TROPICAL CYCLONE) to an inhabited area of the United States. The alert is issued by the National Hurricane Center, in Miami, Florida. Based on observations of the advancing storm and predictions of its future track, a hurricane watch is issued one or two days before the expected arrival of the storm. It is given to people residing in a belt of coastline and its hinterland centered on the point where it is anticipated that the storm center will cross the coast. The affected area extends for a distance equal to about three times the radius of the hurricane to either side of this point. On receipt of a hurricane watch people should prepare for the arrival of the hurricane. This includes preparation to evacuate. A hurricane warning is broadcast when the hurricane is expected to arrive within 24 hours or less. Persons receiving a hurricane warning should prepare for the imminent arrival of the hurricane and should leave a radio or television switched on and tuned to the local station. Updated information and safety instructions will be broadcast and should be obeyed promptly.

A severe THUNDERSTORM watch or warning is issued to local areas. A watch means an area may experience violent storms within the next few hours, although such storms have not yet entered the area. A warning means that violent thunderstorms have entered the area. People within the area should take immediate precautions because of an imminent risk of intense HAIL, possibly with large hailstones, LIGHTNING, winds that may GUST to 140 MPH (225 km/h), and torrential rain that may cause flooding.

A storm watch or warning provides advance information of the likelihood of severe weather. The watch or warning is of sustained winds, lasting for at least one minute, of 48 knots (55 MPH, 89 km/h) or stronger.

A TORNADO watch is issued to people living in an area where tornadoes may occur in the next few hours. Tornadoes have not yet been reported, but the conditions are right for them. A tornado warning means one or more tornadoes have been reported. Everyone in the area should immediately seek shelter.

A tropical storm watch or warning means a tropical storm is approaching an inhabited area of the United States. Based on observations of the advancing storm and predictions of its future track, a tropical storm watch is issued one or two days before the expected arrival of the storm. It is given to people residing in a belt of coastline and its hinterland centered on the point where it is anticipated that the storm center will cross the coast. The affected area extends for a distance equal to about three times the radius of the storm to either side of this point. On receipt of a tropical storm watch people should prepare for the arrival of the storm. A tropical storm warning is broadcast when the storm is expected to arrive within 24 hours or less. Persons receiving a tropical storm warning should prepare for the imminent arrival of the storm and should leave a radio or television switched on and tuned to the local station. Updated information and safety instructions will be broadcast and should be obeyed promptly.

A winter storm watch or warning is issued to alert people in a particular area to the imminent arrival of severe winter weather. A watch provides one or two days' warning of the arrival of severe weather, giving people time to prepare. The warning means conditions have already begun to deteriorate or that they will do so within the next few hours.

A winter weather advisory is a warning that the weather is expected to be bad enough to cause inconvenience and poor, possibly dangerous, driving conditions.

wettability Wettability is the property of a surface that has an affinity for water, because its molecules attract water molecules. The surface becomes wet as water molecules adhere to it and as the moisture penetrates the surface the substance may swell.

Seaweed is the best known wettable material. At low tide, seaweed that is exposed to the air on a beach often becomes so dry it is brittle, but it recovers at once when the incoming tide wets it once more. Small particles with this property act as CLOUD CONDENSATION NUCLEI because of their capacity for capturing molecules of WATER VAPOR.

whirlwind Although many people use the words *whirlwind* and *tornado* as though they were synonymous, in fact there are many differences between them. They are not at all the same things.

Whirlwinds appear in the desert. They rise from the SAND and DUST suddenly, without warning. There is no cloud above them, no dark, menacing sky to warn of their approach, and they are seldom alone. Where there is one, there will be several, sometimes a small army of them. Each individual lasts for only a few minutes, but as one dies another arises. In biblical times they were feared, as much for their mysterious ways as for the damage they could cause. They are much milder than true tornadoes and can do little

A whirlwind, from *The Atmosphere,* translated by James Glaisher from the work of Camille Flammarion, published in 1873. Flammarion described "...gigantic whirlwinds of sand which rise from the earth to the clouds..." *(Historic NWS Collection)*

harm to a solidly constructed building, but a whirlwind can demolish flimsy buildings and tents, and historically desert dwellers have often lived in tents.

Whirlwinds are caused by CONVECTION, as are most tornadoes, but unlike tornadoes their convection is not sustained by the LATENT HEAT of CONDENSATION. They occur in dry air and there is no condensation of WATER VAPOR.

Desert whirlwinds develop on calm days, when there is little or no wind. Wind mixes the air, so that if the ground surface is hotter in one place than it is in another the air in contact with the hot ground will be mixed with cooler air from nearby. This mixing does not happen on still days, however, and because the desert surface is uneven, with some areas exposed to full sunshine and others shaded, and because it is made of a variety of materials, it heats unevenly. By early afternoon patches of exposed rock can be 30°F (17°C) hotter than nearby sand, because of differences in their ALBEDOS and HEAT CAPACITIES.

Air over the hot spots is heated by contact with the surface. The air expands and rises by convection, creating a small region at the base where the AIR PRESSURE is very slightly lower than it is farther away. Air from all sides is drawn into the low-pressure area, and as the air converges (see STREAMLINES) its VORTICITY makes it start to rotate. As it approaches the center, the air accelerates in order to conserve its angular MOMENTUM. Then it is warmed and rises. Air is then spiraling inward and upward. This is very like a tornado, but it is driven from below rather than from a storm cloud overhead.

The spiraling air is made visible by the dust and sand that is carried into its VORTEX and then high into the sky. Although the pressure at its center is low, the relative HUMIDITY is too low for water vapor to condense. The lack of water vapor also limits the lifespan of a whirlwind. It has no source of additional heat above ground level to maintain the upward flow of air. At the same time, relatively cool air flowing into the base of the vortex cools the hot ground, so before long the ground is at the same temperature as the surrounding surface. Air ceases to be drawn into the area and the whirlwind dies. The strongest whirlwinds can last for several hours, but many last for only a few seconds.

Whirlwinds vary in size. Most rise to about 100 feet (30 m). Some reach 300 feet (100 m), and a few grow to 6,000 feet (1,800 m) tall.

Further Reading

Allaby, Michael. *Tornadoes*. Rev. ed. Dangerous Weather. New York: Facts On File, 2004.

whiteout Whiteout is the condition in which the ground, air, and sky are all a uniform white and no landscape features are visible. Persons exposed to a whiteout lose their perception of depth and quickly become disoriented, so a whiteout is extremely dangerous.

There are two ways in which a whiteout can occur. In calm weather, a uniformly white SNOW surface may lie beneath low cloud. The cloud diffuses light passing through it, so light falls on objects evenly from all sides and there are no shadows. Consequently, everything appears white. If dark objects are visible, they appear to float and it is impossible to determine their distance.

Whiteout can also occur in a BLIZZARD. Again the light is diffused by clouds, but in this case there are also SNOWFLAKES between the CLOUD BASE and the ground. The snowflakes are tumbling and turning in all directions, reflecting light in all directions as they do so. A flashlight is useless in the second type of whiteout, because the light is scattered by the falling snowflakes, and so much may be reflected back to the person with the flashlight that it becomes dazzling.

Further Reading

Allaby, Michael. *Blizzards*. Rev. ed. Dangerous Weather. New York: Facts On File, 2004.

Wien's law Wien's law is a physical law that describes the relationship between the TEMPERATURE of a BLACKBODY and the wavelength (see WAVE CHARACTERISTICS) of its maximum emission of radiation. The wavelength varies inversely with the temperature (the higher the temperature the shorter the wavelength), and the law can be stated as:

$$\lambda_{max} = C/T$$

where λ_{max} is the wavelength of maximum emission, C is Wien's constant, and T is the temperature in kelvins. Wien's constant is 2897×10^{-6} m (2,897 μm), so the law becomes:

$$\lambda_{max} = (2897 \times 10^{-6} \text{m})/T$$

Wien's law is valid only for radiation at short wavelengths. The law was discovered in 1896 by the German physicist Wilhelm Wien (1864–1928; *see* APPENDIX I: BIOGRAPHICAL ENTRIES) and for it he was awarded the 1911 Nobel Prize in physics.

wilting Wilting is the limpness that occurs when the cells of a plant contain insufficient water to keep them rigid. The leaves droop and a nonwoody plant may collapse.

Wilting may occur when the rate of TRANSPIRATION exceeds the rate at which water is able to enter the root system of the plant from a soil containing abundant water. In this case the wilting is temporary, and the plant will recover when the transpiration rate decreases.

Wilting may also be due to a deficiency of water in the soil, in which case the plant will not recover unless it is given water. The percentage of water that remains in the soil after a test plant has wilted is known as the permanent wilting percentage (also called the permanent wilting point, wilting coefficient, and wilting point).

Wilting also occurs when the vessels of a plant are blocked, often by a fungal infection, or when water is being taken from the plant vessels by a parasite. Wilting due to infection is called wilt.

wind Wind is the movement of air that results from an uneven distribution of AIR PRESSURE. Air tends to move from regions of high pressure to regions of low pressure at a speed that is proportional to the PRESSURE GRADIENT. It does not flow directly from the center of high pressure to the center of low pressure, however, owing to the CORIOLIS EFFECT caused by the rotation of the Earth (*see* GEOSTROPHIC WIND). Local effects also produce winds such as FÖHN WINDS, ANABATIC WINDS, and KATABATIC WINDS that occur near mountains (*see also* AVALANCHE), LAKE BREEZES, and LAND AND SEA BREEZES. Winds are generally thought of as blowing horizontally, but they can also have a vertical component.

A wind field is a pattern of winds associated with a particular distribution of pressure. Wind stress is the force per unit area that is exerted on the land or water surface by the movement of air.

The wind direction is always given as the direction from which the wind blows (and *not* the direction in which it is blowing). The reason for this is historical (*see* TOWER OF THE WINDS). Wind direction is measured by a WIND VANE or AEROVANE. The wind direction as measured by a magnetic compass is known as the magnetic wind direction. The compass indicates the direction of the magnetic North Pole, which is located in the islands of northern Canada, at approximately 77.3°N 101.8°W (the magnetic South Pole is at approximately 65.8°S 139.0°E). Consequently, the magnetic wind direction is not the same as the wind direction measured in relation to true or geographic north. The difference between the direction of magnetic north and south and true north and south, called the MAGNETIC DECLINATION, varies at different positions on the Earth's surface and at different times (because the magnetic poles move). Wind vanes are usually oriented to geographic north and south, so they indicate the true wind direction. A person who is told the wind direction as it is indicated by a wind vane and who needs to apply this information using a magnetic compass (for example, a sailor at sea or the pilot of an airplane) must remember to make the necessary correction.

The weather shore is the shore from which the wind is blowing, as seen by a ship at sea. The weather side of a ship is the side that faces into the wind or weather. Windward is the side that faces the direction from which the wind is blowing.

A wind-shift line is a boundary that marks a large and abrupt change in the wind direction. Backing is any change in the wind direction (not necessarily a large or abrupt one) that moves in a counterclockwise direction, for example from the northwest to the southwest. If the wind direction is given as the number of degrees counting clockwise from north, a backing wind decreases the number. Veering is a change in the wind direction that moves in a clockwise direction, for example from the southwest to the northwest. If the wind direction is given as the number of degrees from north, a veering wind increases the number.

The surface wind is the speed and direction of the wind that blows at the surface of land or sea. In order to minimize the deflection of the wind by trees, buildings, and similar obstacles, the instrument used to measure the surface wind is mounted on a pole or tower at least 33 feet (10 m) tall.

The spot wind is the wind that is observed or forecast at a specified height over a specified location. The resultant wind is the average speed and direction of the wind at a particular place over a specified period. It is calculated by recording the wind speed and direction at

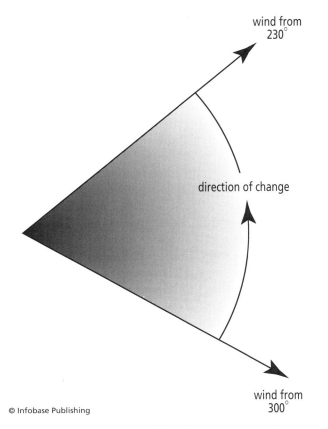

Backing is a change in the wind direction in a counterclockwise sense, in this case from 300° to 230°.

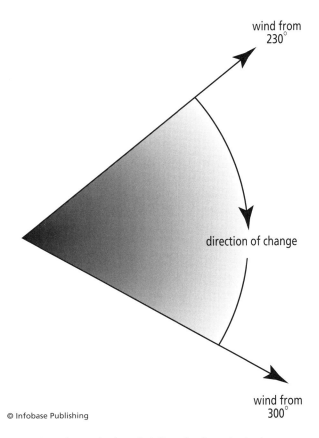

Veering is a change in the wind direction in a clockwise sense, in this case from 230° to 300°.

intervals throughout the period, then calculating their mean values.

A planetary wind is any wind that has a speed and direction caused wholly by the interaction of solar radiation and the rotation of the Earth.

A windbreak is a wall, fence, or other structure that is erected for the purpose of slowing or deflecting the wind.

A windrow consists of loose material that has accumulated naturally to form a line. If it occurs inland or on the surface of the open sea or a lake, the material has been arranged by the wind and the orientation of the line indicates the wind direction. If the windrow occurs on a beach, it has been formed by the action of the TIDES.

windchill Windchill describes the extra feeling of cold that a person experiences when exposed to the wind. The effect can be measured, and its magnitude is usually reported in degrees Fahrenheit (or Celsius).

This gives the misleading impression that when a wind is blowing the air TEMPERATURE actually falls, which is obviously wrong. Wind is simply moving air; that it moves does not alter its temperature.

The confusion arises because of the use of temperature units. These are familiar and therefore easy to understand, but what happens when people are exposed to the wind is that heat energy is removed from their body surfaces. A person's internal body temperature changes very little, and a drop of just a few degrees can be fatal. Heat and temperature are not the same thing and different units are used to measure them. Scientists measure heat in joules, calories, or less commonly in British thermal units (*see* UNITS OF MEASUREMENT). These units are more difficult to understand than degrees, and so temperature units are used.

People keep warm in winter by wearing clothes. These trap a layer of air. The heat of the body warms this air and, once it is warm, the air provides insula-

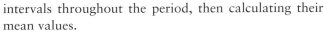

tion. If someone goes outdoors into air that is much colder than body temperature, the layer of warm air inside clothes protects him or her, and the clothes themselves keep the warm air in place.

If the person goes out into a wind, however, the moving air may penetrate the clothing and blow away some of the protective layer of warm air. As warm air disappears, the body must work harder to replace it. Blood vessels near the body surface contract. This reduces the rate at which the blood is cooled by passing through cold tissues. Then stamping the feet, rubbing the hands, beating the arms, eventually shivering all involve rapid muscular movements that generate heat. If the wind is strong enough, however, it may remove warm air faster than the body can either replace it by generating more heat or compensate for its loss. That is when people start to feel cold. How cold they feel depends on the temperature of the air and the speed of the wind.

The rate at which a human body loses heat through windchill increases rapidly as the temperature drops and the wind speed rises, until the wind speed reaches about 40 MPH (64 km/h). At wind speeds faster than this there is only a small increase in windchill.

It is heat that the body is losing, measured in joules or calories, but the effect of the wind is to cool the body at the rate it would chill if the temperature were lower. For anyone going outdoors on a still day when the temperature is 10°F (-12°C), that is how cold it will feel. If there is a wind blowing at 10 MPH (16 km/h), however, the body will lose heat at the same rate as it would on a day with no wind, but with an air temperature of -9°F (-23°C). The effect of windchill does not alter the actual temperature of the air, but it does lower the effective temperature, because it removes some of the insulating layer of warm air and so increases the rate at which the body loses heat.

People need more protection against the cold on a windy day than they do on a still day. When the temperature is below about -21°F (-29°C) on a day with no wind, they need to be well wrapped up, with no bare skin exposed. If there is a wind, that effective temperature will be reached at a much higher actual temperature. With a gentle wind of 5 MPH (8 km/h) the effective temperature will be -21°F (-29°C) when the actual air temperature is -15°F (-26°C). If the wind speed is 25 MPH (40 km/h), that effective temperature is reached when the air temperature is +15°F (-9°C).

Wind chill temperature (°F)

Wind speed (mph)																
0	35	30	25	20	15	10	5	0	-5	-10	-15	-20	-25	-30	-35	-40
5	32	27	22	16	11	6	0	-5	-10	-15	-21	-26	-31	-36	-42	-47
10	22	16	10	3	-3	-9	-15	-22	-27	-34	-40	-46	-52	-58	-64	-71
15	16	9	2	-5	-12	-18	-25	-31	-38	-45	-51	-58	-65	-72	-78	-85
20	12	4	-3	-10	-17	-24	-31	-39	-46	-53	-60	-67	-74	-81	-88	-95
25	8	1	-7	-15	-22	-29	-36	-44	-51	-59	-66	-74	-81	-88	-96	-103
30	6	-2	-10	-18	-25	-33	-41	-49	-56	-64	-71	-79	-86	-93	-101	-109
35	4	-4	-12	-20	-27	-35	-43	-52	-58	-67	-74	-82	-89	-97	-105	-113
40	3	-5	-13	-21	-29	-37	-45	-53	-60	-69	-76	-84	-92	-100	-107	-115

© Infobase Publishing

To calculate the windchill, find the actual air temperature in the top row of figures. Then find the wind speed in the vertical column on the left. Follow the figures along the row and column, and the figure where they intersect is the effective temperature due to windchill. The lightly shaded figures indicate temperatures that are dangerously low. The dark shading indicates temperatures that are extremely dangerous.

Weather reports and forecasts (*see* WEATHER FORE-CASTING) always include wind speeds, and in winter they often include the "windchill factor," but it is simple to work this out using the table at the bottom of page 551. First, though, it is necessary to correct the reported wind speed. Meteorologists measure wind speed well clear of the ground. Close to the ground the wind is slowed by FRICTION, especially in towns, and allowance must be made for this. Unless the report clearly states that the figure quoted refers to the wind speed at ground level, assume the wind speed at ground level will be about two-thirds of the reported speed.

When the effective temperature falls below about -71°F (-57°C), conditions are extremely dangerous. Anyone whose body starts to lose heat in this actual or effective temperature is liable to lose consciousness and die within a short time.

Further Reading
Allaby, Michael. *Blizzards*. Rev. ed. Dangerous Weather. New York: Facts On File, 2004.

wind power Wind power is energy obtained by harnessing the wind to perform useful work. Sails are probably the most ancient example. These were being used to propel boats along the River Nile possibly as early as 4000 B.C.E.

The earliest windmills were built in Persia (Iran) in the seventh century C.E. They were based on designs that were already used for watermills. By the 13th century, windmills were appearing in many parts of Europe, their development having been strongly encouraged by the Mongol leader Genghis Khan.

Until recently, mills were used principally for grinding cereal grains to make flour and meal, or to pump water. The first windmill to generate electrical power was built in Denmark in 1890. One of the first experimental large-scale wind generators was built on the top of Grandpa's Knob, a hill near Rutland, Vermont. Construction of the generator began late in 1940, and it commenced operation in 1941, with a rated generating capacity of 1.25 MW. The generator ran for 1,000 hours over the next 18 months, operating in winds of 70 MPH (113 km/h), and it survived winds of 115 MPH (185 km/h). During this time several repairs and modifications were made to the structure. The generator was taken out of service in

February 1943 due to the failure of a main bearing, and the replacement bearing was not installed until March 3, 1945. On March 26, 1945, one of the 75-foot (23-m), 8-ton (7.3-tonne) rotor blades broke free in a 20-MPH (32-km/h) wind. The blade came to rest on its tip, 700 feet (214 m) down the hillside. The generator was never repaired.

The generation of electrical power is now the principal use for windmills. A modern "windmill," more properly known as a wind turbine, comprises rotor blades or sails. In the horizontal-axis design, which is the one most widely used, the blades are shaped as aerofoils with adjustable pitch—each blade can be turned on its own axis, thus altering the angle at which it meets the air. The horizontal axis to which the rotors are attached is free to turn on a vertical axis. This allows the blades to face into the wind at all times. Vertical-axis mills have sails held on radial arms.

The power that can be derived from the wind depends on the WIND SPEED, the DENSITY of the air, and the area of the circle described by the rotor blades. It can be calculated from the equation:

$$P = 0.5\rho AV^3$$

where P is the power, ρ is the air density (about 0.08 pounds per cubic foot, 1.225 kg/m³ at sea level, but less at higher elevations), A is the area swept by the rotor in square meters, and V is the wind speed in meters per second.

The proportion of this power that a wind turbine is able to extract is given by:

$$P = 0.5\rho AC_p V^3 N_g N_b$$

where C_p is the coefficient of performance of the rotor (a theoretical maximum value of 0.59, but an actual value of 0.35 or less), N_g is the efficiency of the generator (from 0.5 to 0.8), and N_b is the efficiency of the gearing and bearings (up to 0.95). P is in watts (746 W = 1 horsepower; 1,000 W = 1 kilowatt; 1,000 kW = 1 megawatt).

A wind farm comprises a number of wind turbines that are sited at the same location. Each turbine generates electrical power and the combined output from all the turbines at the farm is fed into the public power supply.

Wind farms may comprise as few as 10 turbines or as many as several hundred, and the rated capac-

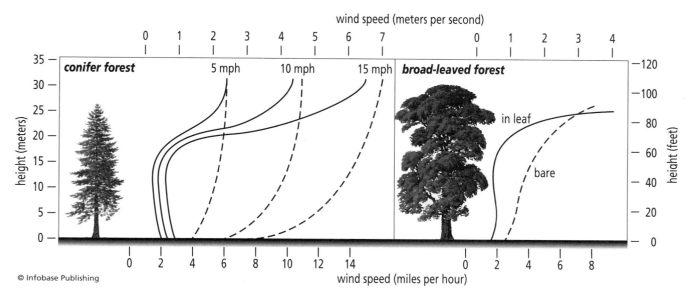

This wind profile diagram compares the change of wind speed with height in a conifer forest (left) and a broad-leaved forest (right). In both cases, wind speed increases sharply above the forest canopy. The broken lines in the diagram of the conifer forest show the wind speed over open ground. The two lines in the diagram of the broad-leaved forest compare the effect of the trees when bare and in full leaf.

ity of the individual turbines varies from about 450 kW to 5 MW. The largest wind farm in Australia is at Esperance, Western Australia. It has nine turbines and a rated capacity of 2 MW. The German Land (state) of Schleswig-Holstein is planning to build the biggest wind farm in Europe. It will be located in North Sea coastal waters near Helgoland, where it will occupy 77 square miles (200 km²) and have a rated capacity of 1,200 MW. On December 1, 2005, the Bundesamt für Schiffahrt und Hydrografie (Federal Authority for Navigation and Hydrography) approved the construction of the first phase of 80 turbines. The farm is scheduled for completion in 2008. Smaller offshore wind farms are also planned for the North and Baltic Seas.

wind profile   A wind profile is a diagram that shows the change in wind characteristics with height and horizontal distance. A wind speed profile, based on measurements, can be used to show the effect on the wind of an obstruction, such as a tree. The profile can then help in designing buildings or SHELTER BELTS. The wind speed at any height in the profile is given by

$$u = (u*/k) \ln (z/z_0)$$

where u is the wind speed, $u*$ is the friction velocity, k is the von Kármán constant, z is the height above the surface, z_0 is the roughness length (*see* AERODYNAMIC ROUGHNESS), and ln means the natural logarithm.

A logarithmic wind profile shows the variation of wind VELOCITY with height throughout the PLANETARY BOUNDARY LAYER, but above the laminar BOUNDARY LAYER. This is expressed by an equation:

$$\overline{u_z} = (u*/k) \ln (z/z_0)$$

where $\overline{u_z}$ is the mean wind speed in meters per second at a height z, $u*$ is the friction velocity, k is the von Kármán constant, and z_0 is the roughness length in meters.

The friction velocity, also known as the shear velocity, is the velocity of air in the planetary boundary layer, but above the laminar boundary layer. It is symbolized by $u*$ and is equal to $(\tau/\rho)^{1/2}$, where τ is the tangential stress on the horizontal surface and ρ is the density of the air.

The von Kármán constant is a value that was discovered by the Hungarian-born American aerodynamicist Theodore von Kármán (1881–1963). Its value is approximately 0.4. The von Kármán constant holds under all circumstances, but the difficulties involved in wind experiments make it impossible to measure it precisely.

A wind profiler is a ground-based instrument that measures the wind speed at different heights and at frequent intervals. Profilers measure the wind through the whole of the troposphere and the lower stratosphere (*see* ATMOSPHERIC STRUCTURE), with a vertical resolution ranging from about 200 feet (61 m) to about 3,300 feet (1 km), taking readings as frequently as every six minutes. The instrument works by Doppler RADAR. It transmits one radar beam vertically and, depending on the instrument type, two to four others at an angle of about 75° with respect to the horizon. The pulses, at a frequency of 900–1,500 MHz, are reflected by turbulent EDDIES. These move with the wind and consequently the reflected beams are Doppler-shifted. The Doppler shift can be interpreted as the speed of the wind crossing from one beam to another, and the time that elapses between the transmission of the beam and reception of its reflection indicates the height above the instrument.

wind rose   A wind rose is a diagram that shows the frequency with which the wind at a particular place blows from each direction. This reveals the direction of the prevailing wind (*see* WIND SYSTEMS). The wind rose is constructed from measurements of the wind direction that are made from the same point at the same time every day.

At the end of the recording period, which is often one calendar month, the daily wind directions are grouped into eight general compass directions: N, NE, E, SE, S, SW, W, and NW, according to which is nearest. They are then drawn as lines originating from a point, with north at the top of the drawing. The direction of each line is that of the wind and it is drawn to a convenient unit of length, such as 0.25 inch (6 mm). If the same direction occurs on another day, the line is made one unit longer. When the rose is complete, it will indicate the comparative frequency of each wind direction and therefore the prevailing wind for that period. The diagram also contains a circle, the radius of which is drawn to the same scale as the wind lines and represents the frequency of days on which the air was calm. The data can also be used to construct wind roses for each season and for the year. Obviously, the longer the record the more reliable the wind roses are likely to be.

Wind roses are very useful aids to planning. Foresters need to take account of prevailing winds when designing plantation forests in order to minimize damage from BLOWDOWN. The design of an urban development also needs to allow for the wind, to avoid the urban canyon (*see* URBAN CLIMATE) effect and exposing open spaces to winds funneled into them (*see* WIND SPEED) along streets.

Wind roses are also used in planning airports so that their main runways can be aligned with the prevailing wind. This reduces the length of takeoff and landing runs. On takeoff it is the speed of the airflow over the wings of an aircraft that determines the amount of lift that is generated. If the aircraft takes off into the wind, that speed is equal to the sum of the forward speed of the aircraft and the speed of the wind, thus reducing the time it takes for the aircraft to attain the speed needed to become airborne. On landing, a headwind has a strong braking effect.

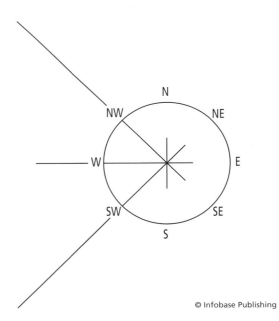

© Infobase Publishing

Each of the straight lines represents a direction from which the wind blows. The length of each line indicates the frequency with which it blows from that direction. In this case, the prevailing winds are from the southwest or northwest.

wind shear   Wind shear is a change of wind VELOCITY with vertical or horizontal distance. FRICTION occurs between adjacent bodies of air that are moving in different directions or at different speeds. Where the difference is in speeds rather than direction, this tends to accelerate the slower air and retard the faster air until both are moving at the same speed. Over land this would cause the wind to cease because of friction between the lowest layer and the ground, and over the sea it would reduce the wind speed to the speed of the surface waves. This rarely happens because the wind is usually driven by a large-scale weather system or by prevailing winds (*see* WIND SYSTEMS) acting at some distance above the surface, and because random movements of air molecules cause air to intermingle at the boundary between the two bodies of air. The molecules also transfer their MOMENTUM, and this distributes the force driving the wind.

Above the level at which friction with the surface exerts an effect the amount of wind shear varies according to the way the air temperature changes and it gives rise to the THERMAL WIND. Where a strong upper-level wind blows across a slower lower-level wind the effect can be to start a column of rising air rotating about a vertical axis. This is the mechanism by which a MESOCYCLONE is formed, in extreme cases leading to a TORNADO.

Anticyclonic wind shear is horizontal wind shear that produces an anticyclonic (*see* ANTICYCLONE) flow in the air to one side of it. Cyclonic wind shear is associated with VORTICITY around CYCLONES. Looking downwind, the winds are stronger on the right in the Northern Hemisphere and on the left in the Southern Hemisphere. This tends to set the air rotating cyclonically along the line of the wind.

wind sock A wind sock is a simple device for indicating the direction of the WIND that was once a feature of every airfield and that is still seen at the smaller airfields used by light aircraft. It consists of a tapering cylinder of fabric that is open at both ends and supported on a circular frame at the larger end. The sock is attached by cords to the top of a tall pole. It fills in the slightest wind and, being free to turn in any direction, it indicates the wind direction. Unlike a WIND VANE or AEROVANE, a wind sock also indicates an absence of wind; when the air is calm the sock hangs limply.

Windsocks are usually colored bright yellow or orange to make them more visible and they can be seen clearly from the air. They were especially useful in the days before airfields had runways and aircraft landed

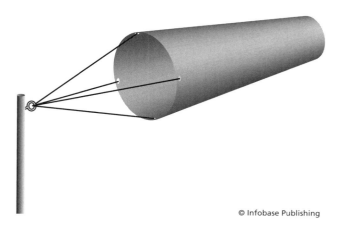

© Infobase Publishing

A wind sock, made of fabric, is free to turn in any direction at the top of its pole. It can be seen clearly from the air.

and took off from a grass surface because they made it possible for the pilot to turn the aircraft into the wind.

wind speed The wind speed is the rate at which the air is moving. This is measured using an ANEMOMETER or AEROVANE and the measurement is of speed only, and takes no account of the wind direction (unlike wind velocity).

Wind velocity is the speed and direction of the wind when these are reported together. VELOCITY is a VECTOR QUANTITY and consequently comprises two values. Wind velocity is reported in the form: direction/speed. Direction is reported in degrees, counting clockwise from north, which is 0°, but usually with the degree sign omitted. East is 090, south is 180, and west is 270. For example, a report to an aircraft might describe the wind as 240/30, meaning that the wind is blowing from 240° at 30 knots (*see* UNITS OF MEASUREMENT). Scientists usually report wind speed in meters per second, so in this example the wind would be 240/15.

There are several ways in which wind speed may be reported, depending on the purpose for which the report is issued. Ships are usually informed of the wind force on the BEAUFORT WIND SCALE. Hurricanes and other types of TROPICAL CYCLONE are allotted categories on the SAFFIR/SIMPSON HURRICANE SCALE that indicate the sustained wind speeds associated with them. TORNADO speeds are re ported using the FUJITA TORNADO INTENSITY SCALE. Some public weather forecasts give the wind as a force on the Beaufort scale, but most use miles per hour (MPH) or kilometers per hour (km/h). Reports to aircraft use knots. Meteorologists usually use meters per second (m/s). To convert these units:

1 MPH = 1.619 km/h = 0.87 knot = 0.45 m/s

1 knot = 1.15 MPH = 1.85 km/h = 0.51 m/s

1 km/h = 0.6214 MPH = 0.54 knot = 0.28 m/s

1 m/s = 2.24 MPH = 1.95 knot = 3.6 km/h

The scientific measurement of wind speed is called anemometry. Speed is the distance a body travels in a unit of time. It is not practicable, however, to label a small volume of air in order to make it visible and then to time its movement over a measured distance. Instead, it is necessary to measure the speed indirectly, by the effect the wind has at different speeds on visible objects. The first successful method was the one devised by Admiral Sir Francis Beaufort (see APPENDIX I: BIOGRAPHICAL ENTRIES). In the earliest version of the Beaufort wind scale, all the wind forces, including gales, were described in terms of their effect on sailing ships and speeds were not allotted to them. The advantage of the Beaufort wind scale is that in its modern version, with wind speeds added, it is based on the response of commonplace objects, such as smoke, flags, trees, and umbrellas. Consequently, it requires no instruments. The Beaufort scale is still used. Weather stations use anemometers to measure wind speed.

The peak gust is the highest wind speed that is recorded at a WEATHER STATION during a period of observation, commonly of 24 hours. Despite the name, a peak gust may be a sustained wind rather than a GUST.

The fastest mile is the greatest wind speed, measured in miles per hour, recorded over a specified period (usually 24 hours) for one mile of wind. A mile of wind is calculated from the number of rotations of a rotating-cups anemometer during a measured interval of time. If each cup is at a distance x from the axis of the anemometer, then in each revolution it travels a distance of $2\pi x$. If the cup completes n rotations in an hour, during that hour it travels a total distance of $n(2\pi x)$. When the distance is converted to miles, it represents the number of miles of wind.

A first gust is a sudden sharp increase in the wind speed that is felt as a cumulonimbus cloud (see CLOUD TYPES) enters its mature stage. The gust is caused by the arrival at the surface of the cold downdraft.

Funneling is an acceleration of the wind that occurs when moving air is forced through a narrow passage. A funneling effect is felt when the wind direction is approximately parallel to the axis of a deep, narrow valley or a street lined by tall buildings (see URBAN CLIMATE). Although the wind is retarded to some extent by FRICTION, the fact that the rate at which air leaves the valley or street must be equal to the rate at which it enters means the flow must accelerate as it passes the constraint. At the same time, the AIR PRESSURE decreases in proportion to the acceleration, due to the BERNOULLI EFFECT.

In the Beaufort wind scale, calm is wind force 0, which is the condition in which the wind speed is 1 MPH (1.6 km/h) or less. The air feels calm and smoke rises vertically.

The threshold velocity is the minimum speed at which wind raises DUST, soil, or SAND particles from the ground. This speed varies according to the size of the particles and the amount of moisture because water adhering to mineral grains holds them together.

In the Beaufort wind scale, light air is force 1. This is a wind that blows at 1–3 MPH (1.6–5 km/h). In the original scale, devised for use as sea, a force 1 wind is just sufficient to give steerage way to a sailing ship. On land, wind vanes and flags do not move, but rising smoke drifts.

A breeze is a light wind. In the Beaufort wind scale, a breeze is a wind blowing at 4–31 MPH (6.4–49.8 km/h). Within this range, breezes are classified as light, gentle, moderate, fresh, and strong. A puff of wind is a breeze that is just strong enough to produce a patch of ripples on the surface of still water.

Funneling occurs when the sides of a valley or buildings lining a street constrict the air flowing parallel to the valley or street, causing it to pass through a smaller passageway in the same amount of time.

In the original Beaufort wind scale, devised for use at sea, light, gentle, and moderate breezes, winds of force 2–4, were defined as "or that in which a man-of-war with all sail set, and clean full would go in smooth water from."

A light breeze is force 2 in the Beaufort wind scale. This is a wind that blows at 4–7 MPH (6.4–11.3 km/h). On land, a light breeze is just strong enough for drifting smoke to indicate the wind direction.

A gentle breeze is force 3 in the Beaufort wind scale. This is a wind that blows at 8–12 MPH (13–19 km/h). On land, a gentle breeze makes leaves rustle, small twigs move, and flags made from lightweight material stir gently.

A zephyr is any very gentle breeze, but especially one that blows from the west. Zephuros was the Greek god of the west wind, and Zephiros, from which *zephyr* is derived, was the Latin version of the name.

A moderate breeze is force 4 in the Beaufort wind scale. This is a wind that blows at 13–18 MPH (21–29 km/h). On land, a moderate breeze causes loose leaves and dry scraps of paper to blow about.

A fresh breeze is force 5 in the Beaufort wind scale. This is a wind that blows at 19–24 MPH (31–37 km/h). In the original scale, devised for use at sea, a force 5 wind was defined as "or that to which a well-conditioned man-of-war could just carry in chase, full and by." On land, a fresh breeze makes small trees that are in full leaf wave about.

A strong breeze is force 6 in the Beaufort wind scale. This is a wind that blows at 25–31 MPH (40–50 km/h). In the original scale, devised for use at sea, a force 6 wind was defined as "or that to which a well-conditioned man-of-war could just carry single-reefed topsails and top-gallant sail in chase, full and by." On land, it is difficult to use an open umbrella in a strong breeze.

A gale is a strong wind, ranging from one that exerts strong pressure on people walking into it to one that breaks and uproots trees. On the Beaufort wind scale there are four categories of gale: moderate gale (force 7), fresh gale (force 8), strong gale (force 9), and whole gale (force 10). A wind stronger than a whole gale is called a STORM, and one weaker than a moderate gale is a strong breeze.

In the original Beaufort wind scale, devised for use at sea, moderate and fresh gales, wind forces 7, were defined as "or that to which a well-conditioned man-of-war could just carry in chase, full and by."

A moderate gale is force 7 in the Beaufort wind scale. This is a wind that blows at 32–38 MPH (51–61 km/h). On land, people walking into the wind feel it exerting a strong pressure in a moderate gale.

A fresh gale is force 8 in the Beaufort wind scale. This is a wind that blows at 39–46 MPH (63–74 km/h). On land, a fresh gale tears small twigs from trees.

A strong gale is force 9 in the Beaufort wind scale. This is a wind that blows at 47–54 MPH (76–87km/h). In the original scale, devised for use at sea, a force 9 wind was defined as "or that to which a well-conditioned man-of-war could just carry close-reefed topsails and courses in chase, full and by." On land, slates and tiles are torn from roofs in a strong gale and chimneys are blown down.

A whole gale is force 10 in the Beaufort wind scale. This is a wind that blows at 55–63 MPH (88–101 km/h). In the original scale, devised for use at sea, a force 10 wind was defined as "or that with which she could scarcely bear close-reefed main topsail and reefed foresail." On land, a whole gale breaks or uproots trees.

A storm is force 11 in the Beaufort wind scale. This is a wind that blows at 64–75 MPH (103–121 km/h). In the original scale, devised for use at sea, a force 11 wind was defined as "or that which would reduce her to storm staysails." On land, a storm uproots trees and blows them some distance, and overturns cars.

A hurricane-force wind is force 12, the highest value on the Beaufort wind scale, with a wind speed in excess of 75 MPH (121 km/h). This wind speed defines a category 1 hurricane on the Saffir/Simpson hurricane scale. In the original scale, devised for use at sea, a hurricane-force wind was described as "or that which no canvas could withstand."

wind systems Many regions of the world experience winds of particular types or from particular directions. These are not LOCAL WINDS, because their effect is felt over large areas. They are prevailing winds, and many of them have names.

A prevailing wind is one that blows more frequently from one direction than from any other. In middle latitudes, for example, the prevailing winds are westerlies, which means they blow from the west. Winds in the TROPICS usually blow from the east, so there the prevailing winds are easterlies. A WIND ROSE is compiled to determine the direction of the prevailing wind.

The trade winds are the most reliable of all prevailing winds. These winds blow toward the equator from either side, from the northeast in the Northern Hemisphere and from the southeast in the Southern Hemisphere. They are extremely dependable, especially on the eastern side of the Atlantic, Pacific, and Indian Oceans, blowing at an average speed of about 11 MPH (18 km/h) in the Northern Hemisphere and 14 MPH (22 km/h) in the Southern Hemisphere.

Air rises over the INTERTROPICAL CONVERGENCE ZONE (ITCZ). When it reaches the tropopause (*see* ATMOSPHERIC STRUCTURE), the air moves away from the equator, and it subsides in the Tropics. As it descends into the SUBTROPICAL HIGHS, some of the air flows away from the equator into higher latitudes, but most of it flows back toward the equator. It is joined by air flowing toward the equator from outside the Tropics. This circulation of air comprises the Hadley cell (*see* GENERAL CIRCULATION), and the air returning to the equator at a low level forms the trade winds.

As the air moves from the subtropical highs and into the equatorial trough, its relative VORTICITY causes the air to start turning about vertical axes centered on the subtropical highs, clockwise in the Northern Hemisphere and counterclockwise in the Southern Hemisphere. This deflects the wind from a direction at right angles to the equator, producing the northeast and southeast trades.

Where the trades cross the equator, vorticity acts in an opposite sense, and the easterly winds can become westerlies. In summer, for example, when the ITCZ moves to about 5–10°N, trade winds blowing from the southeast in the Indian Ocean south of the equator continue across the equator, where they become southwesterly winds that contribute to the onshore winds which produce the summer MONSOON over southern Asia.

There is some seasonal variation in the strength of the trade winds. They are strongest in winter and weakest in summer. This is because they originate in air flowing outward from the subtropical highs, and these are most intense in winter.

The area affected by the trades also changes with the SEASONS. In March, the northeast trades are found between 3°N and 26°N over the Atlantic and between 5°N and 25°N over the Pacific. In September, they occur between 11°N and 35°N over the Atlantic and between 10°N and 30°N over the Pacific. In March, the southeast trades occur between the equator and 25°S

over the Atlantic and between 3°N and 28°S over the Pacific. In September, they are found between 3°N and 25°S over the Atlantic and between 7°N and 20°S over the Pacific.

Weather systems in higher latitudes can also influence the strength of the trades. When these alterations in the distribution of pressure increase the PRESSURE GRADIENT away from the centers of the subtropical highs, the trade winds accelerate in what is called a surge of the trades. They are also affected by the SOUTHERN OSCILLATION. El Niño events (*see* ENSO) cause the trade winds to weaken, fade completely, or even reverse direction. La Niña events cause the winds to strengthen.

Despite these variations, the trade winds are more constant, in both speed and direction, than any other wind system on Earth, and they are especially constant over the ocean. Their constancy was of great importance in the days when goods were traded in sailing ships, and it was noted as soon as vessels were plying regularly through the Tropics.

It is their constancy that gives them their name. In Saxon times, the word *trada* meant footstep or track, and this came into English as *trade*, meaning track. (*Tread* has the same derivation.) Any kind of established track or trail might be called a trade, so a wind that blows almost all of the time and almost always from the same direction was described as a trade wind. The pursuit of a particular occupation was also called a trade, and in time the name came to be attached to the occupation itself.

Once the trade winds had been observed and their reliability verified, scientists began trying to explain them. The first to do so, in 1686, was the British astronomer Edmund Halley (1656–1742). In 1735, the British meteorologist George Hadley (1685–1768) improved on Halley's explanation, but it was not until 1856 that the American meteorologist William Ferrel (1817–91) fully explained the trade winds as part of the general circulation of the atmosphere. (*See* APPENDIX I: BIOGRAPHICAL ENTRIES for information on Ferrel, Hadley, and Halley.)

During the summer, when they extend almost to the tropopause and the westerly winds blowing above them are either nonexistent or too weak to affect the lower troposphere, the trade winds are sometimes called the equatorial easterlies.

The equatorial westerlies are westerly winds that occur in summer between the northeasterly and southeasterly trade winds. These westerlies are most strongly evident over continents, especially over Africa and southern Asia, where heating of the ground surface produces a pressure distribution that shifts the equatorial trough northward. The westerlies extend to a height of about 1–2 miles (2–3 km) over Africa and 3–4 miles (5–6 km) over the Indian Ocean. They are associated with the summer monsoon over Asia, but their cause is quite different from the pressure pattern that produces the monsoon winds. The equatorial trough does not move far enough from the equator to produce westerly winds over the Pacific or Atlantic Oceans.

The antitrade is a wind that blows at high level in the Tropics as part of the Hadley cell circulation. Its direction is opposite to that of the trade winds, so it blows from the southwest in the Northern Hemisphere and from the northwest in the Southern Hemisphere. Air that has been carried away from the equator by the antitrades subsides over the Tropics to produce the subtropical highs.

A meridional wind is a wind, or component of a wind, that blows parallel to the lines of longitude (meridians). The trade winds have a strong meridional component. Those near the surface blow toward the equator at almost 7 MPH (11 km/h). In the tropical upper troposphere, meridional winds blow away from the equator. These meridional winds comprise the horizontal components of the Hadley cell circulation.

The midlatitude westerlies, or prevailing westerlies, are the prevailing winds of the middle latitudes, between about 30° and 60° in both hemispheres. They affect a region that is bounded on one side by subtropical air and on the other by polar air, and they blow throughout the troposphere, so the upper winds blow in the same direction as the surface winds. The winds blow from the southwest in the Northern Hemisphere and from the northwest in the Southern Hemisphere. Their velocity and frequency are both at a maximum at about latitude 35°–40°. The upper winds are centered on a mean latitude of 45°. There is very little meridional flow, except when the westerly flow is interrupted during the index cycle (*see* ZONAL INDEX). A westerly wave is a wavelike disturbance that is embedded within the prevailing westerlies of middle latitudes.

In the days of sailing ships, brave west winds is the name sailors gave to the strong prevailing westerly winds that blow over the oceans in latitudes 40°N–65°N and 35°S–65°S. The brave west winds are more

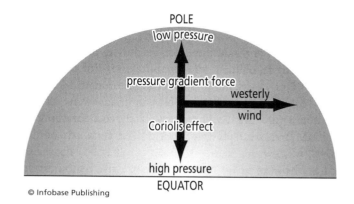

Pressure at any height decreases with distance from the equator. This produces a pressure gradient force directed toward the pole. As soon as air moves in response to this force, the Coriolis effect deflects it to the east. When the two forces balance, the resultant wind blows from west to east, producing the prevailing westerlies of middle latitudes.

persistent in the Southern Hemisphere than in the Northern Hemisphere and are strongest in the roaring forties, in latitudes 40–50°S. They are produced by the strong pressure gradient on the side nearest the equator of the frequent DEPRESSIONS that travel from west to east. Consequently, the generally westerly winds fluctuate between northwest and southwest as weather systems pass.

The roaring forties are fierce winds that blow between latitudes 40°S and 50°S. There is no continental land mass in these latitudes to slow the winds over the Southern Ocean. Terrifying though they are to sailors, the roaring forties are not alone. To their south the furious fifties and shrieking sixties blow in latitudes 50°S to 60°S and to the south of 60°S, respectively.

A stratospheric wind is one that blows in the stratosphere. The little that is known about this air movement has been obtained from satellite observations of the way particles are distributed following a major VOLCANIC ERUPTION. There were two significant eruptions in 1991. Mount Pinatubo, located in Mexico at 15.15°N, erupted in June, and Cerro Hudson, in Chile at 45.92°S, erupted in August. Both volcanoes injected ash and SULFUR DIOXIDE into the stratosphere. The cloud from Mount Pinatubo traveled westward, spreading to the north and south as it did so. The cloud from Cerro Hudson traveled eastward and did not widen. This demonstrated that in the Northern Hemisphere low-latitude stratospheric winds blow predominantly from west to east and those in the middle latitudes of the Southern Hemisphere blow from east to west. Very little material from Cerro Hudson reached Antarctica, illustrating the extent to which stratospheric air over the southern polar region is isolated during the late winter (August and September). This isolation from air in lower latitudes prevents air containing OZONE from entering the polar stratosphere and replenishing the depleted OZONE LAYER.

A country breeze is a light, cool wind that blows into a city from the surrounding countryside, especially on calm nights when the sky is clear. It is produced by the urban HEAT ISLAND effect, and the effect is most pronounced on clear calm nights. Warm air rising above the city produces an area in which the AIR PRESSURE is lower than it is in the surrounding countryside. Cool air then flows towards the city center.

A desert wind is one that blows off the DESERT. It is hot in summer, cold in winter, and very dry.

A firn wind, also called a glacier wind, is a KATABATIC WIND that blows from a GLACIER during the day, especially in summer. Air in contact with the glacier is chilled and becomes denser, its increased DENSITY causing it to flow down the slope.

A valley breeze is an ANABATIC WIND that blows during the day in some mountain areas. It most commonly occurs when conditions are calm and the sky is clear, and it is most frequent in summer. As the ground surface warms in the sunshine, air in contact with it is also warmed. This air expands and becomes less dense. Cooler, denser air that is farther from the surface subsides beneath the warm air, pushing the warm air up the mountain sides as a warm breeze. This sometimes causes the development of cumuliform cloud (see CLOUD TYPES) above the mountain, leading to SHOWERS or THUNDERSTORMS. Mountainsides where mountain and valley breezes occur regularly are often preferred for growing fruit, because the constant air movement prevents the static conditions in which FROST can form.

A mountain-gap wind, also called a canyon wind, gorge wind, or jet-effect wind, is a local wind that occurs where the prevailing wind is funneled through the space between two mountains and accelerated, so it is markedly faster than most winds in the region where it occurs.

A ravine wind is one that blows along a narrow, mountain valley or ravine. The wind is generated by a pressure gradient between the two ends of the valley. Funneling caused by the constricting effect of the valley sides accelerates the airflow, strengthening the wind.

A stowed wind is a wind that is partly blocked by a physical barrier, such as a range of mountains or hills, so it is forced through gaps between them. This increases the speed of the wind.

A tidal wind is a wind produced by the ATMOSPHERIC TIDES. The term is also applied to a wind that

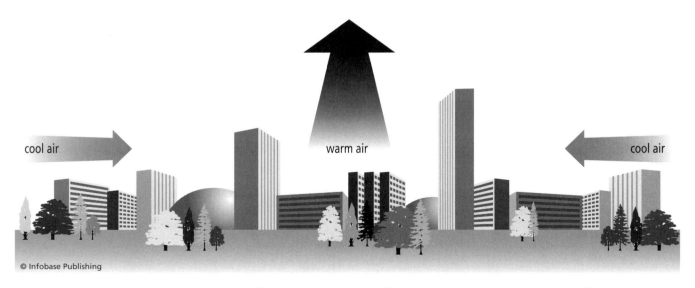

Warm air rises over the city, producing a region of low pressure near ground level. Cool air blows in from the surrounding countryside as a country breeze.

A valley breeze is a warm wind that blows up the side of a mountain during the day, when air is being warmed by contact with the ground surface and rises up the mountainside.

is produced in some tidal inlets by the displacement of air when the TIDE rises strongly. The wind is very light and can be felt only when otherwise the air is calm.

An offshore wind is one that blows across the coast from the land in the direction of the sea. An onshore wind blows across the coast from the sea in the direction of the land.

Perhaps the most famous of all winds, although it occurred only twice, was the kamikaze, the "divine wind" (in Japanese) that saved Japan from invasion in 1274 and again in 1281. The Mongol leader Genghis Khan, who ruled China and Korea, had ordered the Japanese to submit to them. When the Japanese refused, in 1274 a Mongol invasion force of about 40,000 troops sailed in Korean ships for the southernmost Japanese island of Kyushu. Before landing, the invaders had to overcome small Japanese garrisons on the offshore islands of Tsushima and Iki. They did this without difficulty, and advance parties of Mongolian troops landed in several places on Kyushu, but were contained by the Japanese warriors. Had the main Mongol force landed, it is quite possible that it would have overwhelmed the Japanese forces and Japan would have been brought under Mongol rule. Before this could happen, however, a typhoon (*see* TROPICAL CYCLONE) destroyed 200 of the Korean ships and many of the soldiers drowned. The survivors returned to Korea, and the Japanese set about strengthening their defenses. In 1281, the Mongols tried again, this time sending one army of about 40,000 Mongol, Chinese,

and Korean troops from Korea and a second army of about 100,000 troops from southern China. The two armies met, but before they could overwhelm the Japanese forces another typhoon struck, sinking almost all of the invading fleet. Of the force of 140,000, it is said that fewer than 28,000 survived. The Japanese believed the typhoon had been sent by the gods to save them from being conquered by foreigners.

wind vane Wind vanes are the most commonly used device for measuring the direction of the wind. The practice of fastening wind vanes to the tops of church steeples and other tall buildings began in Athens in the first century B.C.E., with the TOWER OF THE WINDS.

The wind vane consists of four fixed arms that point to the cardinal points of the compass (north, south, east, and west) and that are labeled N, S, E, and W. Above them is the vane itself, mounted so that it can move freely around a vertical axis. On one side of its axis, the vane has a flat surface that is aligned by the wind. On the other side of the axis, there is a

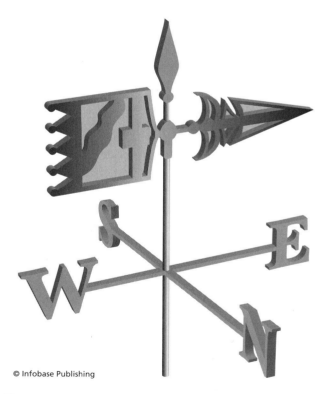

The traditional wind vane, mounted on top of a church steeple, points into the direction from which the wind is blowing. This one is shaped like an arrow, and many are animal figures.

pointer. The wind pushes the vane to the far side of its axis. Consequently, the pointer indicates the direction from which the wind is blowing. That is why winds are named by the direction from which they blow and not by the direction in which they are blowing. For example, a west (or westerly) wind blows from west to east.

Often the vane is shaped as an arrow, with the flight acting as the vane and the head as the pointer. Other, more fanciful designs are also popular. These are usually figures of animals. Roosters are the favorite, and wind vanes of this design are known as weathercocks, but fish and other birds are also used. They all have a large tail or body to act as the vane and their heads act as the pointers.

Directions that lie between the cardinal points have to be judged by eye. For this reason, and also because a wind vane is mounted so high above the ground, the precise wind direction can be difficult to see clearly.

wine harvest   The date when grapes are harvested, the size of the crop, and the quality of the wine that is made from them can be used to infer the weather conditions during the previous year. Wine has always been a product of great commercial importance, and during the Middle Ages many European monasteries relied on it as a source of income. Consequently, harvest records were kept meticulously, and many of them have survived. Crop yields and harvest dates vary according to the variety of grape being grown, but when a grower replaced one variety with another that was also recorded.

Wine harvest records have been used to trace the climatic changes that have taken place since about the year 1000. They show, for example, that prior to 1300, 30–70 percent of wine harvests in southern Germany were described as good, but between 1400 and 1700 the proportion never rose above about 53 percent and sometimes fell to 20 percent. This deterioration coincided with the coldest part of the LITTLE ICE AGE. Similar changes occurred in the French wine-growing regions, and these have also been used in reconstructing the history of climate.

Further Reading

Ladurie, Emmanuel Le Roy. *Times of Feast, Times of Famine: A History of Climate since the Year 1000.* New York: Doubleday, 1971.

wing-tip vortices   Wing-tip vortices are EDDIES that develop around the wing-tips of aircraft and birds and that are sometimes visible as thin streamers of cloud. They result from the way a wing generates lift.

Seen in cross section, the upper surface of a wing is more curved than the lower surface. Air flowing over the wing moves faster over the upper surface than it does over the lower surface because it must cover a greater distance. This reduces the AIR PRESSURE over the upper surface by the BERNOULLI EFFECT, so the pressure is higher on the lower surface than it is on the upper surface. The difference in pressure exerts an upward force on the wing. This is lift, but there is a secondary effect.

Air tends to flow from a region of high pressure to a region of low pressure. The curvature of the upper surface decreases toward the trailing edge (rear) of the wing, so near the trailing edge there is little difference in the curvature of the two surfaces and therefore little difference in the pressure on them. The center of lift, where the maximum lift is exerted, is about one-third of the way back from the leading edge (front) of the wing. Air does not spill around the trailing edge from the high pressure below to the low pressure above, because at the trailing edge there is little difference in the two pressures.

This is not the case at the wing tips, however, because at the tips there is no boundary between the two pressures for the whole cross-sectional distance of the wing. Consequently, air spills over the edge of the wing tip. It then enters the LAMINAR FLOW of air over the main wing surface and is swept to the rear. This

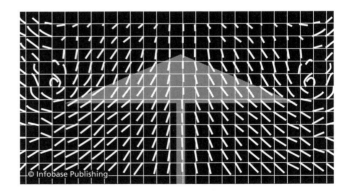

Wing-tip vortices forming around the tips of a delta-shaped wing, seen here in an experimental wind tunnel and indicated by small lengths of wool tied to the intersections of a wire grid

produces an eddy that describes a spiraling path behind the wing tip.

This is the wing-tip VORTEX. Air pressure is a little lower inside the vortex than it is outside. If the air is moist, the lower pressure may be sufficient to cause some WATER VAPOR to condense, producing a streamer of cloud behind the wing tip. This is most often seen behind high-performance aircraft that are turning steeply or pulling out of a dive. It is then that the vortices are strongest and the pressure within them is lowest.

World Climate Program (WCP) The WCP is a program that was established in 1979 by the WORLD METEOROLOGICAL ORGANIZATION to collect and store climate data. The aim of the program is to provide the data necessary for economic and social planning and to improve the understanding of climate processes. Data from the WCP are also used to determine the predictability of climate and to detect climate changes.

The WCP issues warnings to governments of impending climatic changes that may significantly affect their populations. The WCP has four components: the World Climate Data and Monitoring Program, the World Climate Applications and Services Program, the World Climate Impact Assessment and Response Strategies Program, and the World Climate Research Program. The WCP also supports the Global Climate Observing System (GCOS) that covers all aspects of the global climate.

Further Reading
WMO. "World Climate Programme." Available online. URL: www.wmo.ch/index-en.html. Accessed March 7, 2006.

World Meteorological Organization (WMO) The WMO is the United Nations specialized agency that exists to promote the establishment of a worldwide system for gathering and reporting meteorological data, the standardization of the methods used, the development of national meteorological and hydrological services in less-industrialized countries, and the application of meteorological information and understanding to other fields.

The WMO supports and coordinates the work of the national meteorological and hydrological services in its 187 member states. On March 23 each year, the contribution meteorologists and hydrologists make to human welfare and safety is celebrated on World Meteorology Day. Each World Meteorology Day has a theme, and the WMO publishes literature and a film to draw attention to the topic it has chosen for the year. For example, the theme for World Meteorology Day 2006 was "Preventing and Mitigating Natural Disasters."

The WMO was founded in 1947 at the twelfth meeting of its predecessor, the International Meteorological Organization. At their 1947 meeting, the directors of the International Meteorological Organization adopted the World Meteorological Convention, which authorized the creation of the WMO. The convention came into force in 1950, and in 1951 the WMO commenced its operation. Later in 1951, following discussions between the UN and WMO, the WMO became a UN agency.

The first International Meteorological Congress was held in 1874 under the auspices of the International Meteorological Committee, a nongovernmental body consisting of the directors of the national weather services of a number of countries. The committee had been established the previous year with the aim of establishing regular communication among meteorologists throughout the world and developing standards for weather observation and recording. The most important decision made at the congress was to compile and publish an INTERNATIONAL CLOUD ATLAS. The name of the committee was later changed to the International Meteorological Organization, and in 1947 it became the World Meteorological Organization (WMO) of the United Nations. The congress, now called the World Meteorological Congress, still meets at intervals of at least four years and sets the policy for the WMO.

The headquarters of the WMO are in Geneva, Switzerland. It has 187 members, 181 of which are member states and six are member territories that maintain their own weather services. The members are arranged as six regional associations, for Africa, Asia, South America, North and Central America, the Southwest Pacific, and Europe. Each association meets once every four years.

WMO policy is set by the World Meteorological Congress, which meets at least once every four years, and policy is implemented by an executive council with 36 members, including the WMO president and two vice presidents. The council meets at least once every year. The WMO also has eight technical commissions that deal with aeronautical meteorology, agricultural meteorology, atmospheric sciences, basic systems, climatology, hydrology, instruments and methods of observation, and marine meteorology. Each commission meets every four years.

The principal activity of the WMO centers on the WORLD WEATHER WATCH. The WMO also maintains the WORLD CLIMATE PROGRAM and the Atmospheric Research and Environment Program. The Atmospheric Research and Environment Program coordinates and fosters research into the composition and structure of the atmosphere, the physics and chemistry of clouds, weather modification, tropical meteorology, and WEATHER FORECASTING. This program also includes the Global Ozone Observing System that was established in the 1950s and now receives data from 140 ground-based stations. The Tropical Cyclone Landfall Program aims to improve understanding of the rainfall and wind field on the coast and farther inland when a TROPICAL CYCLONE makes landfall.

The international agreement to take steps to halt the depletion of the OZONE LAYER that resulted in the Montreal Protocol on Substances that Deplete the Ozone Layer was based largely on data from the Global Ozone Observing System. The WMO also supports the implementation of the UN Framework Convention on Climate Change, the International Convention to Combat Desertification, and the Vienna Convention on the Protection of the Ozone Layer (*see* APPENDIX IV: LAWS, REGULATIONS, AND INTERNATIONAL AGREEMENTS).

Observing stations throughout the world collect meteorological and climatological data, and disseminate them through the WMO communications network. The WMO devised the system of international index numbers for identifying observing stations. Areas of the world are divided into blocks, each of which is given a two-digit number and a further three-digit number identifies each station within each block.

Stations use the international synoptic code to transmit data. Each element of the data is encoded as a series of five-digit numerals. Synoptic code is one of several recognized codes that is used to transmit SYNOPTIC weather data and observations. It is used widely throughout the world and is the code officially approved by the WMO. It is one of the codes used by the U.S. NATIONAL WEATHER SERVICE (NWS). The NWS also uses the METAR code. The SYNOP code is also used extensively. All of these codes translate observations and data into groups of numbers. This compresses the data and improves the reliability of transmission and reception.

Réseau is the name used by the WMO to describe the global network of weather stations that have been chosen to represent the world climate. The full name is *réseau mondiale,* which is French for "global network."

Further Reading

WMO. "World Meteorological Organisation." Available online. URL: www.wmo.ch/index-en.html. Accessed March 7, 2006.

World Weather Watch (WWW) World Weather Watch is an international program, run by the WORLD METEOROLOGICAL ORGANIZATION, which coordinates the national weather systems of the states that belong to the WMO. The WWW has several components.

The Global Observing System (GOS) gathers data from about 11,000 land-based, surface observing stations, each of which takes readings every three hours and in some cases hourly. There are also approximately 900 upper-air stations collecting data from radiosondes (*see* WEATHER BALLOON). Approximately 40 percent of the 7,000 vessels belonging to the VOLUNTARY OBSERVING SHIP scheme are at sea at any one time, and the GOS receives data from about 900 drifting buoys. About 3,000 aircraft submit data; in 2005, the GOS received almost 300,000 reports from aircraft. The GOS also receives data from five satellites in near-polar ORBIT and six in geostationary orbit.

The Global Telecommunications System (GTS) links meteorological communications centers throughout the world. The main telecommunication network comprises three world telecommunication centers, in Melbourne, Moscow, and Washington, and 15 regional telecommunication centers. These are in Algiers, Beijing, Bracknell, Brasilia, Buenos Aires, Cairo, Dakar, Jeddah, Nairobi, New Delhi, Offenbach, Toulouse, Prague, Sofia, and Tokyo. The regional centers are at the hub of regional networks for Africa, Asia, South America, North America, Central America and the Caribbean, South-West Pacific, and Europe.

The Global Data-processing and Forecasting Systems (GDPFS) analyze data and prepare forecasts and other information. These are distributed to national meteorological and hydrological services through the GTS.

The Public Weather Services Program (PWSP) forms part of the Applications of Meteorology Program. It strengthens and supports national meteorological and hydrological services in providing reliable weather forecasts.

The WWW also includes the Tropical Cyclone Program, which monitors TROPICAL CYCLONES and alerts national services to their approach, and it supports research into Antarctic meteorology.

Further Reading

WMO. "World Weather Watch." Available online. URL: www.wmo.ch/index-en.html. Accessed March 7, 2006.

X, Y

xenon Xenon is a rare gas that accounts for about 0.0000086 percent of the atmosphere, or one part in 10 million by volume. It is heavier than the other atmospheric gases. Atomic number 54; atomic weight 131.30; DENSITY (at sea-level pressure and 32°F, 0°C) 0.059 ounces per cubic foot (5.887 grams per liter). Xenon melts at -169.6°F (-111.9°C) and boils at -160.6°F (-107.1°C).

xerophilous Adapted to living in places that have an arid climate.

xerophyte A plant that is adapted to an arid climate.

xerothermic An adjective that is applied to places, climates, or conditions that are hot and dry.

Younger Dryas (Loch Lomond Stadial) The Younger Dryas was a cold, dry period, affecting the whole of the North Atlantic region and with effects that were felt in the TROPICS, where climates became drier, and as far away as New Zealand. It began about 12,900 years ago and lasted for about 1,300 years, ending abruptly 11,640 years ago when the mean temperature increased by about 13°F (7°C) in the space of about 10 years. The Younger Dryas is recognized by soils of that date containing abundant pollen from mountain avens (*Dryas octopetala*), which is an arctic-alpine plant. Its effects have also been detected in ICE CORES from GISP2 (*see* GREENLAND ICE SHEET PROJECT).

When the Younger Dryas began, ICE SHEETS had probably disappeared from the whole of Scotland as the Devensian Glacial drew to a close. By about 10,800 years ago, however, the ice was hundreds of meters thick over the western Highlands of Scotland, and the climate of Europe was similar to that during the glacial. Many scientists believe that the melting of the LAURENTIDE ICE SHEET triggered the rapid onset of the Younger Dryas by releasing large amounts of freshwater that flowed into the North Atlantic. This shut down the GREAT CONVEYOR and produced a SNOWBLITZ effect. There are difficulties with this explanation, however, because there was another, albeit smaller release of freshwater at the end of the Younger Dryas that did not halt the warming. The cause of the dramatic cooling remains uncertain.

zenith angle The zenith angle is the height of the Sun above the horizon. This is measured as the angle between a line linking the observer to the Sun and the vertical (the zenith). The zenith angle (Z) is calculated from:

$$\cos Z = \sin \theta \sin \delta + \cos \theta \cos \delta \cos h$$

where θ is the latitude, δ is the solar DECLINATION, and h is the hour angle (this is 0 at noon and increases by 15° (360/24) for each hour either side of noon).

As the zenith angle changes through the day, there is a reversal in the relative intensities of light from directly overhead at two wavelengths (*see* WAVE CHARACTERISTICS). The German word for reversal is *Umkehr*, and this is known as the umkehr effect.

One of the two wavelengths is more strongly absorbed by OZONE than is the other, so a series of measurements of the change in their relative intensities can be used to determine the vertical distribution of ozone.

zonal flow A zonal flow is a movement of air in a generally west-to-east direction, approximately parallel to the lines of latitude. When winds are measured over a large area, they may be separated into their zonal and meridional components (*see* MERIDIONAL FLOW). The average zonal winds, known as the zonal average, are strongest in the upper troposphere (*see* ATMOSPHERIC STRUCTURE) over the subtropics during the three winter months (December, January, and February in the Northern Hemisphere and June, July, and August in the Southern Hemisphere). They are then blowing at an average speed of 89.5 MPH (144 km/h).

The movement of air in the TROPOSPHERE in a west-to-east direction that is measured over a specified period or with respect to a particular longitude is known as the zonal circulation.

The zonal kinetic energy is the KINETIC ENERGY of the mean zonal wind. It is calculated by averaging the zonal component of the wind along a specified latitude.

A diagram in which the speed of the ZONAL FLOW is plotted against latitude is called a zonal-wind profile (*see* WIND PROFILE).

zonal index The zonal index is a measure of the strength of the westerly winds between latitudes 33° and 55° in both hemispheres, expressed either as the horizontal PRESSURE GRADIENT, or as the corresponding GEOSTROPHIC WIND. A high zonal index indicates strong westerly winds and a continuous, strongly developed, and almost straight JET STREAM. Weather systems move at a fairly steady pace in the same direction as the jet stream. A low index indicates weak westerlies and the formation of a cellular pattern of air flow. The ROSSBY WAVES in the jet stream and polar FRONT are well developed, so the jet stream follows a sinuous path. Low-pressure polar air (*see* AIR MASS) extends far to the south in some places, and in others tropical air extends far to the north.

Progressive changes in the zonal index are cyclical. The full index cycle typically lasts for between three and eight weeks, at the end of which the original circulation

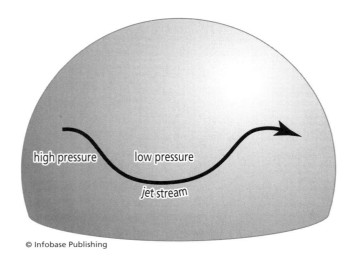

When the zonal index is low, the wave pattern in the jet stream is strongly developed, so the jet stream and polar front follow a very sinuous path.

is restored. The change usually moves westward (from east to west) at a rate of about 60° of longitude a week, but it is very irregular both in the duration of the full cycle and in the speed with which it moves westward. It is especially common in February and March.

The zonal index is a number that represents the difference in AIR PRESSURE between two latitudes, usually 33°N and 55°N. These latitudes mark the boundaries that contain the polar front and polar front jet stream. When the pressure difference is great, the index is said to be high, and when it is small, the index is said to be low.

When the cycle commences, the zonal index is high. The polar front and its jet stream are aligned approximately from west to east, some distance to the north of their mean positions. Polar air lies to the north of the front and tropical air to the south of it, and there is very little mixing between them. A small number of Rossby waves lie along the front and jet stream. The amplitude of the waves is small, and their wavelength is long.

Where the Rossby waves carry the flow northward, that part of the jet stream experiences a stronger CORIOLIS EFFECT (CorF), because the magnitude of CorF increases with latitude. In order to conserve the absolute VORTICITY of the flow, the jet stream develops an equivalent negative vorticity. This turns the jet stream so that it curves cyclonically (*see* CYCLONE). It is then moving back to its original latitude, but it overshoots

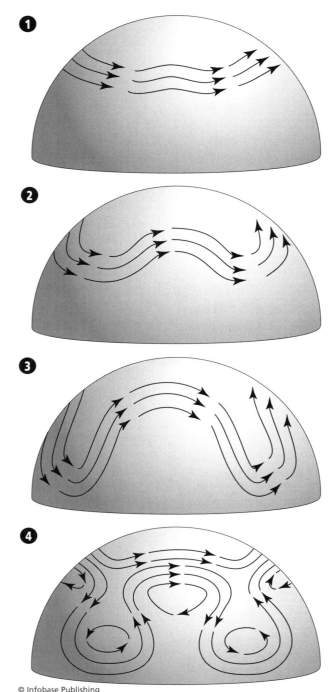

Four stages in the development of the index cycle, during which the pressure difference to each side of the polar front weakens: (1) the initial condition, with a high zonal index; (2) the amplitude of the Rossby waves increases and their wavelength decreases; (3) the undulations in the waves become more and more extreme, carrying polar air a long distance south and tropical air a long distance north; (4) the flow breaks down into a series of cells, with anticyclones in the north and cyclones in the south.

and, to compensate, develops positive vorticity that makes it curve anticyclonically (*see* ANTICYCLONE).

This pattern can remain stable for long periods, with low-amplitude, long-wavelength Rossby waves. During an index cycle, however, each overshoot is slightly greater than the one preceding it. Both the cyclonic and anticyclonic curvatures continue for longer, so the wave amplitude increases and the wavelength decreases. The undulations in the jet stream and polar front become more and more extreme. The original zonal (east to west) flow is then much more meridional (north to south and south to north).

Finally, the pattern breaks down altogether. The flow of air joins on either side of the RIDGES and TROUGHS, forming isolated cells. Cells that form from the troughs lie to the south. The flow around them is cyclonic, and they mark areas of low surface pressure. Cells that form from the ridges are located to the north, and the flow around them is anticyclonic. They mark areas of high surface pressure.

At this stage the pattern may temporarily stabilize. This produces BLOCKING, resulting in prolonged periods of weather. On the southern side of the front, at about 33°N (approximately the latitude of Dallas, Texas, and Little Rock, Arkansas), the weather is associated with low pressure. It is associated with high pressure on the northern side, at about 55°N (a little to the north of Edmonton, Alberta, and Belfast, Northern Ireland).

As the anticyclones weaken and the cyclones fill, the cellular pattern dissipates. The original flow then reestablishes itself, with a high zonal index.

Polar outbreaks sometimes occur in North America during the later stages of the index cycle, when the jet stream follows a deeply undulating path. A polar outbreak is an extension of polar air into lower latitudes. A trough in the middle troposphere usually extends over eastern North America in both summer and winter. It is possibly a LEE trough resulting from the effect of the Rocky Mountains on the high-level westerly winds. The flow of air around the trough tends to bring polar air southward, especially when the trough is strong, and the polar air brings cold weather. When the trough is weak the westerly flow of air is stronger and polar outbreaks are less likely.

APPENDIX I
BIOGRAPHICAL ENTRIES

Abbe, Cleveland
(1838–1916)
American
Meteorologist

Cleveland Abbe was the first person to issue regular daily weather bulletins and forecasts. He is sometimes called the "father of the Weather Bureau." He was also influential in the establishment of standardized time zones across the United States.

Born and educated in New York City, Abbe went to study astronomy at the University of Michigan and privately with Benjamin Apthorp Gould (1824–96) at Cambridge, Massachusetts. Abbe taught at the University of Michigan for several years. Afterward, he spent two years, 1864–66, completing his astronomical studies at the Pulkovo Observatory in Russia. When he returned to the United States, he was appointed director of the Cincinnati Observatory.

Like Joseph Henry, Abbe received telegraphic reports of storms and used them to plot their location and timing across the country. These compilations provided him with the information he issued from Cincinnati Observatory in his daily reports called the *Weather Bulletin,* the first of which appeared on September 1, 1869. On September 22 of the same year, he published his first weather forecast.

The government was immediately interested, and on February 2, 1870, Rep. Halbert E. Paine introduced a Joint Congressional Resolution requiring the secretary of war to establish a national meteorological service. President Ulysses S. Grant signed the resolution on February 9, 1870, and the weather bureau began operations in November 1870. It formed part of the recently instituted Division of Telegrams and Reports for the Benefit of Commerce of the Army Signal Service and was headed by an army general. Abbe was appointed

his scientific assistant, joined the Signal Service, and started work in 1871, issuing the first of his three-day weather forecasts, based on probabilities, on February 19, 1871. The first "cautionary storm signal" (for the Great Lakes region) was issued on November 8, 1871.

In 1891, the national bureau was renamed the United States Weather Bureau and was transferred from the Army Signal Corps to the Department of Agriculture (it was transferred to the Department of Commerce in 1940). Abbe was the meteorologist in charge and retained this position for the rest of his life. At the same time, he conducted research and taught meteorology at Johns Hopkins University.

His advocacy of standardized time zones culminated in a report on the subject that he wrote in 1879. At that time, each community used its own local time, which was accurate enough but meant travelers had to continually adjust their watches as they moved east or west. The railroad companies had devised their own standardization for the purposes of scheduling services, and these were the basis of the Abbe proposal. The government adopted the idea, and in 1883 the country was divided into four time zones. The same time was used throughout each zone and was based on an average value. This system was later extended to the entire world.

Cleveland Abbe died on October 28, 1916, in Chevy Chase, Maryland, where he and his wife are buried.

Agassiz, Jean Louis Rodolphe
(1807–1873)
Swiss–American
Naturalist

Louis Agassiz was born on May 28, 1807, at Motier, not far from Friborg, on the shore of Lake Morat, in Switzerland, where his father was the Protestant pastor. The family was originally French, but was forced to flee from France after Louis XIV revoked the Edict of Nantes in 1685 and Protestants were no longer tolerated in the country.

Louis's mother, Rose Mayor, taught him to love the natural world. His formal education began with four years at the gymnasium (high school) in Bienne (Biel), northwest of Bern, after which he attended a school in Lausanne. In 1824, he enrolled at the University of Zürich, and in 1826 moved from there to the University of Heidelberg, Germany, but he caught typhoid fever in Heidelberg and had to return to Switzerland to recuperate. In 1827, he enrolled at the University of Munich, Germany. He qualified as a doctor of philosophy (Ph.D.) at the University of Erlangen in 1829 and as a doctor of medicine at Munich in 1830.

Two distinguished naturalists from Munich, J. B. Spix and C. P. J. von Martius, had spent 1819 and 1820 touring Brazil, returning to Germany with a large collection of fishes, most of them from the Amazon River. Spix set about classifying the collection, but in 1826 he died and Martius handed the task over to Agassiz. Agassiz completed the classification, and it was published in 1829, when he was only 22 years old, as *Selecta Genera et Species Piscum* (Selection of fish genera and species). This set the course for his major research.

In November 1831, Agassiz went to Paris, where he continued his studies of fishes, for a short time at the Natural History Museum under the supervision of Georges Cuvier (1769–1832), the eminent comparative anatomist. Cuvier, who had read and been greatly impressed by his work on the Brazilian fishes, befriended the young Agassiz. Alexander von Humboldt also helped him. Following the death of Cuvier in May 1832, Agassiz moved to the University of Neuchâtel, Switzerland, where von Humboldt helped him secure the professorship of natural history. Between 1833 and 1844, Agassiz published *Recherches sur les Poissons Fossiles* (Research on fossil fish) in which he classified more than 1,700 species. He also published major works on fossil echinoderms and mollusks.

In 1836, Agassiz turned his attention to a new question. Boulders that were scattered over the plain of eastern France and in the Jura Mountains were different in composition from the solid rock beneath the ground on which they lay. Some scientists thought these erratics might have been transported to their present positions by GLACIERS. If glaciers can push boulders ahead of them, it means that the glaciers flow, and if flowing glaciers pushed boulders deep into France, it means that at one time the Swiss glaciers must have extended much farther than they do today.

With some friends, Agassiz spent his 1836 and 1837 summer vacations on the Aar Glacier. They built a hut on the ice and called it the "Hôtel des Neuchâtelois." From this base they observed the rocks piled to the sides and at the ends of this and other glaciers and the grooves in the solid rock that looked as though they had been made by scouring as harder stones were dragged past them.

Agassiz continued his investigations, and in 1839 he found a hut that had been built on the ice in 1827 and that had moved a mile (1.6 km) from its original position. He drove a line of stakes into the ice across a glacier and found that by 1841 they had moved and the straight line had changed to a U shape, because the stakes at the center had moved faster than the ones at the sides. By then he was convinced that all of Switzerland had been covered by ice in the geologically recent past. He also concluded that all those parts of Europe where erratic boulders and gravel were found had also lain beneath a great sheet of ice resembling that in Greenland. In 1840, Agassiz published the most important of all his works, *Études sur les Glaciers* (Studies of glaciers).

In 1846, with the help of a grant from King Friedrich Wilhelm IV of Prussia, Agassiz visited the United States, partly to continue his studies but immediately to deliver a series of lectures at the Lowell Institute in Boston. He followed these with other popular and technical lectures in various cities. The lectures were popular, and Agassiz extended his stay, studying North American natural history at the same time. In 1848, he was appointed Professor of Zoology at Harvard University. Agassiz became an American citizen and remained in the United States for the rest of his life, for most of the time at Harvard.

Agassiz found evidence that North America had also been covered by ice, and he traced the shoreline of a vast, vanished lake that had once covered North Dakota, Minnesota, and Manitoba. It is now called Lake Agassiz. Agassiz clearly demonstrated that over Europe and North America there had once been what he called a Great Ice Age.

At Harvard in 1858, Louis Agassiz developed the Museum of Comparative Zoology to assist research and teaching. It was built around Agassiz's own collection, and he was its director from 1859 until his death. His scientific research was of great importance, but he was also one of the finest teachers of science America has ever known. He was devoted to his students and treated them as collaborators. During the second half of the 19th century, every well-known and successful teacher of natural history in the country had at one time been a pupil either of Agassiz himself or of one of his former students.

Despite being one of the most knowledgeable biologists of his time, Agassiz remained steadfastly opposed to the Darwinian concept of evolution by natural selection.

He married twice. Cecile Braun, his first wife, died in 1848 in Baden, a few months after Agassiz had taken up his position at Harvard. In 1850, he married Elizabeth C. Cary, who became his valued scientific assistant.

Louis Agassiz died at Cambridge, Massachusetts, on December 12, 1873. He is buried at Mount Auburn, Cambridge, where his grave is marked by a boulder from the Aar glacial moraine. In 1915, Agassiz was elected to the Hall of Fame for Great Americans.

Further Reading
Academy of Natural Sciences, Philadelphia. "Louis Agassiz (1807–1873)." University of California. Available online. URL: www.ucmp.berkeley.edu/history/agassiz.html. Accessed March 9, 2006.

Aitken, John
(1839–1919)
Scottish
Physicist

John Aitken discovered that the air contains large numbers of very small particles, now known as Aitken nuclei (*see* AEROSOL), and invented the Aitken nuclei counter, an instrument for detecting them. He also discovered the part Aitken nuclei play, as CLOUD CONDENSATION NUCLEI, in the formation of clouds.

John Aitken was born and died at Falkirk, in Stirlingshire. He studied at the University of Glasgow, which awarded him an honorary doctorate in 1889.

Due to poor health, he was never able to hold any official position. Instead, he worked from a laboratory he made at his home. There he constructed his own apparatus and conducted experiments. He described

many of his findings in papers published in the journals of the Royal Society of Edinburgh, of which he was a member.

d'Alibard, Thomas-François
(1703–1799)
French
Physicist

Thomas-François d'Alibard performed Benjamin Franklin's famous kite experiment 36 days before Franklin did so.

Franklin had written extensively about electricity, and in 1751 Franklin's friend Peter Collinson (1694–1768), a very distinguished English scientist, assembled his articles and papers on the subject and published them as a short book. Georges-Louis Leclerc, the comte de Buffon (1707–88), obtained a copy of Collinson's book. Buffon was the most eminent French natural scientist of his day—he was made a count (comte) in 1771 and an associate of the French Academy of Sciences and Fellow of the Royal Society of London, both in 1739. Buffon asked d'Alibard to translate the book, which he did.

Other scientists had suggested that clouds might contain electricity and that this was the cause of lightning. Franklin also held this view, and Buffon and d'Alibard devised a way to test it.

They set up their apparatus, which consisted of an iron rod, 40 feet (12 m) long and with a brass tip. They insulated the rod by setting it upon a wooden plank that stood on three wine bottles. On May 10, 1752, a soldier keeping watch on the apparatus heard a clap of thunder. He sent for the village priest. As people from the nearby village stood back, sparks flew from the rod with a crackling sound. The priest wrote to tell d'Alibard what had occurred, and on May 13 d'Alibard submitted a report to the Academy of Sciences. In his report, d'Alibard said: "In following the path that Mr. Franklin has traced for us, I have obtained complete satisfaction." The experiment was repeated in Paris on May 18 and the king of France sent his congratulations to Franklin through Collinson.

When Franklin performed his kite experiment on June 15, he was unaware that d'Alibard had already done so.

Amontons, Guillaume
(1663–1705)
French
Physicist

Guillaume Amontons, one of the most ingenious inventors of his age, was born in Paris on August 31, 1663. His father was a lawyer, originally from Normandy. Guillaume studied the physical sciences, celestial mechanics, and mathematics, as well as drawing, surveying, and architecture, but so far as is known he did not attend a university. He earned his living as a government employee, working on a range of public works.

While still in his teens, Amontons became profoundly deaf. Far from regarding this as a handicap, he considered it a blessing, because it allowed him to concentrate on his scientific work without distraction.

In 1687, he invented a new type of hygroscopic HYGROMETER, based on the expansion and contraction of a substance as it absorbs and loses atmospheric moisture (the modern hair hygrometer was invented in 1783 by Horace de Saussure). The following year he devised an optical telegraph (*see* TELEGRAPHY), which he thought would be of help to deaf people. Messages were transmitted by means of a bright light that was visible to a person with a telescope at the next station. He demonstrated it to the king some time between 1688 and 1695, but it was never adopted.

In 1695, Amontons invented a BAROMETER that did not require a reservoir of mercury. This meant it could be used at sea, where mercury barometers gave unreliable readings because the level of the mercury in the reservoir oscillated with the motion of the ship.

The same year he improved on the air thermoscope (*see* THERMOMETER) that had been invented by Galileo in 1593. Galileo's design used the expansion and contraction of the air in a tube to alter the level of water. The disadvantage of this was that the volume of the water was also affected by changes in AIR PRESSURE. Galileo was unaware of the effect of air pressure, but Amontons knew how to remove it. Instead of water he used mercury, then adjusted the height of the mercury until the air filled a fixed volume. After that, changing the temperature of the air in the tube altered the pressure it exerted on the mercury, and it was this changing pressure that the instrument measured. The Amontons thermometer was more accurate than Galileo's, and he was able to use it to show that, within the limits of his instrument, water always boiled at the same temperature, but it was not accurate enough for most scientific uses.

No one had yet devised a scale by which temperature could be measured. This lack prevented him from discovering Charles's law (*see* GAS LAWS), but his new thermometer allowed him to take the study of gases a step further than Mariotte had done. He noticed that for a particular change in temperature, the volume occupied by a gas always changes by the same amount. This led to Amonton's law, which he described in 1699 and which can be stated as $P_1T_2 = P_2T_1$, where P_1 is the initial pressure, P_2 is the altered pressure, T_1 is the initial temperature, and T_2 is the altered temperature. It also allowed Amontons to visualize a temperature at which gases contracted to a volume beyond which they could contract no further. This was the concept of absolute zero (*see* TEMPERATURE SCALES).

Amontons also invented a type of clock called a clepsydra, operated by the flow of water. He proposed that his clepsydra could be used at sea, although it would not have been accurate enough to measure longitude.

In 1690, Amontons became a member of the French Academy of Sciences. He published a number of papers and one book, *Remarques et expériences physiques sur la construction d'une nouvelle clepsydre, sur les baromètres, thermomètres, et hygromètres* (Observations and physical experiences on the construction of a new clepsydra, on barometers, thermometers and hygrometers), which appeared in 1695.

Amontons died in Paris on October 11, 1705.

Aristotle
(384–322 B.C.E.)
Greek
Philosopher

Aristotle was born in 384 B.C.E. at Stagirus, a Greek colony on the coast of Macedonia. Both his parents were Greek. His father, Nichomachus, was the personal physician to the king of Macedonia, Amyntas III, but Nichomachus died when Aristotle was still a boy. Aristotle was then brought up by a guardian, Proxenus, and in about 367 B.C.E., when he was 17, Proxenus sent him to the Academy in Athens that was led by the philosopher Plato (428 or 427–348 or 347 B.C.E.). Aristotle remained at the Academy for 20 years, first as a pupil and then as a teacher.

In 347 B.C.E., Athens was at war with Macedonia. Amyntas had died and the new king was his son, Philip II. Then Plato died. Perhaps for political reasons, because he was sympathetic to the Macedonian cause, or perhaps because of the change in the leadership of the Academy, Aristotle left Athens. He settled first on the coast of Anatolia, in what is now Turkey, then on the island of Lesbos, where he lived from 345 B.C.E. to 343 B.C.E. Finally, he returned to Macedonia. In the course of his travels he married Pythias, the daughter or niece of Hermias, the ruler of the land where Aristotle first arrived after leaving Athens.

Back in Macedonia, Philip appointed Aristotle to supervise the education of his son, a 13-year-old boy called Alexander, who later became known as Alexander the Great. Later in his life Aristotle was very wealthy, possibly from the money he was paid for teaching Alexander.

Aristotle returned to Athens in about 335 B.C.E., and for the next 12 years he taught at the Lyceum, one of the three most famous schools in the city, established in the grounds of the temple to Apollo Lyceius, hence the name. There Aristotle began to assemble a library of books and maps and a museum of natural history. He also established a zoo with animals that were captured during Alexander's campaigns in Asia.

Alexander died in 323 B.C.E., and the opponents of the Macedonians became powerful in Athens. Aristotle was associated with a renowned Macedonian general, and he was charged with impiety. Rather than face trial, and possibly death, he moved to Chalcis (now called Khalkis) on the Greek island of Euboea, north of Athens. The following year he fell ill and died. He was 62 years old.

Aristotle was one of the most original thinkers who ever lived. Every subject interested him. He wrote about logic, ethics, politics, biology, physics, astronomy, and many other topics. His writings are contained in 47 surviving works. Some of these comprise several volumes and others are very short.

One of his works is called *Meteorologica*. The title means "account (*logos*) of lofty things (*meteoros*)" and from it we derive our word METEOROLOGY.

In *Meteorologica,* Aristotle set out his own explanations for the weather. He had studied Egyptian ideas on the subject and the methods for classifying winds that had been devised by the Babylonians, and in addition to these he drew on a wide variety of sources.

He maintained that the weather is confined to the region between the Earth and the Moon. This region is composed of four elements: earth, air, fire, and water. Earth and water are heavy and sink, while air and fire are light and rise. Aristotle proposed theories to explain the formation of clouds, rain, hail, wind, thunder, lightning, and storms. He argued, for example, that some of the WATER VAPOR formed during the day does not rise very high because the ratio of the fire that is raising it to the water being raised is too small. At night the water cools and descends and is then called DEW, or hoar FROST if it freezes before it has condensed to water. It is dew, he said, when the vapor has condensed to water and the heat is not so great as to dry up the moisture that has been raised, nor the cold sufficient for the water to freeze. Aristotle observed that although hailstones are made from ice, hailstorms are most common in spring and autumn. They are rare in winter and happen when the weather is mild. He suggested that in warm weather the cold forms discrete areas within the surrounding heat and this could cause water vapor to condense rapidly. That is why RAINDROPS are bigger in warm weather than they are when it is cold. If the cold is very concentrated, however, it can freeze the raindrops as they form, producing hail.

Aristotle had no instruments to measure TEMPERATURE, AIR PRESSURE, or HUMIDITY. Nor did he have the facilities to compile a picture of weather conditions over a large area. Lacking any means to validate his ideas, Aristotle was not in a position to develop an accurate understanding of atmospheric processes. The importance of his contribution to scientific thinking arises from his insistence on basing theories on observed facts rather than tradition or unsupported opinion.

Arrhenius, Svante August

(1859–1927)
Swedish
Physical chemist

Svante Arrhenius was born on February 19, 1859, on the estate of Vik, near Uppsala, Sweden, that was owned by the University of Uppsala. In 1860, the family moved to Uppsala.

Svante began his education at the cathedral school in Uppsala. He was brilliantly clever, especially at mathematics, and was accepted as a student at Uppsala

University when he was only 17. He studied chemistry, physics, and mathematics, graduating in 1878, and then stayed on to start working for his doctorate. After a time he grew dissatisfied with the quality of the teaching in physics, and in 1881 he moved to Stockholm to study under the physicist Erik Edlund (1819–1888).

Arrhenius completed his doctoral thesis in 1884 and submitted it to Uppsala University. He wrote it in French: *Recherches sur la conductibilité galvanique des électrolytes* (Investigations on the galvanic conductivity of electrolytes). An electrolyte is a solution of a chemical in water that conducts electricity. The first part of his thesis dealt with ways to measure the electrical conductivity of very weak solutions and the second with the reason the solution is conductive. This, he proposed, is because the dissolved substance dissociates into charged IONS (for example, common salt, which is sodium chloride ($NaCl$) dissociates into sodium (Na^+) and chloride (Cl^-) ions) and the ions move through the solution.

Arrhenius presented his thesis to his professor, a man he greatly admired. According to his own account, Arrhenius said: "I have a new theory of electrical conductivity as a cause of chemical reactions." The professor replied "This is very interesting. Good bye." The professor knew that many theories are formed and almost all of them turn out to be wrong and soon disappear. He concluded, therefore, that Arrhenius's theory was most probably mistaken. Like most of his colleagues, the professor believed in experimentation, and Arrhenius had performed no experiments in the course of developing his idea. The thesis was accepted, but it was awarded only fourth class, the lowest grade. The thesis later earned Arrhenius a Nobel Prize.

Undeterred by the lack of interest at Uppsala, Arrhenius sent copies of his thesis to several of the most eminent chemists of the time. This led to the offer of a job at the University of Riga, Latvia, and the offer persuaded the Uppsala authorities to reconsider their opinion. Toward the end of 1884, he was offered a post at Uppsala, and later a traveling fellowship that allowed him to meet other scientists working in the same field.

In 1891, Arrhenius was offered a professorship at the University of Giessen, Germany, but declined it, because he preferred to remain in Sweden. Instead, he accepted a lectureship at the Stockholms Högskola (High School), where in 1895 he became professor of physics. From 1897 until 1905, Arrhenius was also rec-

tor. The high school was equivalent to the science faculty of a university, but it was not empowered to award degrees or accept doctoral theses. It became the University of Stockholm in 1960.

Arrhenius retired from the professorship in 1905 and refused another invitation from Germany, this time to become a professor at the University of Berlin. In 1905, the Swedish Academy of Sciences decided to establish a Nobel Institute for Physical Chemistry and Arrhenius was appointed director. He remained in this position until shortly before his death.

Arrhenius was a man of wide interests. He applied his knowledge of chemical reactions to the effects on the body of toxins and antitoxins, describing this in a series of lectures he delivered in 1904 at the University of California. He was interested in immunology, and on the origin of life on Earth. Arrhenius was the first scientist is propose the idea of panspermia, according to which life arrived on Earth in the form of spores that had drifted through space. This idea was out of favor for many years, but recent discoveries of extremophiles—single-celled organisms that flourish in extreme environments—and of bacteria that have survived prolonged periods in space have revived interest in it. He wrote about astronomy, especially comets and the possibility of life on Mars.

He also studied what is now called the GREENHOUSE EFFECT. In 1896, he published a paper, "On the Influence of Carbonic Acid in the Air upon the Temperature of the Ground" (*Philosophical Magazine*, vol. 41, pages 237–271). He was not the first scientist to consider the absorption of energy by CARBON DIOXIDE, and it was the French mathematical physicist Jean-Baptiste-Joseph Fourier (1768–1830) who suggested in 1827 that the atmosphere acts like the glass of a greenhouse, allowing light in but preventing heat from leaving. Arrhenius turned earlier speculations into hard numbers.

He calculated the effect carbon dioxide would have if the atmospheric concentration of it were altered. He worked out the resulting change in mean temperature for 13 belts of latitude, each of 10 degrees from 70°N to 60°S, for the four seasons of the year, and the mean for the year. For each of these belts of latitude and seasons, and for the whole year, he worked out what the temperature would be if the carbon dioxide concentration were 67 percent, 150 percent, 200 percent, 250 percent, and 300 percent of the concentration that actually existed in the late 19th century. He calculated that a doubling of atmospheric carbon dioxide would increase the mean annual temperature by 8.91°F (4.95°C) at the equator and by 10.89°F (6.05°C) at 60°N. The task involved thousands upon thousands of calculations, all of which he performed by paper and pencil.

Arrhenius received the 1903 Nobel Prize in chemistry for his work on electrolytes. He also received many other awards and honorary degrees. He married twice, first in 1894 to Sofia Rudbeck, by whom he had a son, and in 1905 to Maria Johansson, by whom he had one son and two daughters.

Arrhenius was a happy, contented, genial man who made many friends and delighted in meeting them. During World War I, he worked successfully to obtain the release of German and Austrian scientists who were prisoners of war.

He was also a popular lecturer and author. In his later years he was in constant demand and traveled widely to attend meetings and deliver lectures. His incessant hard work may have weakened his health, because he was only 68 when he died in Stockholm on October 2, 1927. He is buried in Uppsala.

Further Reading

Arrhenius, Svante. "On the Influence of Carbonic Acid in the Air upon the Temperature of the Ground." (excerpts) *Philosophical Magazine* 41, 237–276 (1896). Available online. URL: http://web.lemoyne.edu/~GIUNTA/Arrhenius.html. Accessed March 9, 2006.

Nobel Foundation. "Svante Arrhenius—Biography." Nobel Foundation. Available online. URL: http://nobelprize.org/chemistry/laureates/1903/arrhenius-bio.html. Accessed March 9, 2006.

Beaufort, Francis

(1774–1857)
Irish
Meteorologist, hydrographer, and naval officer

Francis Beaufort is the scientist who devised the scale of wind force that bears his name (*see* BEAUFORT WIND SCALE). He was born in Navan, County Meath, Ireland, where his father, the Reverend Daniel Augustus Beaufort, was the rector. The family was of Huguenot origin. Francis's father was keenly interested in geography and topography—the art of drawing the natural and built features of a town or area of countryside—and from an early age Francis came to share his enthusiasm.

In 1789, Francis joined the East India Company, the trading company that had been established to administer British commercial interests in India, but eventually came virtually to govern that country. Francis stayed with the company for only a year before leaving to join the Royal Navy, with which he was to spend the rest of his working life. He was 16 years old and began as a cabin boy, but by the age of 22 he had risen to the rank of lieutenant and was serving on HMS *Phaeton*. In 1805, he was given his first command, of HMS *Woolwich*.

Within a very short time of going to sea, Francis recognized the importance of weather conditions and the value of recording them. He began to keep a journal, a habit he maintained for the rest of his life.

By the time of his first command he had become a hydrographer—a scientist who studies, describes, and charts river courses, coastlines, and the depth of the sea—and his task with the *Woolwich* was to survey the Río de la Plata region, in South America. In 1829, he was made the official hydrographer for the Admiralty. He carried out extensive surveying, for example around the Turkish coast in 1812, and was influential in sending out several important voyages of discovery. He also kept up his great interest in meteorology, especially those aspects that affected the operation of sailing ships at sea.

In 1806, Commander Beaufort, as he was then, drew up the chart of wind forces for which he is famous, his *Wind Force Scale and Weather Notation*. Its aim was to provide guidance for sailors, telling them how much sail they should set on a full-rigged warship according to the wind, and in its first version it included no information about the actual speed of the wind. These were not added until long after Beaufort had died. His original scale classified winds from force 0 to force 12—this being a wind "that no canvas could withstand."

In June 1812, during his surveying work in the eastern Mediterranean on HMS *Frederiksteen*, Beaufort led the rescue of some of his men who had been attacked by forces commanded by the local rulers. Beaufort was seriously wounded in the encounter and spent a long time convalescing in Portugal. At the end of the year he was ordered back to Britain. He never went to sea again.

Robert FitzRoy, captain of the *Beagle* on which Charles Darwin sailed, used the Beaufort scale and spoke highly of it, and in 1838 it was introduced throughout the Navy. Captains were required to include details of the wind in their daily logs. In 1874, modified to include details of the state of the sea and the visible effects of the wind on land, the scale was adopted by the International Meteorological Committee for use in international meteorological telegraphy.

Beaufort was knighted for his service in 1848, and by the time he retired in 1855, he had reached the rank of rear admiral. Admiral Sir Francis Beaufort had served in the Royal Navy for 68 years. He died two years later.

Further Reading

Irish Identity. "Navan's Most Famous Son." Hoganstand. com. Available online. URL: www.hoganstand.com/ general/identity/extras/famousgaels/stories/beaufort. htm#top. Accessed March 10, 2006.

Beckman, Ernst Otto

(1853–1923)
German
Chemist

Ernst Beckman, inventor of the Beckman THERMOMETER, was born at Solingen, to the southeast of Düsseldorf, on July 4, 1853. He worked as a pharmacist's assistant before enrolling at the University of Leipzig to study chemistry and pharmacy.

After graduating, Beckman taught at the Technische Hochschule (technical high school) in Brunswick. Later he became professor of physical chemistry at the Universities of Leipzig, Giessen, and Erlangen. He was appointed director of the Laboratory for Physical Chemistry and Electrochemistry at Leipzig, and in 1912 he became the first director of the newly established Kaiser Wilhelm Institute for Physical Chemistry and Electrochemistry in Berlin.

Ernst Beckman died in Berlin on July 13, 1923.

Beer, August

(1825–1863)
German
Mathematician, chemist, and physicist

August Beer was the author of BEER'S LAW and he contributed to LAMBERT'S LAW, which is sometimes known as the Beer–Lambert law.

August Beer was born on July 31, 1825, in Trier. He went to school in Trier, and after leaving school in 1845 he went to Bonn University to study mathematics and natural science, where he worked as an assistant and collaborator to the mathematician and physicist Julius Plücker (1801–1868). Beer was awarded his doctorate in 1848 and was promoted to the faculty of philosophy. In 1850, he began teaching at the university.

Beer conducted research into optics and theories of light. His book *Einleitung in die höhere Optik* (Introduction to higher optics) was published in 1854. Translated into many languages, it became the standard work on theories of light. In 1856, Beer became professor of mathematics at Bonn University.

August Beer died in Bonn on November 18, 1863. At the time he was working on a textbook in which he aimed to bring together the whole of mathematical physics. The book was published posthumously in two volumes in 1865 and 1869.

Bentley, Wilson Alwyn

(1865–1931)
American
Photographer of snowflakes

For 50 years, Wilson A. Bentley studied and then photographed SNOWFLAKES. Eventually, he accumulated an archive of more than 5,000 images and became widely known as "the Snowflake Man."

Bentley was born on February 9, 1865, at the family farm in the village of Jericho, Vermont. His mother had been a schoolteacher prior to her marriage, and she taught Wilson at home until he was 14. She used a small microscope as a teaching aid, and Wilson became fascinated by the world it revealed to him. In particular, he studied the shapes of snowflakes, dewdrops (*see* DEW), FROST, and hailstones (*see* HAIL). He recorded these by drawing what he saw, but this proved unsatisfactory and eventually he acquired the bellows camera and microscope objective that allowed him to photograph them. All of his photomicrographs were taken with this original camera.

Snowflakes consist of ICE CRYSTALS. These have a variety of shapes and can be arranged in an almost infinite number of ways, so that each snowflake is unique. What Bentley discovered, however, is that the temperature and pattern of air circulation within a cloud could be deduced from the form of the ice crystals that fell from it.

Each summer, Bentley turned his attention to the study of rain. He devised a method for measuring the size of RAINDROPS by exposing a dish containing a layer of sifted flour, about 1 inch (2.5 cm) thick. Raindrops formed the flour into little balls of dough. When these were dried, they were approximately the same size as the drops that caused them. The method is still used. From the size of the raindrops he deduced how they had formed. Between 1898 and 1904, he made more than 300 measurements of raindrops.

In 1898, Bentley published his first magazine article, in *Popular Scientific Monthly*. After that he wrote many popular articles and scientific papers, describing many of his most original ideas in the *Monthly Weather Review*. *Snow Crystals,* his only book, was published in 1931 in collaboration with William J. Humphreys, the chief physicist at the UNITED STATES WEATHER BUREAU, who persuaded him to do it. Writing the book involved sifting through his collection of photomicrographs and selecting nearly 2,500 of his favorites. *Snow Crystals* by Wilson A. Bentley and William J. Humphreys (published by McGraw-Hill and republished in 1962 by Dover) contains about 10 pages of text and more than 200 pages of Bentley's illustrations.

In 1924, Bentley received the first research grant ever to be awarded by the American Meteorological Society. The amount was small, but it was a deserved recognition by the scientific community for the work Bentley had been doing for 40 years.

Despite his interest in the weather, Wilson Bentley remained a farmer. After the death of his father, Wilson and his brother worked the farm between them, and Bentley contributed his full share of the physical work. They succeeded and the farm prospered.

Bentley kept detailed meteorological records throughout most of his life, the last on December 7, 1931. Soon after that he fell ill and died at the farm from pneumonia on December 23, 1931.

Further Reading

Blanchard, Duncan C. "The Snowflake Man." *Weatherwise,* 23, 6, 260–269, 1970. Available online. URL: www.snowflakebentley.com/sfman.htm. Accessed March 10, 2006.

Bergeron, Tor Harold Percival
(1891–1977)
Swedish
Meteorologist

Tor Bergeron was born on August 15, 1891, at Godstone, near London, England. He was educated at the universities of Stockholm, Sweden, and Leipzig, Germany, and in 1928 he obtained his Ph.D. from the University of Oslo, Norway.

As part of his education, Bergeron spent the three years from 1918 until 1921 as a student and collaborator of Vilhelm Bjerknes, the Norwegian meteorologist who in 1917 had established what became the most important meteorological research institute in the world, at the Bergen Geophysical Institute. After he qualified, Bergeron joined the staff at the institute.

In 1935, Bergeron proposed a mechanism for the formation of RAINDROPS in cold clouds (*see* CLOUD TYPES). He calculated that ICE CRYSTALS would grow by gathering water at the expense of supercooled water droplets (*see* SUPERCOOLING). This is called the Bergeron process. The crystals would form SNOW-FLAKES, and as the snowflakes fell from the base of the cloud into warmer air they would melt and reach the surface as water droplets. This mechanism was later confirmed experimentally by the German meteorologist Walter Findeisen, and it is now known as the Bergeron-Findeisen mechanism (*see* RAINDROPS).

From 1935 until 1945, Bergeron taught at the University of Stockholm, and in 1946 he moved to the University of Uppsala, in Sweden. He was professor of METEOROLOGY at the University of Uppsala from 1947 until 1961.

Bergeron died in Stockholm on June 13, 1977.

Bernoulli, Daniel
(1700–1782)
Swiss
Natural philosopher

Daniel Bernoulli was born at Groningen, the Netherlands, on February 9, 1700. His father, Johann (sometimes called Jean) Bernoulli (1667–1748) was trained as a physician, but worked as a mathematician. Daniel had an older brother, Nikolaus (1695–1726), who studied law and by the age of 27 was professor of law at the University of Bern, Switzerland, but mathematics was his real passion.

At the time of Daniel's birth, Johann was a professor at the University of Groningen. Johann's elder brother Jakob (sometimes called Jacques) Bernoulli (1654–1705) was also a distinguished mathematician and experimental physicist. Since 1687, he had been professor of mathematics at the University of Basel, Switzerland. After Jakob's death in 1705, Johann succeeded him at Basel and so, at the age of five, Daniel moved to Switzerland. His younger brother Johann (1710–90) was born in Basel and he, too, was a mathematician. In 1743, Johann succeeded his father as professor of mathematics at Basel.

Daniel was educated in Basel. He enrolled at the University of Basel when he was 13 and specialized in philosophy and logic, passing his baccalaureate examination (equivalent to graduating from high school) at the age of 15. When he was 16, he obtained his master's degree. His father wanted him to become a merchant, but commerce did not appeal to Daniel and he refused. He wanted to be a mathematician like the other members of his family, but his father insisted there was no money to be made in that profession. Eventually, they compromised, and in 1717 Daniel began to study medicine at Basel. During his studies he also spent some time at the universities of Heidelberg in 1718 and Strasbourg in 1719, returning to Basel in 1720. The thesis for which he was awarded his doctorate in 1721 was on the action of the lungs.

Having qualified as a doctor and satisfied his father, Daniel sought an academic position, but without success, and he moved to Venice to continue studying medicine. He intended to move to a famous medical school at Padua, but illness kept him in Venice, where he concentrated on mathematics. In 1724, he published a book called *Exercitationes Mathematicae* (Mathematical exercises). He also designed an hourglass that could be used on ships, even in heavy seas. He submitted this to the French Academy of Sciences, and when he returned to Basel in 1725 he learned that it had won a prize.

Daniel also learned that his book had attracted wide attention, and he was offered a professorship in mathematics at the Russian Academy in St. Petersburg. His brother Nikolaus was also offered a professorship in mathematics there, so in 1725 the two men traveled to Russia together. In less than a year Nikolaus died from a fever. This greatly saddened Daniel, and he did not like the Russian climate. Johann, his younger brother, joined him in 1731. Daniel was applying for posts at Basel, and in 1733 he became professor of anatomy and botany. He and Johann left Russia together, visited

Danzig (now Gdansk), Hamburg, the Netherlands, and Paris, and finally reached Basel in 1734. In 1750, Daniel became professor of natural philosophy at Basel, and he retained this post until he retired in 1777. The chair at St. Petersburg was later occupied by Daniel's pupil and nephew Jakob Bernoulli (1759–89).

The prize Daniel won for his hourglass was the first of 10 prizes he received from the French Academy for work on astronomy, magnetism, and a variety of nautical topics. He also wrote on political economy and probability.

Daniel published his most important work, *Hydrodynamica* (Hydrodynamics) in 1738. In it he discussed the theoretical and practical aspects of pressure, velocity, and equilibrium in fluids and showed the link between the pressure of a fluid and its velocity. This is a consequence of the conservation of energy, although more than a century was to pass before that concept was formulated clearly (*see* THERMODYNAMICS, LAWS OF). The relationship between pressure and velocity is now known as the Bernoulli principle, and its consequences as the BERNOULLI EFFECT.

In *Hydrodynamica* Bernoulli also assumed that gases are composed of minute particles. This allowed him to produce an equation of state (*see* GAS LAWS) and to relate atmospheric pressure to altitude.

Daniel Bernoulli was predominantly a mathematician and a friend of many of the most eminent mathematicians of his day. The Swiss mathematician Leonard Euler (1707–83) traveled to St. Petersburg to work with him in 1727, and the French mathematician Jean le Rond d'Alembert (1717–83) was also a close friend.

Bernoulli received many honors and was elected to most of the scientific academies of Europe. He died in Basel on March 17, 1782.

Further Reading
O'Connor, J. J., and E. F. Robertson. "Daniel Bernoulli." St. Andrews University. Available online. URL: www-groups.dcs.st-and.ac.uk/~history/Mathematicians/ Bernoulli_Daniel.html. Accessed March 10, 2006.

Bjerknes, Jacob Aall Bonnevie
(1897–1975)
Norwegian-American
Meteorologist

The son of Vilhelm Bjerknes, Jacob was born in Stockholm, and during World War I he helped his father organize the network of WEATHER STATIONS that supplied them with the data they used to develop their theories of AIR MASSES and polar FRONTS. Jacob also discovered that DEPRESSIONS originate as waves on fronts.

In 1939, Jacob moved to the United States, and in 1940 he became professor of METEOROLOGY at the University of California, Los Angeles. He became a United States citizen in 1946.

After World War II, Bjerknes conducted extensive studies of the upper atmosphere and JET STREAM, in 1952 being among the first to use photographs taken by high-altitude research rockets for this purpose. Bjerknes also studied the climatic consequences of the interaction of the ocean and atmosphere in the tropical Pacific and was the first to propose, in 1969, what became known as the Bjerknes hypothesis. This holds that ENSO arises from changes in sea-surface temperature, leading to changes in wind strength and direction, leading to changes in the ocean circulation, leading to further changes in sea-surface temperature. Bjerknes died in Los Angeles.

Bjerknes, Vilhelm Friman Koren
(1862–1951)
Norwegian
Physicist and meteorologist

Vilhelm Bjerknes was one of the founders of modern METEOROLOGY and scientific WEATHER FORECASTING. He was born in Oslo. His father, Carl Anton Bjerknes (1825–1903), was professor of mathematics at Christiania (now Oslo) University, and Vilhelm helped him with some of his experiments in hydrodynamics before leaving to spend 1890 and 1891 in Germany working as an assistant to and collaborator with the physicist Heinrich Hertz (1857–94). Bjerknes then spent two years as a lecturer at the School of Engineering (Högskola) in Stockholm and in 1895 was appointed professor of applied mechanics and mathematical physics at the University of Stockholm.

In 1897, Bjerknes developed a synthesis of hydrodynamics and thermodynamics that allowed him to propose a system for forecasting weather scientifically. In 1904, he published a scientific paper outlining a method of numerical forecasting. The Carnegie Institution supported this work, allowing Bjerknes to employ a long series of "Carnegie assistants" who joined the "schools" he founded first at Leipzig and later at Bergen.

Bjerknes returned to Norway in 1907 as a professor at Kristiania (the spelling had been changed) University and, in 1910 and 1911, he and three of his assistants (the Swedish meteorologist Johan W. Sandström and the Norwegians Olaf D. Devik and T. Hesselberg) published *Dynamic Meteorology and Hydrography*. This book described their research thus far and proposed many new techniques for weather forecasting as well as suggesting improvements to existing ones. In 1912, Bjerknes was appointed professor of geophysics at the University of Leipzig. While there, he founded the Leipzig Geophysical Institute, his first school.

In 1917, during World War I, Bjerknes returned to Norway to found his second school, the Bergen Geophysical Institute, as part of the Bergen museum. It is now part of the University of Bergen. He joined the staff of the University of Oslo in 1926 and remained there until his retirement in 1932.

Bjerknes did his most important work while at the Bergen Institute. During World War I, he and his colleagues established a network of WEATHER STATIONS throughout Norway. These reported observations and measurements to Bergen, where they were assembled to produce general pictures of weather conditions at particular times over a wide area. Studying these pictures led Bjerknes and other members of the Bergen School to conclude that there exist AIR MASSES that differ from one another and that these masses are separated by distinct boundaries. Likening the masses to opposing armies, they called the boundaries between them fronts and developed a frontal theory to account for their development, disappearance, and the weather associated with them.

The Bergen frontal theory also explained the way CYCLONES form over the Atlantic. Bjerknes described this work in 1921 in a book that became a classic: *On the Dynamics of the Circular Vortex with Applications to the Atmosphere and to Atmospheric Vortex and Wave Motion*. This formed the basis for the modern theory and practice of meteorology.

Vilhelm Bjerknes was an inspired and popular teacher who attracted talented workers and made sure they received full recognition for their work.

Vilhelm Bjerknes died in Oslo on April 9, 1951.

Further Reading
O'Connor, J. J., and E. F. Robertson. "Vilhelm Friman Koren Bjerknes." St. Andrews University. Available online. URL: www-groups.dcs.st-and.ac.uk/~history/

Mathematicians/Bjerknes_Vilhelm.html. Accessed March 10, 2006.

Black, Joseph
(1728–1799)
Scottish
Chemist

Joseph Black discovered the ways in which CARBON DIOXIDE can be released into the air naturally and therefore that this gas is a normal constituent of the air. He also discovered LATENT HEAT.

Black was born in Bordeaux, France, on April 16, 1728. His father was a wine merchant and the family was of Scottish descent, although his father was born in Belfast, Ireland. In 1740, Joseph was sent to Belfast to be educated, and from there he went to Glasgow University, where he studied medicine and natural sciences. His courses included chemistry, for which he displayed aptitude and enthusiasm, becoming more like an assistant to his teacher William Cullen (1712–90) than his student. He moved to Edinburgh University in 1751 to complete his medical studies, and in 1754 he submitted a thesis for his doctor's degree.

The thesis described his researches into the effect of heating magnesia alba (magnesium carbonate, $MgCO_3$). Black found that heating this compound released a gas, which he detected by weighing, that was distinct from the ordinary air. In fact, this gas had been described more than a century earlier by Jan van Helmont (1577–1644), but Black pursued his investigation further and published a fuller account of it in 1756, with the title *Experiments Upon Magnesia Alba, Quicklime, and Some Other Alkaline Substances*. His work showed that what he called "mild alkalis" (carbonates) are causticized (made more alkaline) when they lose this gas and that they become less causticized when they absorb it. This demonstrated that the gas is acid.

Calcium carbonate ($CaCO_3$) was one of the "other alkaline substances" Black studied. He found that when this is heated the same gas is released and the solid substance is converted to quicklime, or calcium oxide (CaO), but that the quicklime can also recombine with the gas. Because it can be "fixed" by being absorbed into a solid substance Black called the gas "fixed air." This is the gas we now know as carbon dioxide, and the reaction he described would now be written as:

$$CaCO_3 \leftrightarrow CaO + CO_2$$

Black used a balance to measure the change when $MgCO_3$ is heated, and he also measured the loss in weight when $CaCO_3$ loses CO_2. He measured how much $CaCO_3$ was needed to neutralize a measured amount of acid. This attention to measurement was new to chemistry, and its importance was recognized some years later in the work of Antoine Lavoisier (1743–94). Investigating the properties of "fixed air," Black discovered that a candle would be extinguished if it was placed in an airtight container. He knew that heat released carbon dioxide, so suspected that this is what was extinguishing the flame, but when he added a substance that would absorb the gas the candle still would not burn. He passed on this problem to one of his students, Daniel Rutherford (1749–1819), who discovered the gas that extinguished the flame is what he called "phlogisticated air," which we know as NITROGEN.

In 1756, the year his book was published, Black returned to Glasgow to succeed William Cullen as lecturer in chemistry, and he was also appointed professor of anatomy at Edinburgh University, although he exchanged the post for that of professor of medicine. Joseph Black was a practicing physician.

Around 1760, Black was becoming interested in a different problem. He found, by careful measurement, that when ice is warmed it melts slowly, but its temperature does not change. This led him to suppose that the intensity of heat is not the same thing as the quantity of heat, and that THERMOMETERS measure only heat intensity. As the ice melted, he concluded it was absorbing a quantity of heat that must have combined with the particles of ice and become latent in its substance. *Latent* means "hidden." He called this latent heat, and at the end of 1761 he verified its existence experimentally. Black introduced the topic of latent heat into his lectures, and he described his work on it to a literary society in Glasgow in April 1762.

In 1764, Black and his assistant William Irvine (1743–97) measured the even larger amount of latent heat that is involved when water boils and WATER VAPOR condenses, although their measurements were not very accurate. Black never published any account of his work on latent heat, with the result that others were able to claim the credit. Jean-André Deluc (1727–1817) also discovered latent heat, independently and at about the same time as Black.

Much of Black's research involved heating substances. In the course of it he noticed that equal masses of different substances require different amounts of heat to raise their temperatures to the same degree. This led to the concept of specific heat (*see* HEAT CAPACITY).

In 1766, Black again succeeded his old teacher and friend William Cullen, this time to become professor of chemistry at Edinburgh University. Joseph Black died in Edinburgh on November 10, 1799. He published very little during his lifetime, but after his death his friend John Robison (1739–1805) published his lecture notes, with some additions from his pupils, together with a biographical preface by his friend. This appeared in 1803 as *Lectures on the Elements of Chemistry, Delivered in the University of Edinburgh.*

Boltzmann, Ludwig
(1844–1906)
Austrian
Theoretical physicist

Ludwig Boltzmann made important contributions to the kinetic theory of gases (*see* KINETIC ENERGY) and thermodynamics (*see* THERMODYNAMICS, LAWS OF). Josef Stefan (1835–93) had observed experimentally that when a substance is heated the radiation it emits increases as the fourth power of the temperature. Boltzmann developed a mathematical explanation for this phenomenon, and it is now known as the STEFAN-BOLTZMANN LAW.

Ludwig Boltzmann was born in Vienna on February 20, 1844. He was educated in Vienna and in Linz and studied at the University of Vienna, where Josef Stefan was one of his teachers. Boltzmann received his Ph.D. from that university in 1866.

In 1867, Boltzmann became an assistant at the Physics Institute (Physikalisches Institut) in Vienna. He then held a series of professorships. He was professor of theoretical physics at the University of Graz (1869–73) and professor of mathematics at the University of Vienna (1873–76). He then held a series of professorships in theoretical physics at the universities of Munich (1889–93), Vienna (1894–1900), where he succeeded Stefan, Leipzig (1900–02), and Vienna (1902–06). In 1904, Boltzmann visited the United States, calling at Stanford University and the University of California at Berkeley, and he lectured at the World's Fair in St. Louis.

Boltzmann showed that the second law of thermodynamics is essentially statistical. A system will approach a state of thermodynamic equilibrium because this is overwhelmingly the most probable state in which to find it. His explanation was based on the idea that matter is composed of atoms. This idea was strongly resisted by certain other physicists who held that the behavior of matter should be described only in terms of energy. A heated debate ensued between the "atomists" led by Boltzmann and the "energists" led by Wilhelm Ostwald (1853–1932).

Boltzmann became depressed by the way other physicists had attacked his work and failed to understand it. On September 6, 1906, he took his own life at Duino, near Trieste—now in Italy but at that time in Austria.

In 1828, the Scottish botanist Robert Brown (1773–1858) had described the erratic movements of pollen grains in water, and in a paper published in 1905 Albert Einstein (1879–1955) showed that this Brownian motion could be explained if the pollen grains were being struck by moving molecules of water. This established that matter is made from atoms and their behavior should be understood statistically. Boltzmann was vindicated, but the news did not reach him in time to prevent his tragic death.

Bouguer, Pierre
(1698–1758)
French
Physicist and mathematician

Pierre Bouguer studied the intensity of sunlight and the effect on it of its passage through the atmosphere.

He was born at Croisic, Britanny, on February 16, 1698. His father was a hydrographer and taught Pierre about the geography of fresh and salt water. Pierre was a prodigy and began teaching hydrography at Le Havre when he was only 15 years old. Bouguer also compiled tables of atmospheric REFRACTION and invented the heliometer, an instrument used to measure the diameter of the Sun and the angular distance between stars. Pierre Bouguer became professor of hydrography at Croisic and, in 1730, at Le Havre.

Pierre Bouguer died in Paris on August 15, 1758.

Boyle, Robert
(1627–1691)
Irish
Natural philosopher

Robert Boyle was an aristocrat. His father, Richard Boyle, was the earl of Cork, and Robert was his seventh

Robert Boyle (1627–1691), the natural philosopher. *(Hulton Archive/Getty Images)*

son and 14th child. He was born on January 25, 1627, at Lismore Castle, in Ireland. He learned to speak Latin and French while still a small child, and he was sent to Eton College, near London, at the age of eight. After three years at Eton, in 1638 he traveled abroad with a French tutor. In 1641, he arrived in Italy and spent the winter in Florence studying the work of Galileo.

Boyle returned to England in 1644 and immediately devoted himself to a life of scientific inquiry. Drawn to others who shared his interests, it was not long before Boyle joined a group of people who called themselves the "Invisible College." They held frequent meetings at Gresham College, in London, and some of the members also met in Oxford. Boyle moved to Oxford in 1654, and there he carried out his most important scientific work.

In 1657, Boyle read of an air pump that had been invented by the German physicist Otto von Guericke (1602–86). The pump was meant to evacuate the air from a chamber, and Boyle enlisted the help of Robert Hooke (1635–1703) to improve it. Boyle and Hooke

became lifelong friends. By 1659, the pump was finished and Boyle began using it to experiment on the properties of air. He published the results of this work in 1660 with the title *New Experiments Physico-Mechanical Touching the Spring of Air and its Effects.* Boyle had discovered that the volume occupied by a gas is inversely proportional to the pressure under which the gas is held. This relationship is known in English-speaking countries as Boyle's law (*see* GAS LAWS). He also found that the weight of a body varies according to the amount of BUOYANCY supplied by the atmosphere.

Boyle and Hooke also studied combustion. They found that neither charcoal nor sulfur will burn when air is excluded, no matter how strongly the vessel containing them is heated, but they will burst into flames as soon as air is allowed into the container. When either charcoal or sulfur is mixed with potassium nitrate (saltpetre), however, the mixture will burn in a vacuum. Boyle concluded from this that both potassium nitrate and air contain some ingredient that is necessary for combustion. Boyle did not identify that ingredient, however. Joseph Priestley (1733–1804) isolated it in 1774, and in 1777 Antoine-Laurent Lavoisier (1743–94) gave it the name oxygen.

In 1661, Boyle published another book, *The Sceptical Chymist,* in which he advanced the idea that matter is composed of "corpuscles." These are of various shapes and sizes, and they are able to combine into groups. Each group of corpuscles makes up a chemical substance. Boyle was the first scientist to use the word *analysis* to describe the separation of a substance into its constituents. He invented a hydrometer for measuring the DENSITY of liquids and made the first match by coating a rough paper with phosphorus and placing a drop of sulfur on the tip of a small stick. The stick ignited when it was drawn along a crease in the paper. He made a portable camera obscura that could be extended or shortened like a telescope in order to focus an image on a piece of paper stretched across the back of the box, opposite the lens.

By a charter granted by King Charles II and passed on August 13, 1662, the Invisible College became the "Royal Society, for the improvement of naturall knowledge by Experiment." The charter named Boyle as a member of the council. Boyle was elected president of the society in 1680, but declined because he was unwilling to take the necessary oath.

In addition to his scientific work, Boyle was deeply interested in theology. He learned Hebrew, Greek, and Syriac in order to be able to read scriptural texts in their original languages. His will provided for the founding of a series of lectures aimed at proving the Christian religion against the views of other religions, but with the proviso that disputes between Christians should not be mentioned.

In 1668, Boyle returned to live in London with his sister. He remained in London for the rest of his life. Boyle died there on December 30, 1691.

Further Reading

Jardine, Lisa. *Ingenious Pursuits: Building the Scientific Revolution.* London: Little, Brown, 1999.

Brückner, Eduard
(1862–1927)
German
Geographer and glaciologist

Eduard Brückner was born at Jena, in Saxony, Germany, on July 29, 1862, the son of Alexander Brückner, a teacher of Russian history, and Lucie Schiele. He was educated at the gymnasium (high school) in Karlsruhe, and from 1881 until 1885 he studied physics and METEOROLOGY at the University of Dorpat (now Tartu, Estonia). He then continued his studies in Dresden and Munich.

In 1885, Brückner joined the staff of the Deutsche Seewarte in Hamburg. Established in 1876, the Seewarte supplied weather information for ships using the port of Hamburg. It developed into the modern German weather service. From 1888 until 1904, he was a professor at the University of Bern, Switzerland, and from 1899 to 1900 he was rector of the university. He married Ernestine Stein in 1888. In 1904, he returned to Germany to become a professor at the University of Halle, and in 1906 was appointed a professor at the University of Vienna, Austria, a post he held until his death.

Brückner was an authority on alpine GLACIERS, and he was especially interested in the effect of glaciers on surface features of the landscape. He was convinced that climate change is of great importance, with direct economic and social implications. He conducted extensive research and made many theoretical studies of changes that have occurred in the past. In the course of

these he discovered the 35-year cycle that now bears his name (*see* BRÜCKNER CYCLE). His interest in the subject is indicated by the titles of some of the papers he published. These include "How Constant is Today's Climate?" (1889), "Climate Change since 1700" (1890, and the paper in which his cycle was first mentioned), "Influence of Climate Variability on Harvest and Grain Prices in Europe" (1895), "An Inquiry About the 35 Year Periods of Climatic Variations" (1902), and "Climate Variability and Mass Migration" (1912).

Brückner died in Vienna on May 20, 1927.

Buchan, Alexander
(1829–1907)
Scottish
Meteorologist

The man who is acknowledged to have been the most eminent British meteorologist of the 19th century was born at Kinnesswood, Kinross, Scotland, on April 11, 1829. He became a schoolteacher, teaching all subjects. His favorite leisure pursuit was botany.

Following a public meeting held in Edinburgh on July 11, 1855, a society was formed with the aim of establishing WEATHER STATIONS throughout Scotland. The society became the Scottish Meteorological Society, and it operated the weather stations from 1856 until 1920, when that task was taken over by the METEOROLOGICAL OFFICE. In December 1860, Buchan was appointed secretary to the society, and he remained in the post until his death in 1907.

Buchan also edited the *Journal of the Scottish Meteorological Society* from its first issue in 1864, and he wrote a great deal of its material. During his editorship, the journal published Thomas Stevenson's description of his louvered screen (*see* STEVENSON SCREEN). The screen is still widely used.

In 1883, the Scottish Meteorological Society opened a meteorological observatory on Ben Nevis, the highest mountain in Britain. Buchan was closely involved with the establishment of the observatory and with the running of it. The observatory remained in operation until 1904.

Buchan established his reputation in 1867, with the publication of his *Handy Book of Meteorology*. This became a standard textbook and remained in use for many years. In 1869, he wrote a paper, "The Mean Pressure of the Atmosphere and the Prevailing Winds

Over the Globe," for the Royal Society of Edinburgh. He also wrote papers on the circulation of the atmosphere and on ocean circulation. It was also in 1869 that he published his paper on "Interruptions in the Regular Rise and Fall of Temperature in the Course of the Year" in the *Journal,* describing what came to be called BUCHAN SPELLS.

Buchan was made a member of the Meteorological Council in 1887 and was elected a fellow of the Royal Society in 1898. In 1902, he was the first person to be awarded the Symons Medal, the greatest honor meteorologists can bestow on one of their colleagues.

Alexander Buchan died in Edinburgh on May 13, 1907.

Budyko, Mikhail Ivanovich
(b. 1920)
Belorussian
Physicist and meteorologist

Budyko was the first scientist to calculate the balance of heat received from the Sun and radiated from the Earth's surface, checking his calculations against observational data from all parts of the world. In 1956, he published his results in his book *Heat Balance of the Earth's Surface.* This work changed CLIMATOLOGY from being a qualitative discipline, based on measuring climatic data from all over the world, into a more physical discipline. Professor Budyko became a pioneer of physical climatology, adding to his 1956 book an atlas, completed in 1963, that shows the Earth as viewed from space with all aspects of the Earth's heat balance displayed. Calculations of climate change are based on this atlas.

By 1960, Professor Budyko was already concerned about the possibility of a general rise in world temperatures caused by human activity. He suggested the day might come when it became necessary to scatter particles in the stratosphere (*see* ATMOSPHERIC STRUCTURE) in order to reflect solar radiation and reduce the rate of temperature increase. In 1972, he was able to confirm a link between past climate changes and changes in the atmospheric concentration of CARBON DIOXIDE. Budyko warned then that his analysis indicated a general warming of the world's climates due to the rise in the carbon dioxide concentration brought about by the increasing consumption of fossil fuels. His 1972 calculations predicted a rise in TEMPERATURE of about 6.3°F (3.5°C) from this cause between 1950 and about 2070.

Budyko's studies of the effects on climate of altering the composition of the atmosphere led him in the early 1980s to ponder the climatic consequences of a large-scale thermonuclear war. He suggested that such a war might inject such a huge quantity of AEROSOLS into the atmosphere that the entire world would be plunged into deep cold, a NUCLEAR WINTER that might threaten human survival.

Mikhail Budyko was born on January 20, 1920, at Gomel, Belarus. He was educated in Leningrad, and from 1942 until 1975 he worked at the Main Geophysical Observatory, Leningrad (St. Petersburg), where he was the director from 1972 to 1975. He was then appointed to his present position, as head of the Division for Climate Change Research, at the State Hydrological Institute, St. Petersburg, the position he still holds. He was elected an Academician of the Russian Academy of Sciences in 1992.

He has been awarded many prizes, including the Lenin National Prize (1958), Gold Medal of the World Meteorological Organization (1987), A. A. Grigoryev Prize of the Russian Academy of Sciences (1995), and the Blue Planet Prize (1999) for his contribution to environmental research.

Further Reading
Asahi Glass Foundation. "Profiles of the 1998 Blue Planet Prize Recipients." Asahi Glass Foundation. Available online. URL: www.af-info.or.jp/eng/honor/hot/enr-budyko.html. Accessed March 10, 2006.

Buys Ballot, Christoph Hendrick Diderik
(1817–1890)
Dutch
Meteorologist

Christoph Buys Ballot was born at Kloetinge, Zeeland, the Netherlands, on October 10, 1817. In 1847, he was appointed professor of mathematics at the University of Utrecht, and in 1854 helped to found and was the first director of the Royal Netherlands Meteorological Institute. He remained in this post until his death.

In 1857, Buys Ballot described the wind circulation around areas of low and high atmospheric pressure. He based his description on his studies of meteorological records, and it quickly became known as a law

Appendix I   589

attributed to him. Buys Ballot did not know that the American meteorologist William Ferrel (1817–91) had reached the same conclusion on theoretical grounds some months earlier. When he learned of this, Buys Ballot acknowledged Ferrel's prior claim to the discovery, but it was too late, and what should be known as Ferrel's law is usually called BUYS BALLOT'S LAW.

Buys Ballot died on February 3, 1890.

Cavendish, Henry
(1731–1810)
English
Physicist

Henry Cavendish discovered that the composition of air is the same everywhere and at all times. Cavendish was born at Nice, France, on October 31, 1731. He was of aristocratic descent. His paternal grandfather was the duke of Devonshire, and his maternal grandfather was the duke of Kent. Henry Cavendish was extremely wealthy, but he was reclusive and made no use of the large fortune he inherited.

Cavendish was educated at Dr. Newcome's Academy in Hackney, London, and in 1749 he went to Peterhouse College of the University of Cambridge. He left Cambridge in 1753 without taking a degree. This was not unusual at the time. His father encouraged his scientific interests and introduced him to the Royal Society. Cavendish became a fellow of the Royal Society in 1760.

Henry Cavendish studied "fixed air" (CARBON DIOXIDE), and in his first paper, published in 1776, he proved the existence of "inflammable air" (HYDROGEN). He also studied what he called "common air," and in 1783, after collecting air samples on 60 days and performing 400 analyses of them, he found that air always has the same composition. He also discovered that a small portion of the air is inert. This was later found to consist mainly of ARGON. Cavendish also studied heat and discovered LATENT HEAT and specific heat before Joseph Black (1728–99), but did not publish his findings.

In the course of his studies of electricity, Cavendish oxidized NITROGEN by sending electric sparks through the air. When he dissolved the resulting gas in water and analyzed it, he found it to be nitric acid (HNO_3). Cavendish also determined the freezing point of mercury.

Henry Cavendish died in Clapham, London, on February 24, 1810.

Celsius, Anders
(1701–1744)
Swedish
Astronomer and physicist

Anders Celsius was born in Uppsala, Sweden, on November 27, 1701. Nils Celsius, his father, was professor of astronomy at the University of Uppsala, and both of Anders' grandfathers were also professors at Uppsala: Magnus Celsius (his paternal grandfather) was professor of mathematics, Anders Spole (his maternal grandfather) preceded Nils as professor of astronomy. Several of his uncles were also scientists.

Anders was educated in Uppsala, and in 1730 he was appointed to succeed his father as professor of astronomy. There was no major observatory in Sweden at that time, so soon after his appointment Celsius embarked on a tour of the major European observatories. His tour lasted five years and in the course of it he met many of the leading astronomers of the day. Between 1716 and 1732, Celsius and his companions made 316 observations of the aurora borealis (*see* OPTICAL PHENOMENA). He published these in Nuremburg in 1733. Celsius and Olof Hiorter, his assistant, discovered that the aurorae are magnetic phenomena.

While visiting Paris in 1734, Celsius met the French astronomer Pierre-Louis Maupertuis (1698–1759), who invited him to join an expedition to Torneå, in Lapland (today on the border between Sweden and Finland, but then in northern Sweden). The purpose of the expedition was to measure the length of one degree of latitude along a meridian (degree of longitude) close to the North Pole and to compare their result with a similar measurement taken in Peru (in a region that is now in Ecuador). The Lapland expedition took place in 1736–37, and it confirmed the opinion of Isaac Newton (1642–1727) that the Earth is flattened at the poles.

His participation in this expedition made Celsius famous in his own country, and he was able to persuade the Swedish government to finance the building at Uppsala of an observatory equipped with instruments Celsius had bought during his European tour. The Celsius Observatory opened in 1741, with Celsius as its first director. Celsius made some of the earliest attempts to measure the magnitude of stars.

In the 18th century, astronomy was not studied purely to obtain information about the stars and planets. Governments were busy delineating the borders of their territories, and to do this accurately their surveyors needed astronomical data to fix positions—inaccurate maps could and did lead to war. Accordingly, Celsius conducted many measurements that were used in the Swedish General Map. He may also have been the first person to observe that the Scandinavian landmass is slowly rising. We now know that this is due to the release of pressure following the melting of the Fennoscandian ice sheet (see glacioisostasy), but Celsius thought the sea level was falling because the sea was evaporating.

In 1742, Celsius presented a paper to the Royal Swedish Academy of Sciences in which he proposed that all scientific measurements of temperature should be made on a scale based on two fixed points that occur naturally. This led to the development of the temperature scale that bears his name (see temperature scales).

Celsius published most of his scientific papers through the Royal Swedish Academy of Sciences, and he was its secretary from 1725 until 1744. He strongly favored the introduction of the Gregorian calendar. This had been tried in 1700 by omitting the leap days between 1700 and 1740, but 1704 and 1708 were declared leap years by mistake, and in 1712 Sweden returned to the Julian calendar. Celsius and his supporters eventually succeeded, and the new calendar was introduced in 1753 and all 11 supernumerary days were dropped together.

By then Celsius was dead. He died from tuberculosis in Uppsala on April 25, 1744.

Further Reading

Astronomical University, University of Uppsala. "Anders Celsius (1701–1744)". University of Uppsala. Available online. URL: www.astro.uu.se/history/Celsius_eng.html. Accessed March 10, 2006.

Charles, Jacques-Alexandre-César
(1746–1823)
French
Physicist and mathematician

The discoverer of Charles's law was born at Beaugency, Loiret, on November 12, 1746. He worked as a clerk in the Ministry of Finance in Paris, and while there he became interested in science. Having heard about the experiments Benjamin Franklin (1706–90) had conducted with electricity, Charles gave popular public lectures in which he popularized Franklin's discoveries, demonstrating them with apparatus he constructed himself.

In June 1783, the Montgolfier brothers made their first experiments with unmanned hot-air balloons at Annonay, in the south of France. When news of this reached Paris, the Academy of Sciences asked Jacques Charles to study the invention. He realized that hydrogen would be a much better lifting gas than hot air. With the help of two friends, Nicolas and Anne-Jean Robert, he successfully launched a hydrogen balloon in August 1783. On December 1, Charles and Nicolas Robert became the first people to ascend in a balloon. In later flights Charles reached a height of nearly 10,000 feet (3,000 m). This made him a popular hero, and the king, Louis XVI, invited him to move his laboratory to the Louvre.

Charles made his most important discovery in about 1787. As long ago as 1699 Guillaume Amontons (1663–1705) had published his finding that different gases expand by the same amount for a given rise in temperature. Using oxygen, nitrogen, and hydrogen, Charles repeated the experiments by which Amontons had reached this conclusion and was able to calculate the precise amount by which the gases expanded. He found that for every 1.8°F (1°C) rise in temperature their volume increased by 1/273 of the volume they had at 32°F (0°C). This meant that if the gas could be cooled to -459.4°F (-273°C) its volume would be zero. This came to be known as absolute zero (see units of measurement).

Charles did not publish the results of these experiments, but he did inform Joseph Gay-Lussac (1778–1850) about them. Gay-Lussac repeated them, and the resulting general rule came to be known in France as Gay-Lussac's law but outside France it is called Charles's law (see gas laws).

In 1785, Charles was elected to the Academy of Sciences. Later he became professor of physics at the Paris Conservatoire des Arts et Métiers. He died in Paris on April 7, 1823.

Clapeyron, Benoit-Paul-Emile
(1799–1864)
French
Mathematician and engineer

In 1834, Emile Clapeyron published a paper explaining the ideas of the French engineer Sadi Carnot

(1796–1832). Carnot had shown that the amount of work a heat engine can do depends entirely on the difference in TEMPERATURE of the working fluid entering the engine and the hot exhaust gases. He also showed that energy can change its form, but it can be neither created nor destroyed. Until Clapeyron's paper, however, few people knew of Carnot's work. Once Clapeyron had explained it, the significance of Carnot's discovery became apparent. William Thomson (Lord Kelvin, 1824–1907) and Rudolf Clausius (1822–88) developed it into the second law of thermodynamics (see THERMODYNAMICS, LAWS OF). Clapeyron also studied the relationship between temperature and SATURATION VAPOR PRESSURE. The Clausius–Clapeyron equation describes the heat of vaporization of a liquid.

Emile Clapeyron was born in Paris on February 26, 1799. He was educated at the École Polytechnique, from which he graduated in 1818, and then at the École des Mines, where he studied engineering.

In 1820, Clapeyron and his friend Gabriel Lamé (1795–1870) went to Russia to lead and train a team of engineers that had been recruited to improve the condition of the country's bridges and roads. Clapeyron taught mathematics at the School of Public Works in St. Petersburg for 10 years. He returned to France in 1830 and became a professor at the École des Mineurs in St. Étienne. Clapeyron and Lamé had proposed building a railroad between Paris and St. Germain. In 1835, the project was approved, and they were asked to head the project, but Lamé had been offered the chair in physics at the École Polytechnique and left to take up the appointment, leaving Clapeyron in charge of the railroad.

Clapeyron became a professor at the École des Ponts et Chaussées in 1844, and in 1848 he was elected to the Paris Academy of Sciences.

He died in Paris on January 28, 1864.

Further Reading

O'Connor, J. J., and E. F. Robertson. "Benoît Paul Emile Clapeyron." University of St. Andrews. Available online. URL: www-groups.dcs.st-and.ac.uk/~history/Mathematicians/Clapeyron.html. Accessed March 10, 2006.

Clarke, Arthur Charles
(b. 1917)
English
Science fiction writer and physicist

Arthur C. Clarke was the first person to suggest that satellites might be used for communications and that they might be placed in geostationary ORBITS—sometimes called Clarke orbits. Clarke was born at Minehead, Somerset, on December 16, 1917. During World War II, he served in the Royal Air Force, where he worked on the experimental trials of ground-controlled approach (GCA). That was the first use of RADAR to allow aircraft to land safely in poor visibility.

After the war Clarke enrolled at King's College, London, from where he graduated in physics and mathematics in 1948. He had already published, in 1945, the technical paper in which he outlined the principles of geostationary orbits and the use of satellites in communications. In 1954, in a letter to Harry Wexler, head of the Scientific Services Division of the UNITED STATES WEATHER BUREAU, Clarke suggested using orbiting satellites to obtain data for use in WEATHER FORECASTING.

Arthur C. Clarke has received many awards and honors. These include the 1982 Marconi International Fellowship, the gold medal of the Franklin Institute, the Lindbergh Award, the Vikram Sarabhai Professorship of the Physical Research Laboratory, Ahmedabad, and a fellowship of King's College, London. He received a knighthood on May 16, 2000.

Since 1956, Sir Arthur has lived in Colombo, Sri Lanka.

Clausius, Rudolf Julius Emmanuel
(1822–1888)
German
Theoretical physicist

Clausius was one of the founders of the study of thermodynamics and the principal originator of its second law (see THERMODYNAMICS, LAWS OF). He coined the word *entropy*.

Rudolf Clausius was born on January 2, 1822, in Köstin, Pomerania (now Koszalin, Poland). His education began at the local school run by his father and continued at the Gymnasium (high school) in Stettin. He enrolled at the University of Berlin in 1840 and obtained his doctorate in 1848 from the University of Halle.

After obtaining his doctorate, Clausius taught in Berlin at the Royal Artillery and Engineering School and in 1855 moved to Switzerland, to become professor of physics at Zurich Polytechnic. He returned to Germany in 1867 to take up an appointment of professor of physics at the University of Würzburg, and in 1869 moved to the University of Bonn as professor of physics.

In 1870, during the Franco–Prussian war, Clausius organized a volunteer ambulance service operated by his students. Wounds received in the war left him in perpetual pain.

Clausius died in Bonn on August 24, 1888.

Coriolis, Gaspard-Gustave de
(1792–1843)
French
Physicist, mathematician, and engineer

The scientist who was the first person to explain why moving bodies, such as winds and ocean currents, are deflected to the right in the Northern Hemisphere and to the left in the Southern Hemisphere was born in Paris on May 21, 1792. His family came from Provence, in the south of France. They were lawyers and were made aristocrats (hence the "de" in the family name) in the 17th century. The French Revolution stripped them of their privileges and wealth, and Gaspard's father became an industrialist, living in the town of Nancy.

In 1808, Gaspard de Coriolis commenced his studies at the École Polytechnique, the school that trained government officials. He completed them at the École des Ponts et Chaussées (School of Bridges and Highways). In the course of his studies there he spent several years in the Vosges Mountains on active service with the corps of engineers.

De Coriolis graduated in highway engineering and was determined to become an engineer, but his health was poor and his father's death left him with the responsibility of keeping the family. In 1816, he joined the staff of the École Polytechnique, first as a tutor and then as an assistant professor of analysis and mechanics. In 1829, he took up a position as professor of mechanics at the École Centrale des Arts et Manufactures, where he remained until 1836, when he became professor of mechanics at the École des Ponts et Chaussées. In 1838, he was made director of studies at the École Polytechnique. He was elected a member of the mechanics section of the French Academy of Sciences in 1836.

De Coriolis was a highly talented scientist, but suffered from poor health, which prevented him from realizing his full potential. Nevertheless, he made several important contributions to science. He succeeded in establishing the term *work* as a technical term, defining it as the displacement of a force through a distance. In dynamics he introduced a quantity, $\frac{1}{2} mv^2$, for which he coined the term *force vive,* now called kinetic energy.

In 1835, he made the contribution to physics for which he is still remembered, describing it in a paper called "Sur les équations du mouvement relatif des systèmes de corps" (On the equations of relative motion of a system of bodies), published in volume XV of the *Journal de l'École Polytechnique.* In his paper he showed that when a body moves in a rotating frame of reference, its motion relative to the frame of reference can be explained only if there is a force of inertia acting upon it. This inertial force causes the body to follow a path that curves to the right if the frame of reference is rotating counterclockwise and to the left if the rotation is clockwise. The inertial force came to be called the Coriolis force and is now known as the CORIOLIS EFFECT. It is of great importance in studies of the movements of air and of ocean currents. It is also relevant to ballistics, because missiles and projectiles traveling a long distance are subject to it.

Coriolis died in Paris on September 19, 1843.

Further Reading
O'Connor, J. J., and E. F. Robertson. "Gaspard Gustave de Coriolis." University of St. Andrews. Available online. URL: www-groups.dcs.st-and.ac.uk/~history/ Mathematicians/Coriolis.html. Accessed March 10, 2006.

Croll, James
(1821–1890)
Scottish
Climatologist and geologist

A self-educated man, in 1864 James Croll proposed that the onset of GLACIAL PERIODS is triggered by changes in the ECCENTRICITY of the Earth's ORBIT and the PRECESSION OF THE EQUINOXES. This developed and extended earlier ideas and aroused great interest.

According to Croll, ice ages occur at intervals of 100,000 years, which is when the orbit reaches its maximum eccentricity. They are experienced alternately in the Northern and Southern Hemispheres, according to which hemisphere is having winter when the

Earth is farthest from the Sun. Although interest in his astronomical theory waned at the end of his life, Croll's work did much to prepare the way for the theory of MILANKOVITCH CYCLES, which is now widely accepted.

James Croll was born near Cargill, Perthshire, on January 2, 1821. The family was poor and James left school at 13. He read avidly, but needed to earn money and so worked at a succession of jobs: as a millwright, then a carpenter, in a tea shop, and as a hotelkeeper. He made and sold electrical goods and sold insurance. Finally, in 1857, he published a book, *The Philosophy of Theism*, which attracted some attention, and after working for a temperance newspaper in 1859 Croll was made the keeper—effectively the janitor—of the Andersonian Museum, in Glasgow. This gave him access to the library, where he continued his self-education. He published several papers on chemistry, physics, and geology, and in 1867 was placed in charge of the Edinburgh office of the Geological Survey of Scotland, a post he held until he retired, in 1880.

After his retirement Croll continued to work on his astronomical theory of climate change and on other topics that interested him. He published several books, including *Climate and Time, in Their Geological Relations* (1875), *Discussions on Climate and Cosmology* (1885), *Stellar Evolution and Its Relations to Geological Time* (1889), and *The Philosophical Basis of Evolution* (1890).

Croll died from heart disease near Perth on December 15, 1890.

Crutzen, Paul
(b. 1933)
Dutch
Atmospheric chemist

Professor Paul Crutzen shared with F. Sherwood Rowland and Mario J. Molina the 1995 Nobel Prize in chemistry, which was awarded for their discovery of the processes that deplete the OZONE LAYER.

Paul Crutzen was born on December 3, 1933, in Amsterdam, the Netherlands. His father, Josef Crutzen, was a waiter and his mother, Anna Gurk, worked in a hospital kitchen. Paul commenced his education in September 1940. In May of that year the Netherlands had been invaded and occupied by the German army, and Crutzen's primary education coincided with the period of occupation. His schooling was interrupted several

times, and conditions were especially severe from the fall of 1944 until the country was liberated in May 1945. Nevertheless, he was able to complete his primary schooling, and in 1946 he entered the middle school, where he prepared to enter university. He specialized in mathematics and physics, languages, and was keen on sports, especially long-distance skating on the Dutch lakes and canals.

Unfortunately, illness meant that Crutzen did not achieve the grades in his final examinations that would have won him a grant to pay for a university education. Instead of embarking on a four-year university course he enrolled at a technical school for a three-year course in civil engineering, lasting from 1951 until 1954. The second year was spent working for a civil engineering company to gain practical work experience, and he managed to live for two years on what he was paid.

On completing his course, Crutzen obtained a job with the Bridge Construction Bureau of the City of Amsterdam. He worked there from 1954 until 1958, with an interruption for his compulsory military service from 1956 to 1958. During his time there he met Terttu Soininen, a student of Finnish history at the University of Helsinki. They were married in February 1958 and made their home in Gävle, a town about 125 miles (200 km) north of Stockholm, Sweden, where Crutzen was then working for a construction company. Their two daughters, Ilona and Sylvia, were born in Gävle.

In 1958, Crutzen saw a newspaper advertisement for a computer programmer in the Department of Meteorology of the Stockholm Högskola (high school, and since 1961 Stockholm University). Despite knowing nothing at all about computer programming, he applied for the post and won it. At the beginning of July 1959, the Crutzen family moved to Stockholm and Paul embarked on a second career. The Department of Meteorology (now the Meteorology Institute) and the International Meteorological Institute that was associated with it were at the forefront of research, and they housed some of the fastest computers in the world.

Until 1966, Crutzen spent much of his time building and running some of the first weather prediction models, including one of a TROPICAL CYCLONE. High-level computer languages, such as Algol and Fortran, had not yet been developed, and all programs had to be written in machine code, using binary notation.

Because he worked at the university, Crutzen was able to attend some of the lectures. He had no opportunity to do laboratory work, however, and so he concentrated on mathematics, statistics, and meteorology. In 1963, he obtained his master of science degree in these subjects and in 1965 began to work for his doctorate. At that time he was assisting another scientist in a project to study the different forms (called allotropes) of OXYGEN in the upper atmosphere, and he chose to make stratospheric chemistry the subject for his doctoral thesis. He received his Ph.D. in 1968 and his D.Sc. (doctor of science, which is a higher degree than a Ph.D.) in 1973, for research into the photochemistry of ozone.

Crutzen left Stockholm in 1969 to work until 1971 as a European Space Research Organization fellow at the Clarendon Laboratory of the University of Oxford, England. From 1974 to 1980, he worked at Boulder, Colorado, on the Upper Atmosphere Project of the National Center for Atmospheric Research and as a consultant in the Aeronomy Laboratory of the Environmental Research Laboratories (NATIONAL OCEANIC AND ATMOSPHERIC ADMINISTRATION). He was an adjunct professor in the Atmospheric Sciences Department of the University of Colorado from 1976 to 1981. In 1980, he was appointed director of the Atmospheric Chemistry Division of the Max-Planck Institute for Chemistry, at Mainz, Germany, and from 1983 to 1985 he was executive director of the Institute. He held a part-time professorship at the University of Chicago from 1987 until 1991, and since 1992 he has been a part-time professor at the Scripps Institution of Oceanography of the University of California.

Professor Crutzen has received many honors and holds honorary degrees from universities at York, Canada, Louvain, Belgium, East Anglia, England, and Thessaloniki, Greece.

Dalton, John
(1766–1844)
English
Meteorologist and chemist

John Dalton was born at Eaglesfield, near Cockermouth, Cumberland (now Cumbria), on September 6, 1766, the son of a weaver and the third of six children. His father, Joseph, was a devout member of the Society of Friends, and in accordance with Quaker practice at the time he did not register the date of his son's birth, so there is some uncertainty about it. John began his education at the Quakers' school in Eaglesfield. The

teacher was John Fletcher, and when Fletcher retired in 1778 John took his place.

Three years later, in 1781, Dalton moved to Kendal, where he continued to earn his living as a teacher and became a headmaster. John Gough, a wealthy Quaker who was also a classicist and mathematician, befriended Dalton. Under his influence Dalton began to write articles on scientific topics for two popular magazines, the *Gentlemen's Diary* and *Ladies' Diary*. Gough also encouraged him to keep a diary of meteorological observations. Dalton began this diary in 1787 and made entries in it regularly for 57 years until his death. It contained more than 200,000 observations.

In the 18th century, only members of the Church of England were accepted as students at the universities of Oxford and Cambridge. As a Quaker, Dalton was excluded, and consequently he was very largely self-taught.

In 1793, Dalton moved to Manchester, where Gough had helped him obtain a position as a teacher of mathematics and natural philosophy at New College. He held this post until 1799, when the college was moved to York, and Dalton remained behind in Manchester as a private teacher of mathematics and chemistry.

Dalton's first publication on weather, *Meteorological Observations and Essays* appeared in 1793, but it did not sell well. In 1794, Dalton was elected to the membership of the Manchester Literary and Philosophical Society, and a few weeks later he delivered his first paper to the society, "Extraordinary Facts Relating to the Vision of Colours." Dalton and his brother were both color blind, and this was the first account of the way the world appears to someone with this condition. For a time color blindness was known as Daltonism. He also lectured to the society on the weather. Dalton became the honorary secretary and then president of the society, and after he resigned his teaching position in 1799 he lived for many years in a house the society bought for him and which he shared with the Reverend W. Johns.

In 1803, Dalton proposed the law of partial pressures, known as Dalton's law (see GAS LAWS). He discovered that the DENSITY of water varies with its TEMPERATURE, reaching a maximum at 42.5°F (6.1°C). In fact, water reaches its maximum density at 39.2°F (4°C), but Dalton was close. He studied what happens when substances dissolve in water and when gases mix, concluding that water and gases must consist of very small particles that intermingle, so the particles of a dissolved substance are located between water particles. In his book *New System of Chemical Philosophy*, published in 1808, he suggested that the particles of different elements have different weights. He compiled a list of these weights (relative atomic masses) and devised a system of symbols for the elements. These could be combined to represent compounds.

Dalton became very famous. He delivered two courses of lectures at the Royal Institution in London, the first in 1804 and the second in 1809–10. He was elected a fellow of the Royal Society in 1822. He became a corresponding member of the French Academy of Sciences, and in 1830 he was elected one of its eight foreign associates. In 1833, the government awarded him an annual pension of £150 and raised it to £300 in 1836. This was a substantial award that would have allowed Dalton to live comfortably at a higher standard than the one he chose for himself. It was also an unusual honor. British governments rarely recognized the worth of individuals in this way. Government support for science took the form then, as now, of providing funds for institutions and paying the salaries of scientists directly employed by government ministries and agencies.

Dalton spent almost all of his time in Manchester, working in his laboratory and teaching. Each year he visited the Lake District, in his native Cumbria, and occasionally he visited London. He made a short visit to Paris in 1822. His only recreation was the game of bowls, which he played every Thursday afternoon. (The English game of bowls is played on an absolutely level surface lawn [or an indoor equivalent]. It is the game Sir Francis Drake allegedly played on Plymouth Hoe as the Spanish Armada approached, and it has no connection to the game of bowling, derived from skittles.)

John Dalton died in Manchester on July 27, 1844, and a statue was erected in his memory. The house in which he lived for so long contained many of his records and other relics, but it was destroyed during a bombing raid in World War II.

Daniell, John Frederic
(1790–1845)
English
Meteorologist, chemist, and inventor

John Daniell was a prolific inventor and became one of the most eminent scientists of his day. He was born in London on March 12, 1790, the son of a lawyer. He was educated privately, mainly learning Latin and Greek. It is not certain whether he earned a degree from Oxford University or was awarded an honorary one. His education completed, he went to work in a sugar refinery and resin factory owned by a relative, where he was able to improve the technology being used.

In his spare time he attended Royal Institution lectures given by William T. Brande, the professor of chemistry. He met Brande, and the two became close friends. Between them they revived the fortunes of the Royal Institution, which were then at a low ebb. Daniell left the factory and became a scientist when, in 1813 at the age of 23, he was appointed professor of physics at the University of Edinburgh. He combined chemistry with the physics he researched and taught at the university, and for a time in 1817 he managed the Continental Gas Company, developing a new process for making gas by distilling resin dissolved in turpentine. This process was used for a time in New York. His interest in gases also extended to the atmosphere.

In 1820, he made his first major contribution to METEOROLOGY with his invention of a DEW point HYGROMETER that measured relative HUMIDITY. It comprised two bulbs made from thin glass that were connected by a tube. One bulb was filled with ether and held a THERMOMETER. When the temperature of the air in the other bulb changed, the ether in the other bulb was warmed or cooled and either evaporated or condensed. This caused WATER VAPOR to condense onto or evaporate from the bulb containing ether. The average temperature at which this happened was the dew point temperature. The Daniell hygrometer remained in use for many years.

In 1823, Daniell published a book describing his meteorological research, *Meteorological Essays*. In the same year he was elected a fellow of the Royal Society.

Daniell moved from Scotland to London in 1831, when he was appointed the first professor of chemistry and meteorology at King's College, which had recently been founded (it is now part of the University of London). Daniell remained in this post until his death.

Daniell was very active in the Royal Society. In 1830, he installed in the entrance hall a BAROMETER that used water to measure pressure. Over the following years, he made many observations with it.

Daniell investigated the climatic influence of solar radiation, the circulation of the atmosphere, and he pointed out the importance of maintaining a humid atmosphere in greenhouses. This revolutionized greenhouse horticulture, and in 1824 Daniell was awarded the silver medal of the Horticultural Society.

His interest in chemistry and physics had not diminished, and in the 1830s Daniell became increasingly interested in electrochemistry. This led to his invention, in 1836, of the Daniell cell, the first reliable source of direct-current electricity.

On March 13, 1845, Daniell suffered a heart attack and died while attending a meeting of the Council of the Royal Society in London.

Further Reading

Corrosion Doctors. "John Frederick Daniell (1790–1845)." Available online. URL: www.corrosion-doctors. org/Biographies/DaniellBio.htm. Accessed March 13, 2006.

King's College, London. "History of the College: John Frederick Daniell." Available online. URL: www.kcl. ac.uk/college/history/people/daniell.html. Accessed March 13, 2006.

Dansgaard, Willi
(b. 1922)
Danish
Geophysicist and paleoclimatologist

Willi Dansgaard is professor emeritus of geophysics at the University of Copenhagen.

Dansgaard was the first scientist to demonstrate that measurements of OXYGEN isotope ratios and the ratio of HYDROGEN and deuterium, also called heavy hydrogen, could reveal information about past climates. In 1966, scientists obtain the first ICE CORE from Greenland, and Dansgaard led a team of paleoclimatologists who analyzed the isotopic composition of the ice. Dansgaard also led the work to perfect the dating of ice layers in ice cores and to measure DUST particles trapped in the ice, and the acidity of the ice. In subsequent years, Dansgaard organized or participated in 19 expeditions to the GLACIERS of Norway, Greenland, and Antarctica.

Professor Dansgaard has received many awards, including the 1996 Tyler Prize for Environmental Achievement.

Darcy, Henri-Philibert-Gaspard
(1803–1858)
French
Civil engineer

Henri Darcy (sometimes called Henry Darcy) improved the PITOT TUBE (invented by the French physicist Henri Pitot, 1695–1771). He also discovered the law (DARCY'S LAW) governing the rate at which water flows through an AQUIFER.

Henri Darcy was born in Dijon, France, on June 10, 1803. His father, Jacques Lazare Darcy, was a civil servant and tax collector, and Henri's younger brother, Hugues, also became a civil servant and prefect (the principal administrative officer of a French district, or département). Jacques Darcy died in 1817, when Henri was 14 years of age, leaving his mother, Agathe Angelique Serdet, with only a small pension. She struggled to obtain the best possible education for her son, borrowing money to hire tutors and gaining a scholarship for Henri to attend college. When he was 18, Henri was admitted to the prestigious École Polytechnique in Paris, and when he was 20, in 1823, he entered the École des Pont et Chaussées (School of Bridges and Highways). After graduating, the Imperial Corps of Bridges and Highways employed Henri in Dijon.

In 1828, Henri married Henriette Carey, an English girl from Guernsey, one of the Channel Islands, who was living in Dijon. They remained together until Henri's death, but had no children.

The same year that he married, Henri was assigned to a project to provide the city with a reliable water supply. They dug a deep well, but found insufficient water, and Henri set about finding a better supply. This led him to supervise the construction of a system of pipes that carried spring water to reservoirs and 17 miles (28 km) of pipes that carried water under pressure to important public buildings and to 142 street pumps. The entire system was driven by gravity and required no pumps or filters. Darcy followed this with many other public works projects in and around Dijon. He also became active in the local government.

By 1848, Henri Darcy had risen to the post of chief engineer for the département of Côte d'Azur. He left Dijon and became chief director for water and pavements in Paris. That is where he conducted the research into liquid flow through pipes that led him to improve the design of the pitot tube.

Failing health compelled him to resign his post in 1855. He returned to Dijon, where he devoted himself to private research into topics that interested him. These established Darcy's law for the flow of water through a column of sand.

During a trip to Paris, Henri Darcy fell ill with pneumonia. He died on January 3, 1858. He is buried in Dijon.

Geer, Gerhard Jacob de
(1858–1943)
Swedish
Geologist

Gerhard (or Gerard) de Geer discovered VARVES (*varv* is the Swedish word for layer), thereby making an important contribution to the study of climate change. He also made other important contributions to QUATERNARY geology and the dating of sedimentary rocks. It was de Geer who coined the term *geochronology,* which is the scientific identification of geological time intervals.

Gerhard de Geer was born in Stockholm on October 2, 1858. His family was descended from Dutch aristocrats who had settled in Sweden early in the 17th century. The de Geer family was prominent in Swedish life. Gerhard was a member of the Swedish parliament (1900–05), and both his father and elder brother were prime ministers of Sweden.

While still a student at Uppsala University, in 1878 de Geer joined the Swedish Geological Survey, marking the commencement of his life's work of studying the Quaternary rocks and landforms of southern Sweden. By the time he graduated in 1879, de Geer had already noted that the laminations in sediments deposited at the edges of lakes as the ice retreated at the end of the last glacial period resembled TREE RINGS. He concluded that both the lakeside laminations and tree rings were probably formed annually. He pursued this study through the 1880s, publishing his first brief account of the work in 1882 and making a more formal presentation to the Swedish Geological Society in 1884. His work attracted international interest following the 11th International Geological Congress in 1910. The congress was held in Stockholm, and de Geer presided over it. It was in the paper he presented to the congress that de Geer first used the term *varves.* De Geer also studied glacial moraines, RAISED BEACHES, and changes in sea level associated with GLACIOISOSTASY.

Gerhard de Geer became professor of geology at Stockholm University in 1897, and from 1902 until 1910 he was president of the university. He still pursued his study of varves, and in 1915 he matched Swedish varves with varves in Norway and Finland. Accompanied by his wife, Ebba Hult de Geer, and two assistants, Ernst Antevs and Ragnar Liden, de Geer traveled to the United States in 1920. When the others returned to Sweden, Antevs remained behind to study North American varve chronology. De Geer retired from teaching in 1924 and became the founder and director of the Geochronological Institute of Stockholm University.

In the 1930s, Antevs showed that de Geer's TELECONNECTIONS were mistaken. De Geer thought he was being deliberately misunderstood, and the dispute became increasingly bitter. Their disagreement coincided with the discovery of RADIOCARBON DATING, and scientific interest in varves diminished.

Gerhard de Geer died in Stockholm on July 24, 1943.

Deluc, Jean André
(1727–1817)
Swiss
Geologist, meteorologist, and physicist

Jean Deluc is principally remembered for having invented the dry pile, a type of electric battery, but his contributions to geology and METEOROLOGY were no less important. He was born in Geneva on February 8, 1727. Until he was 46, Deluc worked in commerce and politics and made scientific excursions among the Swiss mountains. Then, in 1773, he moved to England. He was elected a fellow of the Royal Society and appointed a reader to Queen Charlotte, a post that brought him an income but did not make excessive demands on his time. The post made him free to pursue his scientific interests.

In 1778 and 1779, Deluc published a six-volume work on geology, *Lettres physiques et morales sur les montagnes et sur l'histoire de la terre et de l'homme* (Physical and moral letters on mountains and on the history of the Earth and of man). In this he suggested that each of the six days of the biblical creation was an epoch.

Deluc also discovered that water is more dense at 40°F (4°C) than it is at either higher or lower temperatures. In 1761, he also found that the heat required to melt ice or vaporize liquid water does not raise their temperatures. This was the concept of LATENT HEAT discovered independently by Joseph Black at about the same time. Jean Deluc was also the first scientist to propose that the amount of WATER VAPOR that can be contained in a given space is independent of any other gases that may be present in that space. He invented a HYGROMETER, though not a very successful one, and he was the first to devise a way of measuring height by means of a BAROMETER. He showed that an increase in elevation is proportional to a decrease in the logarithm of the AIR PRESSURE, and that a change in elevation is also inversely proportional to the air temperature.

Jean Deluc died at Windsor, England, on November 7, 1817.

Dobson, Gordon Miller Bourne
(1889–1976)
English
Meteorologist

The scientist who devoted much of his career to studying the OZONE LAYER was born on February 25, 1889. He served in the Royal Flying Corps (precursor of the Royal Air Force) during World War I, attaining the rank of captain, and was director of the experimental department at the Royal Aircraft Establishment, Farnborough. In 1920, Dobson took up the position of lecturer in METEOROLOGY at the University of Oxford.

In collaboration with Professor Frederick Alexander Lindemann (later Lord Cherwell, 1886–1957), Dobson studied meteor trails. They discovered a region of the stratosphere (*see* ATMOSPHERIC STRUCTURE) where TEMPERATURE increases with altitude. Dobson inferred, correctly, that the rise in temperature was due to the absorption of ULTRAVIOLET RADIATION by atmospheric OZONE, and he determined to measure the concentration of stratospheric ozone. In order to do so, in 1924 he built a spectrograph in his workshop at home. He used the instrument through 1925 to monitor the ozone concentration. He found that the concentration changes with the seasons and there is a relationship between the ozone concentration and weather conditions in the upper troposphere and lower stratosphere. In the following years, he built and calibrated five more spectrographs, which were used in other European locations. In 1927 or 1928, Dobson built a spectropho-

tometer that was more sensitive than the earlier spectrographs. He was elected to a fellowship of the Royal Society in 1927.

Early in the 1930s, Dobson became interested in AIR POLLUTION. He helped develop reliable methods for measuring concentrations of SMOKE, SULFUR DIOXIDE, and other pollutants.

During World War II, Dobson studied stratospheric HUMIDITY in order to forecast the height at which aircraft CONTRAILS would form. In the course of this work, Dobson invented the first frost-point HYGROMETER. After the war, Dobson returned to the study of stratospheric ozone. In 1945, the University of Oxford conferred the title of professor on him. Dobson retired from the university in 1956, but continued to study ozone. He wrote his last paper in 1973 and made his last observation the day before he suffered the stroke from which he died six weeks later. Dobson died in Oxford on March 11, 1976.

Further Reading
University of Oxford. "G. M. B. Dobson (25 February, 1889–11 March, 1976)." Available online. URL: www.atm.physics.ox.ac.uk/user/barnett/ozoneconference/dobson.htm. Most recent revision February 1, 2006.

Doppler, Johann Christian
(1803–1853)
Austrian
Physicist

Doppler was an Austrian physicist who discovered the DOPPLER EFFECT. This was first tested in 1845 at Utrecht, in the Netherlands. A train carried a group of trumpeters in an open carriage past a group of musicians located beside the rail track. The stationary musicians reported a change in pitch as the trumpeters approached and receded.

Doppler was born in Salzburg and educated in Vienna as a physicist and mathematician. In 1835, he was appointed professor of mathematics at the State Secondary School in Prague and subsequently held professorships at the State Technical Academy, Prague, and the Mining Academy, Schemnitz. In 1850, he returned to Vienna as director of the Physical Institute and professor of experimental physics at the Royal Imperial University of Vienna.

He died in Venice, Italy, on March 17, 1853.

Ekman, Vagn Walfrid
(1874–1954)
Swedish
Oceanographer and physicist

Vagn Ekman was born in Stockholm on May 3, 1874. He was educated at the University of Uppsala, where he received his doctoral degree in 1902. His doctoral thesis described his research into the cause of a phenomenon first reported in the 1890s by the Norwegian arctic explorer Fridtjof Nansen (1861–1930).

Nansen had noticed that ICE drifting on the surface of the sea did not move in the direction of the wind, but at about 45° to the right of it. Ekman was able to relate the movement of wind-driven ocean currents to FRICTION between different layers of water and to the CORIOLIS EFFECT. Together these produced a change in the direction of currents with depth that became known as the EKMAN SPIRAL. A similar effect occurs in the atmosphere. In 1905, Ekman explained the spiral more fully in a paper titled "On the Influence of the Earth's Rotation on Ocean-Currents," published in the Swedish journal *Arkiv för Matematik, Astronomi och Fysik*.

After receiving his doctorate, in 1902 Ekman moved to Norway (which was still part of Sweden) to take up a post as an assistant at the International Laboratory for Oceanographic Research in Oslo (which was then called Christiania). He remained there until 1908, and while there he came to know Nansen.

Ekman returned to Sweden in 1910, as professor of mechanics and mathematical physics at the University of Lund, a post he held until he retired in 1939. Vagn Ekman died on March 9, 1954.

Elsasser, Walter Maurice
(1904–1991)
German-American
Physicist

Walter Elsasser is the scientist who developed the theory that the Earth's core acts as a dynamo, generating the magnetic field. Elsasser also pioneered the analysis of the magnetic field recorded in the orientation of rock particles as a tool for studying the history of crustal rocks. This has been of great importance in tracing the movements of continents (*see* PLATE TECTONICS) and the history of climate.

Elsasser was born on March 20, 1904, at Mannheim, Germany, and was educated at the Univer-

sity of Göttingen, where he obtained his doctorate in 1927. He then taught at the University of Frankfurt, but he left Germany in 1933, when the Nazis came to power. Elsasser spent three years in Paris and then moved to the United States and joined the staff of the California Institute of Technology. In 1940, he became an American citizen. In the course of his career Elsasser was professor of physics at the University of Pennsylvania, professor of geophysics at Princeton University, and a research professor at the University of Maryland.

Walter Elsasser died on October 14, 1991.

Fabry, Marie-Paul-August-Charles
(1867–1945)
French
Physicist

Charles Fabry was one of the most distinguished physicists of his generation, and one of the most famous. He was born at Marseilles on June 11, 1867, and after commencing his education in Marseilles in 1885 he enrolled at the École Polytechnique in Paris, graduating in 1889. He studied physics and mathematics, but became increasingly drawn toward astronomy and optics. After his graduation, Fabry moved to the University of Paris, where he obtained his doctorate in physics in 1892.

Fabry then spent two years teaching physics at lycées (high schools) in several cities before joining the staff at the University of Marseilles in 1894, where he devoted himself to teaching and pursuing his own research. In 1904, he was appointed professor of industrial physics.

In 1914, the French government called him to Paris to investigate interference in sound and light waves, and in 1921 Fabry moved to Paris permanently as professor of physics at the Sorbonne. Later, he combined this post with that of professor of physics at the École Polytechnique and director of the Institute of Optics. In 1935, he became a member of the International Committee on Weights and Measures. He retired two years later, in 1937.

It is not uncommon for a research scientist to find that further advances are impossible because the available instruments are inadequate. They cannot penetrate the phenomena being investigated deeply enough or provide sufficiently accurate measurements. In this situation the experimenter, who is the only person in a position to know precisely what is needed, often modifies an existing device or invents a new one. That is what Charles Fabry and his colleague the French physicist Albert Pérot (1863–1925) did in 1896. The instruments they invented were based on two perfectly parallel half-silvered plates. If the distance between the plates is fixed, the instrument is known as the Fabry–Pérot INTERFEROMETER and as the Fabry–Pérot etalon if the distance can be varied. These instruments break light into its constituent wavelengths with a much greater resolution than was possible with other devices. Fabry and Pérot spent 10 years designing, improving, and using them. The two colleagues were able to confirm the DOPPLER EFFECT for light in the laboratory, and they applied the instruments to a variety of astronomical questions.

A question of particular interest concerned the absorption of solar ultraviolet radiation (*see* SOLAR SPECTRUM) in the atmosphere. Clearly some atmospheric gas was filtering it, and in 1913 Fabry used the interferometer to discover the presence of abundant OZONE in the upper atmosphere.

Fabry died in Paris on December 11, 1945.

Fahrenheit, Daniel Gabriel (or Gabriel Daniel)
(1686–1736)
Polish–Dutch
Physicist

The scientist whose name is still used every day because of the TEMPERATURE SCALE he devised was born on May 14, 1686, in Danzig (now called Gdansk), an ancient city at the mouth of the Vistula River on the coast of the Baltic Sea. Culturally, the city is Polish, and it now lies deep inside Poland, but at various times in the past changes in frontiers have meant it lay in Prussia or, more recently, in Germany.

Daniel Fahrenheit began his education in Danzig, but in 1701 he moved to Amsterdam in order to learn a business. He became interested in the making of scientific instruments, and in about 1707 he left the Netherlands to tour Europe, meeting scientists and other instrument makers, and learning the craft he had chosen to follow. In the course of his travels, in 1708 he met the Danish astronomer Ole (or Olaus) Christensen Römer (1644–1710). Fahrenheit returned to Amsterdam in 1717, established his own business making instruments, and remained there for the rest of his life.

At that time there was intense scientific interest in studying the atmosphere and the weather it produced, but meteorologists were greatly hindered by the lack of a reliable THERMOMETER. Galileo had made thermometers, and Guillaume Amontons had improved on them, but both of their instruments relied on the expansion and contraction of air to raise and lower a column of liquid and they were highly inaccurate. Fahrenheit turned his attention to the problem.

His first thermometer used alcohol as the liquid. Unlike earlier designs, however, Fahrenheit filled the bulb with liquid, so changes in temperature were indicated by the expansion and contraction of the column of liquid, not changes in the volume of a pocket of air. This was a great improvement, but an alcohol thermometer cannot measure very high temperatures, because of the low boiling point of alcohol (pure ethanol boils at 172.94°F, 78.3°C). So Fahrenheit tried mixing alcohol and water. This raised the boiling point, but the volume of the mixture did not change at a constant rate as the temperature increased or decreased, which made the thermometer very difficult to calibrate. Finally, in 1714, Fahrenheit tried mercury. Mercury boils at a much higher temperature than water (673.84°F, 356.58°C) and freezes at a much lower temperature (-37.97°F, -38.87°C), so a mercury thermometer can be used over a much wider temperature range than can an alcohol thermometer. First, though, Fahrenheit had to devise a way to purify the metal, because impurities caused it to stick to glass surfaces. Once he had achieved this, he found that mercury changed its volume at a fairly constant rate with changing temperature, although it changed by a smaller amount than alcohol.

Having made a thermometer, Fahrenheit then had to calibrate it. Ole Römer had invented a thermometer in about 1701, and during his visit to Copenhagen Fahrenheit had watched him calibrate one. He based his calibration on the Römer temperature scale, using two fixed, or fiducial, points. After some later adjustments, he produced the Fahrenheit temperature scale that is still in use today, on which ice melts at 32°, water boils at 212°, and average body temperature is 98.6°.

Fahrenheit's thermometer was far more reliable and accurate than any that had existed before, and the mercury thermometers in use today are made in the way Fahrenheit devised, although the use of mercury is being phased out because exposure to the metal is harmful to health. Amontons had earlier suggested that water always boils at the same temperature. Fahrenheit set out to check this and found it to be true, but only if the AIR PRESSURE remains constant. He also examined many other liquids and found all of them had characteristic boiling and freezing temperatures.

In 1724, Fahrenheit described his method for making thermometers in a paper he submitted for publication in the *Philosophical Transactions of the Royal Society*. He was elected to the Royal Society in the same year.

Fahrenheit died at The Hague on September 16, 1736.

Ferdinand II
(1610–1670)
Italian
Physicist

A member of the powerful Medici family, Ferdinand was born on July 14, 1610, the son of Cosimo II, grand duke of Tuscany. His father died in 1620. Ferdinand was 10 years old when he became ruler.

Ferdinand was not a strong ruler and was unable to protect Galileo from his trial by the Inquisition in 1633, but he took a keen interest in science, and especially in atmospheric science. One of the scientific challenges of the time was to find a way to measure TEMPERATURE. Galileo had attempted this with his air thermoscope (*see* THERMOMETER), and in 1641 Ferdinand improved on Galileo's instrument by inventing a thermometer consisting of a tube that contained liquid and was sealed at one end. He improved on this with a further design in 1654, his thermometer providing the basis for the improved design that would be made by Daniel Fahrenheit about 60 years later.

Ferdinand also designed one of the earliest accurate HYGROMETERS. It consisted of a tapering cylinder that was filled with ice. WATER VAPOR condensed on the outside of the cylinder and ran down it into a collecting funnel and from there to a flask. The amount of water that accumulated in the flask indicated the HUMIDITY of the air.

In 1657, the year he produced his hygrometer, Ferdinand and his brother Leopold founded the Accademia del Cimento (Academy of Experiments) in Florence. Members of the Accademia were especially interested in studying the atmosphere. Carlo Renaldini was one of those who worked on developing the thermometer. The Accademia itself was the forerunner of other scientific academies, including the Royal Society of London (founded in 1665) and the Royal French Academy of Sciences (founded in 1666). The Accademia ceased to function in 1667.

Ferdinand died on May 24, 1670, and was succeeded by his son, Cosimo III. Ferdinand's grandson, Gian-Gastone, had no male heir, and the Medici family ended with his death in 1737.

Ferrel, William
(1817–1891)
American
Climatologist

William Ferrel was born on January 29, 1817, in Bedford County, Pennsylvania. When he grew up, he earned his living as a school mathematics teacher, but he combined this with an intense interest in the TIDES and weather. His study of the circulation of the atmosphere led him to publish in 1856 a mathematical MODEL of the circulation of the atmosphere. He revised his model in 1860 and again in 1889. The model proposed the existence of a midlatitude cell. In this cell the vertical movement of the air is driven by the Hadley cell on the side nearest the equator and by the polar cell on the side nearest the pole. This third cell is known as the Ferrel cell (*see* GENERAL CIRCULATION).

In 1857, Ferrel's particular interest in and understanding of tides led to an invitation to join the staff of *The American Ephemeris and Nautical Almanac*, which was published in Cambridge, Massachusetts. While working for this publication Ferrel calculated that the combined effect of the PRESSURE GRADIENT FORCE and the CORIOLIS EFFECT must be to cause winds generated by a PRESSURE GRADIENT to flow at 90° to it, parallel to the isobars (*see* ISO-) rather than across them. A few months after he reached this conclusion, the Dutch meteorologist C. H. D. Buys Ballot announced the same phenomenon. Buys Ballot had not known of Ferrel's work, and when he learned of it he readily acknowledged Ferrel's prior claim. It was too late, however, and the discovery came to be known as BUYS BALLOT'S LAW.

On July 1, 1867, Ferrel was appointed to a position at the United States Coast and Geodetic Survey.

His task there was to develop the general theory of the tides to which he had already devoted a considerable amount of work and which he had advanced further than any of his contemporaries.

Winds and atmospheric pressure affect the tides, so Ferrel widened his study of tides to include relevant meteorological phenomena. This led him to a more general investigation of METEOROLOGY, and for some time he alternated between studying tides and studying weather.

William Ferrel wrote extensively on the subject of meteorology. His titles include *Meteorological Researches*, published in three volumes between 1877 and 1882, *Popular Essays on the Movements of the Atmosphere* (1882), *Temperature of the Atmosphere and the Earth's Surface* (1884), *Recent Advances in Meteorology* (1886), and *A Popular Treatise on the Winds* (1889).

On August 9, 1882, Ferrel tendered his resignation from the Coast Survey in order to accept a position in the Army Signal Service. The superintendent accepted his resignation, and Ferrel continued to work for the Signal Service until his retirement in 1886. The Signal Service already had an interest in meteorology, and in November 1870 its newly established Division of Telegrams and Reports for the Benefit of Commerce became a national weather service.

The acceptance of Ferrel's resignation from the Coast Survey was conditional. He was asked to complete the investigations on which he was engaged at the time, and he was also asked to continue supervising the tide-predicting machine he had invented. This was a mechanical device, worked by levers and pulleys, that took account of 19 constituents of the forces affecting tides and gave readings, on five dials, of the predicted times and heights of high and low water. Ferrel had submitted plans and an explanation of the machine to the Coast Survey in the spring of 1880, and in August of that year he described it in Boston at the annual meeting of the American Association for the Advancement of Science. The idea was accepted, but it proved difficult to find a machinist with the adequate skills. Work on constructing the device did not commence until the late summer of 1881, and the machine was not completed until the autumn of 1882. The tide-predicting machine was first used to predict the tides for 1885, and it remained in use until 1991. Computers are now used to predict tides.

After his retirement, William Ferrel moved to Maywood, Kansas, where he died on September 18, 1891.

FitzRoy, Robert
(1805–1865)
English
Naval officer, hydrographer, and meteorologist

In 1860, *The Times* of London became the first newspaper in the world to publish a daily weather forecast. That forecast was prepared by Admiral FitzRoy, the Head of Meteorology at the Board of Trade. This was the department that was to become the British METEOROLOGICAL OFFICE.

Robert FitzRoy was born on July 5, 1805, at Ampton Hall, in Suffolk, an English county to the northeast of London. He was an aristocrat, the grandson on his father's side of the duke of Grafton and on his mother's side of the marquis of Londonderry, and directly descended from King Charles II. FitzRoy was educated at the Royal Naval College, Portsmouth. After graduating, on October 19, 1819, he entered the Royal Navy, and he received his commission as an officer on September 7, 1824.

FitzRoy served in the Mediterranean and was then sent to South America on board HMS *Beagle,* which was conducting a surveying mission. When the captain of the *Beagle* died, FitzRoy assumed command, completing the survey and returning safely to England. He applied to lead a second survey, and in 1831 the Naval Hydrographer, Admiral Sir Francis Beaufort, granted the request. FitzRoy sailed once more as captain of the *Beagle,* this time accompanied by Charles Darwin.

The *Beagle* was well equipped for its scientific mission, and the equipment included several BAROMETERS. FitzRoy used these to prepare short-term weather forecasts. In another innovation, this was the first voyage in which wind observations recorded in the log were based on the BEAUFORT WIND SCALE. A well-trained and experienced sailor, FitzRoy knew how important it was to predict the weather.

With the survey complete, the *Beagle* arrived back in Portsmouth on October 2, 1836, and FitzRoy settled down to write his accounts of the voyage. These were published in 1839 (as two volumes of *Narrative of the Surveying Voyages of His Majesty's Ships* Adventure *and* Beagle *Between the Years 1826 and 1836, Describing Their Examination of the Southern Shores of South*

America, and the Beagle*'s Circumnavigation of the Globe*). FitzRoy was elected a fellow of the Royal Society for his surveying work.

By then an admiral, in 1841 Robert FitzRoy became a Member of Parliament for Durham, and in 1843 he was made governor general of New Zealand. He was recalled from New Zealand in 1845 at the insistence of the British settlers, mainly because he believed the Maori claims to land were as legitimate as theirs. Admiral FitzRoy retired from active service in 1850. In 1854, he took up his post at the Meteorological Office and devoted himself wholly to METEOROLOGY.

At the Meteorological Office, FitzRoy encouraged the collection of weather observations, established barometer stations, and used TELEGRAPHY to gather data. These allowed the Meteorological Office to issue weather forecasts and, in 1861, the first storm warnings. In 1863, FitzRoy published *The Weather Book*, in which he set out principles to guide sailors in WEATHER FORECASTING.

These included 47 "instructions for the use of the barometer to foretell weather." FitzRoy believed a barometer should be installed at every port. Sailors could examine the instrument, use FitzRoy's instructions to interpret the reading, and then decide whether or not to set sail. "FitzRoy" barometers became very popular, and Fitzroy himself invented some versions. There were domestic versions with THERMOMETERS, STORM GLASSES, and various other devices added to them, as well as a set of the admiral's instructions. FitzRoy barometers were still being manufactured in the early 20th century, and reproductions are still made.

FitzRoy's work saved many lives, but criticisms of his forecasting methods by the eminent meteorologist Matthew Maury and of his humanitarian political beliefs by newspapers and politicians greatly troubled him. He also experienced conflict between his strongly held religious views and Darwin's theory of evolution by natural selection, which he had helped to develop. Unable to resolve these difficulties, on April 30, 1865, he committed suicide at his home at Upper Norwood, near London, by cutting his throat.

Flohn, Hermann
(1912–1997)
German
Meteorologist and climatologist

Hermann Flohn, designer of the widely-used climate classification system (*see* FLOHN CLASSIFICATION), was born at Frankfurt-am-Main on February 19, 1912. He was educated at the Universities of Frankfurt and Innsbruck, graduating from both in METEOROLOGY, geography, geophysics, and geology.

Flohn worked at the University of Marburg for a few months, and in 1935 he joined the German state weather service and moved to Berlin. During World War II, Flohn served in the Luftwaffe (air force) meteorological service. He was captured and became a prisoner-of-war.

After the war, Flohn joined the German weather service and was based at Bad Kissingen. He was head of research at the weather service from 1952 to 1961, and from 1961 until 1977 he was a professor at the University of Bonn and director of the Meteorological Institute.

After his retirement in 1977, Hermann Flohn was made professor emeritus and continued to head research projects. The last of these was conducted on behalf of the North Rhine–Westphalia Academy of Science. It investigated large-scale climate forecasting and its environmental significance.

Hermann Flohn died in Bonn on June 23, 1997.

Fortin, Jean-Nicholas
(1750–1831)
French
Instrument maker

Many eminent scientists used precision instruments Jean Fortin had made, but Fortin is best known for the portable mercury BAROMETER that he invented and that bears his name.

Fortin was born at Mouchy-la-Ville, near Paris, on August 8, 1750. No record survives of his education, but as a young man he worked for a time at the Bureau de Longitudes in Paris. Later he worked at the Paris Observatory, where he made astronomical and surveying instruments.

In 1800, Fortin made his first portable barometer. The instrument had a leather bag filled with mercury, a cistern containing a glass cylinder, a pointer to mark the level of the mercury, and a means of adjusting the mercury level to zero. Fortin did not invent any of these features, but he was the first instrument maker to combine all of them in one barometer.

Jean Fortin died in Paris in 1831.

Fourier, Jean-Baptiste-Joseph
(1768–1830)
French
Mathematician and Physicist

Jean Fourier was born on March 21, 1768, at Auxerre, France, a town to the southeast of Paris. His father was a tailor. Jean was orphaned when he was eight, but the bishop exerted his influence to have him admitted to the Auxerre military academy, where boys were educated to become artillery officers. It was there that Jean first encountered mathematics and demonstrated that he had an enthusiasm and great aptitude for the subject.

His humble origin meant he was unable to become an artillery officer, so when he left the academy he went to a Benedictine school in St. Bênoit-sur-Loire. He returned to Auxerre in 1784 and taught mathematics at the military academy. When the French Revolution began in 1787, Fourier took an active part locally, but he did not support the Terror that came later. Despite his initial support, Fourier was arrested in 1794, but was released after a few months, following the execution of Robespierre.

The École Normale opened in Paris in 1795, and Fourier taught there, acquiring such a reputation that it was not long before he was made professor of analysis at the École Polytechnique. In 1798, he was one of the learned advisers, called *savants*, chosen to accompany Napoleon on his campaign in Egypt. Fourier was made governor of lower Egypt and remained there until 1801.

On his return to France, Fourier was made prefect (chief administrator) of the département of Isère and lived at Grenoble, in the south of the country. In 1808, Napoleon conferred the title of baron on him and later made him a count. Fourier rejoined Napoleon in 1815, and after Napoleon's final defeat the following year he settled in Paris.

Fourier was elected to the Academy of Sciences in 1817, and in 1822 he became its joint secretary, sharing the position with the zoologist Georges Cuvier (1769–1832). He was also elected to the Académie Française and to foreign membership of the Royal Society of London.

Fourier had resumed his mathematical studies during the years he lived in Grenoble. He was particularly interested in the conduction of heat and sought to describe this in purely mathematical terms. He explained the theory he had devised to do so in *Théorie Analytique de la Chaleur*. Published in 1822 and translated into English (*Analytical Theory of Heat*) in 1872, this proved to be one of the most influential scientific books of the 19th century.

The rate of conduction varies with the temperature gradient, as well as with the composition of the material and the shape of the conducting body. In order to comprehend this, Fourier developed what came to be known as Fourier's theorem. This is a technique that allows the overall description to be broken down into a series of simpler, trigonometric equations, known as the Fourier series, the sum of which is equal to the original description. The Fourier series can be applied to any complex function that repeats and so it is of value in many branches of physics. It is widely used by meteorologists. Fourier also developed the use of linear partial differential equations for solving boundary-value problems. This, too, is relevant to numerical WEATHER FORECASTING.

His theorem and series were only part of Fourier's contribution to mathematics. He also investigated probability theory and the theory of errors.

Fourier had contracted an illness while he was in Egypt, which may explain his conviction that the heat of the DESERT was beneficial. He lived wrapped up in thick, warm clothes in overheated rooms. He died in Paris on May 16, 1830.

Franklin, Benjamin
(1706–1790)
American
Statesman, physicist, inventor, author, and publisher

One of the best-known and most admired men in the world during the second half of the 18th century, Franklin's achievements were summarized by the French economist Anne-Robert-Jacques Turgot (1727–81) in the words: "He snatched the lightning from the skies and the scepter from tyrants." To Americans he is known as one of the founding fathers, but he is no less famous in Europe. There, in his own day, he was known mainly as a natural philosopher—a person who would nowadays be called a scientist.

Franklin was a complex man, with many sides to his personality, but his attitude to science was expressed in a letter he wrote in 1780 to the English chemist Joseph Priestley (1733–1804): "The rapid progress true science now makes occasions my regretting sometimes that I was born too soon. It is impossible to imagine the height to which may be carried, in a thousand years, the power of man over matter. . . . O that moral science were in as fair a way of improvement, that men would cease to be wolves to one another, and that human beings would at length learn what they now improperly call humanity!"

Benjamin Franklin was born on January 17, 1706, in Boston, Massachusetts. His father, Josiah, a soap and candle maker, had emigrated from Banbury, Oxfordshire, in England. It was a large family—Benjamin was the 15th of 17 children. The family could not afford to send him to college, and he spent only one year at a grammar school. He was educated privately, but was mainly self-taught.

Franklin's greatest scientific achievements arose from his study of electricity. In 1745, the German physicist Ewald Jurgen (George) von Kleist (1700–48) had discovered a way to condense electric charges. His device was first investigated thoroughly by the Dutch physicist Pieter van Musschenbroek (1692–1761) at the University of Leiden (or Leyden) in the Netherlands, and so it became known as a Leyden jar. It consists of a glass jar lined on the inside with metal and sealed by a cork through which a metal rod is inserted. There were already machines that used friction to produce static electrical charges, and the Leyden jar was very good at storing these charges. The stored charge would be discharged as a spark if a hand was brought close to the metal rod, and if enough charge had accumulated, anyone approaching too closely would receive a strong electric shock. If a metal object was brought close to the rod, a spark would leap across the gap and there would be a loud crackling noise.

Like many other scientists, Franklin experimented with a Leyden jar. His observations led him to wonder whether the spark and crackle might not be a tiny demonstration of LIGHTNING and THUNDER and, therefore, whether during a THUNDERSTORM the sky and Earth might become a giant Leyden jar.

It was an idea that needed testing, and to do so Franklin performed the most famous of all his experiments—and by far the most dangerous. In 1752, he flew a kite in a thunderstorm. He had attached a pointed piece of wire to the kite and tied a long silken thread to the wire. At the bottom of the thread he tied a metal key. As the kite flew near the base of the storm cloud and lightning began flashing nearby, Franklin held his hand close to the key and a spark jumped from the key to his hand. Then he held a Leyden jar to the key, and it accumulated an electric charge. Franklin had proved that lightning is a discharge of electricity and thunder is the sound of the spark. In the same year, the French scientist Thomas-François d'Alibard (1703–99) also proved, independently of Franklin, that lightning is an electrical phenomenon.

Franklin was extremely lucky. The next two people to repeat his experiment were killed, and on no account should anyone try to perform it.

Franklin had also noticed that electricity sparks more readily and over a greater distance if there is a pointed surface toward which it can travel. This led him to suggest that buildings could be protected from damage by lightning if pointed metal rods were fixed to their roofs and connected to the ground—earthed—by metal wires. He had invented the lightning conductor, and these were soon being fitted throughout America. Franklin's lightning conductors saved many lives and much damage to property was avoided.

Franklin did make one mistake. It was known that there are two types of electricity and that they repelled or attracted each other, apparently on the principle of like repelling like and opposites attracting each other. This might be explained, Franklin thought, if electricity is some kind of fluid—a gas, perhaps—that can be present either in excess or in deficiency. Then, two bodies each of which contained either an excess or a deficiency of it would repel each other. If one body had an excess and the other a deficiency, on the other hand, they would attract each other and electricity would flow from the excess to the deficiency. This is wrong, of course, because electricity is not a fluid. We do retain a little of the terminology Franklin introduced, however. He proposed that an excess of the fluid be called "positive" electricity and a deficiency "negative" electricity.

At the end of a very long and very distinguished political, diplomatic, and scientific career, Benjamin Franklin retired to Philadelphia at the age of 79. Already ill, he was bedridden for the last year of his life. He died on April 17, 1790. He was given the most impressive funeral there had ever been in Philadelphia; in France, many eulogies were composed to the man the French regarded as the embodiment of freedom and enlightenment.

Fujita, Tetsuya Theodore
(1920–1998)
Japanese–American
Meteorologist

Tetsuya Fujita, who became one of the world's leading authorities on TORNADOES, was born on October 23, 1920, in the town of Sone, on the Japanese island of Kyushu. In April 1939, he enrolled at Meiji College of Technology, graduating in June 1943 with a degree in mechanical engineering. In September of that year he was appointed assistant professor of physics at Meiji College. In 1949, the college became the Kyuhsu Institute of Technology.

A few months after atomic bombs fell on Hiroshima and Nagasaki, Fujita visited those cities to survey the damage and from that to calculate the number of bombs that had been used and the height at which they had detonated. In 1947, he studied DOWNDRAFTS in THUNDERSTORMS, and in 1948 he made his first detailed study of a tornado. Also in 1948, Fujita married Tatsuko Hatano. He conducted his first study of a TROPICAL CYCLONE in 1949.

In May 1951, Fujita commenced his research for a doctorate in science at the University of Tokyo. He completed his research in 1952 and was awarded his D.Sc. in 1953. Fujita then accepted a two-year research appointment at the University of Chicago, where he arrived on August 13, 1953. He returned to Japan in 1955 to obtain an immigrant visa to the United States, and in 1956 he returned with his family to become research professor and senior meteorologist at the University of Chicago.

His first major tornado study was of one that had struck Fargo, North Dakota, in 1957. There were still photographs and movies of that tornado, as well as detailed records of the damage it caused. These allowed Fujita to describe the life cycle of the tornado. In 1962, Fujita was appointed associate professor at the university, and he became a professor of METEOROLOGY in 1965.

Fujita was divorced in 1968, and in the same year he became a United States citizen and added Theodore to his name. From that time he was known as Ted to his friends. In 1969, he married Sumiko Yamamoto, and in February 1971 she collaborated with him and Allen Pearson, formerly the chief tornado forecaster for the NATIONAL WEATHER SERVICE, in developing the FUJITA TORNADO INTENSITY SCALE.

In October 1971, after witnessing a large DUST DEVIL, Fujita proposed the existence of suction vortices (*see* TORNADO). Despite his intense interest in tornadoes and his deep understanding of them, it was not until June 12, 1982, that he saw one for the first time—in fact several of them, in Denver, Colorado. There was a serious air crash in June 1975, and after studying what had happened, Fujita began investigating other crashes in order to find a way of identifying and avoiding MICROBURSTS—at a time when many scientists denied their existence.

Ted Fujita retired in September 1990, but continued to study atmospheric phenomena, and especially tropical cyclones and El Niño (*see* ENSO). In 1995, the first of a series of illnesses weakened his health. He died in his sleep at his home in Chicago on November 9, 1998.

Galilei, Galileo
(1564–1642)
Italian
Physicist and astronomer

One of the most famous scientists who has ever lived, Galileo is usually identified by his given name rather than by his family name, Galilei. His father was Vincenzio Galilei (ca. 1520–91), a musician and mathematician. Galileo was born at Pisa on February 15, 1564.

He received his first lessons from a private tutor. Then, in 1574, the family moved to Vallombrosa, near Florence, and Galileo continued his education at a monastery there. In 1581, he enrolled to study medicine at the University of Pisa, but the family could not afford the expense, and in 1585 he returned home without having taken a degree.

While at the university, Galileo's interest had turned toward mathematics and physics, and by the time he had to leave he had started to study these subjects. A popular story has it that Galileo once watched a lamp that was swinging in the cathedral at Pisa and noticed that no matter how large the range of its swing, the lamp always took the same amount of time to complete an oscillation. He timed the swings by counting his pulse beat. Later in life he confirmed this observation experimentally and suggested that the principle of the pendulum might be applied to the regulation of clocks.

Following his return to Florence, Galileo obtained a post as a lecturer in mathematics and science at the

Florentine Academy, at the same time continuing his studies of Euclid (flourished ca. 300 B.C.E.) and Archimedes (287–212 B.C.E.). In 1586, Galileo invented an improved version of a balance first devised by Archimedes that was used to measure specific gravity. At about this time his father was measuring the lengths and tensions in the strings of musical instruments that produce particular intervals between notes, and this may have helped convince Galileo that mathematical descriptions of phenomena could be tested by experiment.

Galileo became professor of mathematics at the University of Pisa in 1589. The appointment was an honor for him, but it was poorly paid, and in 1592 he applied for and was awarded the better-paid post of professor of mathematics at the University of Padua. He remained at Padua for the next 18 years, and it is there that he did most of his best work.

As well as his experiments with gravity and motion and his astronomical observations and calculations, Galileo maintained the interest in the behavior of fluids that had begun with his studies of the work of Archimedes. In 1593, he invented the first THERMOMETER—called an air thermoscope. It consisted of a bulb filled with air that was connected to a vertical tube containing a column of water. As the temperature rose and fell, the air in the bulb expanded and contracted, pushing the water up the tube or allowing it to fall. Unfortunately, the air thermoscope was highly inaccurate, because no account was taken of changes in atmospheric pressure. Nevertheless, this was one of the earliest attempts to make an instrument for making scientific measurements. Toward the end of his life, Galileo became interested in discovering whether air is a physical substance having mass. A young assistant, Evangelista Torricelli (1608–47), set to work on the problem and the experimental apparatus he devised was the first BAROMETER.

Galileo had little time for those who disagreed with his observations or the arguments he based on them. He was combative and could be sarcastic, but with good reason. He was convinced that natural phenomena can be described mathematically and that observation and experiment can then be used to validate the mathematical description. He was establishing what is now accepted as the basis of scientific procedure, but to do so he had to overthrow the prevailing verbal and nonmathematical approach that was derived from the work of Aristotle.

Galileo's fame rests on three achievements. He was the first person to use a telescope to study the night sky. His observations provided evidence with which he supported the conclusion of Nicolaus Copernicus (1473–1543) that it is the Earth which orbits the Sun and not the reverse and therefore that the Sun, not the Earth, is at the center of what was then thought of as the whole universe. His studies of motion and gravitation outlined the principles Isaac Newton (1642–1727) later formalized as the laws of motion. His final achievement, and perhaps the most important, was his application of mathematics to the study of natural phenomena.

His support for the ideas of Copernicus led Galileo into conflict with the church, and in 1633 he was found guilty of heresy and sentenced to remain for the rest of his life in his villa at Arcetri, near Florence. He continued to study, experiment, and write, summarizing his early experiments and his thoughts on mechanics in *Discorsi e Dimonstrazione Matematiche Intorno a Due Nuove Scienze (Discourses and Mathematical Demonstrations Concerning Two New Sciences)*, a book that was smuggled out of Italy and published in Leiden, the Netherlands, in 1638.

Galileo became blind in 1637, but even this did not stop him working. He finally designed a pendulum-driven clock that was built in 1656 by Christiaan Huygens (1629–95) and he directed the work of his assistants. He was still dictating to them when he fell ill with a fever toward the end of 1641. Galileo died at Arcetri on January 8, 1642.

Galton, Francis
(1822–1911)
English
Scientist, inventor, and explorer

Sir Francis Galton was the first scientist to plot meteorological data onto a WEATHER MAP and to attempt to produce a SYNOPTIC chart showing the weather conditions over a wide area. He played a large part in preparing the daily weather charts that were published by *The Times* of London from data supplied by the METEOROLOGICAL OFFICE. This was one of several contributions he made to the scientific study of weather.

Galton was born into a Quaker family on February 16, 1822, near Sparkbrook, now a suburb of Birmingham, England, and was the youngest of the nine children of a wealthy banker. He was also a first cousin of Charles Darwin (1809–82). Francis was able to read before he was three years old, and by the time he was four he was studying Latin.

In response to his father's wishes Francis Galton studied medicine at Birmingham General Hospital and then at King's College, London, but he interrupted his medical studies to study mathematics at Trinity College, Cambridge. After that he resumed studying medicine at St. George's Hospital, London, but he never completed the course. His father died, and Francis inherited a fortune, so he left the hospital, and for the rest of his life he pursued whatever topic interested him.

First he traveled through the Balkans, Egypt, Sudan, and the countries at the eastern end of the Mediterranean. He spent 1850 and 1851 on a journey of 1,700 miles (2,735 km) through southwestern Africa, after which he visited Spain. His observations in what was then a little known part of Africa led to the award of the Gold Medal of the Royal Geographical Society in 1854, and in 1856 he was elected a fellow of the Royal Society.

His travels ended, in the 1860s Galton began to study the weather. In particular, he wondered whether it might be possible to detect large-scale patterns in the weather and, from these, to forecast weather. He circulated a detailed questionnaire to WEATHER STATIONS in different parts of the British Isles, asking for information about the weather conditions that had prevailed through the month of December 1861. When the replies arrived, he plotted them on a map, using symbols he invented for the purpose. In 1862, he finally succeeded in compiling a detailed weather map. This showed a previously unsuspected relationship between atmospheric pressure and the speed and direction of wind.

Galton was familiar with the work of Matthew Maury (1806–73) and Admiral FitzRoy (1805–65). He also knew that the French astronomer Urbain-Jean-Joseph Leverrier (1811–77) issued daily weather charts of the North Atlantic based on observations from ships and coastal stations, although these were so unevenly distributed that the charts included a great deal of guesswork. Maury, FitzRoy, and Leverrier had established the cyclonic circulation of air around a center of low pressure (see CYCLONE). Galton's questionnaire revealed its opposite: anticyclonic circulation around a center of high pressure. Galton

coined the term ANTICYCLONE, in a paper he submitted to the Royal Society. He published the results of his research in 1863 in a monograph titled *Meteorographica* (published by Macmillan) and summarized them, much later, in his 1908 book *Memories of My Life* (published by Methuen).

As well as preparing weather charts for publication, in *The Times* and in *Meteorographica*, Galton helped find a way to print them using movable type. He had typefaces designed for the purpose and modified a drawing instrument called a pantograph so it used a drill to score curves and arrows in a soft material that could be used to make casts for printing.

Meteorologists were beginning to study the upper air by means of small balloons and kites to which instruments were attached. Galton had the idea of measuring the speed and direction of the wind at a specified location and time by means of the smoke emitted by an exploding shell. Galton was closely associated with the Meteorological Office, and the shell was designed and fired experimentally under their auspices. The experiment was carried out over an area of the Irish coast where no ships might be damaged by falling debris, and it was very successful. The shells exploded consistently at 9,000 feet (2,745 m), releasing a cloud of smoke Galton was easily able to track. On the suggestion of FitzRoy, Galton also invented the WIND ROSE.

In addition to his contributions to METEOROLOGY, Francis Galton was the first person to demonstrate the uniqueness of fingerprints (though that uniqueness has still not been proved) and partly worked out a system for identifying them. He invented a teletype printer and the ultrasonic dog whistle, and also devised new techniques for statistical analysis, as well as a word-association test that was adopted by Sigmund Freud (1856–1939).

Galton's main interest, stimulated by his cousin's book *On the Origin of Species by Means of Natural Selection*, lay in measuring the mental abilities of people and determining the extent to which these were inherited. From this he hoped it might be possible to improve them by means of selective breeding. This project was called eugenics, and although it is now discredited Galton's contribution to it was of great importance to the scientific study of psychology.

Francis Galton was knighted in 1909. He died at Haslemere, Surrey, on January 17, 1911.

Gay-Lussac, Joseph-Louis
(1778–1850)
French
Chemist and physicist

Joseph Gay-Lussac was born on December 6, 1778, at St. Léonard, Haute Vienne, in central France. His father was a judge named Antoine Gay, who added Lussac to the name to avoid confusion with other families called Gay. Lussac was the name of an estate near St. Léonard. Antoine was arrested in 1793 for showing sympathy to the aristocrats.

Joseph began his education locally, and in 1797 he enrolled at the École Polytechnique in Paris, transferring in 1799 to the École des Ponts et Chaussées (School of Bridges and Highways), from which he graduated in 1800. His interest was in engineering, but while still a student Gay-Lussac began to assist the distinguished French chemist Claude-Louis Berthollet (1748–1822), working at Berthollet's home at Arcueil, near Paris. Berthollet was very famous, and his home was a meeting place for many of the leading scientists of the day. For a time, Gay-Lussac worked alongside Bertollet's son in a factory where linen was bleached.

In 1802, Gay-Lussac was appointed a demonstrator in chemistry at the École Polytechnique. He spent 1805 and 1806 on an expedition to measure terrestrial magnetism led by Alexander von Humboldt (1769–1859). In 1808, he married Geneviève Rojet, and on January 1, 1810, he became professor of chemistry at the École Polytechnique. He was also professor of physics at the Sorbonne University in Paris, a position he held from 1808 until 1832, when he resigned to take up the post of professor of chemistry at the Musée National d'Histoire Naturelle, in the Jardin des Plantes. In 1806, he was made a member of the Academy of Sciences.

Gay-Lussac was also a politician. In 1831, he was elected to the chamber of deputies to represent his home département of Haute-Vienne, and in 1839 he was made a peer and entered the chamber of peers, which was then the upper house of the French parliament.

Gay-Lussac published his first research results in 1802. In collaboration with his friend Louis-Jacques Thénard (1777–1857), Gay-Lussac had formulated a law stating that when the TEMPERATURE is increased by a given amount, all gases expand by the same fraction of their volume. Jacques Charles (1746–1823) had discovered this law in 1787, and it is usually known as Charles's law (*see* GAS LAWS), but Charles had not published it.

In 1804, Gay-Lussac and the physicist Jean-Baptiste Biot (1774–1862) were commissioned by the Academy of Sciences to measure the Earth's magnetic field high above the surface. On August 24, they ascended by balloon from the garden of the Conservatoire des Arts and climbed to 13,120 feet (4,000 m). On September 16, Gay-Lussac made a solo ascent in which he reached a height of 23,012 feet (7,019 m) above sea level. This was higher than the tallest peak in the Alps, and it established an altitude record that stood for 50 years. Measurements made during the flights showed not only that the magnetic field remains constant with height, but that the chemical composition of the atmosphere also does so. This discovery makes Gay-Lussac one of the founders of METEOROLOGY. The same year he read a paper describing research on a method of chemical analysis he had done in collaboration with von Humboldt. Using this method, they had found (among a number of other things) that the proportions of the volumes of HYDROGEN and OXYGEN in water were 2:1.

In 1809, Gay-Lussac published what may have been his most important discovery. He had found that when gases combine they do so in simple proportions by volume and that the products of their combination are related to the original volumes. This is known as Gay-Lussac's law and it is used in chemical equations. One of the examples Gay-Lussac used to illustrate it shows that when two molecules of CARBON MONOXIDE (CO) combine with one molecule of oxygen (O_2) the product is two molecules of CARBON DIOXIDE (CO_2):

$$2CO + O_2 \rightarrow 2CO_2$$

From about 1810, Gay-Lussac concentrated increasingly on pure chemistry. He made many important discoveries. These included improving the processes used to manufacture oxalic and sulfuric acids, devising ways to estimate the alkalinity of potash and soda, and the amount of chlorine in bleaching powder. He developed volumetric analysis, and in 1832 he introduced a method for estimating the amount of silver in an alloy by using common salt.

His advice was constantly in demand, and he held a number of official positions. In 1805, he was appointed to the consultative committee on arts and manufactures. In 1818, he was appointed to the department responsible for the manufacture of gunpowder, and in 1829 he became chief assayer to the Mint. Both of these positions were lucrative government appointments.

Joseph Gay-Lussac died in Paris on May 9, 1850.

Hadley, George
(1685–1768)
English
Meteorologist

George Hadley was born in London on February 12, 1685. He studied law and qualified as a barrister (under the English legal system, this is a lawyer who is permitted to appear as an advocate in the higher courts).

Hadley became increasingly interested in physics, however, and in particular in the physics of the atmosphere. He was placed in charge of the meteorological observations that were prepared for the Royal Society, a task he performed for at least seven years.

In 1686, Edmund Halley (1656–1742) had proposed an explanation for the reliability of the trade winds (*see* WIND SYSTEMS). Halley's theory was plausible, but it failed to account for the direction of the trade winds, which blow from the northeast in the Northern Hemisphere and from the southeast in the Southern Hemisphere. George Hadley aimed to complete the theory by explaining the direction of the winds. He agreed with Halley that hot air rises over the equator and is replaced by cooler air flowing toward the equator from higher latitudes, but he noted that the Earth itself is rotating in an easterly direction at the same time. Consequently, the moving air is deflected in a westerly direction, resulting in winds that blow in the directions that are observed. Two centuries passed before this deflection was described in detail by Gaspard de Coriolis. It is now known as the CORIOLIS EFFECT. In 1735, Hadley presented his account of the atmospheric movements that produce the trade winds to the Royal Society as a paper titled "Concerning the Cause of the General Trade Winds."

Hadley had produced the first model for the GENERAL CIRCULATION of the atmosphere. This supposed there is one large CONVECTION cell in each hemisphere. Warm air rises over the equator, moves to the pole where it subsides, and then returns to the equator. Although this model explains the easterly component of the trade winds, however, it fails to account for the

westerly winds that prevail in middle latitudes. Hadley's mistake was to assume there is just a single convection cell in each hemisphere. In fact, three sets of cells are included in the three-cell model of the atmospheric circulation. Also, meteorologists now know that the Coriolis effect is weak in low latitudes and does not exist at the equator, so it cannot be the full explanation for the direction of the trade winds. William Ferrel (1817–91) discovered the true cause of the deflection in the 19th century.

Despite these failings, George Hadley was one of the first people to develop a credible scientific description of the atmospheric circulation. His paper to the Royal Society aroused little interest at the time, and it was not until 1793, long after his death, that John Dalton (1766–1844) recognized Hadley's importance as a meteorologist. That contribution to METEOROLOGY is acknowledged to this day in the name of the Hadley cells.

George Hadley died at Flitton, Bedfordshire, on June 28, 1768.

Halley, Edmond (Edmund)
(1656–1742)
English
Astronomer

Edmond Halley was born at Haggerston, Shoreditch, which was then a village near London, on October 29 according to the Julian calendar, which was then in use, or on November 8 according to the Gregorian calendar, which is in use today, in the year 1656. He said he was born in the year 1656, but there is some doubt about this, partly because in 17th-century England the year began on March 25, and not January 1 as it does today.

The family was from Derbyshire, in the English Midlands, and Edmund's father, who was also called Edmund, had grown wealthy by making soap at a time when the use of soap was increasing throughout Europe. He was murdered in 1684.

Edmund senior could afford a good education for his son, and after employing a tutor to teach him at home, sent the boy to St. Paul's School. Young Edmund excelled in Latin and Greek and in mathematics, and he displayed a talent for devising and making scientific instruments. In 1673, he entered Queen's College at the University of Oxford, where he wrote a book on the laws of Johannes Kepler (1571–1630). This drew him to the attention of John Flamsteed (1646–1719), the Astronomer Royal.

Halley left Oxford without taking a degree, a practice that was not unusual in those days. Flamsteed employed him as an assistant and then helped him launch his career by spending two years on the island of Saint Helena, charting the stars of the Southern Hemisphere. The project was financed partly by Halley's father and partly by King Charles II. In 1678, after his return from Saint Helena, Halley was elected to the Royal Society.

His reputation grew rapidly. On December 3, 1678, Oxford University awarded him a degree without requiring him to take the examination, and in 1679 the Royal Society sent him to Danzig (now Gdansk) to resolve a disagreement between Robert Hooke (1635–1703) and the German astronomer Johannes Hevelius (1611–87). Hevelius, a close friend of Halley's, had made certain astronomical observations without using telescopic sights, and Hooke claimed that the observations could not be accurate.

Halley became friendly with Isaac Newton (1642–1727) and financed the publication of Newton's major work, *Philosophae Naturalis Principia Mathematica* (Mathematical principles of natural philosophy). Halley's success alienated Flamsteed, who became increasingly hostile toward him.

In about 1695, Halley began making a careful study of the ORBITS of comets. He calculated that the comet that had appeared in 1682 was the same object that had been seen in 1531 and 1607, and in 1705 he predicted that it would be seen again in December 1758. When it appeared on December 25, 1758, it was given his name. It is still known as Halley's comet.

Although astronomy was his principal interest, Halley also studied TIDES, winds, and weather phenomena. In 1686, he attempted an explanation of the trade winds (*see* WIND SYSTEMS) by proposing that air is heated more strongly at the equator than it is elsewhere. The warm air rises and draws in cooler air from higher latitudes, flowing toward the equator.

For a time Halley commanded the *Paramore Pink*, a warship, exploring the coasts on both sides of the Atlantic, charting variations in compass readings, and investigating the tides along the English coast. He also inspected harbors in southern Europe on behalf of Queen Anne.

In 1704, Edmund Halley was appointed Savilian professor of geometry at Oxford, and in 1720, he succeeded Flamsteed as Astronomer Royal. Flamsteed's widow was so angry at this that she arranged for all of her husband's instruments to be removed from the Royal Observatory and sold, to prevent Halley from using them.

Edmund Halley died in Greenwich, London, on January 14, 1742, by the Julian calendar and January 25, 1742, by the Gregorian calendar. His tombstone bears an inscription stating that he died in 1741, but this is also correct, because of the change in the date of New Year's Day.

Further Reading
Jardine, Lisa. *Ingenious Pursuits*. London: Little, Brown, 1999.

Helmont, Jan Baptista van
(1577–1644)
Flemish
Physician and alchemist

Jan van Helmont discovered the existence of gases and claimed to have coined the word *gas*. He also identified the gas we know as CARBON DIOXIDE.

Van Helmont was born in Brussels (his date of birth is not known, but it was in the year 1577), and he was educated at Louvain. He studied several sciences, finally concentrating on medicine, in which he graduated in 1599. In 1609, van Helmont moved to Vilvorde, near Brussels, where he spent the rest of his life practicing medicine and conducting chemical experiments.

In his work and ideas van Helmont bridged the medieval and modern worlds. He was a mystic, interested in the supernatural, and an alchemist who believed in the philosopher's stone—the object alchemists believed would change base metals such as lead into gold. He even claimed to have seen it used. Van Helmont believed in spontaneous generation, by which living organisms develop from nonliving materials, and he asserted that mice are produced from dirty wheat.

At the same time, Jan van Helmont was in touch with the scientific developments of his day, and his experiments were performed carefully and accurately. In one experiment he grew a willow tree in a measured quantity of soil to which he added only water. At the end of five years, he found the tree had gained

164 pounds (74.5 kg) in weight, but the soil had lost only two ounces (57 g). Van Helmont concluded that the tree was converting water into its own tissues. This was incorrect, but it was an early example of a carefully quantified experiment in biology, and it did prove that plants do not draw their principal nourishment from the soil.

Some of his experiments produced vapors, and van Helmont was the first person to recognize that these are distinct substances, each with its own properties. Unlike liquids and solids, vapors immediately fill any space they enter. He thought this meant they existed in a state of chaos, a word he spelled *gas*, which is the way it sounded when spoken by a Flemish speaker.

Charcoal gives off a gas when it is burned, and because the gas comes from wood, for which the Latin word is *silva*, van Helmont called it "gas sylvestre." He discovered that the same gas is given off when malted barley is fermented to make beer. It is the gas we call carbon dioxide.

Jan van Helmont continued experimenting and working as a physician until his death, on December 30, 1644.

Henry, Joseph
(1797–1878)
American
Physicist

As a boy, Joseph Henry showed little interest in school. He was born in Albany, New York, on December 17, 1797. His family was poor, and at 13 Joseph left school and was apprenticed to a watchmaker.

Henry might have remained a watchmaker, but for a curious event that happened when he was 16 and spending a vacation on a farm owned by a relative. A rabbit he was chasing ran beneath a church, and Joseph crawled under the building to follow it. Some of the floorboards above him were missing, so he climbed through the gap and into the church, where he came across a shelf of books. One, called *Lectures on Experimental Philosophy*, attracted him. He started to read it and was inspired to return to school.

Joseph Henry enrolled at Albany Academy to study chemistry, anatomy, and physiology, paying his way by teaching in country schools and tutoring. He graduated, hoping to practice medicine, but in 1825 he obtained a job surveying a route for a new road in New York State. This aroused his interest in engineering, but in 1826 he returned to Albany Academy to teach mathematics and science—which was then known as natural philosophy.

Electricity and magnetism were research topics of great interest in Europe, and Henry began experimenting with electromagnets. He was the first American since Benjamin Franklin to undertake important experimental work with electricity. He was the first person to insulate wire for the magnetic coil, using silk from one of his wife's petticoats, and he invented spool winding. In 1829, Joseph Henry made the first electromagnetic motor, and by 1831 he was able to make a range of electromagnets, from very delicate ones to one that could lift 750 pounds (340 kg) and another that lifted more than one ton.

In 1832, Joseph Henry was appointed a professor at the College of New Jersey (now Princeton University). The same year Henry discovered self-induction—the phenomenon in which an electric current flowing through a wire coil induces a second current in the coil, so the current becomes the original and induced currents combined. Henry had read a preliminary account of work on induction by the British physicist and chemist Michael Faraday (1791–1867). Henry was the first to perform the key experiments, but Faraday was the first to publish his results. Nevertheless, at a meeting in Chicago in 1893, the Congress of Electricians named the henry as the unit of inductance. It is now the SI unit of inductance (*see* UNITS OF MEASUREMENT).

This work gave Henry another idea. Suppose there were a very long wire with a battery at one end, an electromagnet at the other, and a key that would open and close the circuit. Every time the key closed the circuit a current would reach the electromagnet, which would attract a small iron bar, causing it to move. When the key was released, opening the circuit, the attraction would cease and a small spring would pull the iron bar back to its original position. When the iron bar moved, it would make an audible click, so if the key were pressed repeatedly to make a pattern of taps, the same pattern would be heard as clicks from the iron bar. By 1831, Henry had made this device and it was working. He had invented the telegraph, and in 1835 he invented the relay. This was a series of similar circuits in which each circuit activated the next. It overcame the diminution in the signal passing through a length of wire that is caused by resistance in the wire itself.

Joseph Henry believed that scientific discoveries should benefit everyone. He did not patent his inventions, and he was happy to describe them in detail. Consequently, it was Samuel Morse (1791–1872) who made the first practical telegraph line and who patented the telegraph.

In December 1846, Joseph Henry was elected the first secretary of the Smithsonian Institution, which had only recently been founded. An excellent administrator, he turned the institution into a major clearinghouse for scientific information.

Henry mobilized scientific effort during the American Civil War. He was elected the second president of the National Academy of Sciences and was active in organizing the American Association for the Advancement of Science and the Philosophical Society of Washington.

One of his first projects at the Smithsonian reflected another of his interests: METEOROLOGY. Henry established a corps of voluntary observers, located all over the United States, who used the telegraph to send weather reports to a central office at the Smithsonian. This was the first use of the telegraph for a scientific purpose. For the next 30 years, Henry supported and encouraged this volunteer corps. When the UNITED STATES WEATHER BUREAU was established in 1891, it used the system that Henry had devised for collecting data.

Joseph Henry died in Washington on May 13, 1878. His funeral was attended by many government officials and by Rutherford B. Hayes, president of the United States.

Hero of Alexandria
(ca. 60 C.E.)
Greek or Egyptian
Engineer

Hero lived in Alexandria, Egypt, and it is not certain whether he was Egyptian or Greek; there was a flourishing Greek community in Alexandria at that time.

Greek or Egyptian, Hero was a gifted mathematician and teacher. He founded a school, part of which was devoted to research. He was also an inventor. He devised systems of gears for lifting heavy objects, and used a suction machine to raise water. Hero also invented the first steam engine, although it was no more than a novelty and not strong enough to perform useful work. The device consisted of a hollow sphere that was made to spin rapidly by the pressure of steam from boiling water.

Hero demonstrated that air is a substance. He believed it is made from minute particles, and he discovered that it can be compressed. This idea had to wait 1,600 years before being developed by Robert Boyle (1627–91) and his contemporaries.

Hooke, Robert
(1635–1703)
British
Physicist

One of the most ingenious experimenters and instrument makers who have ever lived, Robert Hooke was born on July 18, 1635, at Freshwater, in the Isle of Wight, which is an island off the southern coast of England. His father, John Hooke, was a clergyman, employed as the curate at Freshwater. A curate is an assistant to the parish priest; it is a poorly paid occupation.

John wanted his son to enter the church, but as a child Robert was often ill and he was not strong enough to undertake the necessary studies. He spent much of his time alone, amusing himself by making mechanical toys. Later, he attended school in Oxford, and after the death of his father in 1648 Robert was sent to London. At first he seems to have embarked on an apprenticeship, for which he paid the £100 he inherited from his father. Then he enrolled at Westminster School. He learned Latin and Greek, although he never wrote in either, but his best subject was mathematics, in which he excelled. It took him only a week to master the geometry of Euclid. In 1653, he entered Christ Church College, at the University of Oxford. At first he was a chorister and then he became a servitor. This was an undergraduate student who was assisted financially from the college funds, in return for which he performed certain menial duties. Robert was already a highly skilled instrument maker, and he earned a living by selling ideas for modifications and improvements to the owners of the professional workshops where scientific instruments were made.

He did not take a degree, but while he was at Oxford Hooke joined a group of brilliant scientists, one of whom was Robert Boyle (1627–91). Boyle was a wealthy man and employed Hooke as an assistant, paying him generously. The two became close lifelong friends, and although Hooke ceased to be employed by Boyle in 1662, Boyle continued to pay him until 1664, when Hooke was in a better financial position.

Hooke's first task, in 1658, was to help design a pneumatic pump that would remove air from a vessel into which animals and scientific instruments could be placed and the effect on them observed. Hooke succeeded, and Boyle used the pump for his experiments on gases.

In 1659, the group of friends began to separate. Most of them, including Boyle and Hooke, moved to London where, in 1660, they formed a scientific society. In 1662, this became the Royal Society of London. Hooke was appointed curator of experiments for the society, a position that required him to demonstrate new experiments at each weekly meeting. He was paid £30 a year and was provided with accommodation in Gresham College, in London. In 1663, he was elected a fellow of the society, and in the same year he was awarded an M.A. degree from Oxford. In 1664, he was made a lecturer in mechanics at the Royal Society, for which he received £50 a year, and in 1665 he was appointed professor of geometry at Gresham College, also at a salary of £50 a year. He remained in these two posts for the rest of his life. From 1677 until 1683, he was secretary to the Royal Society.

His work for the society, and especially his job as curator of experiments, involved him in every branch of science and allowed him to develop to the full his mechanical skills. It also meant, however, that although he originated many ideas and improved other people's inventions, he left many projects unfinished.

Despite that, his achievements were considerable. He found that the stress placed on an elastic body is proportional to the strain it produces. This is known as Hooke's law. He invented an anchor escapement mechanism for a watch and claimed it was the first, although that led to a dispute over priority with Christopher Huygens (1629–95). He also claimed to have discovered gravitation and the INVERSE SQUARE LAW before Isaac Newton (1642–1727). He greatly improved the design of microscopes, and his book *Micrographia*, published in 1665, contained the first important set of drawings of microscopic observations. He insisted that fossils were the remains of plants and animals that had lived long ago.

Hooke invented the wheel barometer, which indicated the pressure by means of a needle on a dial, and he suggested ways to apply BAROMETER readings to

WEATHER FORECASTING. It was Hooke who first labeled a barometer with the words "change," "rain," "much rain," "stormy," "fair," "set fair," and "very dry." He designed, but did not make, a weather clock to record air TEMPERATURE, PRESSURE, rainfall, HUMIDITY, and WIND SPEED on a rotating drum. He suggested that the freezing point of water be used as the zero reference point on THERMOMETERS.

Hooke died in London on March 3, 1703.

Further Reading

Jardine, Lisa. *Ingenious Pursuits*. London: Little, Brown, 1999.

Howard, Luke
(1772–1864)
English
Chemist, pharmacist, and meteorologist

Luke Howard devised a method for classifying clouds that formed the basis of the system that was adopted internationally and that remains in use to this day. His was not the only attempt at classification, and it was far from being the first. At about the same time as Howard was developing his scheme, the Chevalier de Lamarck (1744–1829) also proposed a scheme (*see* CLOUD CLASSIFICATION) and developed it over several years. Lamarck was a leading scientific figure and an authority on biological classification, but his cloud classification was based on rather vague definitions. Whether for that reason or some other, his scheme was not adopted, and Howard's was.

Luke Howard was born in London into a Quaker family and was educated at a Quaker school at Burford, near Oxford. He was educated as a pharmacist (druggist), and when he grew up he earned his living as a manufacturer of chemicals. He was a businessman and chemist, but he was never a professional scientist.

His interest in METEOROLOGY was kindled in the summer of 1783, when he was 11 years old. That year there were two major volcanic eruptions, one in Iceland and the other in Japan. Volcanic dust formed a haze around the world that produced spectacular skies. Then, on August 18, Luke witnessed a dramatic meteor blaze across the sky. He began to keep a record of his meteorological observations and maintained the habit for more than 30 years.

Howard was a founder member of the Askesian Society, which was one of the many philosophical societies (philosophy was the name given to what we now call science) formed in the late 18th and early 19th centuries. At a meeting of the society in 1800, he presented a paper on "The Average Barometer," and in 1802 he presented another paper on "Theories of Rain."

Howard became interested in biological classification, which was then being strongly influenced by the work of the Swedish naturalist Linnaeus (Carl Linné, 1707–78). In 1735, Linnaeus had published a book, *Systema Naturae,* in which he introduced a system based on a binomial nomenclature, in which every species was given two names, one of its genus and the other of the species itself. In 1803 (although some historians say it was in December 1802), Howard presented a paper to the Askesian Society, "On the Modification of Clouds." He wished to describe the way clouds change from one form to another, but to do so he needed names by which to identify each type. Howard followed Linnaeus in allotting Latin names to cloud types, and he suggested ways of combining the names in a Linnaean, binomial fashion.

Howard proposed that all clouds belong to one of three groups. He called these "cumulus," "stratus," and "cirrus." A fourth group, called "nimbus," denoted a cloud that was producing RAIN, HAIL, or SNOW. *Cumulus,* the Latin word for "heap," Howard described as "convex or conical heaps, increasing upward from a horizontal base—Wool bag clouds." *Stratus,* the past participle of the Latin verb *sternere,* to strew, and meaning "layer," he described as "a widely extended horizontal sheet, increasing from below." *Cirrus,* the Latin word for a "curl," he described as "parallel, flexuous fibers extensible by increase in any or all directions." *Nimbus,* the Latin for "cloud," but to which Howard attached the meaning "rain," he described as "a rain cloud—a cloud or system of clouds from which rain is falling."

Howard maintained that rain could not fall from cumulus, stratus, or cirrus as long as they "retain their primitive forms." Clouds can alter their forms, however. Cumulus clouds could fill the sky to become cumulo-stratus, which is "cirro-stratus blended with cumulus." Similarly, cirrus could become cirro-cumulus and cirro-stratus.

His classification attracted widespread attention, and Howard became a celebrity, his fame increasing even more when all his meteorological papers up to that time were collected by Thomas Forster and published in 1813 as *Researches About Atmospheric Phaenomenae.* Wolfgang von Goethe even dedicated four poems to Howard.

In 1806, Howard began a publication called the *Meteorological Register,* which appeared regularly over several years in the *Athenaeum Magazine.* In 1818–19, he published the first book ever written on urban climate: *The Climate of London,* in two volumes, and in 1833 published as an expanded second edition in three volumes. In it he made what is believed to be the first reference to what is now called a HEAT ISLAND, with temperature records to support it. A heat island is an urban area that is warmer than the surrounding countryside. A series of lectures he delivered in 1817 were published in 1837 as *Seven Lectures in Meteorology* and became the first textbook on the subject. His last book, *Barometrographia,* appeared in 1847.

In recognition of his contributions to meteorology, in 1821 Howard was elected a fellow of the Royal Society of London, the highest honor British scientists can bestow on one of their colleagues.

Luke Howard remained a devout Quaker throughout his life. He died at a great age in London in 1864.

Further Reading

Hamblyn, Richard. *The Invention of Clouds.* New York: Farrar, Straus, and Giroux, 2001.

Humboldt, Friedrich Heinrich Alexander, Baron von
(1769–1859)
German
Geologist, geophysicist, geographer

Alexander von Humboldt was born in Berlin on September 4, 1769. Berlin was then the capital of Prussia, and von Humboldt's father was a Prussian officer who served as an official at the court of the king, Frederick II (Frederick the Great) and who wanted his son to pursue a political career. Alexander was more interested in science, however, and following the death of his father in 1779, he was educated privately before enrolling at the University of Göttingen in 1789 to study science.

While a student at Göttingen, von Humboldt met Georg Forster (1754–94), who had accompanied James Cook (1728–79) on the second of his voyages of exploration. The two became firm friends, and Humboldt spent only one year at Göttingen before he and Forster set off on a journey through the Netherlands and England, where he met many leading scientists.

On his return to Prussia, Humboldt realized he would need a formal qualification if he were to make any useful contribution to science. In 1791, he became a student at the Freiburg Bergakademie (School of Mining). He spent two years there before graduating in geology, and while studying mining he became fascinated by the plants that grow in and around mines.

After graduating from the Bergakademie, Humboldt was appointed assessor of mines and later director of mines in the Prussian principality of Bayreuth. He founded a school of mining, improved conditions for the miners, and also conducted his own research into the magnetic declination of the rocks in the area. He spent the years from 1792 to 1797 on a diplomatic mission that took him to the salt-mining regions of several central European countries. In the course of these travels he met more of the most senior scientists of the day.

Humboldt's mother died in 1796, and Alexander inherited a share of the family fortune. This meant he no longer needed to earn a living and could indulge his passion for travel. He went first to Paris and from there to Marseilles, accompanied by the French botanist Aimé Bonpland (1773–1858). They planned to travel to Egypt, where they hoped to join Napoleon, but instead they went to Madrid, where the prime minister, Mariano de Urquijo, became their patron. With his support they changed their plans and determined to visit the Spanish colonies in South America.

The two friends sailed from Spain in 1799, landed in New Andalusia (modern Venezuela), and early in 1800 they started on a four-month expedition through Latin America. They explored the course of the Orinoco and confirmed that the headwaters of the Orinoco were linked to those of the Amazon. By the time they returned to their base on the coast, at Cumaná, they had traveled 1,725 miles (2,775 km). They then sailed to Cuba and stayed there for several months before returning to South America in March 1801, arriving at Cartagena, Colombia.

They then embarked on a second expedition, this time on a route that crossed the Andes. As they climbed, Humboldt noted the changes in vegetation at different elevations and recorded the decrease in air TEMPERATURE with height. He also made many geophysical observations of the alignment of VOLCANOES and of the Earth's magnetic field. When they reached the Pacific coast, Humboldt measured the temperature

of the water offshore and discovered the existence of the cold current that is sometimes named after him, but nowadays more usually known as the Peru Current (*see* APPENDIX VI: OCEAN CURRENTS).

Humboldt and Bonpland left South America in February 1803, spent a year in Mexico, visited the United States, and sailed for Europe on June 30, 1804. In the course of their explorations the two men had covered about 6,000 miles (9,600 km).

Humboldt spent the following years in Berlin and Paris arranging the vast amount of material he had collected during his travels—including 60,000 plant specimens, many of which were new to science—and writing accounts of his experiences and discoveries. His major work, *Voyage de Humboldt et Bonpland*, appeared in 30 volumes between 1805 and 1834. Most of his fortune had now been spent, and in order to secure an income Humboldt agreed to serve as a Prussian diplomat in Paris.

Humboldt was one of the founders of biogeography, the study of the geographic distribution of plants and animals. He was the first scientist to measure the decrease of temperature with altitude, and he investigated the cause of tropical storms. This supplied information other scientists used later to determine the processes involved in the weather systems of middle latitudes.

His many discoveries and his liberal opinions had made Humboldt a celebrity. He was said to be the second most famous man in Europe, after Napoleon. He died in Berlin on May 6, 1859, at the age of 89 and was given a state funeral.

Keeler, James Edward
(1857–1900)
American
Astrophysicist

James Keeler studied the rings of Saturn, the surface of Mars, and nebulae, but he began his career as an assistant to Samuel Pierpont Langley (1834–1906).

Keeler was born at La Salle, Illinois, on September 10, 1857. He did not attend school between the ages of 12 and 20, but during these years he learned as much as he could about astronomy. A wealthy benefactor made it possible for him to enroll at Johns Hopkins University, where he graduated in 1881.

In 1881, Keeler began to work for Langley and accompanied him on the ascent of Mount Whitney,

California, to measure the intensity of solar radiation at the summit and compare it with measurements they had made at the base of the mountain. In 1883, Keeler went to Germany to study at the Universities of Heidelberg and Berlin. He returned to the United States in 1884. He was appointed astronomer at the newly built Lick Observatory, on Mount Whitney, in 1888, became director of the Allegheny Observatory in 1891, and in 1898 he returned to the Lick Observatory as its director. Keeler was elected a fellow of the Royal Astronomical Society of London in 1898 and a member of the National Academy of Sciences in 1900.

He died unexpectedly on August 12, 1900, in San Francisco.

Keeling, Charles David
(1928–2005)
American
Geochemist

Charles Keeling devised a technique for measuring the atmospheric concentration of CARBON DIOXIDE and used it to show that the concentration was rising. Until that time, the idea that the burning of coal and oil might alter the amount of carbon dioxide in the air was largely supposition. Svante Arrhenius and others had shown this was possible, but scientists believed any surplus would dissolve into the oceans and remain there. Keeling's precise monitoring of carbon dioxide levels demonstrated for the first time that carbon dioxide was accumulating in the atmosphere. The present demand to curb carbon dioxide emissions arises very largely from Keeling's work.

Charles Keeling was born at Scranton, Pennsylvania, on April 20, 1928. He studied chemistry and isotope chemistry at the University of Chicago and was awarded his Ph.D. from Midwestern University in 1954.

Roger Revelle, director of the Scripps Institution of Oceanography, of the University of California, San Diego, offered Keeling a job, and in 1956 Keeling moved to the Scripps, where he remained for the rest of his career, moving away only for short secondments. In 1961–62, Keeling was a Guggenheim Fellow at the

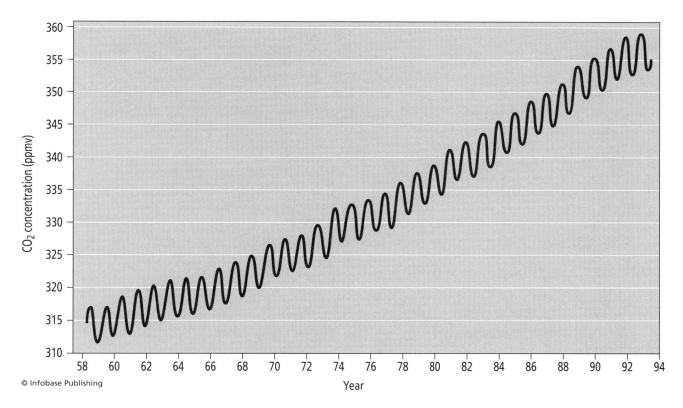

© Infobase Publishing

The Keeling curve, showing carbon dioxide concentrations at Mauna Loa, Hawaii

Meteorological Institute, University of Stockholm, Sweden; in 1969–70, he was a guest professor at the University of Heidelberg, Germany; and in 1979–80, he was a guest professor at the Physical Institute of the University of Bern, Switzerland.

Revelle and his colleague Hans Suess had realized that the surface water of the oceans mix very slowly with deeper water, thus limiting the capacity of the oceans as a sink for dissolved carbon dioxide. Measuring the atmospheric concentration of carbon dioxide suddenly became important. That is the background against which Keeling developed his technique. There is a widely publicized graph showing the steady rise in atmospheric carbon dioxide measured at the Mauna Loa mountain station in Hawaii. The graph is often called the "Keeling curve."

In 2002, President George W. Bush presented Keeling with the National Medal of Science, America's highest award, for a lifetime achievement in scientific research. Keeling received many other awards, including the 1993 Blue Planet Prize from the Science Council of Japan and the Asahi Foundation, and the 2005 Tyler Prize for Environmental Achievement.

Charles David Keeling died from a heart attack at his home in Montana on June 20, 2005.

Further Reading

Helmann, Martin. "Obituary: Charles David Keeling (1928–2005)." *Nature,* 437, 331, September 15, 2005.

Scripps Institution of Oceanography. "Charles David Keeling, Climate Science Pioneer (1928–2005)." Scripps Institution. Available online. URL: http://scrippsnews.ucsd.edu/article_detail.cfm?article_num=687. Accessed March 23, 2006.

Kepler, Johannes
(1571–1630)
German
Astronomer and mathematician

Johannes Kepler calculated the ORBITS of the solar system planets. His three laws of planetary motion showed that: (1) Planets follow elliptical orbits with the Sun at one of the two foci; (2) A line from the Sun to a planet crosses equal areas of space in equal periods of time; and (3) The square of the orbital period of a planet is proportional to the cube of its mean orbital radius from the Sun.

Kepler had less well known claims to fame. He was the author of the first ever science fiction story. He was also fascinated by SNOWFLAKES.

Kepler was born at Weil der Statt, Württemberg, on December 27, 1571. He attended schools in Weil, Leonberg, and Adelberg before enrolling at the University of Maulbronn, where he graduated in theology in 1588. He moved to the University of Tübingen in 1589 to study philosophy, mathematics, and astronomy, obtaining his master's degree in 1591. In 1594, he was appointed professor of mathematics at the University of Graz.

Those were troubled times. Kepler was a Protestant, and in 1598 a religious purge drove him from Graz. He spent a year in Prague before returning, only to be expelled once more, so he went back to Prague. This time he stayed there, in 1600 becoming an assistant to the aging Danish astronomer Tycho Brahe (1546–1601). His meticulous study of Brahe's voluminous records led Kepler to formulate his laws of planetary motion.

In 1611, Johannes Kepler wrote his science fiction story. *Somnium* was about the adventures of a man who travels to the Moon. The story was not published until 1631, after Kepler's death.

It was also in 1611 that Kepler published a description of snowflakes, called *A New Year's Gift, or On the Six-cornered Snowflake.* He dedicated the book to his patron and friend Matthäus Wackher.

Johannes Kepler died at Regensburg (then called Ratisbon), Bavaria, on November 15, 1630.

Kirchhoff, Gustav Robert
(1824–1887)
German
Physicist

Gustav Kirchhoff, the physicist who discovered BLACK-BODY radiation, was born at Königsberg, Prussia (now Kaliningrad, Russia), on March 12, 1824. He was educated at the University of Königsberg, and in 1854 he became professor of physics at the University of Heidelberg.

At Heidelberg Kirchhoff met Robert Wilhelm Bunsen (1811–99), the inventor of the laboratory burner named for him. Bunsen was a gifted inventor, and he and Kirchhoff collaborated in developing the spectroscope. That instrument allowed them to analyze compounds by the light spectra they emitted when heated with the Bunsen burner.

Gustav Kirchhoff also calculated that a perfect blackbody, absorbing all the radiation falling upon it, would emit radiation at all wavelengths if it were heated to incandescence. The Scottish physicist Balfour Stewart (1828–87) reached the same conclusion independently at about the same time, but Kirchhoff is usually credited with the discovery.

Gustav Kirchhoff died in Berlin on October 17, 1887.

Köppen, Wladimir Peter
(1846–1940)
German
Meteorologist and climatologist

Wladimir Köppen was born in St. Petersburg, Russia, on September 25, 1846. His parents were German, and after attending school in the Crimea, he returned to Germany to study at the Universities of Heidelberg and Leipzig. While he was at school in Russia, Wladimir first became interested in the natural environment and especially in the interaction between plants and climate. His student dissertation, which he completed in 1870, dealt with the relationship between TEMPERATURE and plant growth.

Following his graduation, from 1872 to 1873 he was employed in the Russian meteorological service. In 1875, he returned once more to Germany to take up an appointment as chief of a new division of the Deutsche Seewarte, based in Hamburg. His task there was to establish a WEATHER FORECASTING service covering northwestern Germany and the adjacent sea areas. His primary interest lay in fundamental research, however, and he was able to devote himself to it from 1879, once the meteorological service was functioning.

Köppen embarked on a systematic study of the climate over oceans and also investigated the upper air, using kites and balloons to obtain data. In 1884, he published the first version of his map of climatic zones (*see* KÖPPEN CLIMATE CLASSIFICATION). He plotted these on an imaginary continent he called "Köppen'sche Rübe" ("Köppen's beet"). His CLIMATE CLASSIFICATION appeared in full in 1918, and after several revisions the final version of it was published in 1936.

In addition to writing hundreds of articles and scientific papers, Köppen co-authored with Alfred Wegener (1880–1930) *Die Klimate der Geologischen Vorzeit* (The climates of the geological past), published in 1924, and wrote *Grundriss der Klimakunde* (Outline of climate science), which was published in 1931. In 1927, he entered into collaboration with Rudolf Geiger to produce a five-volume work, *Handbuch der Klimatologie* (Handbook of climatology). This was never completed, but several parts, three of them by Köppen, were published.

Köppen had moved to Graz, Austria, to work on the *Handbuch der Klimatologie*, and it was there that he died, on June 22, 1940.

Lamb, Hubert Horace
(1913–1997)
English
Climatologist

One of the first scientists to draw attention to the variability of climates and the social and economic consequences of climate change, Hubert Lamb was arguably the greatest climatologist of the 20th century. In addition to his studies of climate change and the history of climates, Lamb was among the most skillful of weather forecasters.

Hubert Lamb was born at Bedford and educated at Oundle School, in Northamptonshire, and then at Trinity College, Cambridge, where he studied natural sciences and geography. He graduated in 1935 and received a master's degree in the same subjects in 1947. In 1981, he was awarded honorary doctorates from the Universities of Dundee (LL.D.) and East Anglia (D.Sc.), and in 1982 Cambridge University awarded him a doctorate of science.

In 1936, following his graduation from Cambridge, Lamb went to work at the British METEOROLOGICAL OFFICE. As war became increasingly probable, Lamb was asked to study the way clouds of poison gas would be carried by the wind. He refused to do this on moral grounds, and in 1940 he was transferred to the Irish Meteorological Office. In 1941, he was placed in charge of preparing forecasts for transatlantic flights. His forecasts were so accurate that during his period at the Irish Meteorological Office transatlantic flights out of Ireland had a perfect safety record. Eventually, he had a disagreement with the director, however, and in 1945 he returned to the Meteorological Office in Britain.

Lamb sailed as meteorologist on a Norwegian whaling expedition to Antarctica in 1946–47, and during this voyage he came to realize the extent to which climate changes over time. He served as a weather forecaster in Germany from 1951 to 1952, and from 1952 until 1954 he worked in Malta.

During all of this time, from 1945 until 1971, Hubert Lamb was employed by the Meteorological Office and devoted part of his time to long-range WEATHER FORECASTING and studies of global weather and climate change. In 1954, he was placed in charge of the Climatology Division at the Meteorological Office. While there, he undertook the first detailed study of the past climate records held at the office. He used these to trace ways the climate had changed since the middle of the 18th century and devised a classification system, known as LAMB'S CLASSIFICATION, for British weather.

Hubert Lamb also came to realize that DUST injected high into the air by volcanic eruptions can affect surface weather by reflecting incoming solar radiation back into space, thereby shading the surface. This led him to study every eruption since 1500 and to estimate the amount of dust each had released and its effect on weather in the years following. From this he developed a method for estimating the climatic effect of volcanic dust, known as LAMB'S DUST VEIL INDEX.

In 1971, Lamb left the Meteorological Office to establish the Climatic Research Unit at the University of East Anglia, in Norwich. He remained director of the unit until his retirement in 1977, and after retiring he remained at the university as an emeritus professor. Under his direction the unit became one of the foremost centers for the study of climate change.

Hubert Lamb died in Norwich on June 28, 1997.

Further Reading

Lamb, Hubert. *Through All the Changing Scenes of Life, A Meteorologist's Tale*. East Harling, Norfolk: Taverner Publications, 1997.

Langley, Samuel Pierpont
(1834–1906)
American
Astronomer and physicist

Using an instrument of his own invention, Langley was the first scientist to calculate the amount of energy the Earth receives from the Sun and the proportion of that energy absorbed by the atmosphere. He was also the first person to explain clearly how birds are able to soar without flapping their wings, and he came very close to inventing the airplane.

Langley was born at Roxbury, Massachusetts, on August 22, 1834. He was educated at Boston Latin School and Boston High School, graduating in 1851, but he did not go to college, so he was largely self-educated. From 1857 until 1864, he worked as a civil engineer and architect, mainly in Chicago and St. Louis. At the same time, he studied astronomy, and by the time he returned to Boston in 1865 he had attained a sufficiently high standard to be offered a post as an assistant at the Harvard University observatory. In 1866, he left to teach mathematics at the U.S. Naval Academy at Annapolis, Maryland, and in 1867 he was appointed director of the Allegheny Observatory in Pennsylvania and professor of physics and astronomy in the Western University of Pennsylvania. In 1887, he became secretary and then director of the Smithsonian Institution, in Washington, D.C., a post he retained until his death. While there, he established the Washington National Zoological Park and the Smithsonian Institution's astrophysical observatory at Wadesboro, North Carolina.

Throughout his life, Langley was fascinated by all solar phenomena and especially by the task of discovering the amount of solar radiation that reaches the Earth and provides the energy to drive the atmospheric and ocean circulations that produce the world's climates. In 1881, he climbed Mount Whitney, California, accompanied by the American astronomer James Edward Keeler (1857–1900). From the summit the two were able to measure the heat of the solar rays and compare this with the value they measured at sea level. The measurements were made with a BOLOMETER, an instrument Langley had invented for the purpose of studying the SOLAR SPECTRUM.

Langley was also interested in the way that air flows across surfaces, and he made a number of experiments that allowed him to calculate the forces operating on a body moving through the air at a constant speed. He showed how thin wings of a certain shape could support the weight of an airplane, and in 1896 he made a steam-powered airplane. It carried no pilot, but it flew across the Potomac River, a distance of 4,200 feet (1,281 m). Between 1897 and 1903, he made three trials, but failed to achieve a successful flight carrying a pilot. The structural materials available to him were not strong enough and his engines were unreliable. These trials cost $50,000, paid from public funds, and after the third failure he was unable to raise more. The *New York Times* published an editorial berating Langley for wasting public money on a foolish dream, predicting that it would be a thousand years before humans achieved powered flight. The Wright brothers made their flight nine days later, on December 17, 1903.

Samuel Pierpont Langley died at Aiken, South Carolina, on February 27, 1906.

Langmuir, Irving
(1881–1957)
American
Physical chemist

A Nobel laureate, Irving Langmuir was the director of the General Electric research laboratory, at Schenectady, New York, where Vincent Schaefer (1906–93) and Bernard Vonnegut (1914–97) performed the first experiments in CLOUD SEEDING.

Irving Langmuir was born in Brooklyn, New York City, on January 31, 1881. He attended local schools and then a boarding school in a Paris suburb, his parents having moved to Paris for three years. The Langmuirs returned to the United States in 1895, moving to Philadelphia. Irving continued his schooling at Chapel Hill Academy and then at the Pratt Institute, in Brooklyn. In 1903, he graduated in metallurgical engineering from the School of Mines at Columbia University. He obtained his Ph.D. at the University of Göttingen, Germany. After a brief spell teaching, Langmuir joined the staff of the General Electric research laboratories, where he remained for the rest of his career.

Langmuir studied the effect of hot metals on gases, work with direct implications for the development of tungsten filament lamps filled with NITROGEN, which did not blacken as "vacuum-filled" lamps did. He also investigated the properties of liquid surfaces, and during World War II he helped develop more effective smokescreens, using SMOKE particles of an optimum size. This work led to the use of solid CARBON DIOXIDE and silver iodide particles for cloud seeding. For his work on surface chemistry, Langmuir received the 1932 Nobel Prize in chemistry.

Irving Langmuir retired in 1950. He died at Falmouth, Massachusetts, on August 16, 1957.

Laplace, Pierre-Simon, Marquis de
(1749–1827)
French
Mathematician and astronomer

A mathematical and scientific genius, Laplace rose rapidly from humble origins to become very famous and influential. He was born on March 28, 1749, at Beaumont-en-Auge, Normandy, where his father owned a small estate. Neighbors who were better off than the Laplace family recognized that the boy had talent and helped to pay for his education. When he was 16, Pierre enrolled at the University of Caen, where his tutors recognized his genius. He graduated in mathematics after two years, and in 1767, aged 18, he traveled to Paris with a letter of introduction to the famous French mathematician Jean le Rond d'Alembert (1717–83). D'Alembert refused to see him, so the young man sent him a paper on mechanics. This was of such high quality that d'Alembert was delighted with it and sponsored Laplace for a professorship. Only 18 years old, Laplace was appointed professor of mathematics at the École Militaire.

Early in his career, Laplace collaborated with the French chemist Antoine Laurent Lavoisier (1743–94) in determining the specific heats of many substances. In 1780, they were able to show that the amount of heat needed to decompose a compound into its constituent elements is equal to the amount that is released when those elements combine to form the compound. This discovery is regarded as the beginning of the branch of chemistry called thermochemistry. It also developed further the work of Joseph Black (1728–99) on LATENT HEAT and pointed the way toward the concept of the conservation of energy. Sixty years later, this was to become the first law of thermodynamics (*see* THERMODYNAMICS, LAWS OF).

Most of Laplace's work was in mathematics and in the mechanics of the solar system. He published this work between 1799 and 1825 in five volumes called *Mécanique Céleste* (Celestial mechanics), for which he is sometimes known as the French Newton. In 1812, Laplace published *Théorie Analytique des Probabilités* (Analytical theory of probabilities), developing this further in *Essai Philosophique* (Philosophical essay), published in 1814. This work gave the theory of probability its modern form and, with it, helped establish statistics as a branch of mathematics.

In 1799, Napoleon appointed Laplace minister of the interior, but dismissed him after only six weeks and promoted him to the rank of senator. Napoleon later made him a count, and when Louis XVIII was restored to the throne in 1814, far from being penalized for his association with Napoleon, Pierre Laplace was promoted again, this time to the rank of marquis.

Laplace was elected to the Academy of Sciences in 1785. In 1816, he was elected to the far more prestigious French Academy, and in 1817 he became its president. He died in Paris on March 5, 1827.

Leverrier, Urbain-Jean-Joseph
(1811–1877)
French
Astronomer

Urbain Leverrier was an authority on many aspects of astronomy and contributed to the discovery of the planet Neptune. He was also instrumental in establishing a network of meteorological observing stations throughout France and in issuing the first daily weather bulletins for French cities.

Urbain Leverrier was born on March 11, 1811, at St. Lô, Normandy. His father, a civil servant, sold his house to pay for Urbain's education. Leverrier attended local schools and then the Collège de Caen from 1828 until 1830. From Normandy he moved to the Collège de St. Louis in Paris, where in 1831 he won a prize in mathematics, which gained him admission to the École Polytechnique. Given a choice of the branch of public service he would enter, Leverrier opted for the Administration Tobaccos, where he conducted chemical research under the guidance of Joseph-Louis Gay-Lussac (1778–1850). Leverrier then began teaching chemistry privately and at the Collège Stanislas, but when in 1837 he applied for a post as a demonstrator at the École Polytechnique, he was offered a position in astronomy, rather than chemistry. He accepted and, almost by accident, became an astronomer.

Leverrier soon came to be recognized as a highly talented astronomer. His studies of Mercury won him election to the Paris Academy of Sciences in 1846, and later that year he discovered Neptune. In 1847, he became a fellow of the Royal Society.

Urbain Leverrier was also active politically. He sided with the Republicans in the 1848 revolution, becoming a member of the legislative assembly in 1849 and a senator in 1852, but Leverrier changed his allegiance later in 1852 when Louis Napoleon came to power and declared himself Napoleon III.

Leverrier was appointed director of the Paris Observatory in 1854. He was an unpopular director, who kept a tight rein on expenditure and the direction of research, and in 1870 he was dismissed. His successor died unexpectedly, however, and Leverrier was reinstated.

On November 14, 1854, during the Crimean War, a fierce storm devastated the Anglo-French fleet near Balaklava, in the Black Sea. The minister for war in Napoleon's government asked Leverrier to study the storm. Leverrier used records from weather observers to reconstruct the track of the storm and was able to demonstrate that if SYNOPTIC WEATHER MAPS had been available and if information had been telegraphed to commanders, the fleet would have had a day to prepare. Leverrier persuaded the government to use the state telegraph service for this purpose, and in 1855 he began to supervise the establishment of a network of observing stations across France. On January 1, 1858, Leverrier launched a daily weather bulletin with observations from 14 French cities and four cities outside France. From September 1863, the daily *Bulletin International de l'Observatoire de Paris* contained a weather map for western Europe, showing isobars (*see* ISO-), pressures, and wind directions.

Leverrier's health deteriorated steadily over a number of years. He died in Paris on September 23, 1877.

Further Reading
Monmonier, Mark. *Air Apparent: How Meteorologists Learned to Map, Predict, and Dramatize Weather.* Chicago: University of Chicago Press, 1999.

Lorenz, Edward Norton
(b. 1917)
American
Meteorologist

As they develop, weather systems demonstrate chaotic behavior (*see* CHAOS). Edward Lorenz was the scientist who discovered this fact.

Edward Lorenz was born in West Hartford, Connecticut, on May 23, 1917. He was educated at Dartmouth College, Harvard University, and the Massachusetts Institute of Technology (M.I.T.), from where he graduated in mathematics. During World War II, Lorenz served as a weather forecaster in the U.S. Army Air Corps. He continued as a meteorologist after the war, joining the M.I.T. faculty in 1946. He received his doctor of science degree at M.I.T. in 1948. Lorenz was professor of METEOROLOGY at M.I.T. from 1962 until 1981; since then he has been professor emeritus of meteorology.

Lorenz's research centered on ways to improve weather forecasts by programming a computer to track changes in an atmospheric variable, such as TEMPERATURE, over a long period. One day in 1961, he ran his program after he had fed into it the initial conditions printed out from an earlier run of the same program. He expected that the second run would reproduce the results of the earlier run, but to his surprise the weather system described in the second run developed in a markedly different way. On investigating the reason for the difference, Lorenz discovered that in the first run he had entered values with six decimal places, but the printout he used for the second run had rounded these to three decimal places. He had assumed, wrongly, that such a small difference would have no effect.

On December 29, 1979, Lorenz described this result in a paper he presented at the annual meeting of the American Association for the Advancement of Science in Washington, D.C. The title of his paper was: "Predictability: Does the Flap of a Butterfly's Wings in Brazil Set Off a Tornado in Texas?" His discovery was quickly nicknamed the "butterfly effect." In subsequent years, Lorenz continued with his investigation of the mathematical theory of chaos.

Lovelock, James Ephraim
(b. 1919)
English
Atmospheric chemist

James Lovelock achieved fame with his proposal of the GAIA HYPOTHESIS, an idea that was immediately attractive to environmentalists, although scientifically it was controversial. Lovelock was born on July 26, 1919, at Letchworth Garden City, Hertfordshire. His father was a keen gardener, and from him James acquired a deep appreciation of the natural world.

The family was not wealthy, and Lovelock did not move directly from school to university. Instead, he obtained a job in a laboratory. During the day he learned practical laboratory techniques, and in the evenings he attended chemistry classes at night school. In time these gave him the qualifications he needed to become a full-time student. He enrolled at the University of Manchester, graduating in chemistry in 1941.

It was wartime, and as an alternative to military service Lovelock was recruited for war work. He joined the staff at the National Institute for Medical Research (NIMR), in London. After the war, in 1946, he remained with the NIMR and went to work at the Common Cold Research Unit, in Wiltshire, where he remained until 1951, taking part in the (ultimately fruitless) search for a cure for the common cold. He received his Ph.D. in 1948 in medicine, from the London School of Hygiene and Tropical Medicine. In 1959, he received the degree of D.Sc. in biophysics from the University of London.

Still an employee of the NIMR, in 1954 Lovelock was awarded a Rockefeller Traveling Fellowship in Medicine. He spent it at Harvard University Medical School, in Boston, and in 1958 he spent a year working at Yale as a visiting scientist. He resigned from the NIMR in 1961 in order to take up an appointment as professor of chemistry at Baylor University College of Medicine, in Houston, Texas.

In the early 1960s, while living in Texas, Lovelock became a consultant at the Jet Propulsion Laboratory (J.P.L.) of the California Institute of Technology, in Pasadena. He had already helped with some of the instruments that were used to analyze lunar soil, and he was asked to advise on various aspects of instrument design for the team of scientists planning the two Viking expeditions to Mars.

One aim of the Viking expeditions was the search for life on Mars, and although Lovelock was not directly involved in this, he began to speculate about how life on another planet might be detected and recognized. After all, organisms that shared no common ancestor with those living on Earth and that had evolved under radically different conditions might possess no visible feature by which they could be identified as living beings. In discussions with Dian Hitchcock, a philosopher employed to assess the logical consistency of NASA experiments, Lovelock reasoned that any living organism must alter its environment by removing substances (such as food) from it and adding substances (such as body wastes) to it. He thought such changes should be detectable and that the simplest place to look for them would be in the planet's atmosphere. Over the years that followed, this line of reasoning led him to formulate the Gaia hypothesis.

James Lovelock has a talent for devising extremely sensitive instruments. Eventually, he decided it might be possible to earn his living in this way, and in 1964 he became a freelance research scientist. Many of his instruments helped to develop and refine the technique of gas chromatography, and in 1957 Lovelock invented the ELECTRON CAPTURE DETECTOR, which is still one of the most sensitive of all detectors. It revealed the presence of residues of organochlorine insecticides such as DDT throughout the natural environment, a discovery that contributed to the emergence of the popular environmental movement in the late 1960s. Later, the electron capture detector registered the presence of minute concentrations of CFCs (chlorofluorocarbon compounds) in the atmosphere.

Although he has now handed over to younger scientists the task of exploring the implications of the Gaia concept, Lovelock retains a lively interest in environmental science and especially in atmospheric science. He lives in a converted mill in a remote part of southwestern England, surrounded by land he owns and has planted with trees, making it a haven for wildlife.

Further Reading
Lovelock, James. *Homage to Gaia.* New York: Oxford University Press, 2000.

Magnus, Olaus
(1490–1557)
Swedish
Priest and naturalist

Olaus Magnus is the Latinized name of Olaf Mansson, the author of a work on natural history that contained the earliest European drawings of ICE CRYSTALS and SNOWFLAKES.

Olaus was born at Linköping, in southern Sweden, in October 1490, where his father, Magnus Peterson, was a prominent citizen. In those days, Swedish people did not have family names. Olaus Magnus meant "Olaus the son of Magnus." He had an elder brother, Johannes Magnus (Johannes, son of Magnus, 1488–1544). Olaus attended a school in Linköping, and then he and Johannes spent nearly seven years traveling together around Europe to complete their education.

From 1518 to 1520, Magnus served as the deputy to the papal vendor of indulgences. Indulgences were printed forms that pardoned people for sins they had committed (or in some cases for sins they had not yet committed, but might). On receipt of the requisite fee, the official vendor filled in the name of the sinner. The sale of indulgences was an important source of church revenue. Magnus was ordained a priest in 1519. He was a vicar in Stockholm in 1520 and dean of Strengnäs Cathedral in 1522.

In 1523, the Swedish king, Gustav I (Vasa), sent Magnus to Rome on a diplomatic mission and from there on other missions to several Dutch and German cities. Finally, in 1528, Olaus reached Poland, where he was to meet Sigismund, the grandson of Gustav Vasa and later king of both Poland and Sweden. The Swedish Reformation took place while Magnus was in Poland. He remained firmly Catholic, however, and was expelled from Poland for this reason. In 1530, he severed his links with the king of Sweden, and his property there was confiscated.

Magnus took refuge in Danzig (now Gdansk) in 1534 and later the same year moved to Italy, where he and Johannes lived together, for the first three years in Venice and then in Rome. Johannes was then archbishop of Sweden. When Johannes died in 1544, the pope appointed Olaus Catholic archbishop of Sweden in his place. Magnus always hoped to return to Sweden in this capacity and wrote to the king repeatedly, asking for permission, but to no avail.

Olaus had a distinguished ecclesiastical career, associating with several of the leading figures of his day. As well as being archbishop of Sweden, he was the director of the religious house of St. Brigitta, in Rome.

He had other interests as well, all of them related to his homeland. Olaus was a skilled cartographer. In 1539, he published, in Venice, his *Carta marina.* This was the first detailed and reasonably accurate map of Scandinavia.

His drawings of ice crystals and snowflakes—things unfamiliar to his friends in Rome—were in his *Historia de gentibus septentrionalibus.* This was published in 1555 and became very popular throughout Europe. There were many editions and translations. The first English translation of it, called *History of the Goths, Swedes and Vandals,* appeared in 1658.

The *Historia* was his very personal account of the history and daily life of the peoples of the north, told with great affection and pride, together with many details of the natural history of northern Europe. It remains one of the most important sources of information about life in Scandinavia in the early 16th century. Magnus based his book on experiences he had gathered between 1518 and 1520, the two years he spent traveling in his capacity of deputy to the vendor of indulgences.

Olaus Magnus died in Rome on August 1, 1557.

Mariotte, Edmé
(ca. 1620–1684)
French
Physicist

Edmé Mariotte was born at Dijon in about 1620 and spent most of his life in or near that city. He was ordained as a priest, but he was also a scientist. It was as a reward for his scientific work that he was appointed prior of the abbey of Saint-Martin-sous-Beaune, near Dijon.

When the French Academy of Sciences was founded in 1666, Mariotte was one of its first members. His interests were wide and he wrote on many scientific topics, including vision (*Nouvelle découvertes touchant la vue* [New discoveries touching on vision], 1668), color (*Traité des couleurs* [Treatise on colors], 1681), plants (*De la végétation des plantes* [On the growth of plants], 1679 and 1696), and others. But his most important work concerned the behavior of fluids. He wrote about freezing (*Expériences sur la congélation de l'eau* [Findings on the freezing of water], 1682) and in 1679 he wrote *De la nature de l'air* (On the nature of air).

In *De la nature de l'air* Mariotte reported his observation that "The diminution of the volume of the air proceeds in proportion to the weights with which it is loaded." This states the relationship between the volume a gas occupies and the pressure under which it is held. It is the law discovered much earlier by Robert Boyle (1627–91), and Mariotte had discovered it quite independently, but Mariotte had noticed something else. He found that air expands when it is heated and contracts when it is cooled, so that the relationship between pressure and volume remains true only so long as the TEMPERATURE remains constant. Boyle had overlooked this. The relationship is known in English-speaking countries as Boyle's law and in French-speaking countries as Mariotte's law. The importance of Mariotte's contribution justifies supporting the French view.

Many of Mariotte's papers were contained in the first volume of the *Histoire et mémoires de l'Académie*, published in 1733. His collected papers were published in the Netherlands in 1717 and again in 1740.

Edmé Mariotte died in Paris on May 12, 1684.

Marum, Martinus van
(1750–1837)
Dutch
Chemist

Martinus van Marum discovered OZONE. He was born on March 20, 1750, at Delft and was educated at the University of Groningen, receiving a doctorate in medicine in 1773. He practiced as a physician in Haarlem from 1776 until 1780.

Van Marum became active in the Society of Dutch Chemists between about 1790 and 1808. From 1804 until his death in 1837, he was secretary of the Dutch Society of Sciences in Haarlem. He was also director of the society's Cabinet of Natural History and librarian of the Teyler Museum.

As a scientist, van Marum had wide interests. He studied plant breeding, geology, paleontology, and a range of chemical topics. He was also interested in AIR POLLUTION and the ventilation of factory buildings. His most important work was on electricity and the construction of electrostatic machines.

In 1785, while passing electrical discharges through OXYGEN, van Marum noted the "odor of electrical matter" and the accelerated oxidation of mercury. The odor he described was that of ozone, although he failed to recognize it as a different form of oxygen.

Van Marum corresponded with many of the leading scientists of his day, including Benjamin Franklin (1706–90), Joseph Priestley (1733–1804), Allesandro Volta (1745–1827), and Antoine-Laurent Lavoisier (1743–94). He died in Haarlem on December 26, 1837.

Marvin, Charles Frederick
(1858–1943)
American
Meteorologist

Charles Marvin was chief of the UNITED STATES WEATHER BUREAU for 21 years. He was born at Putnam, Ohio (now part of Zanesville), on October 7, 1858, and was educated at Ohio State University, from which he graduated in 1883.

The Office of the Chief Signal Officer of the U.S. Army was the predecessor of the U.S. Weather Bureau, and scientists employed there were styled "professors." In 1884, Marvin was appointed a "junior professor," a post he held until the establishment of the Weather Bureau in 1891. He then became a "professor of meteorology." In 1913, acting on the recommendation of the National Academy of Sciences, President Wilson appointed Marvin chief of the Weather Bureau.

Marvin's principal scientific contributions lay in his work on the compilation of HUMIDITY tables and the design, improvement, and standardization of meteorological instruments. Although he developed instruments to measure and record every type of meteorological phenomenon, he was especially fascinated by the rotating cups ANEMOMETER. For a time the Weather Bureau was assigned the task of collecting seismological data, and Marvin designed a seismograph. He was also interested in the application of statistical methods to the solution of meteorological problems.

Charles Marvin retired in 1934. He died after a brief illness on June 5, 1943, in Washington D.C.

Maunder, Edward Walter
(1851–1928)
English
Astronomer

Edward Maunder was the scientist who first identified the period from 1645 to 1715, now known as the MAUNDER MINIMUM, during which the recorded number of SUNSPOTS and auroras (*see* OPTICAL PHENOMENA) was extremely low.

Maunder was born in London on April 12, 1851, the youngest son of a Methodist minister. He was educated at King's College London (since 1900 a part of the University of London, but then an independent institution). Following his graduation, he went to work at a bank in London.

In 1873, a vacancy occurred at the Royal Observatory, Greenwich, London, for a photographic and spectroscopic assistant. This was a position within the British Civil Service, for which there was an entry examination. Maunder passed the examination and was appointed to the position, although he had no formal qualification as an astronomer.

In 1891, another new member joined the staff. Annie Scott Dill Russell (1868–1947), a brilliant graduate from Girton College, Cambridge, arrived as a "lady computer." A "computer" was a person who performed the mathematical calculations—computations—necessary to astronomical research. She and Maunder married and then collaborated in writing many articles about the Sun and popular articles on astronomy.

Maunder was given the job of photographing sunspots and measuring their areas and positions. As he did so, he discovered that the solar latitudes in which sunspots appear varies in a regular fashion during the course of the 11-year sunspot cycle.

While engaged in photographing and measuring sunspots, his attention was drawn to the work of the German astronomer Gustav Spörer (1822–95), who had identified a period from 1400 to 1510, when very few sunspots were seen. This period is now known as the SPÖRER MINIMUM. Maunder began searching through old records at the Royal Observatory to see whether Spörer was correct and whether there were any other such periods. It was this search that led to his discovery of the Maunder minimum.

Edward Maunder was a keen and accurate observer who experimented to discover the smallest object that he could see without the help of a lens. This demonstrated that objects seen on the surface of the Sun or any other distant object in fact must be very large and contain much fine detail that is invisible from Earth. He used this line of reasoning to challenge the existence of channels, or "canals," on Mars, although at that time these were widely accepted as genuine, and his minority opinion was brushed aside.

Similarly, his discovery of the sunspot minimum was overlooked. Other scientists thought he placed too much reliance on old records that, in their view, were likely to be incomplete or inaccurate. The Maunder minimum is now known to be a real phenomenon.

Edward Maunder was made a fellow of the Royal Astronomical Society in 1873. He died at Greenwich on March 21, 1928.

Maury, Matthew Fontaine
(1806–1873)
American
Naval officer, oceanographer, and meteorologist

Matthew Fontaine Maury was born on January 14, 1806, in Spottsylvania County, Virginia, and spent his youth in Tennessee. His elder brother was a naval officer, and Matthew dreamed of following in his footsteps. In 1825, when he was 18 years old, Maury realized his ambition and joined the U.S. Navy as a midshipman. He was assigned to the USS *Vincennes* and embarked on a four-year cruise that took him around the world. This was followed by other extended voyages to Europe and to the western coast of South America.

Maury returned to the United States in 1834. In the same year, he married Ann Hull Herndon, and the couple settled in Fredericksburg. Matthew spent the years from 1834 until 1841 writing descriptions of ocean voyages and works on navigation. He also wrote essays urging naval reforms.

In 1839, Maury was injured in a stagecoach accident and rendered permanently lame. No longer fit for active service, in 1841 he was appointed superintendent of the Depot of Charts and Instruments. He remained in this post until 1861. It should have given him a quiet, easy life with ample leisure, but Maury threw himself into it. By the time he left, he had transformed this obscure department into the United States Naval Observatory and the Hydrographic Office.

Matthew Maury was especially interested in ocean currents and the winds that drive them. He issued specially prepared logbooks to captains in which they were asked to record their observations of winds and currents. He charted the course of the Gulf Stream and described it as "a river in the ocean." Knowledge of the locations and courses of currents allowed captains to sail with them rather than against them, thus shortening the time their voyages took. In 1850, he charted the depths of the North Atlantic Ocean in order to facilitate the laying of the transatlantic cable.

Maury demonstrated very clearly that the comprehensive study of METEOROLOGY at sea called for international cooperation. He had already achieved international recognition when, in 1853, he was able to play a leading role in organizing a conference on oceanography and meteorology in Brussels, which he attended as the United States representative. Two years later, in 1855, he published *Physical Geography of the Sea,* which was the first textbook in oceanography. Also in 1855, he was promoted to the rank of commander.

When the Civil War broke out in 1861, Maury threw in his lot with the South. He resigned from the navy on April 20, three days after his native Virginia had seceded from the Union. A few days later, he was made a commander in the Confederate States Navy. He became head of coastal, harbor, and river defenses, and he invented an electric torpedo for harbor defense and experimented with electric mines. Because of his international reputation, in 1862 he was sent to England to purchase naval supplies.

At the end of the war Maury went into voluntary exile. He settled for a time in Mexico, where he became commissioner of immigration to the emperor Maximilian and attempted to found a Virginian colony. The emperor abandoned the colonization project in 1866, and Maury moved to England.

By 1868, tempers had cooled, and Maury returned home to take up the post of professor of meteorology at the Virginia Military Institute. He settled in Lexington, where he died on February 1, 1873. His body was buried temporarily in Lexington and then moved to Hollywood Cemetery, Richmond, where it remains.

Maury was completely forgiven for having supported the losing side in the Civil War. There is a Maury Hall at the Naval Academy in Annapolis, Maryland, and in 1930 Maury was elected to the Hall of Fame for Great Americans.

Mie, Gustav
(1868–1957)
German
Physicist

Gustav Mie was the physicist who discovered the way small particles scatter light (*see* SCATTERING). He was born at Rostock, on the coast of the Baltic Sea, on September 29, 1868. He was educated at the gymnasium (high school) in Rostock and went on to study mathematics and geology at the Universities of Rostock and Heidelberg. He obtained his doctorate in 1891 from Heidelberg.

Mie worked for a short time as a teacher at a private school. Then in 1892 he obtained a post as an assistant in the Physics Institute of the Technical University of Karlsruhe. He held this position until 1902, when he was appointed extraordinariat (a special professor) at the University of Greifswald. There, in 1908, Mie wrote his paper on the scattering of light. Mie became a professor at the University of Halle in 1918 and in 1924 at the University of Freiburg.

Gustav Mie died at Freiburg im Bresigau on February 13, 1957.

Milankovitch, Milutin
(1879–1958)
Serbian
Mathematician and climatologist

Milutin Milankovitch was the first person to fully develop an astronomical theory to account for major climatic changes. He was born on May 28, 1879, at Dali, near Osijek, in what was then part of Austria–Hungary. He studied at the Vienna Institute of Technology, graduating in civil engineering in 1902 and with a doctorate in technical sciences in 1904. He worked for a construction company for a short time, and in 1909 he moved to take up the professorship in applied mathematics at the University of Belgrade, where he remained for the rest of his career.

The Balkan Wars of the early 20th century led to World War I, and at the outbreak of war Milankovitch was interned by the Austro–Hungarian army. He was held first at Nezsider but then in Budapest, where his captors permitted him to work in the library of the Hungarian Academy of Sciences. He spent the remainder of his time in captivity studying solar climates and the temperatures of the planets.

Sir John Herschel (1792–1871) was the first of several scientists to suggest a link between changes in climate and in the amount of solar radiation received by the Earth. In 1864, James Croll (1821–90) had suggested that changes in solar radiation trigger ice ages. The idea was fairly vague, however, and Milankovitch determined to test it. Astronomical changes occur with great regularity—think of the accuracy with which solar eclipses can be predicted. This meant that if Milankovitch could identify those factors that alter the amount of radiation Earth receives he would be able to calculate mathematically how they had exerted their influence at various times in the past.

Milankovitch found three cyclical changes that seemed relevant, and he calculated their effects over hundreds of thousands of years. In 1920, he published his results and elaborated on them in subsequent years. Although his theory aroused considerable interest, for a long time there was no firm evidence to support it. Confirmation came in 1976, however. The astronomical changes he identified are now known as the MILANKOVITCH CYCLES, and their influence on the initiation and ending of ice ages is widely accepted.

Milankovitch died in Belgrade on December 12, 1958.

Molina, Mario José
(b. 1943)
Mexican
Atmospheric chemist

Professor Molina was the joint winner of the 1995 Nobel Prize in chemistry, sharing it with Paul Crutzen and F. Sherwood Rowland. The prize was awarded for their work in identifying the threat to the OZONE LAYER from chlorofluorocarbon compounds (CFCs).

Mario Molina was born in Mexico City on March 19, 1943, the son of a lawyer. He was educated at a boarding school in Switzerland and then in Mexico City, where he graduated in 1965 with a degree in chemical engineering from the Universidad Nacional Autonoma de Mexico. In 1967, he obtained a postgraduate degree from the University of Freiburg, Germany, and his Ph.D. from the University of California at Berkeley in 1972, where Sherwood Rowland was one of his supervisors. While at Berkeley Molina met and married Luisa Tan, a fellow graduate student.

Molina held teaching and research posts at the Universidad Nacional Autonoma de Mexico, the University of California at Irvine, and from 1982 to 1989 at the Jet Propulsion Laboratory of the California Institute of Technology, in Pasadena. In 1989, he moved to the Department of Earth, Atmospheric and Planetary Sciences and Department of Chemistry at Massachusetts Institute of Technology. He was named M.I.T. Institute Professor in 1997.

The first warnings about what might happen to the ozone layer appeared in 1970. This attracted scientific attention, and in 1974 Molina and Rowland published a paper in the scientific journal *Nature* describing the results of their studies of the chemistry of CFC compounds. At the time, CFCs were used widely as propellants in spray cans and in refrigerators, freezers, and air conditioners, and in foam plastics. Molina and Rowland calculated that these very stable, and therefore long-lived compounds could cross the tropopause (see ATMOSPHERIC STRUCTURE) and that once they were in the stratosphere CFC molecules would be broken apart by their exposure to ultraviolet radiation (*see* SOLAR SPECTRUM). This chemical degradation would release chlorine, which would engage in reactions that depleted OZONE. Other atmospheric scientists received their paper with some skepticism, but in 1985 stratospheric ozone depletion was detected. It was eventually found that CFCs were involved, just as Molina and Rowland had indicated.

More recently, Mario Molina has studied the pollution of the troposphere. In particular, he is keen to find ways to reduce pollution levels in large cities that suffer from traffic congestion, such as his native Mexico City. He serves on many committees, including the President's Committee of Advisors in Science and Technology, the Secretary of Energy Advisory Board, National Research Council Board on Environmental Studies and Toxicology, and on the boards of U.S.–Mexico Foundation of Science and other nonprofit environmental organizations.

Morse, Samuel Finley Breese
(1791–1872)
American
Artist and inventor

The man responsible for building the first telegraph line in the world and devising a code by which telegraph messages could be sent simply, efficiently, and reliably, was born on April 27, 1791, at Charlestown, Massachusetts. Charlestown is now part of Boston. His father, Jedidiah, was a clergyman and geographer. Samuel studied art at Yale University, where he developed a keen interest in painting miniature portraits. He graduated in 1810 and then persuaded his parents to allow him to travel to England to study historical painting. He stayed in England from 1811 until 1815.

After his return, his talent and gift for making friends brought him enough portrait commissions for him to earn a living, but he never became wealthy, and Americans were not impressed by his historical paintings. He settled in New York City, where his reputation grew, and he was one of the founders of the National Academy of Design and its first president, from 1826 to 1845. Morse taught art at the University of the City of New York (now New York University [NYU]) and twice ran unsuccessfully for election as mayor of New York, on an anti-Catholic, anti-immigrant ticket.

Electricity and electromagnetism were just becoming known and aroused great interest. In 1832, while returning from one of his visits to Europe to study art, Morse fell to discussing electrical experiments with Charles Thomas Jackson (1805–80), a fellow passenger on the ship. Various people had suggested the possibility of transmitting messages electrically, and his conversations with Jackson gave Morse the idea to make a device that would do so. Unfortunately, his knowledge of electricity was inadequate to the task, but a university colleague drew his attention to the work of Joseph Henry (1797–1878), whom he met later. Henry had already given the idea of the telegraph considerable thought, and he was unstinting in his help and advice, answering all of Morse's questions.

By about 1835, Morse had made a telegraph that worked. Henry had designed one earlier, but Morse believed his was the first. Then began his real task. Having failed to build a "Morse line" in Europe, he set out to persuade a reluctant Congress to appropriate the $30,000 it would cost to build a telegraph line linking Baltimore and Washington, a distance of 40 miles (64 km). In 1843, he finally succeeded, and the line was built in 1844. Morse used his own code, the MORSE CODE, which he had developed by 1838, to transmit the first message: "What hath God wrought?" Within a few years TELEGRAPHY was being used to transmit meteorological data, a development that led directly to the first weather reports and forecasts.

After the feasibility of a telegraph line had been demonstrated, Morse was caught up in a succession of legal actions brought by his partners and other inventors, including Jackson, who claimed priority for the invention. Eventually, Morse established his patent rights and was rewarded by many European governments. In old age he became a philanthropist. He died in New York on April 2, 1872, and in 1900, when it first opened, Samuel Morse was made a charter member of the Hall of Fame for Great Americans on the campus of New York University.

Neumann, John von
(1903–1957)
Hungarian–American
Mathematician and physicist

John (originally Janos) von Neumann was born in Budapest, Hungary, on December 28, 1903. He was educated privately until the age of 14, when he entered the gymnasium (high school), but he continued to receive extra tuition in mathematics.

Von Neumann left Hungary in 1919 to escape the chaotic conditions that followed the defeat of the Austro-Hungarian Empire in World War I. He studied at the Universities of Berlin (1921–23) and Zürich (1923–25), graduating from Zürich with a degree in chemical engineering. He received a doctorate in mathematics from the University of Budapest in 1926 and then continued his studies at the University of Göttingen, where he worked with Robert Oppenheimer (1904–67). Von Neumann worked as an unpaid lecturer at the University of Berlin from 1927 until 1929. He then moved to the University of Hamburg, and in 1930 he immigrated to the United States. He was appointed a visiting professor at the University of Princeton, and in 1931 he became a full professor of mathematics at the newly established Institute of Advanced Studies at Princeton, where he remained for the rest of his life. Von Neumann later held a number of advisory posts for the U.S. government.

John von Neumann made many contributions to mathematics and theoretical physics. He invented the study of game theory.

In the 1940s, von Neumann became interested in METEOROLOGY and began to study the methods used in numerical WEATHER FORECASTING. He devised a technique for analyzing the stability of those methods, and in 1946 he established the Meteorology Project at Princeton. In April 1950, the group led by von Neumann made the first accurate numerical forecast, using the ENIAC computer.

In 1952, von Neumann designed and supervised the construction of MANIAC-1, which was the first computer that was able to use a flexible stored program. His work at Princeton on MANIAC-1 influenced the design of all the programmable computers that followed, including those used in weather forecasting.

Von Neumann received many honors and awards, including the Medal of Freedom, the Albert Einstein Award, and the Enrico Fermi Award, all of which were presented to him in 1956.

By 1956, von Neumann was already in poor health. He died from cancer in Washington, D.C., on February 8, 1957.

Nusselt, Ernst Kraft Wilhelm
(1882–1957)
German
Engineer

Wilhelm Nusselt was born at Nuremberg, Germany, on November 25, 1882. He studied machinery at the Technical Universities of Berlin-Charlottenburg and Munich, graduating in 1904. Nusselt continued his studies of mathematics and physics at the Laboratory for Technical Physics in Munich and completed his doctoral thesis in 1907, on the conductivity of insulating materials.

Nusselt worked in Dresden from 1907 until 1909, when he qualified for a professorship with his studies of heat and momentum transfer in tubes. He was a professor at the Technical University of Karlsruhe from 1920 until 1925 and at the Technical University of Munich from 1925 until his retirement in 1952.

In 1915, Wilhelm Nusselt published his most important paper, "The Basic Laws of Heat Transfer." This was the paper in which he proposed the dimensionless groups in the theory of heat transfer (*see* NUSSELT NUMBER).

Wilhelm Nusselt died at Munich on September 1, 1957.

Oeschger, Hans
(1927–1998)
Swiss
Physicist

Hans Oeschger was born at Ottenbach, near Zürich, on April 2, 1927. He trained as a nuclear physicist in Zürich and obtained his doctorate in 1955 at the University of Bern. In collaboration with Professor F. G. Houtermans, in 1955 Oeschger built a device to measure very low levels of radioactivity. Known as the Oeschger counter, this instrument was used to date deep water taken from the Pacific Ocean. In 1963, he was promoted to professor at the University of Bern, where he founded the Laboratory for Low Level Counting and Nuclear Geophysics and the Division of Climate and Environmental Physics. He remained the director of the division until his retirement in 1992.

In 1962, Oeschger began studying GLACIER ice and pioneered research into ICE CORES. He and his team developed drilling and analytical techniques that allowed them to trace changes in climate over the last 150,000 years. In 1979, they showed that during the last GLACIAL PERIOD the atmospheric concentration of CARBON DIOXIDE was a little more than half that of the present day and that its rise during the last 1,000 years was due to the burning of fossil fuels. He was deeply concerned that the rise in carbon dioxide levels might trigger an enhanced GREENHOUSE EFFECT. Hans Oeschger was one of the lead authors of the First Report of the INTERGOVERNMENTAL PANEL ON CLIMATE CHANGE.

With his colleagues Chester C. Langway and Willi Dansgaard, Oeschger discovered that the Greenland ice cores contained records of abrupt climate changes, now known as DANSGAARD-OESCHGER EVENTS.

After a long illness, Hans Oeschger died at Bern on December 25, 1998.

Pascal, Blaise
(1623–1662)
French
Mathematician, physicist, and theologian

Blaise Pascal discovered that the atmosphere has an upper limit and that atmospheric pressure decreases with altitude. He was born on June 19, 1623, at Clermont-Ferrand, in the Auvergne region of France, but the family moved to Paris in 1631.

His mother died when Blaise was three, and his father, a mathematician and civil servant, brought up the family and supervised Blaise's education. The boy was soon recognized as a mathematical prodigy. In 1640, not yet 17 years of age, he wrote *Essai pour les coniques*, about the geometry of conic sections, that made René Descartes (1596–1650) envious. Between 1642 and 1644, the young Pascal designed and made a calculating machine, based on cogwheels, to help his father in the many calculations his work required. This machine was the ancestor of the modern cash register, and the computer programming language Pascal is named after its inventor. Pascal made and sold about 50 of them, and several are still in existence.

Pascal learned of and became interested in the work of Evangelista Torricelli (1608–47). He repeated the Torricelli experiment, but using red wine and water instead of mercury. Wine is even less dense than water, so Pascal's barometers had tubes 39 feet (12 m) tall and were fixed to the masts of ships!

Pascal also reasoned that if air has weight, then it must cover the surface of the Earth rather like an ocean, with an upper surface. This implied that the weight of air measured by a BAROMETER must decrease with height, because the greater the height of the instrument above the surface the smaller must be the mass of air weighing down on it from above.

In 1646, he returned from Paris to Clermont-Ferrand to test this idea on the Puy-de-Dôme, an extinct VOLCANO, 4,806 feet (1,465 m) high, not far from the town. Pascal was never physically strong, and by 1646 he suffered from severe indigestion and insomnia. There was no question that he could make the strenuous ascent of the Puy-de-Dôme—the climb is very steep—so he enlisted the help of his brother-in-law, Florin Périer. Périer carried a barometer up the mountain, at intervals recording the pressure it registered. This showed a steady decrease in pressure with increasing height. Similar records were made from a second barometer, left at the foot of the mountain. These showed no change in pressure throughout the day. The experiment confirmed Torricelli's discovery that air has weight. This achievement is recognized in the name of the derived SI unit of pressure or stress, the pascal (*see* UNITS OF MEASUREMENT and APPENDIX VIII: SI UNITS AND CONVERSIONS).

Pascal made many more contributions to science and mathematics before his death, in Paris, probably from meningitis associated with stomach cancer, on August 19, 1662.

Priestley, Joseph
(1733–1804)
English
Chemist

Joseph Priestley is conventionally credited with the discovery of OXYGEN. He was born on March 13, 1733, at Fieldhead, Yorkshire. His mother died when he was seven, and he went to live with an aunt, who was a Calvinist. Priestley was educated at the Dissenting Academy at Daventry, Warwickshire, and in 1758 he became a Presbyterian minister. In 1761, he became a language teacher at Warrington Academy.

On a visit to London in 1766, Priestley met Benjamin Franklin (1706–90), who aroused in him a keen interest in science. From that time Priestley combined his duties as a minister with scientific research. He discovered "nitrous air" (nitric oxide, NO) in 1772 and reduced it to dinitrogen oxide (N_2O). He isolated ammonia gas (NH_3) in the same year, and it was also in 1772 that he found that plants release a gas that is necessary to animals. In 1774, he discovered that the same gas is released when mercurous oxide (HgO) or red lead (Pb_3O_4) are heated. A mouse thrived in this gas and combustible materials burned brightly in it. He also observed that when the gas is mixed with nitric oxide (his "nitrous air") a red gas is formed, in fact nitrogen dioxide (NO_2).

Priestley concluded that he had discovered air from which the fiery principle phlogiston had been removed. Consequently, this air readily accepted phlogiston released by burning substances. He called the gas "dephlogisticated air." In fact, Karl Wilhelm Scheele (1742–86) had discovered the gas in 1772, but failed to publish his discovery.

Priestley visited Paris in 1774, where he met Antoine Lavoisier (1743–94) and told him about dephlogisticated air. It was Lavoisier who gave it the name oxygen.

Joseph Priestley joined the Lunar Society, a group of eminent scientists based in Birmingham. (The society's name refers to the fact that they met only on moonlit nights, to reduce the risk of being attacked on the unlit streets.) Priestley continued to voice his dissenting views and his support for the French Revolution. This made him increasingly unpopular with the British authorities, and in 1794 he immigrated to Northumberland, Pennsylvania. He refused a professorship at the University of Pennsylvania and died at Northumberland on February 6, 1804.

Ramsay, William
(1852–1916)
Scottish
Chemist

William Ramsay was awarded the 1904 Nobel Prize in chemistry in recognition of his discovery of the rare gases ARGON, KRYPTON, NEON, and XENON.

Ramsay was born in Glasgow on October 2, 1852, the son of an engineer and grandson of the founder of the Glasgow Chemical Society. At first, William's interests were in music and languages, and in 1866, when he was only 14 years old, this hugely talented boy entered the University of Glasgow to study arts. In 1868, he became interested in science and went to work in the laboratory of the Glasgow City Analyst, where he learned chemistry. In 1870, Ramsay traveled to Germany, where he carried out research in organic chemistry, gaining his Ph.D. in 1873 at the University of Tübingen.

Ramsay then returned to Glasgow to work as a teaching assistant at Anderson's College (later called the Royal Technical College) and then at the University of Glasgow. In 1880, he was appointed professor of chemistry at University College Bristol (later the University of Bristol) and the following year, 1881, he became the college principal. In 1887, he moved to London to become professor of chemistry at University College, a post he held until his retirement in 1912.

While at University College Ramsay reorganized the out-of-date laboratory. His research centered on problems posed by Lord Rayleigh, concerning the composition of the atmosphere, which appeared to contain minute amounts of an unknown gas. In 1894, Ramsay discovered argon (the name means "inert"). In 1895, he recovered HELIUM from a uranium mineral. In 1898, Ramsay and his colleagues managed to obtain 26 pints (15 liters) of argon from liquid air. When they allowed the liquid argon to boil, they found it was mixed with other inert gases. The first to be identified they called neon ("new"). It was lighter than argon, but when the argon had completely vaporized, two gases remained that were heavier than argon. One they called krypton ("hidden") and the other xenon ("stranger").

William Ramsay had many talents. As a young man he was athletic. He was also an expert glassblower and made most of his own laboratory glassware.

Ramsay was knighted in 1902. After his retirement in 1912, he moved to High Wycombe, Buckinghamshire, a small town to the north of London, where he continued to conduct research in converted stables. He died at High Wycombe on July 23, 1916.

Rankine, William John Macquorn
(1820–1872)
Scottish
Engineer and physicist

Rankine devoted most of his working life to studying the principles underlying the operation of engines, especially steam engines. He was born in Edinburgh on July 5, 1820. His father was an engineer, and William trained as a civil engineer. In 1855, he was appointed professor of civil engineering and mechanics at the University of Glasgow, a position he continued to hold until his death. In 1853, he was elected a member, and later a fellow, of the Royal Society.

From his work on engines, Rankine's research led him to more abstract studies of thermodynamics. Lord Kelvin (William Thomson, 1824–1907), Rudolf Clausius (1822–88), and Rankine all discovered at about the same time that heat can be converted to work, and work to heat. In the course of his work on thermodynamics, Rankine devised a TEMPERATURE SCALE that is sometimes used (though not by scientists) as an alternative to the Kelvin scale.

William Rankine died at Glasgow on December 24, 1872.

Raoult, François-Marie
(1830–1901)
French
Physical chemist

The author of RAOULT'S LAW was born at Fournes-en-Weppes, in the Nord département of France, on May 10, 1830. He obtained his doctorate at the University of Paris in 1863 and held a post at the University of Sens before moving to a faculty position at the University of Grenoble, where he remained for the rest of his life.

Raoult was one of the founders of physical chemistry, together with Jacobus Henricus Van't Hoff (1852–1911), Friedrich Wilhelm Ostwald (1853–1932), and Svante Arrhenius (1859–1927). His studies of solutions led him to propound the law for which he is famous.

François-Marie Raoult died at Grenoble, in the département of Isère, on April 1, 1901.

Rayleigh, Lord
(**John William Strutt**)
(1842–1919)
English
Physicist

The scientist who explained, in 1871, why the sky is blue (*see* RAYLEIGH SCATTERING) spent much of his life studying the properties of light and sound waves, waves in water, and earthquake waves, but his interests were much wider. His most famous discovery was in the field of chemistry, not physics. Rayleigh had become interested in the densities of different gases and found that when he measured the DENSITY of NITROGEN in air it was always 0.5 percent greater than the density of that gas obtained from any other source. He eliminated all the reasons he could think of for this, but the disparity remained, and in desperation in 1892 he published a short note in the scientific journal *Nature* asking for suggestions. He received a reply from the Scottish chemist William Ramsay (1852–1916) who solved the problem in 1894 by discovering the previously unknown gas ARGON. For

their discovery, in 1904 Rayleigh was awarded the Nobel Prize in physics and Ramsay received the Nobel Prize in chemistry.

Rayleigh was born as John William Strutt at Terling Place, Langford Grove, Essex, on November 12, 1842. He was educated by a private tutor until he entered Trinity College, Cambridge in 1861, graduating in mathematics in 1865. He then visited the United States, returning to England in 1868 and establishing a private laboratory at his home, Terling Place.

In 1873, when Strutt was 31, his father died and he acceded to the title, becoming the third Baron Rayleigh. Lord Rayleigh, the name by which he is usually known, continued with his research, an occupation that was considered somewhat eccentric for an English aristocrat. Rayleigh was elected a fellow of the Royal Society in the same year he acceded to his title, and in 1879 he succeeded James Clerk Maxwell (1831–79) as Cavendish Professor of Experimental Physics at Cambridge University. After 1884, Rayleigh spent most of his time working in his private laboratory. He held the post of professor of natural philosophy at the Royal Institution, London, from 1887 until 1905, but this allowed him to remain at home for most of the time. Rayleigh was secretary to the Royal Society from 1885 until 1896 and its president from 1905 until 1908. From 1908, until his death, he was chancellor of Cambridge University.

Lord Rayleigh died at Terling Place on June 30, 1919.

Réaumur, René-Antoine-Ferchault de
(1683–1757)
French
Physicist

As well as making a THERMOMETER and devising a TEMPERATURE SCALE with which to calibrate it, Réaumur contributed to many branches of science and technology. His most important work concerned the process of digestion.

Réaumur was born at La Rochelle, in Charente-Maritime on the French Atlantic coast, on February 28, 1683. In 1703, when he was 20, he moved to Paris, where he lived under the protection of a relative who was an important civil servant. Réaumur was admitted to the Academy of Sciences in 1708, on the strength of some mathematical work. In 1710, the authorities commissioned him to prepare a report describing all of the nation's useful arts and manufactures.

His own achievements are considerable. Réaumur invented a kind of white glass that is still known as Réaumur porcelain. He showed that certain curious stones found in southern France were in fact the fossil teeth of animals that are now extinct. He helped devise new methods of steel manufacture, wrote a six-volume work on insects, and was the first person to show that corals are animals. His work on digestion resolved a long-standing debate as to whether food is digested chemically or physically, by being ground up in some way. Réaumur showed experimentally that the process is chemical. He developed his thermometer in about 1730, apparently unaware of the work of Daniel Fahrenheit (1686–1736).

René Réaumur spent much of his time at La Bermondière, his country house in Maine et Loire, where he died on October 17, 1757.

Renaldini, Carlo
(1615–1698)
Italian
Physicist

An aristocrat from Ancona, on the Adriatic coast of northern Italy, Renaldini became professor of philosophy at the University of Pisà. He was also a member of the Accademia del Cimento (Academy of Experiments), in Florence. The Accademia was the principal institution studying the atmosphere.

In 1694, Renaldini proposed a method for calibrating THERMOMETERS. This was causing difficulty, because when heated from freezing to boiling temperature water does not expand at a constant rate and there seemed no reason to suppose that any other liquid would do so. It was agreed that the freezing and boiling temperatures of water should mark two ends of the temperature scale, and Renaldini suggested that the intermediate points might be identified by mixing boiling and ice-cold water in varying proportions. Equal weights of water at 32°F (0°C) and 212°F (100°C) would reveal the point at which to mark the halfway point, 122°F (50°C). Twenty parts of boiling water mixed with 80 parts of freezing water would indicate 68°F (20°C) and so on, with each degree rise in temperature corresponding to the addition of the same amount of heat. The method proved difficult to put into practice and did not overcome the difficulty of depending on the behavior of a particular liquid.

Reynolds, Osborne
(1842–1912)
English
Physicist and engineer

Osborne Reynolds was born on August 23, 1842, in Belfast, Northern Ireland. His father was a mathematician who became a schoolteacher at Dedham Grammar School, in Suffolk, England. This necessitated the family moving to England, but in a sense the family was returning home. Osborne's great-grandfather and great-great grandfather had both held the position of rector in the parish of Debach-with-Boulge, to the northeast of Ipswich, only a few miles from Dedham. Later, Osborne's father also became rector there.

Reynolds began his education at Dedham Grammar School. A proficient mathematician, he left school at 19 and went to work for an engineering company, where he was able to apply his knowledge of mathematics. Having gained some practical experience, Reynolds enrolled at Queens' College, Cambridge, to study mathematics. He graduated in 1867 and was immediately awarded a fellowship at Queens'. He then moved to London to work for John Lawson, a civil engineer, but left in 1868 because he had been elected the first professor of engineering at Owens College, Manchester (now the University of Manchester). Reynolds was elected a fellow of the Royal Society in 1877 and received the Society's Royal Medal in 1888. This was followed by many more medals and honorary degrees.

Reynolds conducted research into the movements of comets and atmospheric phenomena caused by electricity, but his most important work concerned the flow of water through channels, including wave and tidal movements in rivers and estuaries. His discoveries led to his formulation of what is now known as the REYNOLDS NUMBER. The Reynolds number is widely used by meteorologists and atmospheric physicists in calculations of the TURBULENT FLOW of air. Account must be taken of turbulence when designing buildings, aircraft, or any other objects that move through the air or that the air moves around. Turbulence also affects the rate at which atmospheric pollutants disperse.

Reynolds retired in 1905 and died at Watchet, Somerset, on February 21, 1912.

Richardson, Lewis Fry
(1881–1953)
English
Mathematician and meteorologist

In 1922, a book was published with the title *Weather Prediction by Numerical Process*. It described a system for numerical WEATHER FORECASTING, which is the preparation of weather forecasts by the use of mathematical techniques for solving equations. Lewis Fry Richardson, its author, had been developing the system since 1913, and his book contained a worked example to show how it might be done.

Unfortunately, Richardson was half a century ahead of his time. His book demonstrated the principle of numerical forecasting, but some of his equations were incorrect, and the data he used for his example were inaccurate. Only 750 copies of the book were printed, and 30 years later not all of them had been sold. Accurate WEATHER FORECASTING requires detailed information about conditions in the upper atmosphere. Modern meteorologists have access to these data, but they were not available in the 1920s. Nor did Richardson and his contemporaries have access to modern computers or even pocket calculators. His scheme involved so many calculations, all of which had to be performed with paper and pencil, that by the time they were completed the conditions they were forecasting would have come and gone.

Richardson estimated that 64,000 individuals equipped with slide rules would be needed. He imagined them seated in a vast hall, each person dealing with data from a particular WEATHER STATION and everyone occupying a place corresponding to the location of the stations, so the hall was a kind of map of the world. The operation would be coordinated by a person standing at a higher level and pointing a red light at any station that was running ahead of the field and a blue light at a station that was behind. His estimate was very conservative. To produce a forecast for the whole world, the hall would have had to contain more than 200,000 workers, and even so the forecasting would have been able to do no more than keep up with the weather as it developed. If the forecast was to advance five times faster than the weather, more than 1 million workers would have been required.

Had it been possible to produce them fast enough, the forecasts would not have been very reliable. The equations Richardson used work only under certain atmospheric conditions, a fact of which he was unaware. These deficiencies made the method impractical, and so little attention was paid to it. Today, upper-air data and adequate computing power are available. Richardson's book was republished in 1965, and this time it sold thousands of copies. A modified version of his method is now used for making large-scale, long-range weather forecasts.

Lewis Fry Richardson was born into a family of tanners on October 11, 1881, at Newcastle-upon-Tyne, an industrial city in the northeast of England. It was a Quaker family, and Richardson remained a committed Quaker throughout his life. He was educated at a school in York and then studied natural science (physics, mathematics, chemistry, biology, and zoology) at King's College, at the University of Cambridge. In 1927, at the age of 47, he was awarded a doctorate in mathematical psychology by the University of London. (Mathematical psychology is the name given to any theoretical work that uses mathematical methods, formal logic, or computer simulation.)

From 1903 to 1904, Richardson worked for the National Physical Laboratory, a government research institution, and in 1912, following the sinking of the *Titanic,* he conducted experiments with echo-sounding. In 1913, he went to work at the METEOROLOGICAL OFFICE.

Richardson's mathematical work, concerned mainly with the calculus of finite differences, showed great originality. Later in life, he even sought to use mathematics to identify and clarify the causes of war, an interest that arose from his religious beliefs.

During the First World War, Richardson served among fellow Quakers in the Friends Ambulance Unit. He was sent to France, where he tried to practice Esperanto, another interest of his, on German prisoners of war. After the war he returned to the Meteorological Office. It was while serving in France that Richardson worked out the equations needed to forecast the weather. His notes were lost in the chaos of war, but were found later beneath a pile of coal.

As a member of the Society of Friends, Richardson was a pacifist, and in 1920 he resigned from the Meteorological Office when it was absorbed into the Air Ministry. He became head of the Physics Department at Westminster Training College, and in 1929 he was appointed principal of Paisley Technical College (now the University of Paisley) in Scotland. He remained at Paisley until his retirement in 1940.

Retirement gave him time to develop his ideas on eradicating sources of conflict. His two books on the subject, *Arms and Insecurity* and *Statistics of Deadly Quarrels* were published posthumously in 1960, attracting interest among religious and pacifist groups.

During his lifetime, Richardson was best known for his studies of atmospheric turbulence. This led to his proposal of what came to be called the Richardson number (*see* TURBULENT FLOW) for predicting whether turbulence would increase or decrease.

In 1926, Richardson was elected a fellow of the Royal Society of London. He died at Kilmun, Argyll, Scotland, on September 30, 1953.

Further Reading

Collins, Paul. "The slide-rule orchestra." *New Scientist,* 22 January 2005, 48–49.

Monmonier, Mark. *Air Apparent: How Meteorologists Learned to Map, Predict, and Dramatize Weather.* Chicago: University of Chicago Press, 1999, 88–93.

University of Paisley: "Lewis Fry Richardson." University of Paisley. Available online. URL: http://maths.paisley. ac.uk/LfR/home.htm. Accessed March 28, 2006.

Römer, Ole (or Olaus) Christiansen
(1644–1710)
Danish
Astronomer

Ole (sometimes spelled Olaus) Römer was born at Århus, in Jutland, Denmark, on September 25, 1644. He was educated at the University of Copenhagen. In 1671, the French astronomer Jean Picard (1620–82) visited Denmark to ascertain the precise location of Tycho Brahe's observatory. Picard hired Römer as an assistant and when the project was complete took him back with him to Paris. Römer worked as an assistant at the French Academy of Sciences and within a year had been elected a full member.

Römer observed the movements of the satellites of Jupiter. He noted that Jupiter eclipsed the satellites earlier when Jupiter and Earth were approaching each other and later when the two planets were moving apart. He deduced from this that light must travel at a finite speed, contradicting most scientists of his day, who believed light propagated instantaneously. Römer

calculated the speed of light, announcing this at a meeting of the academy in 1676 as 141,060 miles per second (227,000 km/s). The true speed is 186,291 miles per second (299,792 km/s), but Römer had made an impressive first attempt at it.

Römer visited England in 1679, where he met Isaac Newton (1642–1727) and Edmund Halley (1656–1742). In 1681, King Christian V recalled him to Denmark, where Römer was appointed astronomer royal and professor of astronomy at the University of Copenhagen. In 1705, he became mayor of Copenhagen.

As well as his astronomical skills, Ole Römer was highly practical and had a talent for making scientific instruments. In about 1701, he made a THERMOMETER and devised a method for calibrating it. Daniel Fahrenheit (1686–1736) met Römer during his visit to Copenhagen in 1708, and the two instrument makers discussed Römer's thermometer. This discussion strongly influenced Fahrenheit in the design of his own thermometer.

Ole Römer died at Copenhagen on September 23, 1710.

Rossby, Carl-Gustav Arvid
(1898–1957)
Swedish-American
Meteorologist

Carl-Gustav Rossby was born in Stockholm, Sweden, on December 28, 1898. He was educated at the University of Stockholm, and in 1918, after graduating with a degree in theoretical mechanics, he moved to the Bergen Geophysical Institute. There he worked with Vilhelm Bjerknes (1862–1951) on oceanographic as well as meteorological problems. When Bjerknes moved to Germany in 1921 to take up a position at the University of Leipzig, Rossby followed him and spent a year there. In 1922, Rossby returned to Stockholm to join the Swedish Meteorological Hydrologic Service.

During the next three years, Rossby traveled as the meteorologist on several oceanographic expeditions. He also studied mathematics at the University of Stockholm, graduating in 1925 with a licentiate. This is a European degree one rank below a doctorate.

In 1926, Rossby visited the United States with a scholarship from the Scandinavian–American Foundation. He joined the staff of the UNITED STATES WEATHER BUREAU in Washington, D.C. At the time this was the only meteorological center in the United States. While working there, Rossby wrote several papers on turbulence in the atmosphere and on the dynamics of the stratosphere (*see* ATMOSPHERIC STRUCTURE).

In 1927, Rossby moved to California. Sponsored by the Daniel Guggenheim Fund for the Promotion of Aeronautics, Rossby established experimentally the first weather service that was designed expressly for the benefit of aviators. It became the model on which later aviation weather services were based.

The following year Rossby received his first important academic appointment when he was made the country's first professor of METEOROLOGY, at the Massachusetts Institute of Technology (M.I.T.). He devised the first university meteorological program, and during the 11 years he spent at M.I.T. he continued to pursue his research interests in the thermodynamics of AIR MASSES, turbulence in the atmosphere and oceans, and on BOUNDARY LAYERS. Later, he became increasingly interested in large-scale atmospheric movements.

In 1939, Rossby was appointed assistant chief of research and education at the U.S. Weather Bureau, but he left in 1940 to become chairman of the Institute of Meteorology at the University of Chicago. Soon after his arrival in Chicago Rossby developed his theory describing the long atmospheric waves that now bear his name (*see* ROSSBY WAVES).

During World War II, Rossby organized training for military meteorologists, while continuing his research on long waves. His wartime work took him to many parts of the world and brought him into personal contact with many British as well as American meteorologists. After the war, he was able to bring many of these scientists to the University of Chicago, where together they played an important part in developing the mathematics needed to introduce numerical WEATHER FORECASTING using electronic computers.

In 1947, Rossby was invited to establish a department of meteorology at the University of Stockholm. This was funded by American and Swedish foundations as well as by international bodies including UNESCO, and it attracted students from many countries. Appropriately, the institute was called the International Institute of Meteorology. Rossby divided his time between working at the institute in Stockholm and at the outpost of the University of Chicago that was opened at the Woods Hole Oceanographic Institute. His work at Stockholm was concerned mainly with the development

of numerical forecasting methods for European weather services.

Rossby died in Stockholm on August 19, 1957.

Rowland, Frank Sherwood
(b. 1927)
American
Chemist

F. Sherwood Rowland shared the 1995 Nobel Prize in chemistry with Mario Molina and Paul Crutzen. The prize was awarded to these three scientists for having shown first that the amount of OZONE in the OZONE LAYER might decrease as a result of reactions with chemicals released by human activity, and later that CFCs (chlorofluorocarbons) were the chemical compounds primarily involved.

F. Sherwood Rowland was born on June 28, 1927, in Delaware, Ohio, where his parents had moved the previous year when his father took up the post of professor of mathematics at the Ohio Wesleyan University in that city. As a small boy, Sherwood developed a keen interest in naval history.

He was educated at schools in Delaware, and in 1943 he enrolled at the Ohio Wesleyan University, studying chemistry, physics, and mathematics, but in June 1945, before completing his course, Rowland enlisted in the navy. He was discharged 14 months later in California, hitchhiked the 2,000 miles (3,200 km) back to Delaware, and resumed his studies. These were combined with sport, an interest that began when he was in high school. He played tennis, basketball, and baseball for university teams.

Rowland graduated in 1948 and the same year entered graduate school in the chemistry department at the University of Chicago. He obtained his Ph.D. in 1952, and in June of that year he married Joan Lundberg, a fellow graduate student.

In September 1952, Rowland became an instructor in the chemistry department at Princeton University, where he remained until 1956. He spent three summers, 1953–55, working on the use of tracer chemicals in the chemistry department of the Brookhaven National Laboratory. From 1956 until 1964, he was an assistant professor at the University of Kansas, and in 1964 he was appointed professor of chemistry at the University of California, Irvine, where he is now Donald Bren Research Professor of Chemistry and Earth

System Science. His link with Brookhaven continued until 1994.

In January 1972, Rowland attended a workshop on chemistry and METEOROLOGY in Fort Lauderdale, Florida, where he heard the English chemist James Lovelock describe how he had detected minute concentrations of a CFC in the air. Lovelock thought their great chemical stability would make CFCs useful for tracing the movement of AIR MASSES, but Rowland realized that no molecule can remain inert for ever. At high altitudes it will be broken apart by sunlight. He began to wonder what would happen to CFCs when they decayed. In 1973, Mario Molina joined his group as a research associate, and the two men set about studying the fate of airborne CFC molecules.

Paul Crutzen had already drawn attention to the possibility of depleting stratospheric ozone. Rowland and Molina published the results of their research as a paper in *Nature* in 1974. They had calculated that the breakdown of CFC molecules would release chlorine in a form that would destroy ozone molecules. Their paper stimulated a federal investigation of the situation, and in 1978 the use of CFCs as propellants in spray cans was banned in the United States. The phasing out of the use of CFCs was agreed internationally in 1987 under the terms of the Montreal Protocol on Substances That Deplete the Ozone Layer (*see* APPENDIX IV: LAWS, REGULATIONS, AND INTERNATIONAL AGREEMENTS). It was for this work that Rowland, Molina, and Crutzen were awarded their Nobel Prize.

Rutherford, Daniel
(1749–1819)
Scottish
Chemist

Daniel Rutherford, the chemist who discovered NITROGEN, was born in Edinburgh on November 3, 1749. He studied medicine at the University of Edinburgh, where one of his teachers was Joseph Black (1728–99).

In 1772, for his final thesis, Black set Rutherford the task of examining the portion of air that will not support combustion. Rutherford kept a mouse in an airtight container until the mouse died. Then he burned a candle and some phosphorus in the same air until they would no longer burn. He assumed that the air contained some CARBON DIOXIDE from respiration

and burning the candle, so he passed the gas through a strongly alkaline solution to remove the carbon dioxide.

The remaining gas was still noxious. A mouse could not survive in it, and it would not support combustion. Rutherford believed in the phlogiston theory. This proposed that phlogiston, a substance present in all combustible materials, was released when the material burned. Consequently, Rutherford believed the air in the container had accepted all the phlogiston it could hold. He called it "phlogisticated air." Joseph Priestley (1733–1804) and Karl Wilhelm Scheele (1742–86) also discovered the gas at about the same time. In 1790, the French chemist Jean-Antoine Chaptal (1756–1832) gave it the name nitrogen.

Daniel Rutherford became professor of botany at the University of Edinburgh in 1786. He died in Edinburgh on November 15, 1819.

Saussure, Horace-Bénédict de
(1740–1799)
Swiss
Physicist

Horace-Bénédict de Saussure was born at Conches, near Geneva, Switzerland, on February 17, 1740. His education began in 1746, when he enrolled at the public school in Geneva. He entered the Geneva Academy in 1754 and graduated in 1759, having presented a dissertation on the physics of fire. He was then 19 years old.

It was in the following year that he paid his first visit to Chamonix, a small resort in southeastern France standing at the foot of Mont Blanc. This is the tallest mountain in Europe, its peak 15,771 feet (4,810 m) above sea level, and when de Saussure visited it in 1760, no one had ever managed to climb it. De Saussure was fascinated by the mountain and toured the neighboring parishes offering a considerable reward to anyone who discovered a practicable route to the summit. In fact, Mont Blanc was first climbed 26 years later, on August 8, 1786, by Jacques Balmat and Gabriel Paccard. De Saussure climbed it himself in the summer of 1787, reaching the summit at 11 A.M. on August 3, accompanied by a number of guides and his personal valet.

De Saussure returned to the mountain many times over the succeeding years. His interest was not primarily in mountaineering as a sport, but in alpine plants, geology, and METEOROLOGY. From 1773, he began to spend increasing amounts of time in the area and climbed many of the mountains. The first volume of his most famous book, *Voyages dans les Alpes* (Journeys in the Alps) was published in 1779. The remaining three volumes were published between then and 1796. In them he described seven of his alpine journeys.

In 1761, de Saussure applied unsuccessfully for the vacant professorship of mathematics at the Geneva Academy. The following year he applied again, this time for the professorship of philosophy, and was successful. The new professor delivered his inaugural lecture in October. In 1772, he was elected a fellow of the Royal Society of London, and in the same year he founded the Society for the Advancement of the Arts in Geneva.

By this time de Saussure's reputation as a physicist was well established. He is credited with having constructed the first solar collector in 1767. This was a box with a glass top and heavily insulated sides, and he used it to discover why it is always cooler in the mountains than it is at lower levels. He took his box to the top of Mont Cramont. There the outside TEMPERATURE was 43°F (6°C), but the temperature inside the box rose to 190°F (88°C). Then he repeated the experiment 4,852 feet (1,480 m) lower down, on the Plains of Cournier. The air temperature there was 77°F (25°C), but the temperature inside the box was almost the same as it had been at the higher elevation. De Saussure concluded that the Sun shines just as warmly in the mountains as it does on the plains, but that the more transparent mountain air is unable to trap and hold so much warmth.

In 1783, de Saussure invented the hair HYGROMETER, based on his observation that human hairs increase in length as the HUMIDITY rises and grow shorter as the air becomes drier. Hair hygrometers are still the most widely used instruments for measuring relative humidity.

De Saussure married Albertine Boissier in May 1765. In February 1768, in the company of his wife and sister-in-law, he visited Paris, the Netherlands, and England, returning to Geneva in January 1769. In 1771, he visited Italy and in the autumn of 1772, with his wife and six-year-old daughter, he toured Italy, visiting Sicily, climbing Mount Etna, and calling at Rome, Rimini, and Venice before returning to Switzerland over the Brenner Pass. He also had an audience with Pope Clement XIV.

These were turbulent times in Geneva, and de Saussure became involved in politics. He drew up plans in 1776 for the reform of city institutions, and during the troubles of 1782 he was arrested and spent two days in prison. During revolutionary riots in July of the same year, he was besieged in his home for several days, suspected of harboring armed men and concealing weapons. The Terror that began in France after the Revolution of 1789 spread to Geneva, and in 1792 Geneva had its own revolution aimed at introducing a measure of democracy. The following year de Saussure was a member of the commission appointed to draft a constitution for the city. This led to an invitation, which he refused, to stand as a candidate for the governing council. The new constitution failed, and in 1794 the Terror returned.

In 1787, de Saussure resigned from his position at the Geneva Academy, and he then spent some time in the south of France, where he could live at sea level and collect measurements of atmospheric pressure that he could compare with those he had taken in the Alps. His health had begun to deteriorate in 1772, and by 1794 de Saussure was a sick man. He was also experiencing financial difficulties and was compelled to return with his family to the country house at Conches, where he had been born. News of his poverty spread, and he received offers of help from abroad. Thomas Jefferson himself considered offering de Saussure a position at the newly founded University of Virginia. It was not to be. De Saussure remained at Conches, and died there on the morning of January 22, 1799.

Schaefer, Vincent Joseph
(1906–1993)
American
Physicist

Vincent Schaefer was born at Schenectady, New York, on July 4, 1906. After leaving school, he worked for a time in the machine shop at the General Electric Corporation (G.E.C.) in Schenectady. Then, thinking outdoor work would suit him better, he attended classes at Union College, New York, and enrolled at the Davey Institute of Tree Surgery, from where he graduated in 1928 and became a tree surgeon. He was a keen skier and loved the snow, but unable to earn an adequate salary at tree surgery, had to abandon this profession.

In 1933, Schaefer returned to G.E.C. There he came to the notice of Irving Langmuir (1881–1957). In 1932, Langmuir had become the first American industrial scientist to win the Nobel Prize in chemistry. He was at the peak of his fame, and he recruited Schaefer as a research assistant. Schaefer became a research associate in 1938, and he remained at G.E.C. until 1954.

During World War II, Langmuir and Schaefer studied the problem of icing on the wings and other external surfaces of aircraft. This was extremely dangerous and caused many aircraft to crash, but before remedies could be found the scientists had to discover what was causing ice to form.

Working with his colleague Bernard Vonnegut (1914–97), Schaefer studied the formation of ICE and SNOW using a refrigerated box with an inside TEMPERATURE that remained at a constant -9°F (-23°C). He hoped to be able to induce WATER VAPOR to be deposited as ice around DUST particles. This work continued for some years until July 1946, when there was a spell of unusually hot weather. It became difficult to maintain the temperature inside the box, so on July 13 Schaefer dropped some dry ice (solid CARBON DIOXIDE) into the box to chill the air. The result was dramatic. The moment the dry ice entered the air in the box, water vapor turned into ICE CRYSTALS and there was a miniature snowstorm.

This suggested a way to make PRECIPITATION fall from a cloud that otherwise would not have released it. By November 13, Schaefer was ready for a full-scale trial. An airplane flew him above a cloud at Pittsfield, Massachusetts, about 50 miles (30 km) southeast of Schenectady. Schaefer dropped about 6 pounds (2.7 kg) of dry ice into the cloud and started the first artificially-induced snowstorm in history. This discovery led to the development of other techniques for CLOUD SEEDING.

Schaefer received the degree of doctor of science (Sc.D.) in 1948 from the University of Notre Dame, and in 1959 he joined the faculty of the State University of New York at Albany, where he founded the Atmospheric Sciences Research Center. He was appointed professor of atmospheric science at the State University of New York in 1964 and held the position until 1976. He was appointed a fellow of the American Academy of Arts and Sciences in 1957, and in 1976 he received a special citation from the American Meteorological Society.

Schaefer died at Schenectady on July 25, 1993.

Scheele, Karl (or Carl) Wilhelm
(1742–1786)
Swedish
Chemist

Karl Scheele discovered the elements OXYGEN, chlorine, and NITROGEN, as well as a long list of compounds. He was born at Stralsund, Pomerania, on December 9, 1742. Pomerania is now part of Germany, but then it belonged to Sweden. Scheele spoke and wrote in German, however.

Karl was the seventh of the 11 children of a poor family. Until he was 14 he received little education, but then he was apprenticed to an apothecary (druggist) in the city of Göteborg. He was a keen observer, and, determined to learn chemistry, he read and performed experiments. His growing skill and experience allowed him to advance to increasingly prestigious positions as an apothecary in Malmö (1765), Stockholm (1768), and Uppsala (1770). Other Swedish scientists, including Johann Gahn (1745–1818) and Torbern Bergman (1735–84), recognized his talent and helped build his reputation.

In 1775, Scheele was elected to the Swedish Royal Academy of Sciences, and in the same year he moved to Köping, in Västmanland. He remained in Köping for the rest of his life, practicing as an apothecary. He refused offers of academic appointments in England and Germany, and of an invitation to become court chemist to Frederick II of Prussia.

As well as chlorine, oxygen, and nitrogen, Karl Scheele discovered a long list of chemical compounds, yet not one of his discoveries was fully attributed to him. In some cases other scientists genuinely preceded him, and in other cases he failed to complete the research, allowing others to claim the credit. He also made mistakes. When he discovered chlorine, he thought it was an oxygen compound. Nevertheless, Scheele was arguably the greatest chemist of the 18th century.

Scheele had a habit of tasting the substances he discovered. One of these was hydrogen cyanide. This dangerous habit, together with overwork, may have contributed to the decline in his health that began in early middle age. He married while on his deathbed, and died at Köping on May 21, 1786, aged 43.

Schönbein, Christian Friedrich
(1799–1868)
German-Swiss
Chemist

Christian Schönbein, the chemist who discovered OZONE, was born at Metzingen, Württemberg, Germany, on October 18, 1799. He was educated at the Universities of Tübingen and Erlangen, and in 1828 he joined the faculty of the University of Basel, Switzerland. He became a professor at Basel in 1828.

In 1840, Schönbein began to experiment with the peculiar odor that for about half a century people had been noticing in the vicinity of electrical equipment. Schönbein found that he could produce the same smell by allowing phosphorus to oxidize and by electrolyzing water. He traced the smell to a gas that he called ozone, from the Greek *ozon,* meaning "smell."

Christian Schönbein also discovered guncotton— by accident. It seems that he was strictly forbidden to experiment in the kitchen at home, but one day in 1845 his wife was out and he was feeling brave. Working with a mixture of sulfuric and nitric acids, he accidentally spilled some and panicked, grabbing the nearest piece of cloth to mop up the acid before it could do any damage. The cloth happened to be his wife's apron, and he hung it by the kitchen stove to dry. When it was thoroughly dry, the apron disappeared with a pop. Schönbein realized something significant had happened, and when he analyzed the process, he found that the acid had added nitro groups (NO_2) to the cellulose in the cotton to form nitrocellulose, which he found was highly flammable, burning with no smoke and leaving no residue.

Schönbein died at Sauersberg, Baden, Germany, on August 29, 1868.

Schwabe, Heinrich Samuel
(1789–1875)
German
Astronomer and chemist

Heinrich Schwabe, who discovered the SUNSPOT cycle, was born at Dessau, to the southwest of Berlin, on October 25, 1789. His father was a doctor, and his mother ran a pharmacy. Heinrich was educated in Berlin, and at the age of 17 he entered the pharmacy business. After three years he returned to Berlin to study pharmacy at the university. His course lasted two years, and when it ended, in 1812, Schwabe returned to the pharmacy.

While at the university, Schwabe became keenly interested in astronomy and chose a research topic he could pursue during the daytime, during quiet periods in the pharmacy. He thought he might be able to discover a new planet orbiting the Sun inside the orbit of Mercury. Such a planet would be visible as it crossed in front of the Sun's disc. He began observing the Sun in about 1825, using a 2-inch (5-cm) telescope, and what he saw were sunspots. He became fascinated by the appearance and disappearance of the sunspots, and eventually he forgot about the search for a new planet and began drawing the sunspots. This became his life's work. Every day when the sky was clear he would record the sunspots.

In 1829, Schwabe sold the pharmacy and became a full-time astronomer. In 1831, he drew a picture of Jupiter that showed the Great Red Spot for the first time. In 1843, while recognizing the limitations of his equipment, Schwabe was able to announce that the number of sunspots waxes and wanes over a 10-year cycle (in fact the cycle is 11 years). The announcement was ignored, however, until Alexander von Humboldt (1769–1859) referred to it in his book *Kosmos.*

By the end of his life, Heinrich Schwabe had published 109 scientific papers, and the data he had collected filled 31 volumes, which were presented to the Royal Astronomical Society after his death. The Royal Astronomical Society presented him with its gold medal in 1857, and in 1868 he became a fellow of the Royal Society.

Schwabe died at Dessau on April 11, 1875.

Shackleton, Nicholas
(1937–2006)
English
Paleoclimatologist

Sir Nicholas Shackleton was one of the founders of paleoclimatology (*see* CLIMATOLOGY). He pioneered the interpretation of OXYGEN isotope ratios and confirmed that major climate changes were caused by variations in the Earth's solar ORBIT, thus validating the MILANKOVITCH CYCLES.

Nicholas Shackleton was born on June 23, 1937. He studied physics at Clare College, University of Cambridge, graduating in 1961. He received his Ph.D. from Cambridge in 1967 for a thesis entitled "The measurement of palaeotemperatures in the Quaternary era." From 1972 until 1987, he was assistant director of research at the sub-department of QUATERNARY research at the University of Cambridge, and from 1988 he was director of the sub-department. Shackleton was a visiting research fellow at the Lamont–Doherty Geological Observatory of Columbia University 1974–75 and a senior research associate 1975–2004. He was the director of the Godwin Institute of Quaternary Research at Cambridge from 1995 until his retirement in 2004. He was appointed to the academic rank of professor at Cambridge in 1991, and on his retirement he became an emeritus professor.

Shackleton discovered that the OXYGEN isotope ratios found in marine organisms are controlled not by temperature, as had previously been thought, but by the preferential removal of the lighter ^{16}O isotope by EVAPORATION and its accumulation in ICE SHEETS. This made it possible to measure the change in the volume of ice between GLACIAL PERIODS and INTERGLACIALS, while other scientists were able to convert the isotope ratios to temperatures. This work led to the CLIMATE LONG-RANGE INVESTIGATION MAPPING AND PREDICTION (CLIMAP) project. Arising from this work, in 1973 Shackleton found the link between variations in ice volume and the Milankovitch cycles. Shackleton also studied changes in carbon isotopes in order to understand the cause of variations in the amount of CARBON DIOXIDE stored in ICE CORES between glacial and interglacial periods.

As well as his climatological research, Professor Shackleton was a keen clarinet player and an authority on the history of the instrument. He taught a course at Cambridge on the physics of music.

Nicholas Shackleton received many awards and honors. He became a fellow of the Royal Society in 1985, a fellow of the American Geophysical Union in 1990, and a foreign associate of the National Academy of Sciences in 2000. He was awarded the Crafoord Prize by the Royal Swedish Academy of Science in 1995, the Ewing Medal by the American Geophysical Union in 2002, the Royal Medal by the Royal Society in 2003, and the Founder's Medal by the Royal Geographical Society in 2005. He received a knighthood in 1998.

Professor Sir Nicholas Shackleton died on January 24, 2006.

Further Reading

Asahi Glass Foundation. "Profiles of the 2005 Blue Planet Prize Recipients: Professor Sir Nicholas Shackleton." The Asahi Glass Foundation. Available online. URL: www.af-info.or.jp/eng/honor/hot/enr-shackleton.html. Accessed March 29, 2006.

Haug, Gerald H., and Larry C. Peterson. "Obituary: Nicholas Shackleton (1937–2006)." *Nature*, 439, 928, February 23, 2006.

Shaw, William Napier
(1854–1945)
English
Meteorologist

Shaw contributed greatly to the establishment of METEOROLOGY as a scientific discipline. He was born in Birmingham on March 4, 1854, and educated at a school in Birmingham at Emmanuel College, University of Cambridge, from which he graduated in 1876. He was elected a fellow of the Emmanuel College in 1877 and appointed a lecturer in experimental physics at the Cavendish Laboratory, which is part of the university. In 1898, he was appointed assistant director of the Cavendish Laboratory, but he resigned in 1900 to take up an appointment as secretary of the Meteorological Council. He was made director of the METEOROLOGICAL OFFICE in 1905 and remained there until his retirement in 1920. Shaw was also a reader in meteorology at the Royal College of Science of the Imperial College of Science and Technology, in London, from 1907 until 1920, and from 1920 until 1924 he was the college's first professor of meteorology.

Shaw introduced the millibar (*see* UNITS OF MEASUREMENT), in 1909, as a convenient unit of atmospheric pressure. It was adopted internationally in 1929. Some time about 1915, Shaw devised the TEPHIGRAM. He pioneered the use of instruments carried beneath kites and balloons to study the upper atmosphere and wrote several books on WEATHER FORECASTING, as well as one, *The Smoke Problem of Great Cities* (1925), on AIR POLLUTION. He received many honors, including the

1910 Symons Gold Medal of the Royal Meteorological Society, and in 1915, he received a knighthood.

Shaw died in London on March 23, 1945.

Simpson, George Clark
(1878–1965)
English
Meteorologist

Simpson studied atmospheric electricity and the effect of radiation on polar ice. He also standardized the wind speeds in the BEAUFORT WIND SCALE.

George Simpson was born in Derby on September 2, 1878, the son of a prominent tradesman. He went to school in Derby, leaving in 1894 to work in his father's business, but his reading of popular books about science aroused his interest, and he began attending night school. His father persuaded him to continue his education at Owens College, Manchester. He was coached for the entrance examination, entered the college, and graduated in 1900. He became an unsalaried tutor at Owens College until he won a traveling scholarship in 1902. This allowed him to continue his studies at Göttingen University, in Germany. He then visited Lapland to study atmospheric electricity.

On his return to his college in 1905, which by then had become the University of Manchester, he was appointed to head a newly formed METEOROLOGY department. He was the first lecturer in meteorology at any British university. Also in 1905 Simpson was appointed assistant director of the METEOROLOGICAL OFFICE. He was its director from 1920 until he retired in 1938.

Simpson traveled widely. He spent some time in India inspecting meteorological stations, and in 1910 he traveled as meteorologist on Robert Scott's last expedition to Antarctica. Between 1916 and 1920, he worked in the Middle East and Egypt.

George Simpson assigned wind speeds to the Beaufort scale based on measurements made by an ANEMOMETER standing 36 feet (11 m) above the ground (the international standard is now 20 feet [6 m] elevation). He studied the effect of increasing solar radiation on the polar ice caps, concluding that initially this increases PRECIPITATION, causing the ICE to advance, but that further radiation causes the ice to retreat. As solar radiation decreases, more precipitation falls as SNOW, producing a second advance of the ice.

On the outbreak of World War II in 1939, he came out of retirement, studying electrical storms, and retired for a second time in 1947.

Simpson was awarded the Symons Gold Medal of the Royal Meteorological Society in 1930. In 1935, he was knighted. He died in Bristol on January 1, 1965.

Spörer, Friedrich Wilhelm Gustav
(1822–1895)
German
Solar astronomer

Gustav Spörer identified a period of low SUNSPOT activity that now bears his name. Spörer was born in Berlin on October 23, 1822. He studied mathematics and astronomy at the University of Berlin. After graduating, he worked as a schoolteacher.

Spörer began his solar observations in 1858 and quickly established his reputation. He joined the staff of the Potsdam Astrophysical Laboratory in 1874, and in 1882 he was appointed its chief observer. By 1887, his studies of sunspot activity had convinced him that during the 17th century there had been a period during which there were very few sunspots. He also found a similar lack of sunspots between 1400 and 1510. This is the SPÖRER MINIMUM, which coincided with a sequence of very cold winters.

Gustav Spörer retired in 1894 and died at Potsdam on July 7, 1895.

Stefan, Josef
(1835–1893)
Austrian
Physicist

Josef Stefan discovered the relationship between the TEMPERATURE of a body and the amount of radiation it emits. Ludwig Boltzmann (1844–1906) explained the underlying theory, and the result is now known as the STEFAN-BOLTZMANN LAW.

Josef Stefan was born at St. Peter, a village near Klagenfurt, on March 24, 1835. He attended the gymnasium (high school) at Klagenfurt and enrolled at the University of Vienna in 1853. After graduating, he became a lecturer at the university in 1858, and in 1863 he was appointed professor of higher mathematics and physics. Stefan held this post for the rest of his life. From 1866, he was also director of the Institute for Experimental Physics, and from 1885 he was vice president of the Imperial Academy of Sciences.

Stefan discovered his radiation law in 1879. Isaac Newton (1642–1727) had maintained that the rate at which a body cools is proportional to the difference in temperature between the body and its surroundings. Physicists had known for some time, however, that this was not true if the temperature difference is very large. John Tyndall (1820–93) had measured the radiant heat emitted by a platinum wire heated to several high temperatures, and Stefan followed up this work. Eventually, he found that the radiation emitted by a hot body is proportional to the fourth power of its absolute temperature. In 1884, Boltzmann showed that this could be deduced from thermodynamic principles.

Josef Stefan died in Vienna on January 7, 1893.

Stevin, Simon
(1548–1620)
Flemish
Mathematician

In 1586, Simon Stevin published a book, *Statics and Hydrostatics,* in which he reported his discovery that the pressure a liquid exerts on the surface beneath it depends only on the area of the surface on which it presses and the height to which the liquid extends above the surface. Contrary to what seemed obvious to most people at the time, it has nothing whatever to do with the shape of the vessel holding the liquid. This finding paved the way for Evangelista Torricelli (1608–47), who found a way to weigh air and in doing so invented the BAROMETER.

Stevin, often known by the Latin version of his name as Stevinus, was born in Bruges in 1548. He worked as a clerk in Antwerp and then entered the service of the Dutch government, becoming director for the department of roads and waterways and later quartermaster-general to the Dutch army. He devised for military purposes a scheme for opening sluices in the dikes protecting the polders—cultivated land reclaimed from the sea. This would flood the land in the path of any invading army. He also invented a carriage propelled by sails that ran along the beach and could carry 26 passengers, and he is said to have dropped two objects of different weights from a tall tower and found they reached the ground simultaneously—an experiment usually attributed to Galileo

(1564–1642). Stevin established the use of decimal notation in mathematics—representing ½ as 0.5, for example, and ¼ as 0.25. He maintained that decimal weights, measures, and coinage would eventually be introduced.

Stevin wrote in Flemish and advised all scientists to describe their work using their own native language. This was unusual, and his works were all translated into Latin later.

Simon Stevin died either at The Hague or Leiden, two cities in the Netherlands that are 10 miles (16 km) from each other.

Stewart, Balfour
(1828–1887)
Scottish
Physicist

Balfour Stewart discovered the properties of BLACK-BODIES independently of Gustav Robert Kirchhoff (1824–87), who was more famous and who is usually credited as the sole discoverer. Stewart is best known, however, for his studies of the Earth's magnetic field.

Balfour Stewart was born in Edinburgh on November 1, 1828. He was educated at the Universities of Dundee and Edinburgh, and after graduating he joined the staff of the Kew Observatory, near London, becoming its director in 1859. In 1870, he joined the faculty of Owens College, Manchester (later the University of Manchester).

Stewart studied the theory of heat exchange and found that if the TEMPERATURE of a body remains constant, the amount of radiation the body absorbs is equal to the amount it emits. His studies of magnetism led him to suggest in 1882 that the daily variations in the orientation of the Earth's magnetic field might be caused by electric currents flowing horizontally in the upper atmosphere. Most of his colleagues thought the idea absurd, but it was confirmed in 1902 by the British-American electrical engineer Arthur Edwin Kenneally (1861–1939) and the English physicist and electrical engineer Oliver Heaviside (1850–1925). They proposed the existence of a layer of the atmosphere that was known for a time as the Kenneally–Heaviside layer and is now called the ionosphere (*see* ATMOSPHERIC STRUCTURE).

Balfour Stewart died near Drogheda, Ireland, on December 19, 1887.

Stokes, George Gabriel
(1819–1903)
Irish
Physicist

One of the most eminent physicists of his generation, George Stokes was born at Skreen, Sligo, Ireland, on August 13, 1819. He went to school in Dublin and to a college in Bristol, England, before enrolling at the University of Cambridge to study mathematics. He graduated in 1841 and became a fellow of Pembroke College.

In 1849, Stokes was appointed Lucasian Professor of Mathematics at the University of Cambridge, a position he held until his death. He was elected secretary of the Royal Society in 1854 and in 1885 its president. The last individual to hold all three of these posts was Isaac Newton (1642–1727).

It was between 1845 and 1850, while working on the theory of viscous fluids, that Stokes developed the law bearing his name. STOKE'S LAW relates the force required to move a body through a fluid to the size and VELOCITY of the body and the VISCOSITY of the fluid. This makes it possible to calculate the FRICTION on a moving body and the TERMINAL VELOCITY of a body falling through the air. It explains why clouds float in the sky, but it can also predict the resistance a ship will experience as it moves through the water.

Stokes also studied sound, light, and fluorescence. He was the first person to show that quartz is transparent to ultraviolet radiation (*see* SOLAR SPECTRUM) but glass is not.

Stokes received a knighthood in 1889. He died at Cambridge on February 1, 1903.

Teisserenc de Bort, Léon-Philippe
(1855–1913)
French
Meteorologist

The scientist who discovered the stratosphere (*see* ATMOSPHERIC STRUCTURE) was born in Paris on November 5, 1855, the son of an engineer. His career began in 1880 when he went to work in the meteorological department of the Central Bureau of Meteorology, in Paris. He undertook three expeditions to North Africa, in 1883, 1885, and 1887, to study the geology and geomagnetism. During the same period, he was growing increasingly interested in the distribution of atmospheric pressure at an altitude of about 13,000 feet (4,000 m).

In 1892, Teisserenc de Bort was made chief meteorologist at the bureau, but he resigned in 1896 in order to establish a private meteorological observatory. This was located at Trappes, near Versailles, and Teisserenc de Bort used it primarily to study clouds and the upper air. He was one of the pioneers in the use of balloons to take soundings of the upper atmosphere, and he described their use and the results he had obtained with them in a paper he published in 1898 in the journal *Comptes Rendus*. As well as working from his own observatory, Teisserenc de Bort conducted investigations in Sweden, over the Zuider Zee and Mediterranean Sea, and over the tropical Atlantic.

The balloon soundings Teisserenc de Bort made revealed that the air temperature decreased with height as expected, but only to an altitude of about 7 miles (11 km). Above this height, the temperature remained constant as far as the greatest altitude his balloons could reach. In 1900, he proposed that the atmosphere comprises two layers. He called the lower layer, in which temperature decreases with height and the air is constantly moving, the troposphere. The Greek word *tropos* means "turning." He found that the upper boundary of the troposphere is marked by a layer he called the tropopause.

Above the tropopause there is an isothermal layer. Teisserenc de Bort believed that because the temperature in this layer remains constant, there is no mechanism such as CONVECTION to make the air move vertically. Consequently, he thought the air might separate into its constituent gases, which would then form layers, with the heaviest gases at the bottom and the lightest at the top. Immediately above the tropopause there would be a layer of OXYGEN, above that a layer of NITROGEN, then HELIUM, and a layer of HYDROGEN at the top. This stratified arrangement led him to name the isothermal layer the stratosphere. The stratosphere is not, in fact, layered in the way Teisserenc de Bort supposed, but his name for it has survived.

Teisserenc de Bort died at Cannes on January 2, 1913.

Theophrastus
(371 or 370–288 or 287 B.C.E.)
Greek
Philosopher

Theophrastus is often called the "father of botany" for his studies of plants and attempts at plant classification.

His interests embraced the entire field of contemporary knowledge, however, and they included atmospheric phenomena. Two of his books, *On Weather Signs* and *On Winds* are among the earliest accounts of natural signs and their meteorological significance (*see* WEATHER LORE).

Theophrastus was the nickname Aristotle (384–322 B.C.E.) gave him; it means "divine speech." He was born Tyrtamus in 371 or 370 B.C.E. at Eresus on the island of Lesbos. His education began in Lesbos, and as a young man he went to Athens to study at the Academy headed by Plato (428 or 427–348 or 347 B.C.E.), where Aristotle also taught. Aristotle left Athens soon after Plato died, and Theophrastus may have accompanied him. When Aristotle returned to Athens in 335 B.C.E., Theophrastus continued studying under him at the Lyceum.

Aristotle retired to Chalcis in 322 or 321 B.C.E., appointing Theophrastus as his successor and bequeathing him his library. Theophrastus built a covered walkway, in Greek a *peripatos*, at the Lyceum, and his style of philosophy, based mainly on Aristotelian teaching, became known as the Peripatetic School. He was the most successful of all the peripatetic philosophers. At one time he had 2,000 students. His popularity is attested by the fact that an attempt to charge him with the capital offence of impiety collapsed.

Theophrastus presided over the Peripatetic School for 35 years, until his death in 288 or 287 B.C.E. He was honored with a public funeral, and many Athenians followed the funeral procession to his grave.

Thornthwaite, Charles Warren
(1899–1963)
American
Climatologist

C. W. Thornthwaite was one of the most eminent climatologists of his generation, with an international reputation. He held many important positions, but his enduring fame rests on the CLIMATE CLASSIFICATION he devised. This remains in widespread use, especially among agricultural scientists, because of its emphasis on thermal efficiency and PRECIPITATION efficiency, two concepts that Thornthwaite introduced.

Major climate classifications do not appear all at once, in their complete and final form. Their authors revise and amend them over the years, and Thornth-waite was no exception. The first version of his scheme applied only to North America. It appeared in October 1931, in an article titled "The Climates of North America According to a New Classification" that appeared in the *Geographical Review* (21: 633–55). He expanded the classification to cover the world in another article, "The Climates of the Earth," that appeared in the *Geographical Review* in July 1933 (23:433–40). In January 1948, he published a second version of his classification, in which he introduced the concept of potential EVAPOTRANSPIRATION. He described the new scheme in a third article, "An Approach Toward a Rational Classification of Climate" (*Geographical Review* 38:55–94).

Charles Warren Thornthwaite was born on March 7, 1899, near Pinconning, in Bay County, Michigan. He graduated as a science teacher in 1922 from Central Michigan Normal School and received his doctorate from the University of California, Berkeley, in 1929. Thornthwaite held faculty positions from 1927 to 1934 at the University of Oklahoma, from 1940 to 1946 at the University of Maryland, and from 1946 to 1955 at the Johns Hopkins University. He headed the Division of Climatic and Physiographic Research of the U.S. Soil Conservation Service from 1935 to 1942. From 1946 until his death, he was the director of the Laboratory of Climatology, at Seabrook, New Jersey, and professor of climatology at Drexel Institute of Technology, Philadelphia.

From 1941 to 1944, Thornthwaite was president of the section of METEOROLOGY of the American Geophysical Union. In 1951, he became president of the Commission on Climatology of the WORLD METEOROLOGICAL ORGANIZATION, a post he held until his death. He died from cancer on June 11, 1963.

Torricelli, Evangelista
(1608–1647)
Italian
Physicist and mathematician

The Italian physicist and mathematician who discovered the principle of the BAROMETER was born at Faenza, near Ravenna, on October 15, 1608. In 1627, he went to Rome to study science and mathematics at the Sapienza College. Having read some of the works of Galileo (1564–1642), he was inspired to write a treatise developing some of Galileo's ideas on mechanics.

Torricelli's teacher, a former student of Galileo, sent the treatise to Galileo, who was impressed. Galileo invited Torricelli to Florence, and in 1641 the two met. Galileo was then old and blind. Torricelli became his assistant, but Galileo died three months later. It was Galileo who suggested the problem that Torricelli solved.

The question was: Why does water rise up a cylinder when a piston in the cylinder is raised, but only so far? The conventional explanation for water rising in this way at all, which Galileo supported, was that raising the piston produces a vacuum in the cylinder and that nature abhors a vacuum, so the water is compelled to rise. It can be made to rise only about 33 feet (10 m) above its ordinary level, however, and that was the puzzle. Galileo suspected that nature's abhorrence of a vacuum was limited, so vacuums can be tolerated under certain conditions. He asked Torricelli to investigate.

Torricelli rejected the idea of "vacuum abhorrence." He considered what might happen if the air possessed weight. In his day most scientists believed air had a property of "levity," making it tend to rise, so Torricelli's idea was fairly radical. If air possessed weight, however, then the weight of air would press down on the surface of the water outside the cylinder. As the piston was raised, there would be a space containing no air between the bottom of the piston and the surface of the water. The air would not press down inside the cylinder below the piston, but it would still do so outside. Consequently, the pressure outside the cylinder would cause water to rise inside the cylinder as the piston was raised.

Suppose, though, that the pressure due to the weight of air was sufficient only to raise the water by 33 feet? In that case, 33 feet would be as far as the water could be made to rise, and withdrawing the piston further would have no effect.

In 1643, Torricelli tested this idea, choosing to use mercury, which is 13.6 times denser than water. He partly filled a dish with mercury and completely filled a glass tube 4 feet (1.2 m) long and open at one end. Placing his thumb over the open end, he inverted the tube of mercury and placed the open end below the surface of the mercury in the dish, holding the tube absolutely upright. When he removed his thumb, mercury flowed out of the tube and into the dish, but some mercury remained in the tube. The level of mercury in the tube was about 30 inches (760 mm) higher than the surface of the mercury in the dish.

Because the liquid had fallen from a full tube, rather than being drawn upward by the withdrawal of a piston, Torricelli had proved it was the pressure exerted by the weight of air on the mercury in the dish that supported the column of mercury in the tube. He had proved that air has weight, a finding with implications that were explored by Blaise Pascal (1623–62).

Above the mercury, the tube was empty except for a small amount of mercury vapor. Torricelli was the first person to make a vacuum, and a vacuum produced in this way is still known as a Torricellian vacuum.

Then Torricelli noticed that the height of the column of mercury varied slightly from day to day. He attributed this, correctly, to variations in the weight of the air. He had invented a device for measuring these small changes—the first barometer.

After Galileo's death, Torricelli was appointed mathematician to the grand duke as well as professor of mathematics in the Florentine Academy. He died in Florence on October 25, 1647.

Travers, Morris William
(1872–1961)
English
Chemist

Morris Travers collaborated with William Ramsay (1852–1916) in the work that led to the discovery of rare gases.

Travers was born in Kensington, London, on January 24, 1872, the second of four sons of a London physician. He was educated at schools in Ramsgate, Kent (1879–82) and Woking, Surrey (1882–85), before being sent to Blundell's School at Tiverton, Devon, chosen because it had a good chemistry laboratory. He enrolled at University College, London, in 1889, graduating in chemistry in 1893.

In 1894, he began to conduct research in organic chemistry at the University of Nancy, France, but after a few months he returned to University College and became a demonstrator, working under Ramsay. This was the start of the collaboration that led to their discoveries of ARGON, NEON, KRYPTON, and XENON. He received his D.Sc. in 1898 and was made an assistant professor. Travers became a full professor in 1903.

In 1906, Travers was appointed director of the Indian Institute of Scientists, in Bangalore, India. He returned to England at the outbreak of World War

I in 1914 to direct glass production at a factory in Walthamstow, London. He later became president of the Society of Glass Technology. He became interested in fuel technology, the gasification of coal, and in high-temperature furnaces. In 1927, he was appointed an honorary professor of chemistry at the University of Bristol. Travers retired in 1937.

During World War II, Travers was a consultant and adviser to the explosives section of the Ministry of Supply. He died at his home in Stroud, Gloucestershire, on August 25, 1961.

Tyndall, John
(1820–1893)
Irish
Physicist

John Tyndall was born at Leighlin Bridge, County Carlow, in southwestern Ireland, on August 2, 1820, the son of a police officer. He was educated at the school in the nearby town of Carlow, where he acquired a sound background in mathematics and English. He left school at 17, and two years later, in 1839, he joined the Ordnance Survey to train as a surveyor.

The Ordnance Survey is the British government agency responsible for mapping the country. In the 19th century, it was part of the military establishment. Tyndall worked on the survey of Ireland (which was then part of Britain), and when that task was completed, in 1842, he transferred to the English survey. He was dismissed in 1843, however, because he had complained about the efficiency of the Ordnance Survey and about its treatment of Irish people. He returned to Ireland, but then obtained a position with a firm of surveyors in England, so he returned and conducted surveys for the companies that were rapidly expanding the railroad network.

The railroad boom ended in 1847, and Tyndall found a post as a mathematics teacher at Queenwood College, Hampshire. There he became friendly with the science teacher, the chemist Edward Frankland (1825–99) and, through Frankland's influence, he developed an interest in science. The two men decided to further their education, and in October 1848 they enrolled together at the University of Marburg, Germany. Tyndall was then 28 years old.

Tyndall studied physics, calculus, and chemistry at Marburg. His chemistry teacher was Robert Bunsen (1811–99), after whom the laboratory Bunsen burner is named. Bunsen helped and encouraged Tyndall, who began the course with only a limited knowledge of German. Despite this, he qualified for a doctorate in 1850. He specialized in the study of magnetism and optics and continued his research for an additional year in Germany. Short of funds, in 1851 Tyndall returned to his post at Queenwood College, where he spent the next two years, supplementing his salary by translating and reviewing scientific articles from foreign journals.

By this time, Tyndall was beginning to earn the respect of other scientists. He was elected a fellow of the Royal Society in 1852, and in 1853 it was arranged that he should give a lecture at the Royal Institution in London. This was so successful that he was invited to give a second lecture, and then a whole course of lectures. A few months later, he was appointed professor of natural philosophy at the Royal Institution. Michael Faraday (1791–1867) was the superintendent of the Royal Institution. The two men were close friends, and when Faraday died Tyndall succeeded him as superintendent.

John Tyndall acquired a reputation as a brilliant and entertaining lecturer. In 1872 and 1873, he made a lecture tour of the United States. Tyndall paid the proceeds from those lectures into a trust for the advancement of American science. He was also a talented writer and journalist and did much to popularize science in Britain and the United States. From 1859 to 1868, Tyndall was also professor of physics at the Royal School of Mines, where he gave a series of popular lectures on science, working in collaboration with another close friend, Thomas Henry Huxley (1825–95). The success of those lectures prompted the British Association for the Advancement of Science to hold similar lectures at its annual meetings.

In addition to lecturing, teaching, and writing, Tyndall was also conducting his own research in the laboratories in the basement of the Royal Institution. At first these were on diamagnetism. Then, starting in January 1859, he studied radiant heat. He found that although OXYGEN, NITROGEN, and HYDROGEN are completely transparent to radiant heat (infrared radiation, *see* SOLAR SPECTRUM), other gases, especially WATER VAPOR, CARBON DIOXIDE, and OZONE, are relatively opaque. Tyndall said that without water vapor the surface of the Earth would be permanently frozen. Later, he speculated about the way changing the concentrations of these gases might affect the climate. This was the first suggestion of what is now known as the GREENHOUSE EFFECT.

In 1869, Tyndall discovered what is now called the Tyndall effect. He was investigating the way light passes through liquids and found that light passes unimpeded through a solution or pure solvent, but that the light beam becomes visible in a colloidal solution (*see* COLLOID). This suggested that although the colloidal particles cannot be seen, they are big enough to scatter light. From this he reasoned that particles in the atmosphere should also scatter light, and that air molecules should scatter blue light more than red light, causing the blue color of the sky. Lord Rayleigh (1842–1919) confirmed this in 1871. Tyndall used the effect to measure the pollution of London air. It also led him to suspect that the air may contain microscopic living organisms, and he was able to show in 1881 that bacterial spores are present in even the most carefully filtered air. This confirmed the rejection by Louis Pasteur (1822–95) of the idea that life is generated spontaneously from nonliving matter, and it provided added support to the germ theory of disease.

Tyndall had many interests. He was a keen mountaineer. He climbed Mont Blanc (15,781 feet, 4,813 m) several times and was the first person to climb the Weisshorn (14,804 feet, 4,515 m). In 1849, he and Huxley began to take annual vacations in the Alps. This led them to publish a treatise, *On the Motion and Structure of Glaciers,* about glacial movement. He also invented many scientific instruments. He led an active civic life, acting as a government adviser and helping to investigate the causes of mining and other industrial accidents. He was especially concerned about safety at sea.

Tyndall also devoted considerable energy to championing the cause of those he believed had been badly treated. In the 1860s, for example, he lectured widely on the work of the German physicist Julius Robert Mayer (1814–78). Mayer was years ahead of all his contemporaries in calculating the mechanical equivalent of heat. He proposed the conservation of energy (*see* THERMODYNAMICS, LAWS OF), and he suggested that solar energy is the ultimate source of all the energy on Earth and that solar energy is produced by the conversion of KINETIC ENERGY into radiant energy. Yet Mayer's work aroused little interest, and more famous scientists, including James Prescott Joule (1818–89),

Hermann Ludwig Ferdinand von Helmholtz (1821–94), and Lord Kelvin (1824–1907) were given the credit for discoveries Mayer had made first. Mayer became so depressed at the lack of recognition and his complete failure to claim priority for his discoveries that in 1849 he attempted suicide, and in 1851 he was admitted to a mental institution that would be judged harsh and cruel by modern standards. Tyndall worked tirelessly to put right the wrongs Mayer had suffered, and he finally succeeded in having Mayer's achievements recognized.

In 1876, when he was 56, Tyndall married Louisa Hamilton. They had no children, but lived together devotedly until Tyndall's death.

Despite his energy, throughout his adult life Tyndall slept badly and was often unwell. By the 1880s, his health was deteriorating. He retired from the Royal Institution in 1887 and went to live at Hindhead, Surrey. He gave up most of his scientific work, but took an active part in political campaigns, including the campaign against the Home Rule Bill, which would have made Ireland independent of Britain. In 1891, for the first time in 30 years, he was too ill for his annual climbing vacation in the Alps. His insomnia became steadily worse, and he experimented with a variety of drugs to treat it. He died on December 4, 1893, after his wife had accidentally given him an overdose of the sedative chloral hydrate.

Further Reading

Earth Observatory. "On the Shoulders of Giants: John Tyndall (1820–1893)." NASA. Available online. URL: http://earthobservatory.nasa.gov/Library/Giants/Tyndall/. Accessed March 29, 2006.

Irish Scientists. "John Tyndall (Physicist, Mountaineer, and Author)." Available online. URL: http://members.tripod.com/~Irishscientists/scientists/JOHN.HTM. Accessed March 29, 2006.

Von Guericke, Otto
(1602–1686)
German
Physicist

Otto von Guericke carried out one of the most famous of all experiments, demonstrating beyond doubt that the atmosphere exerts pressure. He also demonstrated that Aristotle's assertion that a vacuum is impossible, summed up by medieval scholars in the saying "nature abhors a vacuum," was false. Vacuums can and do exist.

Von Guericke was born at Magdeburg on November 20, 1602. His family was wealthy, and Otto was able to study law, science, and engineering at the Universities of Leipzig (1617–20), Helmstedt (1620), Jena (1621–22), and Leiden (1622–25). He traveled in France and England. A Protestant city, Magdeburg suffered severe damage during the Thirty Years' War and most of its 40,000 population were killed. Guericke and his family fled to Brunswick and the city of Erfurt, where he worked as an engineer for the Swedish and later Saxon governments. On his return to Magdeburg in 1627, Guericke was made an alderman. As the war continued to rage, he represented Magdeburg in negotiations with various occupying powers. He was mayor of Magdeburg from 1646 until 1676, and in 1646 he was ennobled, becoming von Guericke.

Aristotle had proposed that a moving body will accelerate if the medium through which it travels becomes less dense, because it will encounter less resistance. This being so, a moving body entering a vacuum will accelerate to infinite speed. Aristotle did not believe infinite speed was possible, so he concluded that a vacuum cannot exist. In 1643, Evangelista Torricelli made his device to demonstrate that air has weight—the first BAROMETER—using a tube filled with mercury. When the level of the mercury in the tube fell, the empty space above the mercury contained nothing (apart from a small amount of mercury vapor). It was a vacuum.

Von Guericke determined to conduct a more convincing demonstration. In 1647, he built an air pump worked by human muscle and attached it to a copper vessel. Once the seals were airtight, he pumped air out of the vessel until it imploded. He followed this demonstration by others showing that after the air has been removed from a vessel a bell ringing inside the vessel cannot be heard clearly, that candles will not burn in the vessel, and that animals cannot survive in it.

In 1654, he constructed a cylinder and piston. He had 50 strong men pull on a rope attached to the piston. For all their strength they could pull it no more than halfway up the cylinder, and as von Guericke's pump removed the air, the men were unable to prevent the piston from moving back into the cylinder.

Von Guericke performed his most famous and most dramatic demonstration in 1657. This time he used two metal hemispheres that fitted together snugly along a greased flange. These came to be called the Magdeburg hemispheres. A team of horses was attached by ropes to each hemisphere, and von Guericke evacuated the air from inside. Try as they might, the two teams of horses could not pull the hemispheres apart. When von Guericke allowed air to enter the hemispheres, they immediately fell apart of their own accord.

Von Guericke had shown not simply that the atmosphere exerts pressure, but that it exerts very great pressure. He also experimented with the elasticity of air. Blaise Pascal had shown in 1648 that AIR PRESSURE decreases with height. Von Guericke explored the link between air pressure and weather conditions. Using a water barometer to monitor changes in air pressure, von Guericke began issuing weather forecasts, and in 1660 he proposed the establishment of a string of weather stations that would contribute data to a WEATHER FORECASTING system.

This remarkable man also demonstrated the magnetization of iron by hammering it in a north–south direction, and he devised a machine for producing static electricity by friction. He was interested in astronomy and suggested that comets return at regular intervals.

Von Guericke retired in 1681 and went to live with his son in Hamburg. He remained there for the rest of his life and died at Hamburg on May 11, 1686.

von Kármán, Theodore
(1881–1963)
Hungarian
Aerodynamicist

Theodore von Kármán studied the flow of fluids around cylinders, discovering that the wake separates into two rows of vortices, known as von Kármán vortex streets (*see* VORTEX). These vortices can cause vibrations that are serious enough to damage structures. The spiral fluting around many factory chimneys are designed to prevent vortex streets forming.

Von Kármán was born in Budapest on May 11, 1881. His father, Mór, was a professor at the University of Budapest. Theodore studied engineering at the Budapest Royal Polytechnic University, graduating in 1902. In 1903, he was appointed an assistant professor there. He studied for his doctorate at the University of Göttingen, Germany, where he helped design a wind tunnel for airship research. He obtained his doctorate in 1909.

During World War I, von Kármán served in the Austro-Hungarian army, but apart from that interlude he spent the years between 1913 and 1930 directing aeronautical research at the Aachen Institute, Germany. He visited the United States in 1926, and in 1928 he began to divide his time between Aachen and the Guggenheim Aeronautical Laboratory at the California Institute of Technology. He became director of the Guggenheim Laboratory and concentrated his research on aerodynamics.

Theodore von Kármán died at Aachen, while on vacation visiting a friend, on May 7, 1963.

Vonnegut, Bernard
(1914–1997)
American
Physicist

Bernard Vonnegut was born at Indianapolis, Indiana, on August 29, 1914. He was educated at the Massachusetts Institute of Technology (M.I.T.), graduating in 1936. He obtained his Ph.D. from M.I.T. in 1939 for research into the conditions that produce icing on aircraft (*see* FLYING CONDITIONS). From 1939 until 1941, he worked for the Hartford Empire Company, and from 1941 until 1945 Vonnegut was a research associate at M.I.T.

In 1945, Bernard Vonnegut moved to the laboratories of the General Electric Corporation in Schenectady, New York, where he continued his research into icing in collaboration with Vincent Schaefer (1906–93). Following the discovery that dry ice (solid CARBON DIOXIDE) was effective at CLOUD SEEDING, Vonnegut turned his attention to the search for other materials that might perform the same task. Deciding that crystals of silver iodide were the right size and shape to act as FREEZING NUCLEI, he experimented with them and was proved correct. Silver iodide largely replaced dry ice as a seeding medium. Unlike dry ice, it can be stored indefinitely at room temperature and it does not have to be released from an airplane flying above the target cloud. Silver iodide can be released at ground level and will be carried into the cloud by vertical air currents. If dry ice were released in this way, it would vaporize before it could cause ice DEPOSITION.

Vonnegut moved to the Arthur D. Little Corporation in 1952, and in 1967 he was appointed Distinguished Research Professor of the State University of New York, a position he held until his death at Albany, New York, on April 25, 1997.

Wegener, Alfred Lothar
(1880–1930)
German
Meteorologist and geologist

Alfred Wegener is best known for having formulated the theory of CONTINENTAL DRIFT that developed later into the theory of PLATE TECTONICS. He was primarily a meteorologist, however, and studied the formation of RAINDROPS and the circulation of air over polar regions.

Alfred Wegener was born in Berlin on November 1, 1880. His father was a minister and director of an orphanage. Alfred was educated at the Universities of Heidelberg, Innsbruck, and Berlin. In 1905, he received a Ph.D. in planetary astronomy from the University of Berlin, but immediately switched to METEOROLOGY, taking a job at the Royal Prussian Aeronautical Observatory, near Berlin. He used kites and balloons to study the upper atmosphere and also flew hot air balloons. In 1906, Alfred and his brother Kurt remained airborne for more than 52 hours, breaking the world endurance record.

In 1906, he joined a Danish two-year expedition to Greenland as the official meteorologist. Wegener studied the polar air, using kites and tethered balloons, and on his return to Germany in 1909 he became a lecturer in meteorology and astronomy at the University of Marburg. He collected his lectures into a book published in 1911, *Thermodynamik der Atmosphäre* (Thermodynamics of the atmosphere). This became a standard textbook throughout Germany. In it, Wegener pointed out that where ICE CRYSTALS and supercooled droplets (*see* SUPERCOOLING) are present together, the crystals will grow at the expense of the droplets, because the equilibrium vapor pressure (*see* WATER VAPOR) is lower over the crystals. He suggested that this might lead to the formation of crystals that were large enough to sink through the cloud and melt at lower levels to become raindrops. Wegener never had an opportunity to test this idea in real clouds. Tor Bergeron and Walter Findeisen finally tested it in the 1930s. Bergeron acknowledged his debt to Wegener, and although this type of raindrop formation is usually known as the Bergeron-Findeisen mechanism, it is sometimes called the Wegener-Bergeron-Findeisen mechanism.

Wegener was also the first person to explain two kinds of arc that are occasionally seen in the Arctic opposite the Sun. The arcs are caused by ice crystals that form in the very cold air. They are now called Wegener arcs.

Since 1910, Wegener had been intrigued by the apparent fit of the continental coastlines on either side of the Atlantic Ocean. In 1912, he published a short book, *Die Entstehung der Kontinente und Ozeane* (*The Origin of the Continents and Oceans*), drawing together various strands of evidence to support the idea that he called "continental displacement." This proposed that the continents had once been joined and that they have moved slowly to their present positions.

It was also in 1912 that Wegener married Else Köppen, the daughter of Wladimir Köppen (1846–1940), the most eminent climatologist in Germany. Wegener and Köppen collaborated in a book about the history of climate, *The Climates of the Geological Past*. Wegener then returned to Greenland to take part in the four-man 1912–13 expedition. The team crossed the ice cap and was the first to spend the winter on the ice.

On the outbreak of war in 1914, Wegener was drafted into the German army, but was wounded almost at once. He spent a long time recuperating, during which he elaborated on his theory of continental drift, publishing an expanded version of *The Origin of the Continents and Oceans* in 1915 (it was not translated into English until 1924). The book received a hostile reception from German scientists, and Wladimir Köppen strongly disapproved of his son-in-law's digression from meteorology into geophysics. Wegener spent the remainder of the war employed in the military meteorological service.

After the war, Wegener returned to Marburg. In 1924, he accepted a post created especially for him and became the professor of meteorology and geophysics at the University of Graz, in Austria.

In 1930, he returned to Greenland once more, this time as the leader of a team of 21 scientists and technicians planning to study the climate over the ice cap. They intended to establish three bases, all at 71°N, one on each coast and one in the center, but they were delayed by bad weather. On July 15, a party left to establish the central base, called Eismitte, 250 miles (402 km) inland. The weather then prevented

necessary supplies from reaching them, including the radio transmitter and hut in which they were to live. On September 21, Wegener, accompanied by 14 others, set off with 15 sleds to carry supplies to Eismitte. The appalling conditions forced all but Wegener, Fritz Lowe, and Rasmus Villumsen to give up and return. These three finally reached Eismitte on October 30. Lowe was exhausted and badly frostbitten. They stayed long enough to celebrate Wegener's 50th birthday on November 1, then Wegener and Villumsen began their return, leaving Lowe to recover. They never reached the base camp. At first it was assumed they had decided to overwinter at Eismitte, but when they had still not appeared in April a party went in search of them. They found Wegener's body on May 12, 1931. He appeared to have suffered a heart attack. Villumsen had carefully buried the body. They marked the site with a mausoleum made from ice blocks, later adding a large iron cross. Despite a long search, Villumsen was never found.

There is now an Alfred Wegener Institute for Polar and Marine Research at Bremerhaven, Germany.

Further Reading

University of California. "Alfred Wegener (1880–1930)." University of California, Berkeley. Available online. URL: http://www.ucmp.berkeley.edu/history/wegener. html. Accessed March 30, 2006.

Watson, J. M. "Alfred Lothar Wegener: Moving Continents." U.S. Geological Survey. Available online. URL: http://pubs.usgs.gov/gip/dynamic/wegener.html. Last updated May 5, 1999.

Wien, Wilhelm
(1864–1928)
German
Physicist

WIEN'S LAW describes the relationship between the TEMPERATURE of a BLACKBODY and the wavelength (*see* WAVE CHARACTERISTICS) of its maximum emission of radiation. Its author, Wilhelm Wien, received the 1911 Nobel Prize in physics for his work on blackbody radiation.

Wilhelm Wien was born at Fischhausen, in East Prussia (now in Poland), on January 13, 1864. His father, Carl Wien, was a landowner. Wilhelm might have become a gentleman farmer, but he was determined to obtain a good education. In 1866, the family moved to Rastenburg, East Prussia, and in 1879 Wilhelm enrolled at a school in that town. From 1880 until 1882, he studied at the City School in Heidelberg, and in 1882 he entered the University of Göttingen to study mathematics and natural sciences. Later in the same year, he moved to the University of Berlin. While still a student, from 1883 until 1885, he worked as an assistant to Hermann von Helmholtz (1821–94). He obtained his doctorate in 1886, with a thesis on the diffraction of light on sections of metals and the influence of different materials on the color of refracted light.

Carl Wien then fell ill, and between 1886 and 1890 Wilhelm had to spend most of his time helping to manage the family estate, although he did manage to spend one semester working with Helmholtz. Wien returned to Helmholtz's laboratory when the estate was sold, remaining there until 1896, when he was appointed professor of physics at the University of Aix-la-Chapelle (also called Aachen). In 1899, he became professor of physics at the University of Giessen, and in 1900 professor of physics at Würzburg. In 1902, he succeeded Ludwig Boltzmann (1844–1906) as professor of physics at the University of Leipzig, and in 1906 he became professor of physics at the University of Berlin.

It was in 1893 that Wien published the law relating radiation emission to temperature, known as the law of displacement. The following year he defined an ideal body that absorbs all the radiation falling upon it. He called such a body a blackbody. He published Wien's law in 1896. It was shown later that Wien's law applies only to short-wave radiation. Lord Rayleigh (1842–1919) had published an equation that applied to long-wave radiation, and the attempt to devise an equation that was valid at all wavelengths drew the German physicist Max Planck (1858–1947) to develop quantum theory.

In later years, Wien turned his attention to X-rays and cathode rays, confirming that cathode rays are streams of electrons.

Wilhelm Wien died at Munich, Bavaria, on August 31, 1928.

Appendix II
Tropical Cyclones and Tropical Storms

Abby A typhoon that struck Taiwan on September 19, 1986; it killed 13 people and caused damage estimated at $80 million.

Abe A typhoon that struck Zhejiang Province, China, on August 31, 1990; it killed 48 people.

Aere A typhoon that struck northern Taiwan on August 24, 2004, killing at least 24 people. Aere then moved on to the Philippines, where five persons lost their lives.

Agnes A hurricane, rated category 1 on the SAFFIR/SIMPSON HURRICANE SCALE, that struck Florida and New England in 1972; it caused $2.1 billion of damage.

Agnes is also the name of two typhoons. The first struck South Korea on September 1, 1981; it delivered 28 inches (711 mm) of rain in two days and left 120 people either dead or missing.

The second Typhoon Agnes, rated category 5 on the SAFFIR/SIMPSON HURRICANE SCALE, struck the central Philippines in November 1984; it generated winds of 185 MPH (297 km/h). At least 300 people were killed and 100,000 rendered homeless. The damage was estimated at $40 million.

Alex A typhoon that struck Zhejiang Province, China, on July 28, 1987. It triggered a huge landslide. There was widespread damage, and at least 38 people were killed.

Alicia A hurricane, rated category 4 on the SAFFIR/SIMPSON HURRICANE SCALE, that struck southern Texas on August 18, 1983. It generated winds of 115 MPH (185 km/h) and caused extensive damage in Galveston and Houston. At least 17 people were killed, and the damage was estimated at $1.6 billion.

Allen A hurricane, rated category 5 on the SAFFIR/SIMPSON HURRICANE SCALE, that struck islands in the Caribbean and the southeastern United States in August 1980. It brought winds of 175 MPH (280 km/h) gusting to 195 MPH (314 km/h). Barbados, St. Lucia, Haiti, the Dominican Republic, Jamaica, and Cuba were affected. More than 270 people died, most of them in Haiti.

Allison A tropical storm that moved across the southeastern United States from June 6 to 17, 2001. At least 20 people lost their lives in Texas, two in Louisiana, nine in Florida, nine in North Carolina, six in Pennsylvania, and one in Virginia. The damage was estimated to cost $5 billion.

Alpha A tropical storm that formed over the Caribbean on October 22, 2005. On October 23, it reached Haiti and the Dominican Republic, bringing heavy rain and killing at least 26 people. Alpha was the 22nd named storm of the 2005 season, breaking a record set in 1933 and making 2005 the most active hurricane season ever recorded.

Ami A category 3 cyclone that struck Fiji in January 2003 with winds of up to 125 MPH (200 km/h) and waves up to 98 feet (30 m) high. It killed 11 people.

Amy A typhoon that struck southern China on July 20 and 21, 1991. It killed at least 35 people and injured 1,360.

Andrew A hurricane, rated as category 5 on the SAF-FIR/SIMPSON HURRICANE SCALE, that struck the Bahamas, Florida, and Louisiana from August 23 to 26, 1992. It generated winds of up to 164 MPH (264 km/h). It was the costliest hurricane in United States history up to that time. In Florida, Homestead and Florida City were almost destroyed, 38 people were killed, 63,000 homes were destroyed, and damage was estimated at $20 billion. In Louisiana, 44,000 people were rendered homeless and damage was estimated at $300 million.

Angela Two typhoons, the first of which struck the Philippines in October 1989. It killed at least 50 people.

The second Typhoon Angela, rated at category 4 on the SAFFIR/SIMPSON HURRICANE SCALE, struck the eastern Philippines on November 3, 1995, generating winds of 140 MPH (225 km/h). It killed more than 700 people, rendered more than 200,000 homeless, and destroyed 15,000 homes. The cost of damage to crops, roads, and bridges was estimated at $77 million.

Audrey A hurricane, rated category 3 on the SAFFIR/SIMPSON HURRICANE SCALE, that struck the U.S. Gulf Coast near the border between Louisiana and Texas on June 17, 1956. It killed nearly 400 people and generated winds of 100 MPH (160 km/h) and a STORM SURGE of 12 feet (3.6 m).

Babs A typhoon that struck the Philippines in late October 1998. It caused floods and landslides in which at least 132 people died and about 320,000 people were rendered homeless.

Bebinca A typhoon that struck the Philippines in November 2000. It left 43 people dead and forced more than 630,000 people to leave their homes in metropolitan Manila and in 14 northern provinces.

Benedict One of five cyclones that struck Madagascar between January and March 1982. The others were called Frida, Electra, Gabriel, and Justine. Together they killed more than 100 people and rendered 117,000 homeless.

Bertha A hurricane that struck the U.S. Virgin Islands on July 8, 1996, bringing torrential rain and winds of up to 103 MPH (166 km/h) and triggering FLASH FLOODS and mudslides. It then moved to the British Virgin Islands, St. Kitts and Nevis, and Anguilla. It reached the Bahamas on July 10, where it caused waves 20 feet (6 m) high and winds of 100 MPH (160 km/h). On July 12, it crossed the Carolina coast then traveled north, eventually reaching Delaware and New Jersey. Bertha was initially rated as category 1 on the SAFFIR/SIMPSON HURRICANE SCALE, but was then upgraded to 2 and on July 9 it reached category 3. It was a large hurricane, with a diameter of 460 miles (740 km), and produced hurricane-force winds (see BEAUFORT WIND SCALE) 115 miles (185 km) from the eye. It killed seven people.

Bilis A typhoon that struck Taiwan and the coast of mainland China in August 2000. It killed 11 people in Taiwan, including seven farmers and a six-year-old girl who were buried in a mudslide and a woman who was killed by a falling power line. The typhoon had weakened by the time it reached the mainland province of Fujian, but it destroyed more than 1,000 homes.

Bob A hurricane that struck the eastern coast of the United States from August 18 to 20, 1991. It killed at least 20 people.

Brenda A typhoon that struck southern China in May 1989, killing 26 people.

Bret A tropical storm that struck Venezuela on August 8, 1993, causing floods and mudslides in which at least 100 people died.

Bret is also the name of a hurricane that struck the coast of Texas in 1999. With winds of 125 MPH (201 km/h), it struck Kenedy County at about 6 P.M. on August 22, traveling westwards at about 7 MPH (11.25 km/h). Its winds later weakened to 105 MPH (169 km/h). The following day it was downgraded to a tropical storm and its winds abated. The area affected was sparsely populated, but four people lost their lives in floods caused by the hurricane.

Calvin A hurricane that struck Mexico on July 6 and 7, 1993. It killed 28 people.

Camille A hurricane, rated category 5 on the SAFFIR/SIMPSON HURRICANE SCALE, that struck Mississippi and

Louisiana on August 17 and 18, 1969, killing about 250 people along the coast and causing $1.42 billion of damage. Camille then weakened and headed south and then east, over the Blue Ridge Mountains, Virginia, and was funneled through the narrow valleys of the Rockfish and Tye Rivers, where it encountered an advancing cold FRONT with associated THUNDERSTORMS. This resulted in 18 inches (457 mm) of rain that flooded 471 square miles (1,220 km²). FLASH FLOODS damaged or destroyed 185 miles (298 km) of road, and 125 people died, either drowned or crushed by boulders. After 80 days the floods subsided.

Carlotta A hurricane that formed off the Mexican coast on June 19, 2000. It headed out to sea, but generated heavy rain that caused mudslides, killing at least six people and causing more than 1,000 people to leave their homes.

Catarina A hurricane that developed in the South Atlantic Ocean during the week of March 22–28, 2005, and made landfall on the coast of the Brazilian state of Santa Catarina on March 28, with winds estimated at nearly 90 MPH (145 km/h). Catarina caused extensive damage, and some lives were lost. Catarina is believed to be the first tropical cyclone ever recorded in the South Atlantic.

Cecil Two typhoons, the first of which struck South Korea in August 1982. It killed at least 35 people and caused damage estimated at more than $30 million.

The second Typhoon Cecil struck Vietnam on May 25 and 26, 1989. It destroyed 36,000 homes and killed at least 140 people.

Charley A hurricane that reached the British Isles, where it arrived on August 25, 1986. It caused at least 11 deaths.

A second hurricane Charley struck southwestern Florida on August 13, 2004, devastating Punta Gorda and Port Charlotte, and killing 27 people.

Chata'an A typhoon that struck Japan in July, 2002, killing five people and causing widespread flooding.

Chebi A typhoon that struck Taiwan and Fujian Province, China, on June 23–24, 2001. It killed at least 73 in Fujian and nine in Taiwan.

Clara A typhoon that struck Fujian Province, China, on September 21, 1981. It destroyed 130 square miles (337 km²) of rice crops.

Clare A category 2 cyclone that struck Western Australia at about midnight on January 9, 2006. The storm made landfall near Dampier, with winds of up to 87 MPH (140 km/h), causing minor damage and no injuries.

Cobra A typhoon, generating winds up to 130 MPH (208 km/h) and waves up to 70 feet (21 m) high, that occurred in the Philippine Sea in December 1944. It struck a U.S. naval fleet, causing three destroyers to sink and destroying 150 carrier-borne aircraft. It caused the deaths of 790 sailors.

Connie A hurricane that struck the United States in August, 1957, producing rain that saturated the ground, shortly before the arrival of Hurricane Diane.

Damrey A typhoon that crossed East Asia for a week in September 2005. It killed 36 people in Vietnam, 16 in the Philippines, 16 in southern China, and at least three in Thailand. On September 28, Damrey was downgraded to a tropical depression.

Dan A typhoon that struck the Philippines on October 10, 1989. It killed 43 people and rendered 80,000 homeless.

Dan was also the name of a typhoon that struck the Philippines in October 1999 and then weakened to a tropical storm as it approached the outlying Taiwanese islands of Penghu and Kinmen. In the Philippines at least eight people were killed, thousands of homes were flooded, and the cost of damage to crops was estimated as $2 million.

David A hurricane, rated category 5 on the SAFFIR/SIMPSON HURRICANE SCALE, that struck islands in the Caribbean and the eastern coast of the United States in late August and early September 1979. It brought winds of up to 150 MPH (240 km/h), killed more than 1,000 people, and caused damage estimated at billions of dollars. The hurricane affected the Dominican Republic, Dominica, Puerto Rico, Haiti, Cuba, the Bahamas, Florida, Georgia, and New York State.

Dawn A typhoon that struck central Vietnam on November 19 to 23, 1998. It was the worst storm to affect the country in 30 years. More than 100 people were killed, and about 200,000 were forced to leave their homes.

Dennis   A hurricane that struck Haiti on July 7, 2005, killing at least 60 people. On July 8, Dennis reached Cuba, where 16 people lost their lives.

Diane   A hurricane that struck the United States in August 1957. It killed more than 190 people and caused $1.6 billion of damage.

Dolly   A tropical storm that struck islands in the Caribbean on August 20, 1996, strengthening to hurricane force as it reached Punta Herrero, Mexico. It then weakened but strengthened to hurricane force again as it moved across the sea toward northeastern Mexico and Texas.

Dolores   A hurricane, with extremely heavy rain, that struck an area housing poor people on the outskirts of Acapulco, Mexico, on June 17, 1974. At least 13 people died, and 35 were injured.

Domoina   A cyclone that struck Mozambique, South Africa, and Swaziland from January 31 to February 2, 1984. It caused severe flooding in which at least 124 people died and thousands were rendered homeless.

Dot   A typhoon that struck Luzon, Philippines, on October 19, 1985. It destroyed 90 percent of the buildings in the city of Cabanatuan, killed at least 63 people, and caused damage estimated at $1 billion.

Eline   A cyclone that struck Mozambique on February 22, 2000. It produced winds of up to 160 MPH (257 km/h) and torrential rain. There were reports of coastal villages being swept away by the STORM SURGE.

Eloise   A hurricane, rated category 4 on the SAFFIR/SIMPSON HURRICANE SCALE, that struck several Caribbean islands and the eastern United States in September 1975. It reached Puerto Rico on September 16, bringing winds of up to 140 MPH (225 km/h) and heavy rain, and killing 34 people. It then moved to Hispaniola, where 25 people died. It crossed Haiti and the Dominican Republic and from there reached Florida, where 12 people were killed. Eloise then moved north-

ward along the coast as far as the northeastern United States, where a state of emergency was declared.

Elsie   A typhoon that struck the Philippines on October 19, 1989; it killed 30 people and left 332,000 homeless.

Eric   One of two cyclones that struck Viti Levu, Fiji, on January 22, 1985. The other was Nigel. Between them they caused the deaths of 23 people. It is most unusual for two cyclones to occur so close together.

Eve   A typhoon that struck the southern tip of Kyushu, Japan, on July 18, 1996, bringing winds of 199 MPH (191 km/h).

Faith   One of two tropical storms that struck central Vietnam in December 1998. Between them, Faith and Gil brought rain that caused floods in which at least 22 people died and thousands had to leave their homes.

Faye   A typhoon that struck South Korea in July 1995; it killed at least 16 people.

Fifi   A hurricane, rated category 3 on the SAFFIR/SIMPSON HURRICANE SCALE, that struck Honduras on September 20, 1974. It generated winds of 130 MPH (209 km/h) and heavy rain. An estimated 5,000 people were killed, and tens of thousands were rendered homeless.

Firinga   A cyclone, rated category 4 on the SAFFIR/SIMPSON HURRICANE SCALE, that struck the island of Réunion, in the Indian Ocean, on January 28 and 29, 1989. It generated wind speeds of more than 125 MPH (200 km/h). At least 10 people died and 6,000 were rendered homeless.

Flo   A typhoon that struck Honshu, Japan, on September 16 and 17, 1990; it killed 32 people.

Floyd   A hurricane that formed in September 1999 and at its peak generated winds of 155 MPH (249 km/h), making it category 4 bordering category 5 on the SAFFIR/SIMPSON HURRICANE SCALE. It was also unusually large, with a diameter of about 600 miles (965 km).

Floyd struck the Bahamas on September 14 and then moved northward along the eastern coast of the

United States, reaching New York and New Jersey on September 16, by which time it had been downgraded to a tropical storm, with winds of 65 MPH (105 km/h). It brought heavy rain, causing flooding, and STORM SURGES of up to 7 feet (2.1 m). More than 2.3 million people were evacuated from their homes in Florida, Georgia, and the Carolinas. Crop damage in North Carolina was estimated to cost more than $1 billion. A total of 49 people lost their lives, one of them in the Bahamas and the remainder in the United States, 23 of them in North Carolina.

Forest A typhoon that struck the Japanese islands of Honshu, Kyushu, and Shikoku on September 19, 1983. It delivered up to 19 inches (483 mm) of rain, causing flooding that inundated 30,000 homes and resulted in the deaths of 16 people.

Fran A typhoon, rated category 3 on the SAFFIR/SIMPSON HURRICANE SCALE, that struck southern Japan from September 8 to 13, 1976. It generated winds of 100 MPH (160 km/h) and brought 60 inches (1,524 mm) of rain. It caused the deaths of 104 people and made an estimated 325,000 homeless.

Hurricane Fran, rated as category 3, struck the eastern United States in September 1996. It triggered TORNADOES and produced a STORM SURGE with waves of up to 12 feet (3.6 m) and in some places 16 feet (4.8 m). It passed Cape Fear, North Carolina, on September 6 and the following day crossed North and South Carolina, Virginia, and West Virginia. More than 500,000 people were evacuated from the coastal areas of South Carolina, and areas of North Carolina were also evacuated. It killed 34 people.

Frankie A tropical storm that crossed from the Gulf of Tonkin to the Red River delta, Vietnam, on July 23, 1996. It killed 41 people.

Fred A typhoon that struck Zhejiang Province, China, on August 20 and 21, 1994. It killed about 1,000 people and caused damage estimated at more than $1.1 billion.

Frederic A hurricane that struck the eastern coast of Florida, Alabama, and Mississippi in September 1979. About eight people died, and 500,000 were evacuated.

Gafilo A cyclone that struck Madagascar on February 24, 2004, killing approximately 200 people and leaving thousands homeless.

Gavin A cyclone that struck Fiji in March 1997. It killed at least 26 people, 10 of whom died when a fishing trawler sank.

Gay A typhoon that struck Thailand on November 4 and 5, 1989. It killed 365 people and destroyed or damaged 30,000 homes.

Geralda A cyclone, rated as category 5 on the SAFFIR/SIMPSON HURRICANE SCALE, that struck Madagascar from February 2 to 4, 1994. It generated winds of up to 220 MPH (354 km/h) and destroyed 95 percent of the buildings in the port of Toamasina. It killed 70 people and rendered 500,000 homeless.

Gert A hurricane that struck Bermuda in September 1999. At its peak it generated winds of 145 MPH (233 km/h) and was classed as category 4 on the SAFFIR/SIMPSON HURRICANE SCALE. By the time the edge of the hurricane reached Bermuda the wind speed had fallen to 110 MPH (177 km/h). Gert generated a STORM SURGE of 5 feet (1.5 m) and waves 10 feet (3 m) high, but caused no deaths or serious injuries.

Gilbert A hurricane, rated category 5 on the SAFFIR/SIMPSON HURRICANE SCALE, that struck islands in the Caribbean and the Gulf Coast in the United States from September 12 to 17, 1988. It caused widespread damage in Jamaica before moving toward the Yucatán Peninsula, Mexico. It killed about 200 people and caused an estimated $10 billion of damage in Monterrey, Mexico. In Texas, it killed at least 260 people and generated nearly 40 TORNADOES. The core pressure in Gilbert fell to 888 mb, making it the strongest hurricane recorded until that time. Its record was broken in 2005 by hurricane Wilma.

Gladys A typhoon that struck South Korea on August 23, 1991. It brought 16 inches (406 mm) of rain to Pusan and Ulsan, killing 72 people and rendering 2,000 homeless.

Glenda A category 4 cyclone that struck the northwestern coast of Australia on March 30, 2006, with winds of up to 155 MPH (250 km/h). The cyclone caused damage in the town of Onslow, but there were no reports of injuries. The cyclone was later downgraded to category 2.

Gloria   A typhoon that struck the Philippines on July 25, 1996. It killed at least 30 people. It was then downgraded to a tropical storm. It reached Taiwan and the southeastern coast of China on July 26, where it killed three people.

Gordon   A typhoon that struck Luzon, Philippines, on July 16, 1989. At least 200 people were killed.

Tropical Storm Gordon struck islands in the Caribbean, Florida, and South Carolina from November 13 to 19, 1994. It killed 537 people and caused damage estimated as at least $200 million.

Harvey   A tropical storm, with 60-MPH (96-km/h) winds, that brought more than 10 inches (254 mm) of rain to southwestern Florida in September 1999.

Hazel   A hurricane that struck islands in the Caribbean and the eastern United States in October 1954. It formed in the Lesser Antilles, then struck Haiti, South Carolina, North Carolina, Virginia, Maryland, Pennsylvania, New Jersey, and New York State, then continued into Canada. On October 18, the hurricane moved out into the Atlantic, eventually causing strong winds and heavy rain in Scandinavia. It killed an estimated 1,000 people in Haiti, 95 in the U.S., and 80 in Canada, and caused $250 million of damage in the United States and $100 million in Canada.

Herb   A typhoon that struck Taiwan on July 31, 1996. It brought 44 inches (1,118 mm) of rain in 24 hours to Mount Ali, Taiwan, flooding thousands of homes. It killed at least 41 people. Herb then crossed the Taiwan Strait, reaching Pingtan, in Fujian Province, China, on August 1.

Higos   A typhoon that struck Japan on October 2, 2002. It passed across Tokyo, then moved northward to Hokkaido. The storm caused at least four deaths and widespread flooding.

Honorinnia   A cyclone that struck Madagascar on March 17, 1985. It destroyed 80 percent of the buildings in Toamasina and left 32 people dead and 20,000 homeless.

Hortense   A tropical storm that struck Martinique on September 7, 1996 and then headed for the British Virgin Islands, Puerto Rico, and the Dominican Republic. By September 10, it had strengthened to a category 1 hurricane on the SAFFIR/SIMPSON HURRICANE SCALE. It then moved to the Bahamas and Turks and Caicos Islands. It killed 16 people in Puerto Rico and the Dominican Republic.

Hugo   A hurricane, rated category 5 on the SAFFIR/SIMPSON HURRICANE SCALE, that struck islands in the Caribbean and the eastern coast of the United States from September 17 to 21, 1989, generating winds of up to 140 MPH (224 km/h). On September 11, Hugo reached Guadeloupe, where it killed 11 people. It reached Dominica in the Lesser Antilles (Leeward Islands) on September 17. Then it moved to St. Croix, St. John, St. Thomas, and the smaller islands of the U.S. Virgin Islands and Puerto Rico, arriving on September 19.

Hugo killed 10 people on Montserrat, six in the Virgin Islands, and 12 in Puerto Rico. Almost the entire population of Montserrat was made homeless, as well as 90 percent of the population of St. Croix and 80 percent of the people of Puerto Rico, and 99 percent of all homes were destroyed in Antigua.

Hugo then turned north and weakened to category 4. On September 21, it struck Charleston, South Carolina, where one person died when a house collapsed. At Folly Beach, South Carolina, 80 percent of the population was left homeless. The following day Hugo reached Charlotte, North Carolina, where one child was killed. It crossed the Blue Ridge Mountains the same day and in the afternoon moved across Virginia, where two people died and winds of 81 MPH (130 km/h) were recorded. There was a STORM SURGE at Awendaw, South Carolina, and the storm triggered several TORNADOES that caused damage on several islands and on the mainland in North Carolina. In all, the hurricane caused damage in the United States estimated at $10.5 billion.

Hyacinthe   A cyclone that struck the island of Réunion, in the Indian Ocean, in January 1980. At least 20 people were killed.

Ike A typhoon that struck the Philippines on September 2 and 3, 1984. It killed more than 1,300 people and left 1.12 million homeless. It then struck the coast of Guangxi Zhuang, China, on September 6, where it caused widespread damage and killed 13 fishermen whose boats were lost at sea.

Imbudo A typhoon that struck Guangdong Province and the Guanxi region of China on July 26, 2003, killing at least 20 people.

Irene A hurricane that struck Caribbean islands and then moved to Florida and North Carolina in October 1999. It formed as a tropical storm over the northwestern Caribbean and reached hurricane force, category 1 on the SAFFIR/SIMPSON HURRICANE SCALE, as it approached Cuba. Irene reached Florida, moved from there to North Carolina, and then traveled out into the Atlantic. Although its winds were not strong, Irene brought very heavy rain and triggered several TORNADOES. Two people were killed in Cuba, five in Florida, and in North Carolina one person died in a traffic accident caused by the hurricane.

Iris A hurricane that struck southern Belize on October 8–9, 2001, destroying at least 3,000 houses, leaving 12,000 people homeless, and killing 22.

Irma Two typhoons, the first of which, rated category 5 on the SAFFIR/SIMPSON HURRICANE SCALE, struck the Philippines on November 24, 1981. It generated winds of 140 MPH (225 km/h) and caused great destruction in the coastal towns of Garchitorena and Caramoan. More than 270 people were killed, and 250,000 were rendered homeless. The damage was estimated at $10 million.

The second Typhoon Irma struck Japan on July 1, 1985. It caused the deaths of 19 people and extensive damage in Numazu and Tokyo.

Irving A typhoon that struck Thanh Hoa Province, Vietnam, on July 24, 1989. At least 200 people were killed.

Isabel A hurricane that struck the United States on September 18, 2003. It caused serious damage, especially in North Carolina and Virginia, and approximately seven persons lost their lives in seven states.

Ismael A hurricane that struck Mexico on September 14, 1995. It killed at least 107 people in the northwestern states, many of them fishermen who were lost at sea.

Ivan A hurricane that crossed the Caribbean from September 7 to 17, 2004. It killed 39 people in Grenada and destroyed the island's crops, then killed 18 people in Jamaica. It strengthened before striking the Gulf Coast of the United States, where approximately 52 people lost their lives in several states.

Jeanne A tropical storm that struck the Gonaïve region of Haiti on September 18, 2004. The area had already been devastated by floods in May, and the flooding due to Jeanne killed more than 3,000 persons.

Joan A hurricane that struck the Caribbean coast of Central and South America from October 22 to 27, 1988. It caused severe damage and killed at least 111 people in Nicaragua, Costa Rica, Panama, Colombia, and Venezuela. It then weakened and was renamed Tropical Storm Miriam. Miriam struck El Salvador, where it rendered 3,000 people homeless.

Joe A typhoon that struck northern Vietnam on July 23, 1980. More than 130 people were killed and about 3 million were made homeless.

John A cyclone from the Indian Ocean that struck the sparsely populated northwestern coast of Australia on December 15, 1999. At its full strength it brought sustained winds of 130 MPH (209 km/h) and gusts of up to 185 MPH (298 km/h), making it the most powerful storm ever recorded in Australia up to that time. It crossed the coast near Whim Creek (estimated population 12), Western Australia, weakening as it did so, but still with sustained winds of 106 MPH (170 km/h). The cyclone brought STORM SURGES of up to 20 feet (6 m).

Jose A hurricane, rated category 2 on the SAFFIR/SIMPSON HURRICANE SCALE, with winds of 100 MPH (160 km/h), but later downgraded to a tropical storm with 65-MPH (105-km/h) winds, that struck Caribbean islands in October 1999. At its maximum extent its diameter was about 300 miles (480 km). It caused heavy rain, but no deaths and few serious injuries.

Judy   Three typhoons, the first of which struck South Korea on August 25 and 26, 1979. It caused severe flooding in which 60 people died and 20,000 were rendered homeless.

The second Typhoon Judy, rated category 3 on the SAFFIR/SIMPSON HURRICANE SCALE, struck Japan on September 11 and 12, 1982. It killed 26 people and caused extensive damage.

The third Typhoon Judy struck South Korea in July 1989 and killed at least 17 people.

Kaemi   A tropical storm that struck Vietnam on August 22, 2000. It killed 14 people.

Kai Tak   A typhoon that struck the Philippines, Japan, and Taiwan in July 2000, with winds of 93 MPH (150 km/h) and heavy rain. Kai Tak arrived soon after typhoon Kirogi. Between them, the two typhoons killed 42 people in the Philippines, five in Japan, and one in Taiwan.

Kate   A hurricane that struck Cuba and Florida from November 19 to 21, 1985. It killed at least 24 people.

Katrina   The hurricane that brought the worst natural disaster to strike the United States for a century made landfall early on August 29, 2005, near Buras, Louisiana. By the time the hurricane dissipated two days later, Louisiana, Mississippi, and Alabama had suffered severe devastation. The storm had also caused damage in Florida, and its effects were felt in Texas, Arkansas, Georgia, and Tennessee. New Orleans was left flooded and largely abandoned, a ghost town. Biloxi, Mississippi, was almost totally demolished by the wind and 30-foot (9-m) STORM SURGE, and most of the buildings in nearby Gulfport were damaged and many destroyed. There was extensive flooding in Mobile, Alabama. The hurricane then headed northward. It was downgraded to a tropical storm, and it reached Clarksville, Tennessee, as a tropical depression. Its last known position was over southeastern Quebec and northern New Brunswick, Canada, where what remained of Katrina merged with an ordinary frontal weather system. By that time it brought nothing more serious than moderate rainfall.

New Orleans was battered by the winds, but suffered principally from the rain, which fell over the region draining into the Mississippi River and Lake Pontchartrain. Early on August 30, the pressure of water breached the levees protecting New Orleans in three places on the Lake Pontchartrain side of the city. Floodwater poured into New Orleans, rendering most of the city uninhabitable.

Katrina triggered TORNADOES in Pennsylvania, Virginia, Georgia, and Alabama. These caused several injuries, although no fatalities, but damage to poultry houses in Carroll County, Georgia, led to the death or liberation of 500,000 chickens.

The surface atmospheric pressure in the eye of the storm is the principal measurement used to evaluate hurricanes. The eye pressure in Katrina fell to 90.2 kPa (902 millibars), making it the fourth most intense hurricane on record (after Gilbert 1988, the Labor Day storm 1935, and Allen 1980). This pressure was recorded at sea. When Katrina reached land, the central pressure was 91.8 kPa (918 mb), making Katrina the third most intense storm to make landfall in the United States (after the Labor Day storm 1935 and Camille 1969).

The monetary cost was predicted to be between $20 billion and $100 billion, making Katrina the most costly hurricane ever. The final death toll was 972 in Louisiana and 221 in Mississippi.

Keith   A hurricane that struck Central America in 2000. On September 30, it killed one person in El Salvador, and the following day it struck Nicaragua, where a 16-year-old boy was swept away by a swollen river as the hurricane reached Category 3 on the SAFFIR/SIMPSON HURRICANE SCALE, with winds of 135 MPH (217 km/h). Keith then struck Belize and the Yucatán peninsula of Mexico, its winds dropping to 90 MPH (145 km/h). It weakened to a tropical storm as it moved out over the Gulf of Mexico, but then strengthened to category 2 as it headed back toward the Mexican coast. By the time it died, Keith had caused at least 12 deaths and had dropped 22 inches (559 mm) of rain on Belize.

Kelly   A typhoon that struck the Philippines on July 1, 1981. It caused floods and landslides in which about 140 people died.

Khanun   A typhoon that struck Zhejiang Province, China, on September 11, 2005, destroying more than 7,000 homes and killing at least 14 people.

Kina A cyclone, rated as category 4 on the SAFFIR/ SIMPSON HURRICANE SCALE, that struck Fiji on January 2 and 3, 1993. It generated winds of up to 115 MPH (185 km/h) and killed 12 people.

Kirk A typhoon that crossed southwestern Honshu, Japan, on August 14, 1996, bringing winds of 130 MPH (209 km/h) and up to 12 inches (300 mm) of rain. It returned to the northern part of the island on August 15, but by then it had weakened. From there it moved to northeastern China where it caused floods that inundated 845 villages along the Yellow River.

Kirogi A typhoon that struck the Philippines, Japan, and Taiwan in July 2000. Kirogi was closely followed by another typhoon, Kai Tak. Kirogi generated sustained winds of 89 MPH (143 km/h) and brought heavy rain. Between them, the two typhoons killed 42 people in the Philippines, five in Japan, and one in Taiwan.

Kyle A typhoon that struck Vietnam on November 23, 1993. It killed at least 45 people.

Larry A category 5 cyclone that struck the coast of Queensland, Australia, on March 20, 2006, with winds of 180 MPH (290 km/h). The storm made landfall at Innisfail, about 60 miles (100 km) south of Cairns, uprooting trees, destroying sugar and banana crops, and wrecking homes, leaving thousands homeless.

Lenny A hurricane that struck islands in the Caribbean in November 1999. It was upgraded from a tropical storm on November 14, passed to the south of Jamaica with winds of 100 MPH (160 km/h), and its wind speeds increased to 125 MPH (201 km/h) on November 17. The following day its winds increased to 150 MPH (241 km/h) as it struck St. Croix in the U.S. Virgin Islands, where the storm continued for 12 hours. Other islands in the region also suffered, and a total of 13 people were killed. On November 19, it was downgraded to a tropical storm.

Lili A hurricane that struck the Caribbean and Louisiana in late September and early October 2002. The storm caused seven deaths and widespread damage in Jamaica and St. Vincent, and tore roofs from buildings in the Cayman Islands. It made landfall in Louisiana on October 2, with winds of 90 MPH (145 km/h).

Linda A typhoon that struck southern Vietnam, Cambodia, and Thailand in November 1997, destroying thousands of homes and sinking hundreds of fishing vessels. It killed 464 people in Vietnam and more than 20 in Cambodia and Thailand.

Linfa A tropical storm that struck Luzon, Philippines, on May 27, 2003, bringing torrential rain. At least 25 people were killed.

Lingling A tropical storm that struck the Philippines on November 7, 2001, with winds up to 56 MPH (90 km/h). It killed at least 68 people.

Liza A hurricane, rated category 4 on the SAFFIR/ SIMPSON HURRICANE SCALE, that struck La Paz, Mexico, on October 1, 1976, killing at least 630 people and leaving tens of thousands homeless. It brought winds of 130 MPH (160 km/h) and 5.5 inches (140 mm) of rain. This destroyed an earth dam 30 feet (9 m) high, sending a wall of water 5 feet (1.5 m) high through a shantytown.

Longwan A typhoon that struck Fujian Province, China, causing floods that swept away a military school at Fuzhou on October 2, 2005, killing at least 80 people.

Luis A hurricane, rated as category 4 on the SAFFIR/ SIMPSON HURRICANE SCALE, that struck Puerto Rico and the U.S. Virgin Islands from September 4 to 6, 1995. Its winds gusted to more than 140 MPH (225 km/h). At least 15 people were killed. Luis was shortly followed by Marilyn.

Lynn A typhoon that struck Taiwan on October 24, 1987, and destroyed 200 homes.

Maemi A typhoon that struck South Korea on September 12, 2003. It was said to have been the worst storm to strike the country for 100 years. Maemi caused severe damage to the port of Pusan and killed at least 124 people.

Maria A typhoon that struck the southern Chinese provinces of Guangdong and Hunan on September 1, 2000. A cargo ship sank in Shanwei harbor, more than 7,000 homes were damaged, and at least 45 people were killed. The cost of the damage was estimated at $223 million.

Marilyn A hurricane that struck the U.S. Virgin Islands and Puerto Rico on September 15 and 16, 1995, not long after Luis. It generated winds of more than 100 MPH (160 km/h) and destroyed 80 percent of the houses on St. Thomas. It killed nine people.

Martin A cyclone that struck the Cook Islands in November 1997. It killed nine people.

Maury A typhoon that struck northern Taiwan on July 19, 1981. It caused floods and landslides in which 26 people died.

Meli A cyclone that struck Fiji on March 27, 1979. It killed at least 50 people and destroyed about 1,000 homes.

Mike A typhoon that struck the Philippines on November 14, 1990. It killed 190 people and rendered 120,000 homeless.

Mindulle A typhoon that struck Luzon, Philippines, on June 29, 2004, where it killed 31 people. It then moved on to Taiwan, where on July 1 it killed 15 people.

Mireille A typhoon, rated as category 4 on the SAF-FIR/SIMPSON HURRICANE SCALE, that struck Kyushu and Hokkaido, Japan, on September 27, 1991. It generated winds of up to 133 MPH (214 km/h) and killed 45 people.

Mitch A hurricane, which weakened to a tropical storm, that struck Central America in late October 1998. It formed on October 21 in the southwest Caribbean, then moved toward Honduras and intensified to category 5 hurricane status on the SAFFIR/SIMPSON HURRICANE SCALE. On October 29, it was downgraded to a tropical storm and moved inland and southward. The storm was declared over on November 1. The high rainfall associated with it caused appalling damage from flooding, in which an estimated 11,000 people died, the highest death toll from a hurricane for 200 years. During a period of 41 hours, from 3 P.M. on October 29 to 7 A.M. on October 31, a total of 27.5 inches (698 mm) of rain fell on Honduras, and between 6 P.M. on October 27 and 9 P.M. on October 31 the rainfall was 35.3 inches (896 mm).

Muroto II A typhoon that struck Japan in September 1961. It caused a STORM SURGE of 13 feet (3.9 m) that produced floods in Osaka in which 32 people died.

Nabi A typhoon that struck southern Japan on September 6, 2005. Approximately 250,000 people had to be evacuated, and at least 18 were killed.

Namu A typhoon that struck the Solomon Islands on May 19, 1986. It killed more than 100 people and rendered more than 90,000 homeless.

Nanmadol A typhoon that struck the Philippines in December 2004, during the cleanup following typhoon Winnie. After the two storms had passed, at least 1,000 persons were dead or missing.

Nari A typhoon that struck Taiwan from September 16 to 19, 2001. It caused floods, mudslides, and extensive damage, killing at least 94 people.

Nell A typhoon that struck the Philippines on December 25 and 26, 1993. It killed at least 47 people.

Nina A typhoon that struck the Philippines on November 26, 1987, and caused a STORM SURGE. It killed 500 people in Sorsogon Province, Luzon.

Ofelia A typhoon that struck the Philippines, Taiwan, and China on June 23 and 24, 1990. It killed a total of 57 people.

Olga A typhoon that struck Luzon, Philippines, in May 1976. It brought rains so heavy that they caused widespread flooding in which 215 people died and at least 600,000 were rendered homeless. The floods caused $150 million of damage.

Olga is also a category 1 typhoon on the SAFFIR/SIMPSON HURRICANE SCALE that struck Japan and South Korea on August 2 and 3, 1999. In Japan it brought 23 inches (584 mm) of rain to the area around Kochi and up to 1 inch (25 mm) of rain an hour to other places. One woman died when she was crushed by a landslide caused by the heavy rain. The heavy rain in South Korea caused 29 deaths and left 22 people missing. It destroyed more than 74,000 acres (30,000 ha) of farmland, more than 8,000 homes, and left nearly 20,000 people homeless.

Olivia A hurricane that struck Mazatlán, Mexico, on October 24, 1975. It killed 29 people.

Opal A hurricane, rated as category 4 on the SAF-FIR/SIMPSON HURRICANE SCALE, that formed over the Yucatán Peninsula, Mexico, on September 27, 1995. A few days later, it weakened to a tropical storm, but by October 2 had strengthened and was once again classed as a hurricane. By the time it reached Florida, on October 4, it had weakened to category 3. From Florida it moved into North Carolina, Georgia, and Alabama. Most of the damage it caused, estimated at more than $2 billion, was due to a STORM SURGE that produced breaking waves and waves 12 feet (3.6 m) high. Opal killed 50 people in Guatemala and Mexico and 13 in the United States.

Orchid A typhoon that struck South Korea on September 11, 1980. It killed seven people, and more than 100 fishermen were lost at sea.

Pat A typhoon, rated category 4 on the SAFFIR/SIMP-SON HURRICANE SCALE, that struck Kyushu, Japan, on August 30, 1985. It generated winds of up to 124 MPH (200 km/h) and killed 15 people.

Paul A hurricane, rated category 4 on the SAFFIR/SIMPSON HURRICANE SCALE, that struck Sinaloa, Mexico, on September 30, 1982. It generated winds of 120 MPH (193 km/h) and left 50,000 people homeless.

Pauline A hurricane that struck southern Mexico from October 8 to 10, 1997. It generated winds of up to 115 MPH (185 km/h) and waves up to 30 feet (9 m) high, and caused extensive damage in the city of Acapulco and in coastal villages in the states of Oaxaca and Guerrero. The hurricane killed 217 people and rendered 20,000 homeless.

Peggy A typhoon that struck the northern Philippines on July 9, 1986. It caused floods, landslides, and mudslides, bringing extensive damage to property. More than 70 people were killed. It then moved to south-

eastern China, where it arrived on July 11 and caused widespread flooding. More than 170 people were killed, at least 1,250 injured, and more than 250,000 homes were destroyed.

Phyllis A typhoon that struck the Japanese island of Shikoku in August 1975. It killed 68 people. A week later, the island was struck by Typhoon Rita.

Polly A tropical storm that caused a STORM SURGE with waves 20 feet (6 m) high on August 30 and 31, 1992, at Tianjin, China. It killed 165 people along the southeastern coast and rendered more than 5 million homeless.

Pongsona A category 4 typhoon that struck Guam and the Mariana Islands on December 9, 2002, with winds of up to 150 MPH (240 km/h).

Rananim A typhoon that made landfall in Zhejiang Province, China, on August 12, 2004, then moved inland, killing at least 164 people. It was the most powerful typhoon to strike China in seven years.

Rex A typhoon that struck northern Japan in August 1998. It caused floods and landslides in which at least 11 people died, and 40,000 were forced to evacuate their homes.

Rita Two typhoons and one hurricane, the first of which struck the Japanese island of Shikoku in August 1975, one week after Typhoon Phyllis. Rita killed 26 people and injured 52.

The second typhoon Rita, rated category 4 on the SAFFIR/SIMPSON HURRICANE SCALE, struck the Philippines on October 26, 1978. Nearly 200 people were killed, and about 10,000 homes were destroyed.

Hurricane Rita struck Louisiana on September 23, 2005, breaching levees at New Orleans that had recently been repaired following hurricane Katrina, and flooding parts of the city once more. Rita reached the Gulf Coast on September 24, making landfall close to the Texas–Louisiana border, but causing little damage and few injuries.

Roxanne A hurricane, rated at category 3 on the SAFFIR/SIMPSON HURRICANE SCALE, that struck the island of Cozumel, off the Mexican coast, in October 1995, generating winds of 115 MPH (185 km/h). It killed 14 people, and tens of thousands were forced to flee their homes.

Ruby A typhoon that struck the Philippines on October 24 and 25, 1988. It caused floods and landslides in which about 500 people died. The damage was estimated at $52 million.

Rumbia A tropical storm that struck southern Mindanao, Philippines, on November 30, 2000. It brought heavy rain and high waves that caused flooding. More than 1,600 people were forced to leave their homes.

Rusa A typhoon that struck South Korea on August 31 and September 1, 2002, with winds of up to 124 MPH (200 km/h). It killed more than 180 people and caused damage costing more than $1 billion.

Ruth Two typhoons, the first of which struck Vietnam on September 15, 1980. It killed at least 164 people.

The second Typhoon Ruth, rated as category 5 on the Saffir/Simpson hurricane scale, struck Luzon, Philippines, on October 27, 1991. It generated winds of up to 143 MPH (230 km/h) and killed 43 people.

Sally A typhoon that passed Hong Kong on September 10, 1996, then crossed the coast of Guangdong, China, bringing winds of up to 108 MPH (174 km/h). It killed more than 130 people and destroyed nearly 400,000 homes.

Sarah A typhoon that struck Taiwan on September 11, 1989. It broke a Panamanian-registered freighter in half and killed 13 people.

Skip A typhoon that struck the Philippines on November 7, 1988. High winds and heavy rain caused mudslides and floods in which at least 129 people died. It was the second typhoon to strike the Philippines in two weeks.

Stan A hurricane that made landfall on the Gulf Coast of Mexico on October 4, 2005, then moved through Central America. It caused floods and landslides that killed at least 71 people in El Salvador, at least 654 in Guatemala, and more than 60 in Nicaragua, Honduras, Mexico, and Costa Rica.

Steve A tropical cyclone that struck Cairns, Queensland, Australia, on February 27, 2000. It brought winds of 105 MPH (169 km/h). There were no reported injuries or deaths.

Sybil A tropical storm that struck the Philippines on October 1, 1995. It triggered floods, landslides, and volcanic mudflows, causing damage in 29 provinces and 27 cities. More than 100 people were killed.

Tad A typhoon, rated category 1 on the Saffir/Simpson hurricane scale, that struck central and northern Japan on August 23, 1981, killing 40 people and leaving 20,000 homeless. It brought winds of up to 80 MPH (128 km/h).

Talim A typhoon that struck Anhui Province, China, on September 1, 2005, causing flooding and landslides, and killing 53 people.

Teresa A typhoon that struck Luzon, Philippines, on October 23, 1994. It killed 25 people.

Thelma Two typhoons, the first of which, rated category 4 on the Saffir/Simpson hurricane scale, struck Kaohsiung, Taiwan, on July 25, 1977. It generated winds of up to 120 MPH (193 km/h). Nearly 20,000 homes were destroyed and 31 people died.

The second Typhoon Thelma struck South Korea on July 15, 1987. It caused floods, landslides and mudslides in which at least 111 people died.

Tico A hurricane that struck the coast of Mazatlán, Mexico, on October 10, 1983. It killed 105 fishermen, whose boats were lost at sea.

Tip A typhoon that struck Japan on October 19, 1979, with winds of up to 55 MPH (88 km/h). At least 36 people died. On October 12, surface pressure in the eye of Tip was 870 mb; this is the lowest surface pressure ever recorded.

Tokagi A typhoon that struck Japan on October 30, 2004, killing at least 83 people.

Toraji A typhoon that struck Hualien and Nantou Provinces, Taiwan, on July 30, 2001. It caused floods and landslides in which 77 people died and 133 were missing and presumed dead.

Tracy A cyclone that struck Darwin, Australia, on December 25, 1974. It destroyed 90 percent of the city and killed more than 50 people.

Uma A cyclone that struck Vanuatu on February 7, 1987. It killed at least 45 people.

Utor A typhoon that occurred in early July 2001. Utor killed one person in Taiwan, at least 121 people in the Philippines, and 23 in Guangdong Province, China.

Val A typhoon, rated as category 5 on the SAFFIR/SIMPSON HURRICANE SCALE, that struck Western Samoa from December 6 to 10, 1991. It generated winds of up to 150 MPH (241 km/h) and killed 12 people and rendered 4,000 homeless.

Vera Two typhoons, the first of which struck Honshu, Japan, in September 1959. It killed nearly 4,500 people, destroyed about 40,000 homes, and left 1.5 million people homeless.

The second Typhoon Vera struck Zhejiang Province, China, on September 16, 1989. It killed 162 people and injured 692.

Victor A typhoon that struck Guangdong and Fujian Provinces, China, in August 1997. It killed 49 people and destroyed 10,000 homes.

Violet A typhoon that crossed Japan on September 22, 1996, with winds of 78 MPH (125 km/h). It killed at least seven people, most of them in the Tokyo area, and caused about 200 landslides in Honshu, where it destroyed more than 80 homes and flooded more than 3,000. By the following day, as it moved out into the Pacific, Violet had weakened to a tropical storm.

Wally A cyclone that struck Fiji in April 1980. It caused floods and landslides in which at least 13 people died and thousands were left homeless.

Willie A typhoon that struck the island of Hainan, China, in September 1996. It caused floods that affected 70 percent of the streets of the capital, Haikou, and inundated 95,000 acres (38,000 ha) of farmland. At least 38 people were killed.

Wilma A category 4 hurricane that crossed the Caribbean in October 2005, producing winds of up to 150 MPH (240 km/h). It killed 13 people in Haiti and Jamaica before making landfall on the Mexican coast on October 21. Wilma then remained stationary for a full day, devastating Cancún, Cozumel, and Playa del Carmen and killing 6 people. On October 24 Wilma crossed Florida, killing 22 people. The core pressure of Wilma fell to 882 mb, making this the strongest hurricane ever recorded, with a lower pressure than that of Gilbert in 1988 (though not so low as that of typhoon Tip).

Winnie A typhoon that struck Taiwan, eastern China, and the Philippines on August 18 and 19, 1997. It generated winds of up to 92 MPH (148 km/h) and caused widespread flooding, especially in Taiwan and the Philippines. At least 37 people were killed in Taiwan, 140 in the Chinese provinces of Zhejiang and Jiangsu, and 16 in the Philippines. Tens of thousands of homes were destroyed in China, and in the Philippines 60,000 people were forced to leave their homes.

A second typhoon Winnie struck the Philippines on November 29, 2004, closely followed by typhoon Nanmadol. After the two storms had passed, at least 1,000 persons were dead or missing.

Wukong A typhoon that struck southern China on the weekend of September 9–10, 2000. Five people were killed, more than 1,000 houses collapsed, and 49,420 acres (30,710 ha) of rubber trees, bananas, rice, and pepper plants were destroyed.

Xangsane A typhoon that struck Taiwan on November 1 and 2, 2000. It brought winds of 90 MPH (145 km/h) and severe flooding. A Panamanian cargo ship, the *Spirit of Manila*, sank in the storm after breaking into three pieces. The storm killed 58 people, and 31 were missing, including 23 of the crew of the *Spirit of Manila*. Xangsane (pronounced Chang*sharn*) caused more than $2 billion of damage to crops and farms. The name means "elephant" in Thai.

Yancy Two typhoons, the first of which struck the Philippines and China in August 1990. It killed 12 people in the Philippines and 216 people in Fujian and Zhejiang Provinces, China.

The second Typhoon Yancy struck Kyushu, Japan, in September 1993. Rated category 4 on the SAFFIR/

Simpson hurricane scale, it generated winds of up to 130 MPH (209 km/h) and killed 41 people.

Yanni A tropical storm that struck South Korea in late September and early October 1998. It flooded about one-quarter of the country's cropland and caused the deaths of at least 27 people.

York A typhoon that struck Hong Kong on September 16, 1999, crossing the territory and reaching Guangzhou, in mainland China, later the same day. The maximum wind speed was 93 MPH (150 km/h). York brought heavy rain to the province of Fujian. At least one person was killed in Hong Kong and six in Fujian.

Zack A tropical storm that struck the Philippines in October 1995. It capsized a ship sailing between islands, killing 59 people. It caused severe flooding on land. A total of at least 100 people died, and 60,000 were forced to leave their homes.

Zane A typhoon that crossed Taiwan on September 28, 1996. It triggered mudslides and killed two people, then moved away to Okinawa.

Zeb A typhoon that struck the Philippines, Taiwan, and Japan in October 1998. It killed at least 74 people in the Philippines, 25 in Taiwan, and 12 in Japan.

Zoe A category 5 cyclone that struck the Solomon Islands on December 28, 2002, with winds of up to 220 MPH (350 km/h). It caused widespread devastation on the islands of Tikopia and Anuta.

APPENDIX III
CHRONOLOGY OF TORNADOES

Many tornadoes occur in remote areas, far from human habitations, but if they strike in populated areas, they can cause great devastation.

1140: Warwickshire, England. Extensive damage.

July 1558: Nottingham, England. Extensive damage and some deaths.

October 1638: Devon, England. Between five and 50 deaths.

May 1840: Natchez, Mississippi. 317 killed.

June 1865: Viroqua, Wisconsin. More than 20 killed.

December 1879: Scotland. The Tay Bridge destroyed by two tornadoes that struck simultaneously. Between 75 and 90 killed.

February 1884: Mississippi, Alabama, North Carolina, South Carolina, Tennessee, Kentucky, Indiana. More than 800 killed.

May 1896: Missouri and Illinois. 300 killed.

June 1899: New Richmond, Wisconsin. 117 killed, at least 150 injured.

May 1902: Goliad, Texas. 114 killed.

March 1925: Missouri, Illinois, and Indiana. 689 killed by up to seven tornadoes. A separate tornado at Annapolis, Maryland, overturned passenger trains and lifted 50 motorcars, carried them over houses, and dropped them in fields.

March 1932: Alabama. 268 killed.

April 1936: 216 killed at Tupelo, Mississippi, and 203 at Gainesville, Georgia, by two separate tornadoes.

June 1944: Ohio, Pennsylvania, West Virginia, and Maryland. 150 killed.

April 1947: Texas, Oklahoma, and Kansas. 169 killed.

March 1952: Arkansas, Missouri, and Texas. 208 killed.

May 1953: Texas. 114 killed.

June 1953: 143 killed in Michigan and Ohio and 90 around Worcester, Massachusetts, by two separate tornadoes.

May 1955: Kansas, Missouri, Oklahoma, and Texas. 115 killed.

April 1965: Indiana, Illinois, Ohio, Michigan, and Wisconsin. 271 killed.

February 1971: Mississippi Delta. 110 killed.

January 1973: San Justo, Argentina. 60 killed.

April 1974: More than 300 killed during the super outbreak.

June 1974: Oklahoma, Kansas, and Arkansas. 24 killed by several tornadoes.

January 1975: Mississippi. 12 killed, 200 injured when a tornado struck a shopping mall.

April 1976: Bangladesh. 19 killed, more than 200 injured.

April 1977: Bangladesh. More than 600 killed, 1,500 injured.

May 1977: Moundou, Chad. 13 killed, 100 injured.

March 1978: Delhi, India. 32 killed, 700 injured.

April 1978: Orissa State, India. Nearly 500 killed, more than 1,000 injured. 100 believed killed in West Bengal by another tornado.

April 1979: Texas and Oklahoma. 59 killed, 800 injured.

August 1979: Irish Sea. 18 killed when tornadoes struck yachts taking part in the Fastnet Race between England and Ireland.

May 1980: Kalamazoo, Michigan. Five killed, at least 65 injured.

April 1981: Bangladesh. About 70 killed, 1,500 injured. Orissa State, India. More than 120 killed.

April 1982: Kansas, Oklahoma, and Texas. Seven killed.

May 1982: Marion, Illinois. 10 killed.

April 1983: Fujian Province, China. 54 killed.

April 1983: Bangledesh. 12 killed, 200 injured.

May 1983: Texas, Tennessee, Missouri, Georgia, Louisiana, Mississippi, and Kentucky. 24 killed by at least 59 tornadoes.

May 1983: Vietnam. 76 killed.

March 1984: North and South Carolina. More than 70 killed.

April 1984: Water Valley, Mississippi. 15 killed. Oklahoma. 14 killed. Kentucky, Louisiana, Tennessee, Ohio, Maryland, and West Virginia. 14 killed.

June 1984: Russia. Hundreds killed.

October 1984: Maravilha, Brazil. 10 killed.

May 1985: Ohio, Pennsylvania, New York, and Ontario. 90 killed.

May 1987: Saragosa, Texas. 29 killed.

July 1987: Heilongjiang Province, China. 16 killed, more than 400 injured.

July 1987: Edmonton, Alberta, Canada. 25 killed.

April 1989: Bangladesh. Up to 1,000 killed, 12,000 injured.

May 1989: Texas, Virginia, North Carolina, Louisiana, South Carolina, and Oklahoma. 23 killed, more than 100 injured.

November 1989: Huntsville, Alabama. 18 killed.

June 1990: Indiana, Illinois, and Wisconsin. 13 killed.

May 1991: Bangladesh. 13 killed.

January 1993: Bangladesh. 32 killed, more than 1,000 injured.

April 1993: West Bengal, India. 100 killed.

March 1994: Alabama, Georgia, and North and South Carolina. 42 killed.

May 1996: Bangladesh. More than 440 killed, more than 32,000 injured.

March 1997: Arkansas. 25 killed.

May 1997: Central Texas. 30 killed, 24 in Jarrell.

July 1997: Kiangsu Province, China. 21 killed, more than 200 injured.

July 1997: Southern Michigan. 16 killed, more than 100 injured.

October 1997: Bangladesh. 25 killed, thousands injured.

May 1999: Kansas, Oklahoma, Texas. 46 people killed and about 900 injured by more than 76 tornadoes, at least one classified F-5.

February 2000: Georgia. 18 killed, about 100 injured.

July 2000: Alberta, Canada. 10 killed when a tornado swept through a trailer park near Edmonton.

October 2000: Bognor Regis, England. Four people injured and hundreds of homes damaged.

December 2000: Tuscaloosa, Alabama. 11 people killed.

February 2001: Mississippi. Five killed. Arkansas. One person killed.

May 2001: Ellicott, Colorado. 18 killed.

November 9–11, 2002: United States. Nearly 90 tornadoes along a storm front from Gulf of Mexico to the Great Lakes. 17 killed in Tennessee, 12 in Alabama, five in Ohio, one in Mississippi, one in Pennsylvania; more than 200 injured.

May 4–12, 2003: United States. More than 300 tornadoes sweep through the midwestern and southern United States. Hundreds of homes damaged, at least 42 people killed.

April 14, 2004: Northern Bangladesh. Thousands of homes destroyed, at least 66 people killed.

March 20, 2005: Northern Bangladesh. At least 56 people killed in Gaibandha district, thousands homeless.

November 6, 2005: United States. Southern Indiana and northern Kentucky. 24 people killed, most in a trailer park outside Evansville, Indiana.

January 2, 2006: Missouri. Three persons injured.

March 9–13, 2006: United States. Outbreak of at least 105 tornadoes sweeps across the south-central United States, killing at least 11 people. One tornado is rated F-5 on the Fujita tornado intensity scale. Two other persons are also killed: one in an automobile accident and the other in a fire caused by lightning. These deaths are due to the weather but not directly to the tornadoes. March 12 is the most active day, with 62 confirmed tornadoes and storms producing hailstones the size of softballs.

Appendix IV
Laws, Regulations, and International Agreements

Air Quality Act 1967 A United States federal law that empowers the Department of Health, Education and Welfare to define areas within which air quality should be controlled, to set ambient air standards, to specify the methods and technologies to be used to reduce AIR POLLUTION, and to prosecute offenders if local agencies fail to do so.

Clean Air Act A law designed to improve air quality that was passed in 1956 in the United Kingdom, and a law with the same name that was passed in 1963 in the United States.

The British legislation was drafted in response to the London smog incidents (*see* AIR POLLUTION INCIDENTS) and empowered local governments to designate "smokeless zones" in which it became an offense to emit black smoke. As the act was implemented over the succeeding years, this had the desired effect of eliminating the domestic burning of coal in most cities in favor of smokeless fuels such as coke. Industrial plants continued to burn coal, but were required to fit devices to remove the smoke.

The U.S. act was introduced in order to strengthen the provisions of the Air Quality Act of 1960, and it was subsequently revised in 1970, with amendments passed in 1977, 1987, and 1990. The Clean Air Act increased the powers of the Environmental Protection Agency, especially in stipulating the emissions permitted from particular industrial installations. This proved difficult in practice, because it would have forbidden any further industrial development in states where existing emissions exceeded the permitted limits. This difficulty

was overcome by allowing companies to agree with the regulatory authority to accept stricter emission standards for one part of their operation in return for a relaxation in the standards for another part. This was called a "bubble policy." It increased the effectiveness of pollution control while reducing its cost.

Clean Development Mechanism A procedure that is included in the Kyoto Protocol under which countries and companies are permitted to offset their own carbon emissions by paying for a project in another country that would reduce carbon emissions.

Framework Convention on Climate Change A United Nations agreement that was reached at the United Nations Conference on Environment and Development held in Rio de Janeiro, Brazil, in June 1992 (and sometimes called the Rio Summit or the Earth Summit). The Framework Convention aims to address the issue of GLOBAL WARMING by seeking the agreement of national governments to promote relevant research and to reduce emissions of greenhouse gases (*see* GREENHOUSE EFFECT). Its most direct achievement is the Kyoto Protocol, which sets targets for reduced emissions.

Geneva Convention on Long-Range Transboundary Air Pollution A legally binding international agreement that was drafted by scientists after a link had been established between sulfur emissions in Europe and the acidification of Scandinavian lakes (*see* ACID DEPOSITION). The convention was drawn up under the auspices of the United Nations Economic Commission for Europe, was signed in Geneva, Switzerland, in 1979, and came into force in 1983. It lays down the general principles that form the basis for international cooperation to reduce the emission of air pollutants that drift across international frontiers and also establishes an institutional framework for research and the development and implementation of policy. The executive body issues an annual report. Since it came into force, eight protocols have been added to the convention. These are:

- The 1984 Protocol on Long-term Financing of the Cooperative Programme for Monitoring and Evaluation of the Long-Range Transmission of Air Pollutants in Europe
- The 1985 Protocol on the Reduction of Sulphur Emissions or their Transboundary Fluxes by at least 30 percent
- The 1988 Protocol Concerning the Control of Nitrogen Oxides or their Transboundary Fluxes
- The 1991 Protocol Concerning the Control of Emissions of Volatile Organic Compounds or their Transboundary Fluxes
- The 1994 Protocol on Further Reduction of Sulphur Emissions
- The 1998 Protocol on Heavy Metals
- The 1998 Protocol on Persistent Organic Pollutants
- The 1999 Protocol to Abate Acidification, Eutrophication, and Ground-level Ozone

Kyoto Protocol An international agreement that was drawn up in 1997 under the auspices of the United Nations to provide guidelines for the implementation of the United Nations Framework Convention on Climate Change. The protocol committed the industrialized nations to reducing their emissions of six greenhouse gases (*see* GREENHOUSE EFFECT) by an average of 5.2 percent (measured against their 1990 levels) by 2012. The European Union agreed to an 8 percent reduction, the United States to 7 percent, Japan to 6 percent, and 21 other countries to varying reductions. Less-developed countries were not required to make binding commitments.

The protocol was adopted by about 170 nations, but not the United States, at Bonn, Germany, on July 23, 2001. The 5.2 percent target was retained, but countries were permitted to offset their emission reductions by counting the absorption of carbon dioxide by forest planting, changes in forest management, and improved management of croplands and grassland. Nations emitting less than their target amounts were allowed to sell the surplus, up to a cap of 10 percent of their total emission entitlement, as credits to nations unable to meet their targets.

Montreal Protocol on Substances That Deplete the Ozone Layer An international agreement that was reached in 1987 under the auspices of the United Nations Environment Program. The protocol aims to reduce and eventually eliminate the release into the atmosphere of all man-made substances that deplete stratospheric ozone (*see* OZONE LAYER). The provisions made in the protocol for achieving this have been strengthened through amendments adopted in London in 1990, Copenhagen in 1992, Vienna in 1995, and Montreal in 1997.

APPENDIX V
THE GEOLOGIC TIMESCALE

Eon/Eonothem	Era/Erathem	Sub-era	Period/System	Epoch/Series	Began Ma
Phanerozoic		Quaternary	Pleistogene	Holocene	0.11
				Pleistocene	1.81
	Cenozoic	Tertiary	Neogene	Pliocene	5.3
				Miocene	23.3
			Paleogene	Oligocene	33.9
				Eocene	55.8
				Paleocene	65.5
	Mesozoic		Cretaceous	Late	99.6
				Early	145.5
			Jurassic	Late	161.2
				Middle	175.6
				Early	199.6
			Triassic	Late	228
				Middle	245
				Early	251
	Paleozoic	Upper	Permian	Late	260.4
				Middle	270.6
				Early	299
			Carboniferous	Pennsylvanian	318.1
				Mississipian	359.2
			Devonian	Late	385.3
				Middle	397.5
				Early	416
		Lower	Silurian	Late	422.9
				Early	443.7
			Ordovician	Late	460.9
				Middle	471.8
				Early	488.3
			Cambrian	Late	501
				Middle	513
				Early	542

(table continues on next page)

(continued from previous page)

Eon/Eonothem	Era/Erathem	Sub-era	Period/System	Epoch/Series	Began Ma
Proterozoic	Neoproterozoic		Ediacaran		600
			Cryogenian		850
			Tonian		1,000
	Mesoproterozoic		Stenian		1,200
			Ectasian		1,400
			Calymmian		1,600
	Palaeoproterozoic		Statherian		1,800
			Orosirian		2,050
			Rhyacian		2,300
			Siderian		2,500
Archaean	Neoarchaean				2,800
	Mesoarchaean				3,200
	Palaeoarchaean				3,600
	Eoarchaean				3,800
Hadean	Swazian				3,900
	Basin Groups				4,000
	Cryptic				4,567.17

Source: International Union of Geological Sciences, 2004.

Note: Hadean is an informal name. The Hadean, Archaean, and Proterozoic Eons cover the time formerly known as the Precambrian. Tertiary has been abandoned as a formal name, and Quaternary is likely to be abandoned in the next few years, although both names are still widely used.

APPENDIX VI
OCEAN CURRENTS

Agulhas Current A current that flows in a south-westerly direction at the surface of the Indian Ocean, between the eastern coast of Africa and Madagascar, between latitudes 25°S and 40°S. Its speed is 0.7–2.0 feet per second (0.2–0.6 m/s).

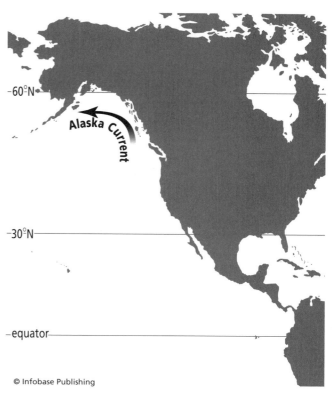

© Infobase Publishing

The Alaska Current is a branch of the North Pacific Current carrying warm water to the coasts of northwestern Canada and southeastern Alaska.

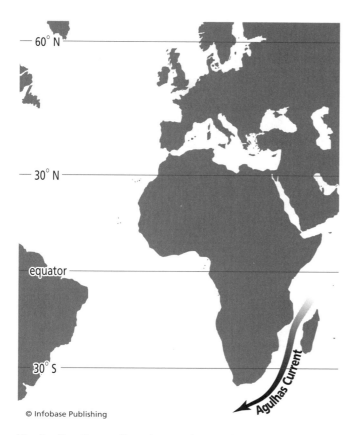

© Infobase Publishing

The Agulhas Current flows in a southwesterly direction, parallel to the coast of East Africa.

Alaska Current A warm ocean BOUNDARY CURRENT that flows northwestward and then westward along the coast of Canada and southeastern Alaska. It results from the deflection of the North Pacific Current as this current approaches the North American continent. It is sometimes called the Aleutian Current, although the two are usually regarded as distinct.

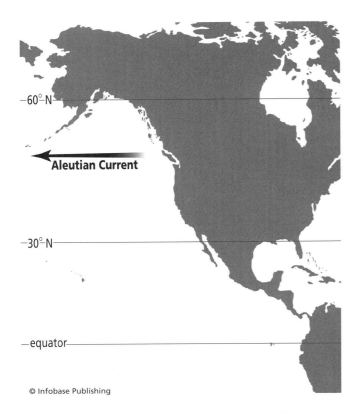

© Infobase Publishing

The Aleutian Current flows in a westerly direction to the south of the Aleutian Islands.

Aleutian Current (Sub-Arctic Current)   A warm ocean current that flows in a westerly direction to the south of the Aleutian Islands. It runs parallel to the North Pacific Current, but to the north of it, and carries a mixture of warm water from the Kuroshio Current and cold water from the Oyashio Current. The Alaska Current is sometimes called the Aleutian Current, although the two are usually considered distinct.

Antarctic Circumpolar Current (West Wind Drift)   An ocean current that flows from west to east around the coast of Antarctica. It is driven by the prevailing winds (see WIND SYSTEMS), which are from the west, and it is the only ocean current that flows all the way around the world. It carries water that is cold, with a temperature of 30–40°F (-1–+5°C), and has a low salinity, of less than 34.7 per mil. This current is not to be confused with the Antarctic Polar Current.

Antarctic Polar Current   An ocean current that flows from east to west, parallel to the coast of Antarctica.

It is driven by easterly winds that blow from off the ice cap. The current affects only surface waters. This current is not to be confused with the Antarctic Circumpolar Current, also known as the West Wind Drift.

Antilles Current   An ocean current that branches from the North Equatorial Current and carries warm water along the northern coasts of the Great Antilles, in the Caribbean.

Benguela Current   An ocean current that flows northward from the Antarctic Circumpolar Current and along the western coast of Africa, from about 35°S to 15°S. It carries cold water, with many UPWELLINGS, and flows fairly slowly, at less than 0.6 MPH (0.9 km/h).

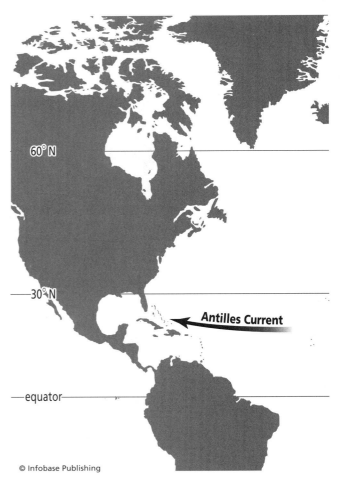

© Infobase Publishing

The Antilles Current is a warm ocean current that flows past the Antilles, in the Caribbean.

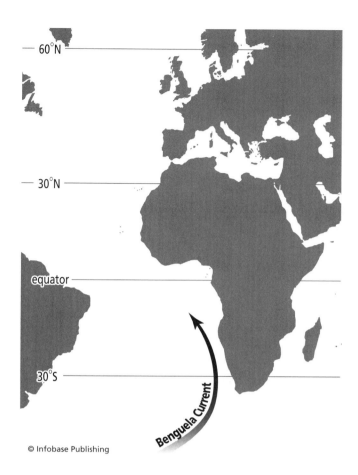

The Benguela Current is a cold current flowing parallel to the coast of southern Africa.

Bering Current An ocean current that flows southward through the Bering Strait separating Alaska from eastern Siberia, bringing cold water into the North Pacific Ocean.

Brazil Current An ocean current that carries warm water southward from the South Equatorial Current, along the eastern coast of South America, to where it joins the Antarctic Circumpolar Current. The current moves very slowly, is no more than 330–660 feet (100–200 m) deep, and the salinity of its water is 36–37 per mil, which is saltier than the average for seawater.

California Current A slow-moving, somewhat diffuse ocean current that conveys cold water southward par-allel to the western coast of North America. In the latitude of Central America, it turns westward to become the North Equatorial Current.

Canary Current A slow-moving ocean current that conveys cold water southward parallel to the coasts of Spain, Portugal, and West Africa. It is the cause of frequent sea FOGS off northwestern Spain and Portugal.

Caribbean Current An ocean current that flows westward through the Caribbean Sea. As it passes the coast of Florida, it joins the Florida Current, and it then becomes part of the Gulf Stream. The Caribbean Current carries warm water and flows at an average of 0.85–0.96 MPH (0.38–0.43 m/s).

Cromwell Current (Equatorial Undercurrent) An ocean current that flows from west to east beneath the

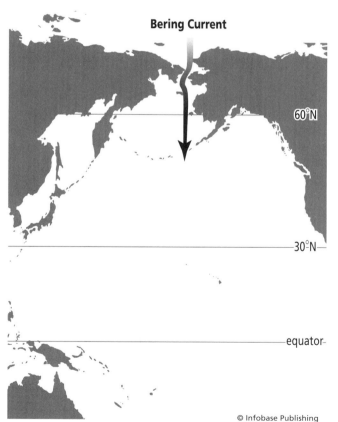

The Bering Current carries cold water from the Arctic into the North Pacific.

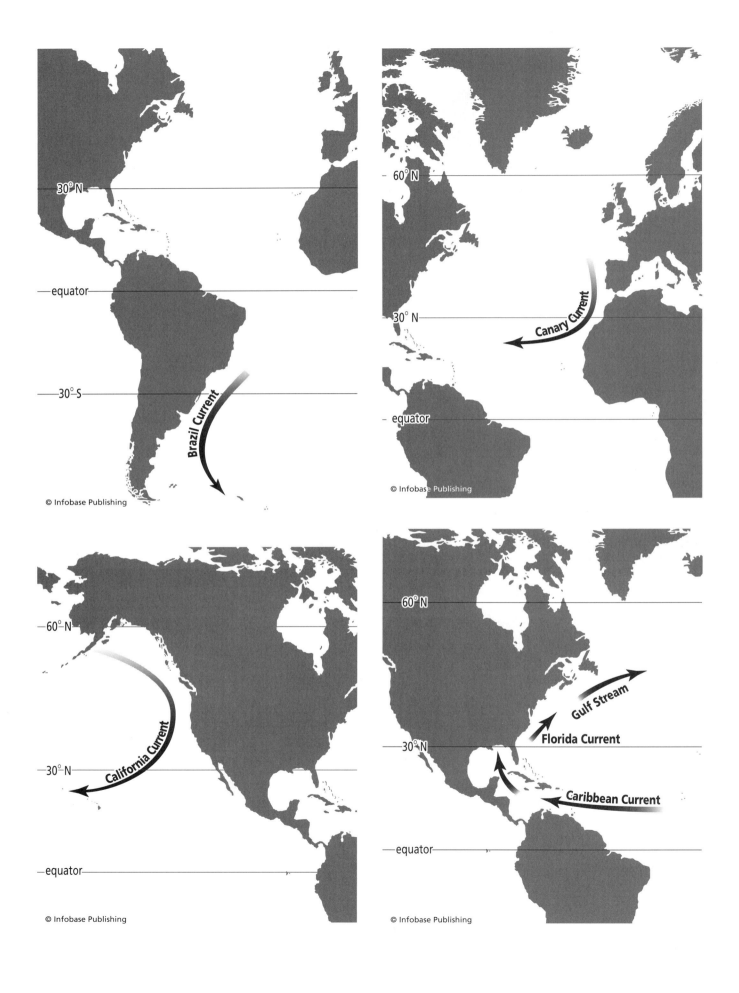

(*Opposite page: Top left*) The Brazil Current is a warm current flowing parallel to the coast of Brazil.
(*Bottom left*) The California Current is a cold current flowing parallel to the coast of northwestern North America.
(*Top right*) The Canary Current is a cold current flowing parallel to the western coast of North Africa.
(*Bottom right*) The Caribbean Current flows westward through the Caribbean Sea, then joins the Florida Current before joining the Gulf Stream.

surface of the Pacific Ocean between latitudes 1.5°N and 1.5°S and at a depth of 165–1,000 feet (50–300 m). The current is about 185 miles (300 km) wide and flows at up to 3.4 MPH (5.5 km/h). The surface Equatorial Current is driven by the trade winds (see WIND SYSTEMS) and carries warm surface water from east to west. The Cromwell Current counterbalances the Equatorial Current by conveying water below the surface in the opposite direction.

East Australian Current   An ocean current that flows southward carrying warm water parallel to the eastern coast of Australia. The current is narrow, being only 330–660 feet (100–200 m) wide, and slow-moving, flowing at 0.6–1.2 MPH (1–2 km/h).

East Greenland Current   An ocean current that flows southward from the Arctic Ocean into the North Atlantic Ocean, parallel to the northeastern coast of Greenland. In about the latitude of Iceland it merges with a branch of the North Atlantic Drift. The East Greenland Current carries cold water with a low salinity.

Equatorial Countercurrent   A narrow ocean current that flows from west to east between the North and South Equatorial Currents.

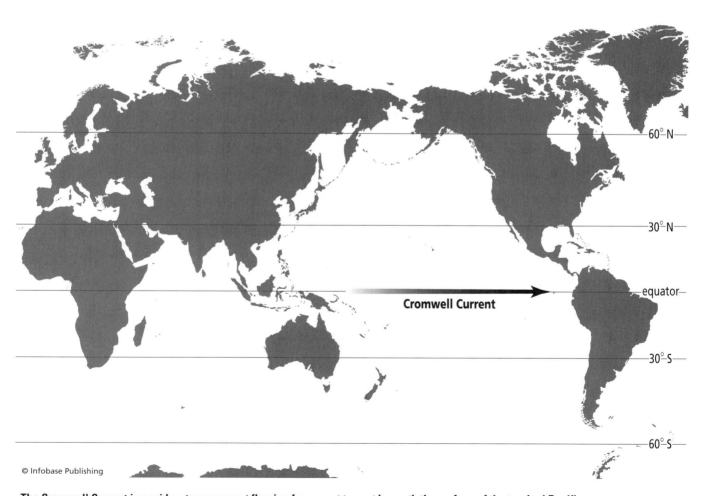

© Infobase Publishing

The Cromwell Current is a wide, strong current flowing from west to east beneath the surface of the tropical Pacific.

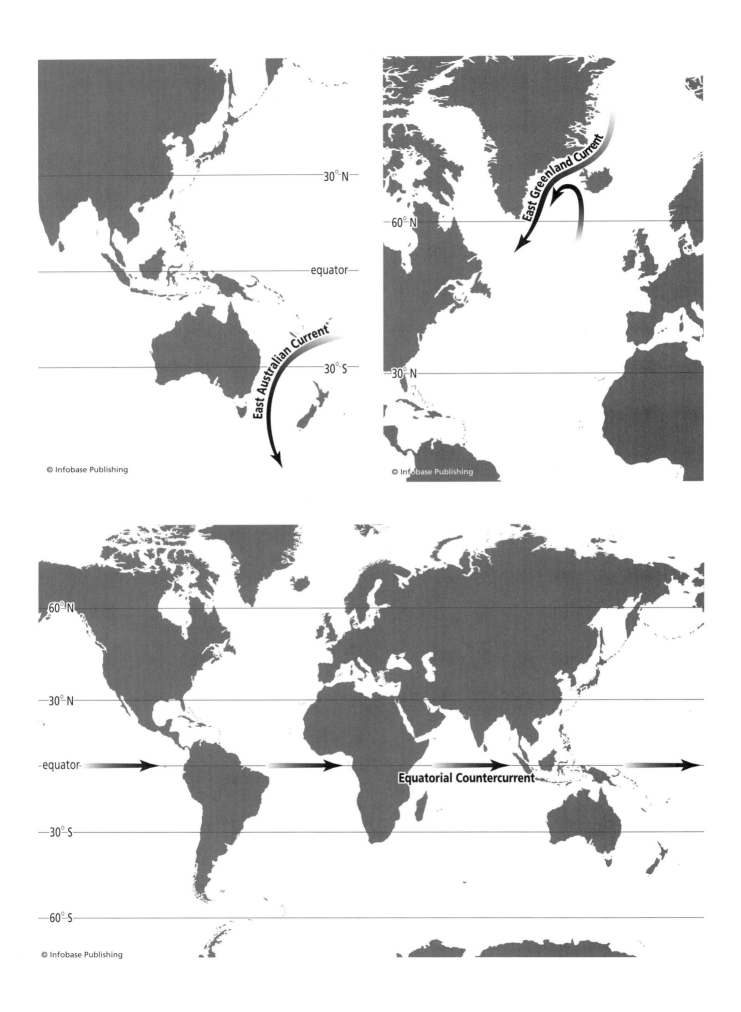

(Opposite page: Top left) The East Australian Current is a warm current flowing parallel to the eastern coast of Australia. *(Opposite page: Top right)* The East Greenland Current is a cold current flowing parallel to the eastern coast of Greenland. *(Opposite page: Bottom)* The Equatorial Countercurrent flows close to the equator and from west to east in all oceans.

Falkland Current   An ocean BOUNDARY CURRENT that carries cold water northward past the Falkland Islands (Malvinas) and parallel to the coast of Argentina as far as about latitude 30°S and rather farther north in winter. Its influence greatly reduces the effect ordinarily produced by boundary currents outside the TROPICS, of bringing warm water to western coasts.

Florida Current   An ocean current that flows northward parallel to the coast of Florida and that forms part of the Gulf Stream. It extends from the southern tip of Florida to Cape Hatteras, North Carolina. It is narrow, being 30–47 miles (50–75 km) wide, and fast, flowing at 2.2–6.7 MPH (3.6–11 km/h).

Guinea Current   An ocean current that flows from west to east along the West African coast and into the

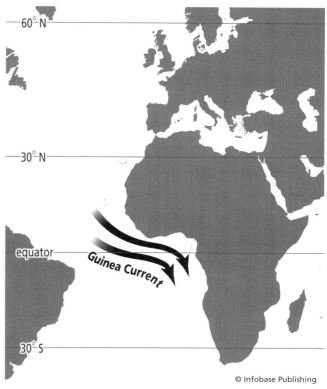

The Guinea Current is a warm current flowing past the West African coast into the Gulf of Guinea.

The Falkland Current is a cold current flowing northward, parallel to the coast of Argentina.

Gulf of Guinea. It is part of the Equatorial Countercurrent and carries warm water with a temperature that often exceeds 80°F (27°C). The current brings hot, humid weather to coastal regions in the west, but farther east, off the coast of western Nigeria, the current produces UPWELLINGS. These frequently produce FOG and a fairly low rainfall of about 30 inches (762 mm) a year, compared with the 100 inches (2,540 mm) a year along the coast to the east of Lagos.

Gulf Stream   A system of ocean currents that convey warm water from the Gulf of Mexico to the center of the North Atlantic Ocean. It begins as the Florida Current, where the North Equatorial Current enters the Gulf and ends in the latitude of Spain and Portugal,

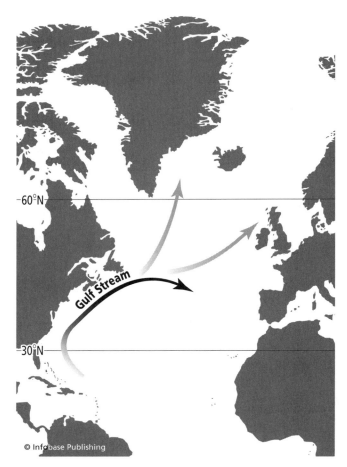

The Gulf Stream is a warm current that follows an approximately circular, clockwise path in the North Atlantic.

Kuroshio Current   An ocean current that flows northward from the Philippines, along the coast of Japan, and then eastward into the North Pacific Ocean, carrying warm water northward. It is a narrow current, less than 50 miles (80 km) wide, and flows rapidly, at up to 7 MPH (11 km/h).

Labrador Current   An ocean current that conveys cold water from the Arctic Ocean into the North Atlantic Ocean. It flows in a southeasterly direction between the coasts of eastern Canada and western Greenland, often carrying ICEBERGS south into the North Atlantic. Sea FOGS are common off Newfoundland, where the cold water of the Labrador Current meets the warm water of the Gulf Stream.

where it turns south, becoming the Canary Current and rejoining the North Equatorial Current. A branch from it becomes the North Atlantic Drift. Its influence on climates makes it the most important current system in the Northern Hemisphere. It is clearly defined, as a belt of water at a fairly constant temperature of 64–68°F (18–20°C) and salinity of 36 per mil.

Irminger Current   An ocean current that flows past the southern coast of Iceland and continues past the southern cape of Greenland. It is a branch of the Gulf Stream that breaks away from the North Atlantic Drift in about latitude 50°N and carries warm water northwestward. As it passes Greenland, its water mixes with cold water from Baffin Bay, but it can still be detected by its higher salinity as far as 65°N.

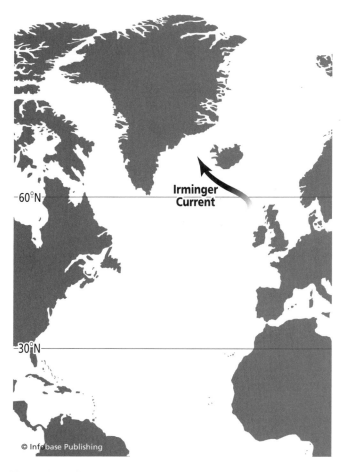

The Irminger Current breaks away from the Gulf Stream and flows past the southern coast of Iceland and Greenland.

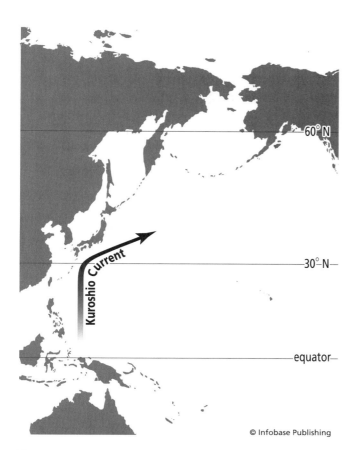

The Kuroshio Current is a warm current flowing parallel to the eastern coast of Japan.

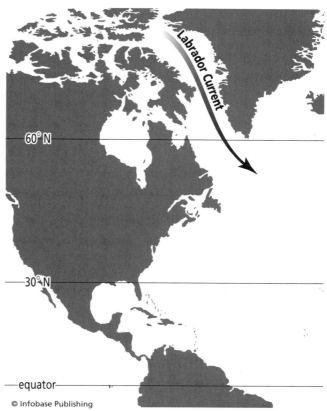

The Labrador Current is a cold current flowing parallel to the northeastern coast of Canada.

Monsoon Drift   An ocean current that flows through the Arabian Sea. It breaks away from the North Equatorial Current off the southernmost tip of India, flows parallel to the western coast of India, then turns in about latitude 15–20°N to flow parallel to the southern coast of the Arabian Peninsula, finally joining the Somalia Current. This current flows past the eastern coast of Africa, through the Mozambique Channel separating Africa and Madagascar, and joins the Agulhas Current.

Mozambique Current   A branch of the South Equatorial Current that flows around the northern end of Madagascar and continues as a warm ocean current in a southwesterly direction through the Mozambique Channel off the eastern coast of Africa. To the south of Madagascar, it becomes the Agulhas Current.

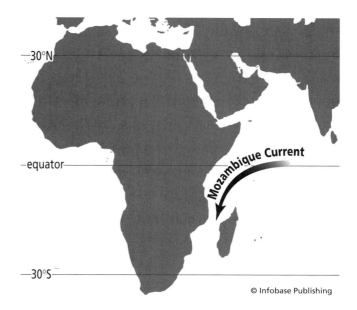

The Mozambique Current carries warm water into the Mozambique Channel, between Mozambique and Madagascar.

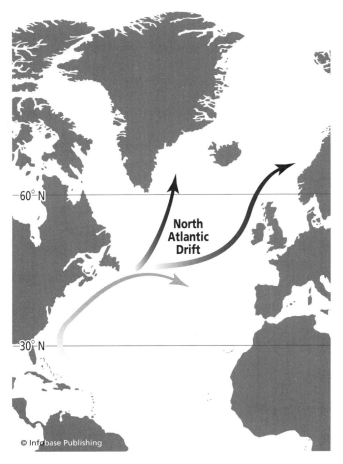

The North Atlantic Drift is a warm current that breaks away from the Gulf Stream to form two streams, one flowing to the east of Iceland and past the coasts of northwestern Europe, the other flowing to the west of Iceland toward Greenland.

North Atlantic Drift (North Atlantic Current) A broad, shallow, warm, surface current that is an extension of the Gulf Stream. It flows in a northeasterly direction from the middle of the North Atlantic Ocean and divides into two branches south of Iceland. The westerly branch flows northward and then turns south to join the East Greenland Current. The other branch passes to the east of Iceland in a northeasterly direction, approaching close to the coasts of the British Isles and continuing to become the Norwegian Current. The North Atlantic Drift exerts a strong influence on the climate of the coastal regions of northwestern Europe, making it milder and wetter than it would be otherwise.

North Equatorial Current Two ocean currents, one in the North Atlantic and the other in the North Pacific, that flow from east to west parallel to and just north of the equator. The current flows within the upper 1,600 feet (500 m) of water at a speed of 0.6–2.5 mph (1–4 km h^{-1}) and is separated from the South Equatorial Current by the Equatorial Countercurrent.

North Pacific Current An ocean current that flows from west to east across the North Pacific Ocean, carrying warm water toward California. It is an extension of the Kuroshio Current.

Norwegian Current (Norway Current) An ocean current that flows in a northeasterly direction, parallel to the northern coast of Norway. It is a continuation of the North Atlantic Drift and carries relatively warm water into the Arctic Ocean.

Oyashio Current (Kamchatka Current) An ocean current that flows southward from the Bering Sea, past the Kuril Islands, to the northeast of Japan, where it meets the Kuroshio Current. It flows at less than 1.5 MPH (2.4 km/h) and carries cold water, with a low salinity of 33.7–34.0 per mil.

Peru Current (Humboldt Current) An ocean current that flows northward from the Antarctic Circumpolar Current, past the western coast of South America, to join the South Equatorial Current. It carries cold water and is broad and slow-moving. The current is noted for the many UPWELLINGS along its course, which bring nutrients near to the surface.

South Equatorial Current Two ocean currents, one in the South Atlantic and the other in the South Pacific, that flow from east to west parallel to and just south of the equator. The current flows within the upper 1,600 feet (500 m) of water at a speed of 0.6–2.5 MPH (1–4 km/h) and is separated from the North Equatorial Current by the Equatorial Countercurrent.

(Opposite page: Top) The North Equatorial Current flows just north of the equator and parallel to it, in the North Atlantic and North Pacific Oceans. Where it encounters continents, the current is deflected into a more northerly path.
(Bottom left) The North Pacific Current carries warm water from Asia toward California.
(Bottom right) The Norwegian Current is a warm current flowing parallel to the coast of Norway.

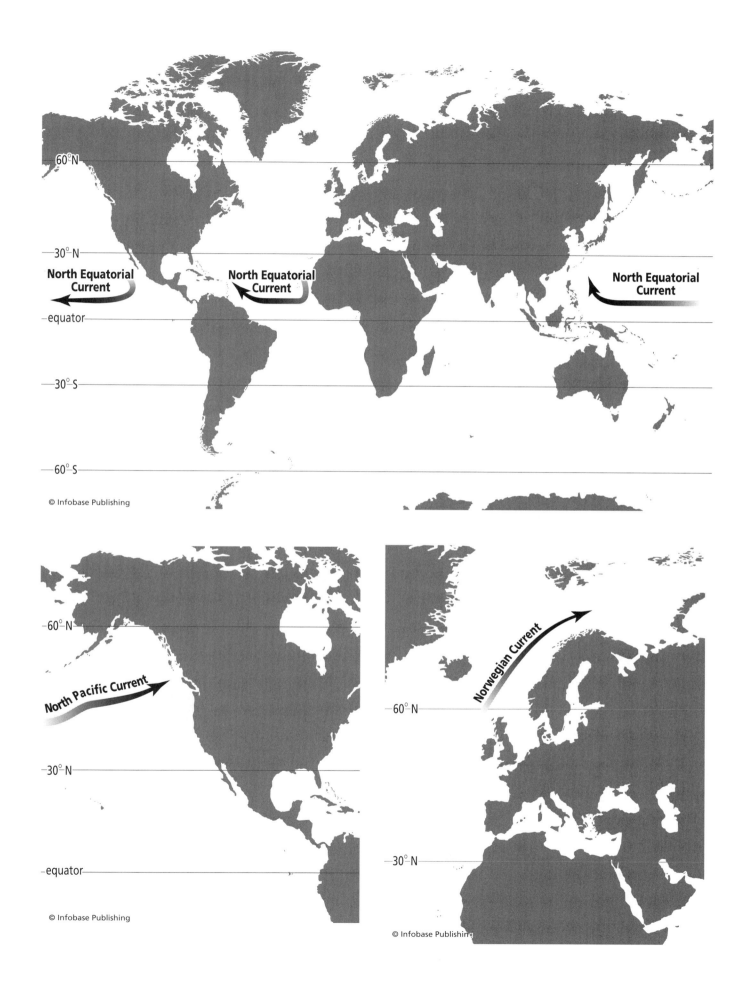

North Equatorial Current

North Equatorial Current

North Equatorial Current

© Infobase Publishing

North Pacific Current

© Infobase Publishing

Norwegian Current

© Infobase Publishing

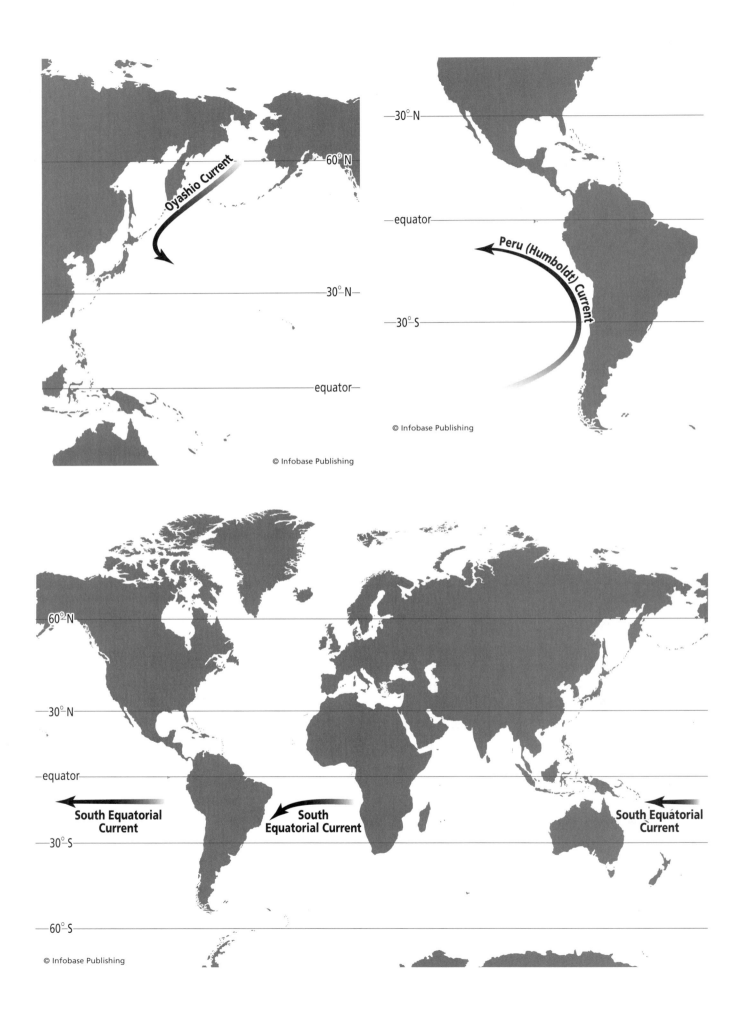

Oyashio Current

30° N

60° N

30° N

equator

© Infobase Publishing

30° N

equator

Peru (Humboldt) Current

30° S

© Infobase Publishing

60° N

30° N

equator

30° S

60° S

South Equatorial
Current

South
Equatorial Current

South Equatorial
Current

(Opposite page: Top left) The Oyashio Current is a cold current flowing southward from Alaska toward Japan.
(Top right) The Peru (or Humboldt) Current is a cold current flowing northward from the Southern Ocean, parallel to the western coast of South America.
(Bottom) The South Equatorial Current flows from east to west close to the equator in the South Atlantic and South Pacific Oceans. Where it encounters continents, the current is deflected into a more southerly path.

West Australian Current   An ocean current that flows northward from the Antarctic Circumpolar Current, parallel to the western coast of Australia. The current flows strongly and steadily in summer, but weakens in winter. Its water is cold, at 37–45°F (3–7°C), and its salinity, of 34.5 per mil, is below the average for seawater.

West Greenland Current   An ocean current that flows to the west of Greenland and that is an extension of the western branch of the North Atlantic Drift. It carries relatively warm water northward parallel to the western coast of Greenland and into the Davis Strait. The current then divides, part of it continuing into Baffin Bay and part joining the Labrador Current.

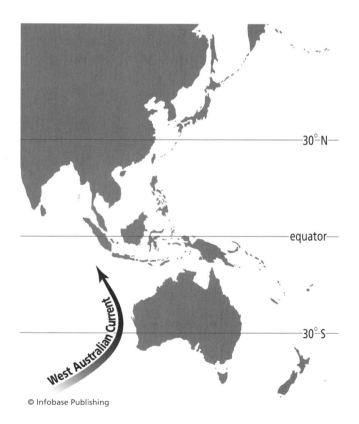

The West Australian Current is a cold current flowing from the Southern Ocean and passing close to the western coast of Australia.

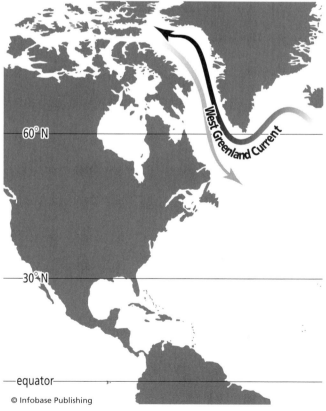

The West Greenland Current is a warm current flowing parallel to the western coast of Greenland.

APPENDIX VII
PLIOCENE, PLEISTOCENE, AND HOLOCENE GLACIALS AND INTERGLACIALS

Approximate date ('000 years BP)	N. America	Great Britain	N.W. Europe
Holocene			
10–present	*Holocene*	*Holocene (Flandrian)*	*Holocene (Flandrian)*
Pleistocene			
75–10	Wisconsinian	Devensian	Weichselian
120–75	*Sangamonian*	*Ipswichian*	*Eeemian*
170–120	Illinoian	Wolstonian	Saalian
230–170	*Yarmouthian*	*Hoxnian*	*Holsteinian*
480–230	Kansan	Anglian	Elsterian
600–480	*Aftonian*	*Cromerian*	*Cromerian complex*
800–600	Nebraskan	Beestonian	*Bavel complex*
740–800	*Pastonian*		
900–800		Pre-Pastonian	Menapian
1,000–900		*Bramertonian*	*Waalian*
1,800–1,000		Baventian	Eburonian
Pliocene			
1,800		*Antian*	*Tiglian*
1,900		Thurnian	
2,000		*Ludhamian*	
2,300		Pre-Ludhamian	Pretiglian

BP means "before present" (present is taken to be 1950). Names in italic refer to interglacials. Other names refer to glacials (ice ages). Dates become increasingly uncertain for the older glacials and interglacials, and prior to about 2 million years ago evidence for these episodes has not been found in North America; in the case of the Thurnian glacial and Ludhamian interglacial, the only evidence is from a borehole at Ludham, in eastern England.

APPENDIX VIII
SI UNITS AND CONVERSIONS

Unit	Quantity	Symbol	Conversion
Base units			
meter	length	m	1 m = 3.2808 feet
kilogram	mass	kg	1 kg = 2.205 pounds
second	time	s	
ampere	electric current	A	
kelvin	thermodynamic temperature	K	1 K = 1°C = 1.8°F
candela	luminous intensity		
mole	amount of substance	cd	mol
Supplementary units			
radian	plane angle	rad	$\pi/2$ rad = 90°
steradian	solid angle	sr	
Derived units			
coulomb	quantity of electricity	C	
cubic meter	volume yards^3	m^3	1 m^3 = 1.308
farad	capacitance	F	
henry	inductance	H	
hertz	frequency	Hz	
joule	energy	J	1 J = 0.2389 calories
kilogram per cubic meter	density	kg m^{-3}	1 kg m^{-3} = 0.0624 lb. ft.$^{-3}$
lumen	luminous flux	lm	
lux	illuminance	lx	
Unit	Quantity	Symbol	Conversion
meter per second	speed	m s^{-1}	1 m s^{-1} = 3.281 ft s^{-1}
meter per second squared	acceleration	m s^{-2}	
mole per cubic meter	concentration	mol m^{-3}	
newton	force	N	1 N = 7.218 lb. force
ohm	electric resistance	Ω	
pascal	pressure	Pa	1 Pa = 0.145 lb. in^{-2}
radian per second	angular velocity	rad s^{-1}	
radian per second squared	angular acceleration	rad s^{-2}	
square meter	area	m^2	1 m^2 = 1.196 yards^2
tesla	magnetic flux density	T	
volt	electromotive force	V	
watt	power	W	1W = 3.412 Btu h^{-1}
weber	magnetic flux	Wb	

Prefixes used with SI units

Prefix	Symbol	Value
atto	a	$\times 10^{-18}$
femto	f	$\times 10^{-15}$
pico	p	$\times 10^{-12}$
nano	n	$\times 10^{-9}$
micro	μ	$\times 10^{-6}$
milli	m	$\times 10^{-3}$
centi	c	$\times 10^{-2}$
deci	d	$\times 10^{-1}$
deca	da	$\times 10$
hecto	h	$\times 10^{2}$
kilo	k	$\times 10^{3}$
mega	M	$\times 10^{6}$
giga	G	$\times 10^{9}$
tera	T	$\times 10^{12}$

Prefixes attached to SI units alter their value.

APPENDIX IX
CHRONOLOGY OF DISASTERS

1246–1305

DROUGHT in what is now the southwestern United States.

1281

A typhoon destroys a fleet of Korean ships carrying Mongol troops on their way to invade Japan. This is the kamikaze wind.

1703

On November 26 and 27, hurricane-force winds in the English Channel destroy 14,000 homes and kill 8,000 people in southern England.

1762

In February, a BLIZZARD in England lasts for 18 days and kills nearly 50 people.

1831

A hurricane strikes Barbados, killing 1,477 people.

1865

In June, a TORNADO moves through Viroqua, Wisconsin, destroying 80 buildings and killing more than 20 people.

1875

On November 15, the River Thames rises, possibly by more than 28 feet (8.5 m), causing extensive flooding in London.

1876

A cyclone coinciding with high, MONSOON, river levels floods islands in the Ganges Delta and on the mainland, drowning about 100,000 people in half an hour.

1879

On December 28, the TAY BRIDGE DISASTER causes 70–90 deaths in Scotland.

1887

In September and October, the Yellow River, China, floods about 10,000 square miles (26,000 km²). Between 900,000 and 2.5 million people die.

1888

On March 11–13 BLIZZARDS with winds up to 70 MPH (113 km/h) strike the eastern United States. More than 400 people die, including 200 in New York City.

1925

In March, a series of possibly seven TORNADOES develop over Missouri and cross Illinois and Indiana, killing 689 people.

1931

The Yangtze River, China, rises 97 feet (29.6 m) following heavy rain. About 3.7 million people die, some in the floods but most from the famine that follows.

1954

On October 12, Hurricane Hazel kills 1,175 people (1,000 of them in Haiti).

1956

On June 27, Hurricane Audrey kills near 400 people.

1957

In August, Hurricane Diane kills more than 190 people.

1959

In September, Typhoon Vera kills nearly 4,500 people and leaves 1.5 million homeless.

1966

On November 3, the River Arno floods Florence, Italy, causing extensive damage to historic buildings and works of art, killing 35 people, and leaving 5,000 homeless.

1969

On August 17–18, Hurricane Camille kills about 275 people.

1970

In November, a cyclone kills about 500,000 people in Bangladesh.

1973

On January 10, a TORNADO kills 60 people and injures more than 300 in San Justo, Argentina.

1974

On September 20, Hurricane Fifi kills about 5,000 people in Honduras. Cyclone Tracy strikes Darwin, Australia, on December 25.

1976

On September 8–13, Typhoon Fran kills 104 people and makes 325,000 homeless in Japan.

1977

On November 19, a cyclone and STORM SURGE washes away 21 villages and damages 44 more in Andhra Pradesh, India, killing an estimated 20,000 people and making more than 2 million homeless.

1978

On April 16, a TORNADO kills nearly 500 people and injured more than 1,000 in Orissa, India. On October 26, Typhoon Rita kills nearly 200 people and destroys 10,000 homes in the Philippines. On November 23, a cyclone kills at least 1,500 people and destroys more than 500,000 buildings in Sri Lanka and southern India.

1979

On April 10, a TORNADO kills 59 people and injures 800 at Wichita Falls, Texas. On May 12–13, a cyclone kills more than 350 people in India. On August 11, heavy rain causes a dam to break, flooding the town of Morvi, India, and killing up to 5,000 people. In August, Hurricane David kills more than 1,000 people in the Caribbean and eastern United States.

1980

In August, Hurricane Allen kills more than 270 people. A heatwave kills 1,265 people in the United States; in Texas, temperatures exceed 100°F (38°C) almost every day.

1981

On July 12–14, MONSOON rains cause the Yangtze River, China, to flood, killing about 1,300 people and leaving 1.5 million homeless. On September 1, typhoon Agnes kills 120 people in South Korea. On November 24, typhoon Irma kills more than 270 people and leaves 250,000 homeless in the Philippines.

1982

On January 23–24, floods kill at least 600 people and leave 2,000 missing in Peru. On June 3, MONSOON floods in Sumatra, Indonesia, kill at least 225 people and leave 3,000 homeless. In September, monsoon floods in Orissa, India, kill at least 1,000 people.

1983

In June, floods killed at least 935 people in Gujarat, India.

1984

Between January 31 and February 2, Cyclone DOMOINA kills at least 124 people. On September 2–3, Typhoon Ike kills more than 1,300 people in the Philippines. In November, typhoon Agnes kills at least 300 people in the Philippines and leaves 100,000 homeless.

1985

On May 25, a cyclone and STORM SURGE kill an estimated 2,540 people, but possibly as many as 11,000, on islands off Bangladesh. On May 31, TORNADOES kill 88 people and cause extensive damage in Pennsylvania, Ohio, New York, and Ontario.

1986

On March 17, Cyclone Honorinnia kills 32 people and leaves 20,000 homeless in Madagascar. On September 4, a typhoon kills 400 people in Vietnam.

1987

In August, floods kill more than 1,000 people in Bangladesh. On November 26, typhoon Nina kills 500 people in the Philippines.

1988

In August and September, monsoon floods inundate 75 percent of Bangladesh, killing more than 2,000 people and leaving at least 30 million homeless. On September 12–17, Hurricane Gilbert kills at least 260 people in the Caribbean and Gulf of Mexico and generates nearly 40 TORNADOES in Texas. On October 24–25, Typhoon Ruby kills about 500 people in the Philippines. On November 7, Typhoon Skip kills at least 129 people in the Philippines. On November 29, a cyclone kills up to 3,000 people in Bangladesh and eastern India.

1989

On April 26, a TORNADO in Bangladesh kills up to 1,000 people and injures 12,000. On September 17–21, Hurricane Hugo kills more than 40 people in the Caribbean and eastern United States. On November 4–5, typhoon Gay kills 365 people in Thailand.

1990

On May 9, a cyclone kills at least 962 people in Andhra Pradesh, India. In August, typhoon Yancy kills 228 people in the Philippines and China. On August 28, a TORNADO kills 29 people and injures 300 at Plainfield, Illinois.

1991

On March 10, floods kill more than 500 people and leave 150,000 homeless in Mulanje, Malawi. On April 26, more than 70 TORNADOES kill 26 people and injure more than 200 in Kansas. On April 30, a cyclone kills at least 131,000 people on coastal islands off Bangladesh. On May 7, a tornado kills 100 people at Tungi, Bangladesh. In June, FLASH FLOODS kills up to 5,000 people in Jowzjan Province, Afghanistan.

1992

In July, floods kills more than 1,000 people in Fujian and Zheijiang Provinces, China. On August 23–26, Hurricane Andrew kills 38 people and causes extensive damage in the Bahamas, Florida, and Louisiana. On September 11–16, MONSOON rains cause the Indus River to flood, killing at least 500 people in India and more than 2,000 in Pakistan.

1993

On January 8, a TORNADO kills 32 people and injures more than 1,000 in Bangladesh. On March 12–15, a BLIZZARD kills at least 238 people in the eastern United States, four in Canada, and three in Cuba. Between October 31 and November 2, mudslides kill 400 people and destroy 1,000 homes in Honduras.

1994

On February 2–4, Cyclone Geralda kills 70 people and leaves 500,000 homeless in Madagascar. On August 20–21, Typhoon Ted kills about 1,000 people in Zheijiang Province, China. On November 13–19, Tropical Storm Gordon kills 537 people in the Caribbean, Florida, and South Carolina.

1995

Beginning in July, floods affect 5 million people, nearly 25 percent of the population, in North Korea. On November 3, Typhoon Angela kills more than 700 people and leaves more than 200,000 homeless in the Philippines.

1996

On May 13, a TORNADO in Bangladesh destroys 80 villages in less than half an hour, killing more than 440 people and injuring more than 32,000. On September 10, Typhoon Sally kills more than 130 people and destroy nearly 400,000 homes in Guangdong, China.

1996–1997

In December and January, floods in California, Idaho, Nevada, Oregon, and Washington cause at least 29 deaths and force 125,000 people to leave their homes.

1997

In January, a COLD WAVE crosses Europe kills at least 228 people. On May 2, a SANDSTORM kills 12 people and injures 50 in Egypt. On May 27, TORNADOES in central Texas destroy about 60 homes and kill 30 people. On July 2, THUNDERSTORMS and tornadoes in southern Michigan destroy 339 homes and business premises, kill 16 people, and injure more than 100. On August 18–19, Typhoon Winnie kills nearly 200 people in China, Taiwan, and the Philippines. On October 8–10, Hurricane Pauline kills 217 people and leaves 20,000 homeless in southern Mexico. On October 12, a tornado kills at least 25 people and injures thousands who had gathered for a religious ceremony at Tongi, Bangladesh. In November, Typhoon Linda kills at least 484 people in Vietnam, Cambodia, and Thailand.

1998

On February 23, TORNADOES in Florida kill at least 42 people, injure more than 260, and leave hundreds homeless. On March 20, tornadoes kill at least 14 people and injure 80 in Georgia and kill two and injure at least 22 in North Carolina. In March, a cyclone kills at least 200 people and makes 10,000 homeless in West Bengal and Orissa, India. On April 8–9, tornadoes kill 39 people in Mississippi, Alabama, and Georgia. In May and early June, a heat wave kills at least 2,500 people in India. From June to August, the Yangtze River, China, floods, killing 3,656 people and affecting an estimated 230 million. On September 21–28, Hurricane Georges kills more than 330 people in the Caribbean and along the U.S. Gulf coast. In September and early October Tropical Storm Yanni kills 27 people in South Korea. In October, typhoon Zeb kills 111 people in the Philippines, Taiwan, and Japan. In late October, Hurricane Mitch kills more than 8,600 people, leaving 12,000 unaccounted for, and makes more than 1.5 million homeless in Central America. In late October, Typhoon Babs kills at least 132 people and makes about 320,000 homeless in the Philippines. On November 19–23 typhoon Dawn causes floods in Vietnam that force 200,000 people from their homes and kill more than 100.

1999

In August, typhoon Olga causes extensive flooding in South Korea. In September, Hurricane Floyd kills 49 people in the Bahamas and the eastern coast of the United States.

2000

In February, the worst floods in 50 years devastate Mozambique, destroying about 200,000 homes. Shortly after midnight on February 14, TORNADOES sweeping through southwestern Georgia kill 18 people and injure about 100. On February 22, Cyclone Eline strikes Mozambique, with winds of up to 162 MPH (260 km/h). Eline moves to Madagascar, which is also struck by Tropical Storm Gloria on March 4–5. The two storms leave at least 500,000 people homeless on the island and kills at least 137. In May, severe flooding combine with a tidal surge, killing at least 140 people and leaving about 20,000 homeless on West Timor, Indonesia. Between late July and early October, the Mekong Delta, in Vietnam, Laos, and Cambodia, experiences the worst flooding in 40 years, killing at least 315 people. In September and October, flooding kills more than 900 people in India and about 150 in Bangladesh, and leaves some 5 million homeless in the two countries. On November 1–2, typhoon Xangsane causes severe flooding on Taiwan.

2001

Floods early in the year kill at least 52 people in Mozambique and leave more than 80,000 homeless. In January, a BLIZZARD in northern China kills 20 people and leaves thousands cut off, with no access to food supplies. In February, TORNADOES kill five people in Mississippi and one in Arkansas. In April, the Mississippi bursts its banks, flooding parts of Minnesota, Iowa, Illinois, and Wisconsin. A 165-foot (50-m) dam of ice blocks cause the River Lena, in Siberia, to flood in May, washing away thousands of homes and killing at least five people. On May 28, 18 people are injured and many buildings damaged by a tornado in Ellicott, Colorado. Weekend storms on July 7–8 cause widespread flooding in West Virginia. In July, Typhoon Utor causes floods and a mudslide in which 23 people die in the Philippines and one person dies in Taiwan. Two days of rain in South Korea, also in July, cause 40 deaths. On July 29 and 30, Typhoon Toraji swept through Taiwan, causing at least 72 deaths and leaves more than 130 unaccounted for. MONSOON floods in late July trap nearly 50,000 people in inundated villages in Bangladesh.

2002

Heavy rain and extreme cold in Mauritania in January kill at least 25 people and an estimated 80,000 head of livestock. Winds of almost 120 MPH (200 km/h) batter Europe in late January, killing at least 18 people. Heavy rain causes floods and landslides in Java, Indonesia, in February. At least 150 people die. On February 19, the most destructive storm ever experienced in La Paz, Bolivia, triggers FLASH FLOODS and mudslides that kill 69 people and leave hundreds homeless. From May 9 to 15, an intense heat wave in Andhra Pradesh, India, kills at least 1,030 people. In early June, a heat wave kills more than 60 people in Nigeria. In June, floods kill more than 200 people in northwestern China. Floods in June inundate approximately 70 villages in southern Russia, killing at least 53 people and rendering 75,000 homeless. MONSOON floods lasting from June until the middle of August in southern Asia kill at least 422 people in Nepal, nearly 400 in India, and at least 157 in Bangladesh. Approximately 15 million people are made homeless in Bihar and Assam States, India, and about 6 million in Bangladesh. On June 4–5, the Zeyzoun Dam near the town of Hama, Syria, collapses following prolonged heavy rain. Several villages were flooded and at least 28 people are killed. In July, a heat wave in Algeria produces temperatures up to 133°F (56°C) and kill at least 50 people. Severe cold weather in July kills at least 59 people in Peru, and about 80,000 head of livestock die. Heavy rain causes flooding and triggers landslides in southern China in August. At least 133 people die. In mid-August, prolonged rain causes extensive flooding in central Europe and southern Russia. At least 100 people die. At least 113 people are killed in South Korea on August 31 and September 1, when Typhoon Rusa brings winds of more than 125 MPH (200 km/h) and widespread flooding. In September, floods cause heavy rain, killing 23 persons around Sommières, France. On October 26–27, a storm crosses northern Europe, killing seven people in Britain, six in France, at least 10 in Germany, five in Belgium, four in the Netherlands, and one in Denmark. From November 9 to 11, a storm front sweeps through the southeastern and midwestern United States, generating almost 90 TORNADOES and killing 36 people. An ICE STORM strikes North and South Carolina on December 4–5, disrupts power supplies to 1.8 million people, and is probably responsible for at least 22 deaths. On December 8–9, heavy rain triggers mudslides that bury many homes in Angra dos Reis, Brazil, and kill at least 34 people. During late December, extremely cold weather claims at least 100 lives in northern Bangladesh.

2003

On January 16, mudslides due to heavy rain kill at least 14 people in Minas Gerais State, Brazil. Widespread flooding in northern Mozambique in February destroys about 6,000 homes and kills at least 47 people. In mid-February, a huge snowstorm dumps about two feet (60 cm) of snow along the eastern seaboard of the United States, killing 59 people. On February 17, storms bringing heavy rain and snow in southern Pakistan, Kashmir, and Afghanistan cause floods and destroy houses and a bridge, killing a total of 86 people. On March 31, a mudslide triggered by heavy rain engulfs the gold-mining town of Chima, Bolivia, killing at least 14 people. FLASH FLOODS and mudslides wash away homes and kill at least 29 people in Flores Island, Indonesia, on April 1. On April 20, a mudslide destroys the town of Kurbu-Tash, Kyrgyzstan. At least 38 people die, and the area is declared a grave because it proves impossible to recover the bodies. THUNDERSTORMS kill at least 33 people in Assam State, India, on April 22. Heavy rain in May cause floods in the Horn of Africa in which more than 160 people die. A TORNADO outbreak, with more than 300 tornadoes, sweeps through several southern and midwestern states of the United States from May 4 to May 12. The tornadoes destroy entire towns and kill at least 42 people. On May 4, a tropical storm causes a landslide that engulfs a village in Bangladesh, killing at least 23 people. A heat wave and drought lasting from mid-May until June 10 causes a shortage of drinking water that results in the deaths of 1,522 people in India, 40 in Pakistan, and more than 60 in Bangladesh. On May 16, mudslides and FLASH FLOODS wash away factories and bury the homes of coal miners in Wanshui, Hunan Province, China, killing at least 12 people. On May 17, more than 300 people die in Sri Lanka during floods and landslides caused by heavy rain. Tropical Storm Linfa kills at least 25 people in Luzon, Philippines, on May 27. At least 32 people lose their lives when they are either swept away or buried by floods and mudslides in Bangladesh on June 26. The MONSOON rains deliver 4.5 inches (120 mm) in 24 hours. On July 7, monsoon rains swell the Jamuna River, which breaks its banks and sweeps away several villages. An intense heat wave and drought lasting from mid-July until mid-August in western Europe causes the deaths of approximately 14,800 people in France, 4,200 in Italy, 1,400 in the Netherlands, 1,300 in Portugal, 900 in Britain, and 100 in Spain. More than 100 people die in Himachal Pradesh State, India, on July 16 when a cloudburst causes FLASH FLOODS that sweep away a camp housing migrant workers at a hydroelectric project. At least 88 people die in late July in Sind, Pakistan, and 100,000 lose their homes due to floods caused by monsoon rains. Floods in Kassala Province, Sudan, kill 20 people and make 250,000 homeless in early August. Floods in Haiti leave 20 people dead in early September. Typhoon Maemi kills at least 124 people in South Korea on September 12. About 40 people die in the eastern United States when Hurricane Isabel strikes on September 18. On November 2, flash floods destroy a tourist village in Sumatra, Indonesia, killing about 200 people. In mid-November, floods kill at least 50 people in Vietnam. A rare winter cyclone on December 16 destroys the homes and crops of about 8,000 people in Andhra Pradesh State, India, and kills at least 50. At least 200 people die on December 19 in Leyte Province, Philippines, when mudslides engulf entire villages and towns. Extreme cold weather late in December kills more than 200 people in northern India.

2004

On March 7, Cyclone Gafilo kills about 200 people in Madagascar and leaves hundreds of thousands homeless. FLASH FLOODS kill at least 34 people in Piedras Negras, Mexico, on April 5. On April 14, TORNADOES destroy thousands of homes in northern Bangladesh and kill at least 66 people. Floods in April kill at least 16 people in Kenya and 30 in Djibouti. On April 23, a mudslide engulf a bus in Sumatra, Indonesia, killing at least 37 people. A heat wave in May kills at least 17 people in Bangladesh. On May 19, typhoon Nida destroys several villages and kills 19 people in Catanduanes Province, Philippines. Nida was classed as a supertyphoon (see TROPICAL CYCLONE). On May 24, heavy rain cause floods and mudslides that kill almost 2,000 people in Haiti and the Dominican Republic. Extreme MONSOON rain and storms across southern Asia kills almost 2,000 people between June and August. Typhoon Mindulle kills 31 people in Luzon, Philippines, on June 29 and 15 in Taiwan on July 1. Floods and landslides kill nearly 400 people in southern and central

China in early July. Taiwan suffers its worst floods for 25 years in early July; at least 21 people die. Typhoon Rananim kills at least 164 people in Zhejiang province, China, on August 12. On August 12, FLASH FLOODS in Adamawa state, Nigeria, drown at least 23 people while they sleep. Hurricane Charley strikes Florida on August 13, killing at least 27 people. Typhoon Aere kills at least 24 people in Taiwan and five people in the Philippines on August 24. Hurricane Ivan crosses the Caribbean and the U.S. Gulf coast from September 7 to September 17, destroying crops in Grenada and killing at least 109 people, including 52 in the United States. At the end of the three-month rainy season in early September, floods kill more than 1,000 people in China. On September 18, tropical storm Jeanne strikes Haiti, causing floods in which more than 3,000 people die. Flash floods kill at least 44 people on September 21 in Uttar Pradesh State, India. On October 9, flash floods kill more than 100 people in Assam State, India, and at least 44 in Bangladesh. Typhoon Tokage kills at least 83 people in Japan on October 20. On November 29, Typhoon Winnie causes floods and landslides in the Philippines in which at least 412 people die. On December 2, while rescuers are still searching for survivors from Typhoon Winnie, Typhoon Nanmadol strikes the Philippines, leaving more than 1,000 people dead or missing. On December 26, a Richter magnitude 9.0 earthquake beneath the seabed off the coast of Sumatra, Indonesia, causes TSUNAMIS that kill more than 250,000 people in Sumatra's Aceh Province, Sri Lanka, India, the Maldives, and Thailand.

2005

On January 8–9, storms across northern Europe kill at least 11 people. Prolonged heavy rain and snow in Southern California kill approximately 20 people in early January, and on January 10 a hillside collapses at La Conchita, burying four blocks and killing at least 10 people. A storm on January 22, the last day of the Hajj, causes FLASH FLOODS that kill approximately 29 people in Medina, Saudi Arabia. In late January, Georgetown, Guyana, suffers its worst flooding for a century. Thousands of people are forced to evacuate their homes, and 34 die, mainly from diseases. In early February, flooding kills at least 53 people in Venezuela and 33 in Colombia. Heavy rainfall causes the failure of the Shadi Khor Dam, in Balochistan Province, Pakistan, on February 10; at least 60 people die. Heavy rain and snow cause the deaths of 65 people in North-West Frontier Province, Pakistan, in early February. At least 278 die on February 20 when AVALANCHES destroy several villages in Indian-administered Kashmir. Approximately 120 people are dead or missing in Java, Indonesia, when on February 21 heavy rains cause a municipal waste dump sited on the top of a hill to collapse, triggering a landslide that partly buries the village of Cimahi. A TORNADO kills 56 people and leaves thousands homeless on March 20 in the Gaibandha district of northern Bangladesh. At least 123 people die on April 23 in the Somali region of Ethiopia when the Shebeli River overflows its banks, flooding the surrounding area. Storms and flash floods kill approximately 30 people in Jiddah, Saudi Arabia, on April 28. On May 18, a BLIZZARD kills at least 26 soldiers on a training march in the Andes Mountains, Chile. At least 92 people drown at Shalan, in Heilongjiang Province, China, on June 10, when about eight inches (200 mm) of rain fall in 40 minutes, causing flash floods. A mudslide buries homes killing at least 23 people in Senahú, Guatemala, on June 16. Flooding in China kills 536 people between June 10 and June 24. MONSOON floods kill at least 94 people in Gujarat, India, in late June. Hurricane Dennis strikes Haiti on July 7, killing at least 60 people, and on July 8 it kills 16 people in Cuba. Heavy rain in early July destroys 26,000 homes in southern China and kills at least 29 people. On July 26, 37.1 inches (942 mm) of rain fall on Mumbai, India, in 24 hours, paralyzing the city and causing at least 736 deaths. Category 4 hurricane Katrina strikes the U.S. Gulf coast on August 29, causing devastation in New Orleans and Slidell, Louisiana, and Gulfport and Biloxi, Mississippi. On August 30, the New Orleans levees are breached in three places, flooding about 80 percent of the city. A total of 972 lives are lost in Louisiana and 221 in Mississippi. Rain trigger a landslide on September 1 in Sumatera Province, Indonesia, killing at least 10 people and leaving 34 buried in rubble. Landslides and flooding caused by Typhoon Talim kill 53 people in Anhui Province, China, on September 1. Typhoon Nabi strikes southern Japan on September 6, forcing 250,000 people to evacuate their homes and killing at least 18. Typhoon Khanun strikes Zhejiang Province, China, on September 11, killing at least 14 people and destroying more than 7,000 homes. Between about September 20 and 28, when it is downgraded to a tropical storm, Typhoon Damrey kills 36 people in Vietnam, 16 in

the Philippines, 16 in southern China, and at least three in Thailand. At least 80 students are killed on October 2 at a military school in Fuzhou, Fujian Province, China, by floodwaters released by Typhoon Longwang. Hurricane Stan strikes Central America on October 4, triggering landslides and floods that kill at least 71 people in El Salvador, 654 in Guatemala, and 60 in Nicaragua, Honduras, Mexico, and Costa Rica. Hurricane Wilma strikes Mexico on October 21, devastating the resorts of Cancún, Cozumel, and Playa del Carmen and killing six people. Wilma makes landfall on the U.S. coast on October 24, killing approximately 22 people. On October 23, tropical storm Alpha becomes the 22nd named storm in the 2005 season, making this the most active hurricane season ever recorded in the Atlantic and Caribbean region. Alpha causes floods that kill at least 26 people in Haiti and the Dominican Republic. More than 100 people are reported dead in southern India on October 27 following five days of heavy rain. A TORNADO kills 24 people in Indiana on November 6. An AVALANCHE kills 24 gemstone miners in Pakistan on December 28.

2006

On January 2, heavy snow causes the roof of an ice-skating rink at Bad Reichenhall, Germany, to collapse, killing 15 people. Heavy snowstorms dump two feet (60 cm) or more of snow across the northern United States on February 11–12, closing airs in Washington, D.C., and New York and bring road traffic almost to a standstill; 26.9 inches (1) of snow fall in New York City. On February mud-slide caused by the collapse of a mountain gulfs the town of Guinsaugon, Philippines, more than 1,000 people, few of whom su tween March 9 and March 13, at least 105 T affect five southern and midwestern U.S. sta least 11 people. In early April, melting espread rain cause rivers to overflow, pr lows its flooding in central Europe. The thousand banks and breaches flood defen omania, people have to leave their hom Danube and Bulgaria. Flooding final people die delta, in Ukraine. On April ois, Iowa, in storms that cross Arkans triggering Kentucky, Ohio, Mississipp ide caused at least 63 tornadoes. On naventura, by heavy rain kills 31

Colombia. On May 11–13, Typhoon Chanchu crosses the Philippines; 41 people are killed and thousands are made homeless. Chanchu makes landfall in China on May 17, causing at least 29 deaths; 28 Vietnamese fisherman also die, and 150 are reported missing. On June 19, heavy rain triggers a landslide in Shiji village, Fujian Province, China, killing 11 people. On June 19–20 torrential rain causes flooding in eastern Sulawesi, Indonesia; at least 216 people die. On June 25, a FLASH FLOOD kills 11 people in Hunan Province, China. Between July 16 and 25 a heat wave across the United States, affecting California most severely, brings temperatures in excess of 104°F (40°C); 140 persons die.

APPENDIX X
CHRONOLOGY OF DISCOVERIES

ca. 340 B.C.E.

Aristotle (*see* APPENDIX I: BIOGRAPHICAL ENTRIES) writes *Meteorologica,* the oldest known work on meteorology and possibly the first. It gives us the word *meteorology.*

140–131 B.C.E.

Han Ying, in China, writes *Moral Discourses Illustrating the Han Text of the* Book of Songs. This contains the first known description of the hexagonal structure of SNOWFLAKES.

first century B.C.E.

The TOWER OF THE WINDS is built in Athens. It is possibly the first attempt to forecast the weather systematically on the basis of observations.

ca. 55 B.C.E.

Lucretius Carus (ca. 94–55 B.C.E.) proposes that THUNDER is the sound of great clouds crashing together. Although he is wrong, he may have been the first person to notice that thunder is always associated with big, solid-looking clouds.

first century C.E.

Hero of Alexandria (*see* APPENDIX I: BIOGRAPHICAL ENTRIES) demonstrates that air is a substance.

1555

Olaus Magnus (*see* APPENDIX I: BIOGRAPHICAL ENTRIES) publishes a book containing the first European depictions of ICE CRYSTALS and snowflakes.

1586

Simon Stevinus (*see* APPENDIX I: BIOGRAPHICAL ENTRIES) shows that the pressure a liquid exerts on a surface depends on the height of the liquid above the surface and the area of the surface on which it presses, but it does not depend on the shape of the vessel containing the liquid.

1593

Galileo (*see* APPENDIX I: BIOGRAPHICAL ENTRIES) invents an air thermoscope (*see* THERMOMETER).

1611

Johannes Kepler (*see* APPENDIX I: BIOGRAPHICAL ENTRIES) publishes *A New Year's Gift, or On the Six-cornered Snowflake* in which he described snowflakes.

1641

Ferdinand II (*see* APPENDIX I: BIOGRAPHICAL ENTRIES) invents a thermometer consisting of a sealed tube containing liquid.

1643

Evangelista Torricelli (*see* APPENDIX I: BIOGRAPHICAL ENTRIES) invents the BAROMETER.

1646

Blaise Pascal (*see* APPENDIX I: BIOGRAPHICAL ENTRIES) demonstrates that atmospheric pressure decreases with height.

1654

Ferdinand II (*see* APPENDIX I: BIOGRAPHICAL ENTRIES) improves on his thermometer, producing the design that will lead to the mercury thermometer invented by Daniel Fahrenheit in 1714.

1660

Robert Boyle (*see* APPENDIX I: BIOGRAPHICAL ENTRIES) publishes his discovery of the relationship between the volume occupied by a gas and the pressure under which the gas is held.

1686

Edmund Halley (*see* APPENDIX I: BIOGRAPHICAL ENTRIES) proposes the first explanation for the trade winds (*see* WIND SYSTEMS).

1687

Guillaume Amontons (*see* APPENDIX I: BIOGRAPHICAL ENTRIES) invents the HYGROMETER.

1714

Daniel Fahrenheit (*see* APPENDIX I: BIOGRAPHICAL ENTRIES) invents the mercury thermometer and the temperature scale that bears his name.

1735

George Hadley (*see* APPENDIX I: BIOGRAPHICAL ENTRIES) proposes his model of the circulation of the atmosphere to explain the direction from which the trade winds blow (*see* GENERAL CIRCULATION).

1738

Daniel Bernoulli (*see* APPENDIX I: BIOGRAPHICAL ENTRIES) demonstrates that when the velocity of a flowing fluid increases, its internal pressure decreases.

1742

Anders Celsius (*see* APPENDIX I: BIOGRAPHICAL ENTRIES) proposes the temperature scale that bears his name.

1752

Benjamin Franklin (*see* APPENDIX I: BIOGRAPHICAL ENTRIES) performs his experiment with a kite, demonstrating that storm clouds carry electric charge and that a lightning stroke is a giant spark.

1761

Joseph Black (*see* APPENDIX I: BIOGRAPHICAL ENTRIES) demonstrates that when ice melts it absorbs heat with no rise in its own temperature. He later shows that heat is absorbed or released when water vaporizes and condenses. He calls this LATENT HEAT.

1783

Horace Bénédict de Saussure (*see* APPENDIX I: BIOGRAPHICAL ENTRIES) invents the hair hygrometer.

1806

Admiral Sir Francis Beaufort (*see* APPENDIX I: BIOGRAPHICAL ENTRIES) proposes a scale for classifying wind forces.

1820

John Daniell (*see* APPENDIX I: BIOGRAPHICAL ENTRIES) invents the dewpoint hygrometer.

1824

John Daniell shows the importance of maintaining a humid atmosphere in hothouses growing tropical plants.

1827

Jean-Baptiste Fourier (*see* APPENDIX I: BIOGRAPHICAL ENTRIES) writes what is possibly the first account of the GREENHOUSE EFFECT.

1835

Gaspard de Coriolis (*see* APPENDIX I: BIOGRAPHICAL ENTRIES) explains why anything moving over the surface of the Earth, but not attached to it, is deflected by inertia acting at right angles to its direction of motion.

1840

Louis Agassiz (*see* APPENDIX I: BIOGRAPHICAL ENTRIES) discovers that GLACIERS move and that they had once covered a much larger area than they do now.

1842

Matthew Maury (*see* APPENDIX I: BIOGRAPHICAL ENTRIES) discovers the shape of storms from data gathered from ships at sea.

Johann Christian Doppler (*see* APPENDIX I: BIOGRAPHICAL ENTRIES) discovers the effect bearing his name, that the pitch of a sound rises and light becomes bluer if the source is approaching, and the pitch of a sound falls and light becomes redder if the source is receding.

1844

The world's first telegraph line opens between Baltimore and Washington.

1846

Joseph Henry (*see* APPENDIX I: BIOGRAPHICAL ENTRIES) is elected secretary of the Smithsonian Institution and uses his position to obtain weather reports from all over the United States.

1851

The first WEATHER MAP is exhibited at the Great Exhibition in London, England.

1855

Urbain-Jean-Joseph Leverrier (1811–77; *see* APPENDIX I: BIOGRAPHICAL ENTRIES) begins supervising the installation of a network to gather meteorological data from observatories across Europe.

1856

William Ferrel (*see* APPENDIX I: BIOGRAPHICAL ENTRIES) finds that winds blowing close to the equator are deflected by VORTICITY, not the CORIOLIS EFFECT, and that once established they continue to rotate in order to conserve angular MOMENTUM. He also proposes that in the Northern Hemisphere winds blow counterclockwise around areas of low pressure.

1857

C. D. H. Buys Ballot (*see* APPENDIX I: BIOGRAPHICAL ENTRIES) proposes the law bearing his name (but discovered earlier by William Ferrel).

1858

The first weather bulletins are issued in France on January 1, containing observations from 14 French cities and four cities outside France.

1861

The Meteorological Department of the Board of Trade issues the first British storm warnings for coastal areas on February 6 and for shipping on July 31.

John Tyndall (*see* APPENDIX I: BIOGRAPHICAL ENTRIES) shows that certain atmospheric gases absorb heat, and therefore, that the chemical composition of the atmosphere affects climate.

1863

The first network of meteorological stations to be linked to a central point by telegraph open in France. From September the *Bulletin International de l'Observatoire de Paris* includes a daily WEATHER MAP.

Francis Galton (*see* APPENDIX I: BIOGRAPHICAL ENTRIES) devises a method for mapping weather systems and coined the term *ANTICYCLONE*.

1869

The first daily weather bulletins begin to be issued from Cincinnati Observatory on September 1.

1871

The first three-day weather forecasts are issued by the Weather Bureau.

1874

The International Meteorological Congress is founded.

1875

A weather map appears in a newspaper, *The Times* of London, for the first time.

1884

S. P. Langley (*see* APPENDIX I: BIOGRAPHICAL ENTRIES) publishes a paper on the climatic effect of the absorption of heat by atmospheric gases.

1891

The U.S. Weather Bureau is founded.

1893

Edward Maunder (*see* APPENDIX I: BIOGRAPHICAL ENTRIES) discovers the link between solar activity and the LITTLE ICE AGE.

1896

Svante Arrhenius (*see* APPENDIX I: BIOGRAPHICAL ENTRIES) links climatic changes to the atmospheric concentration of carbon dioxide.

The International Meteorological Congress publishes the first edition of the *INTERNATIONAL CLOUD ATLAS*.

1902

L. P. Teisserenc de Bort (*see* APPENDIX I: BIOGRAPHICAL ENTRIES) discovers the stratosphere (*see* ATMOSPHERIC STRUCTURE).

Vilhelm Bjerknes (*see* APPENDIX I: BIOGRAPHICAL ENTRIES) publishes one of the first scientific studies of weather forecasting.

1905

V. W. Ekman (*see* APPENDIX I: BIOGRAPHICAL ENTRIES) discovers that the deflection of winds and ocean currents changes with vertical distance from the surface.

1913

Charles Fabry (*see* APPENDIX I: BIOGRAPHICAL ENTRIES) discovers the OZONE LAYER.

1918

W. P. Köppen (*see* APPENDIX I: BIOGRAPHICAL ENTRIES) publishes a system for classifying climates.

1918

Vilhelm Bjerknes establishes the existence of AIR MASSES.

1922

L. F. Richardson (*see* APPENDIX I: BIOGRAPHICAL ENTRIES) describes a method for numerical WEATHER FORECASTING.

1923

Gilbert Walker describes the high-level flow of air from west to east close to the equator. He also describes the SOUTHERN OSCILLATION that is linked to El Niño events (*see* ENSO).

1930

Milutin Milankovitch (*see* APPENDIX I: BIOGRAPHICAL ENTRIES) proposes a link between the onset and ending of GLACIAL PERIODS and variations in the Earth's orbit and rotation.

1931

Wilson Bentley (*see* APPENDIX I: BIOGRAPHICAL ENTRIES) publishes more than 2,000 photographs of snowflakes.

C. W. Thornthwaite (*see* APPENDIX I: BIOGRAPHICAL ENTRIES) publishes a system for classifying climates.

1940

Carl-Gustav Rossby (*see* APPENDIX I: BIOGRAPHICAL ENTRIES) discovers large-wavelength undulations in the westerly winds of the upper atmosphere.

1946

Vincent Schaeffer (*see* APPENDIX I: BIOGRAPHICAL ENTRIES) discovers that pellets of dry ice (solid carbon dioxide) can trigger the formation of ice crystals.

1949

RADAR is used for the first time to obtain meteorological data.

1951

An international system for the classification of snowflakes is adopted.

1959

The U.S. Weather Bureau begins publishing a temperature-humidity index as an indication of how comfortable the air will feel on a warm day.

1960

The first weather satellite, *Tiros 1*, is launched.

1964

The *Nimbus 1* weather satellite is launched.

1966

The first satellite to be placed in a geostationary ORBIT is launched on December 6.

1971

The FUJITA TORNADO INTENSITY SCALE is published.

1973

Doppler RADAR is used successfully for the first time to study a TORNADO.

1974

The first of the GOES (GEOSTATIONARY OPERATIONAL ENVIRONMENTAL SATELLITE) is launched.

F. Sherwood Rowland and Mario Molina (*see* APPENDIX I: BIOGRAPHICAL ENTRIES for information about Rowland and Molina) propose that CFCs might deplete stratospheric OZONE.

1977

Meteosat-1, the first European meteorological satellite, is launched on November 23. It remains operational until 1985.

1979

Edward Lorenz (*see* APPENDIX I: BIOGRAPHICAL ENTRIES) presents a paper describing what came to be called the BUTTERFLY EFFECT.

1981

Meteosat-2 is launched in June.

1985

Depletion of the ozone layer over Antarctica is discovered by J. C. Farman, B. G. Gardiner, and J. D. Shanklin of the British Antarctic Survey.

1988

The INTERGOVERNMENTAL PANEL ON CLIMATE CHANGE (IPCC) is founded.

1989

Meteosat-4 is launched.

1990

The IPCC publishes its first report in June, summarizing scientific understanding of climate change at that time.

1992

The TOPEX/POSEIDON satellite is launched, with instruments to measure sea level.

The IPCC publishes its second report in February.

1993

Using powerful computers and advanced climate models, the National Weather Service is able to predict a major storm five days in advance.

1995

F. Sherwood Rowland, Mario Molina, and Paul Crutzen (*see* APPENDIX I: BIOGRAPHICAL ENTRIES) share the Nobel Prize in chemistry for their work on the ozone layer.

1996

Technological advances mean five-day forecasts are as accurate as three-day forecasts were in 1980.

1999

The Drought Monitor program is launched to provide weekly updates on drought conditions in the United States.

2001

The IPCC publishes its third report.

2002

Preparations are made to expand the Drought Monitor program to cover Canada and Mexico as well as the United States; the new service will be called the North American Drought Monitor.

2007

The INTERGOVERNMENTAL PANEL ON CLIMATE CHANGE (IPCC) publishes its fourth report.

BIBLIOGRAPHY AND FURTHER READING

Abrahamson, John, and James Dinniss. "Ball lightning caused by oxidation of nanoparticle networks from normal lightning strikes on soil." *Nature,* 403, 519–521.

Allaby, Michael. *Deserts.* Rev. ed. Ecosystem. New York: Facts On File, 2007.

———. *Temperate Forests.* Rev. ed. Ecosystem. New York: Facts On File, 2007.

———. *Deserts.* Biomes of the Earth. New York: Facts On File, 2006.

———. *Blizzards.* Rev. ed. Dangerous Weather. New York: Facts On File, 2004.

———. *A Change in the Weather.* New York: Facts On File, 2004.

———. *A Chronology of Weather.* Rev. ed. New York: Facts On File, 2004.

———. *Tornadoes.* Rev. ed. Dangerous Weather. New York: Facts On File, 2004.

———. *Fog, Smog, and Poisoned Rain.* New York: Facts On File, 2003.

———. *Droughts.* Rev. ed. Dangerous Weather. New York: Facts On File, 2003.

———. *Floods.* Rev. ed. Dangerous Weather. New York: Facts On File, Inc., 2003.

———. *Hurricanes.* Rev. ed. Dangerous Weather. New York: Facts On File, 2003.

———. *Air, the Nature of Atmosphere and the Climate.* New York: Facts On File, 1992.

———. *A Guide to Gaia.* New York: E. P. Dutton, 1989.

Allen, Philip A., and Paul F. Hoffman. "Extreme winds and waves in the aftermath of a Neoproterozoic glaciation." *Nature* 433, (January 13, 2005): 123–127.

Ashman, M. R. and G. Puri. *Essential Soil Science.* Oxford: Blackwell Science, 2002.

Barry, Roger G., and Richard J. Chorley. *Atmosphere, Weather & Climate.* New York: Routledge, 7th ed. 1998.

Bluestein, Howard B. *Tornado Alley.* New York: Oxford University Press, 1999.

Bryant, Edward. *Climate Process & Change.* Cambridge: Cambridge University Press, 1997.

Burroughs, William, ed. *Climate Into the 21st Century.* Cambridge: World Meteorological Organization and Cambridge University Press, 2003.

———. *Climate Change: A Multidisciplinary Approach.* Cambridge: Cambridge University Press, 2001.

Conrad, V. "Usual formulas of continentality and their limits of validity." *Transactions of the American Geophysical Union* 17: 663–664.

Cox, John D. *Storm Watchers.* Hoboken, New Jersey: John Wiley & Sons, 2002.

Eddy, John A. "The Case of the Missing Sunspots." *Scientific American,* 236, no. 5, vol. 3 (May 1977): 80–92.

Fischer, Hubertus, et al. "Ice Core Records of Atmospheric CO_2 around the Last Three Glacial Terminations." *Science* 283 (March 12, 1999): 1,712–14.

Flasar, F. M. et al. "Titan's Atmospheric Temperatures, Winds, and Composition." *Science,* 308 (May 13, 2005): 975–978.

Flohn, H. "Neue Anshcauungen über die allgemeine Zirkulation der Atmosphäre und ihre klimatische Bedeutung" ("New Views of the General Circulation of the Atmosphere and its Climatic Significance"). *Erdkunde (Earth Science),* 3:141–62.

Gillet, Nathan P. "Northern Hemisphere circulation," *Nature:* 437, 498 (September 22, 2005): 498.

Gleick, James. *Chaos: Making a New Science.* London: William Heinemann, 1988.

González, Frank I. "Tsunami!" *Scientific American* 21, no. 12, (May 1999): 56–65.

Guilderson, Tom P., Paula J. Reimer, and Tom A. Brown. "The Boon and Bane of Radiocarbon Dating." *Science*, 307 (January 21, 2005): 362–364.

Hamblyn, Richard. *The Invention of Clouds*. New York: Farrar, Straus, and Giroux, 2001.

Herzog, Howard, Baldur Eliasson, and Olav Kaarstad, "Capturing Greenhouse Gases." *Scientific American*, 282, (February 2000): 72–79.

Hoffman, Paul F., and Daniel P. Schrag. "Snowball Earth." *Scientific American*, 282, (January 2000): 68–75.

Hoffman, Ross N. "Controlling Hurricanes." *Scientific American*, 291, (October 2004): 68–75.

Houghton, J. T., Y. Ding, D. J. Griggs, M. Noguer, P. J. van der Linden, X. Dai, K. Maskell, and C. A. Johnson, eds. *Climate Change 2001: The Scientific Basis*. Cambridge: Intergovernmental Panel on Climate Change and Cambridge University Press, 2001.

Hubler, Graham K. "Fluff balls of fire." *Nature* 403: 487–488.

Jardine, Lisa. *Ingenious Pursuits: Building the Scientific Revolution*. London: Little, Brown, 1999.

Joseph, Lawrence E. *Gaia: The Growth of an Idea*. New York: St. Martin's Press, 1990.

Kendrew, W. G. *The Climates of the Continents*. Oxford: Oxford University Press, 1st ed. 1922, 5th ed., 1961.

Kent, Michael. *Advanced Biology*. Oxford: Oxford University Press, 2000.

Kerr, Richard A. "Scary Arctic Ice Loss? Blame the Wind," *Science* 307 (January 14, 2005): 203.

———. "A New Force in High-Latitude Climate," *Science* 284 (April 9, 1999): 241–42.

———. "Confronting the Bogeyman of the Climate System." *Science* 310 (October 21, 2005): 432–433.

Kurchella, Charles E., and Margaret C. Hyland. *Environmental Science*. Boston: Allyn and Bacon, 2nd ed. 1986.

Ladurie, Emmanuel Le Roy. *Times of Feast, Times of Famine: A History of Climate Since the Year 1000*. New York: Doubleday, 1971.

Lamb, H. H. *Climate, History and the Modern World*. London: Routledge, 2nd edition 1995.

Lovelock, James. *Gaia: A New Look at Life on Earth*. Oxford: Oxford University Press, 1979.

———. *The Ages of Gaia*. New York: Oxford University Press, 1989.

Lutgens, Frederick K., and Edward J. Tarbuck. *The Atmosphere*. Upper Saddle River, N.J.: Jersey: Prentice Hall, 7th ed. 1998.

MacKenzie, James J., and Mohamed T. El-Ashry, eds. *Air Pollution's Toll on Forests and Crops*. New Haven, Conn.: Yale University Press, 1989.

McIlveen, Robin. *Fundamentals of Weather and Climate*. New York: Chapman and Hall, 1992.

Monmonier, Mark. *Air Apparent: How Meteorologists Learned to Map, Predict, and Dramatize the Weather*. Chicago: University of Chicago Press, 1999.

Morgan, J. J., and H. M. Liljestrand. *Final Report, Measurement and Interpretation of Acid Rainfall in the Los Angeles Basin*. Sacramento: California Air Resources Board, 1980.

Morton, Oliver. "Storms Bow Out, But Boughs Remember." *Science* 309 (August 6, 2005): 1,321.

Oke, T. R. *Boundary Layer Climates*. New York: Routledge, 2d. ed. 1987.

Oliver, John E., and John J. Hidore. *Climatology, An Atmospheric Science*. Upper Saddle River, N.J.: Prentice Hall, 2d. ed. 2002.

Page, Robin. *Weather Forecasting The Country Way*. London: Penguin Books, 1981.

Palmer, W. C. "Meteorological Drought," Washington, D.C.: U.S. Weather Bureau. *U.S. Department of Commerce Research Paper No. 45, 1965.*

Parkinson, Claire L. *Earth From Above: Using Color-coded Satellite Images to Examine the Global Environment*. Sausalito, Calif.: University Science Books, 1997.

Penman, H. L. "Natural evaporation from open water, bare soil and grass." *Proceedings of the Royal Society* 193: 120–145.

Petit, J. R., et al. "Climate and Atmospheric History of the Past 420,000 Years from the Vostok Ice Core, Antarctica." *Nature* 399 (June 3, 1999): 429–36.

Robinson, Peter J., and Ann Henderson-Sellers. *Contemporary Climatology*. Upper Saddle River, N.J.: Prentice Hall, 2d ed. 1999.

Ruddiman, William F. *Earth's Climate, Past and Future*. New York: W. H. Freeman, 2001.

Saunders, Mark A., and Adam S. Lea. "Seasonal prediction of hurricane activity reaching the coast of the United States." *Nature* 434 (April 21, 2005): 1,005–1,007.

Schilling, Govert. "Volcanoes, Monsoons Shape Titan's Surface." *Science* 309 (September 23, 2005): 1,985.

Stephanou, Euripedes G. "The decay of organic aerosols," *Nature* 434 (March 3, 2005): 31.

Strahler, A. N. *Physical Geography.* New York: J. Wiley, 1969.

Sutton, Rowan T., and Daniel L. R. Hodson, "Atlantic Ocean Forcing of North American and European Summer Climate," *Science* 309 (July 1, 2005): 115–118.

Volk, Tyler. *Gaia's Body: Toward a Physiology of Earth.* New York: Springer-Verlag, 1998.

Wettlaufer, John S. and J. Greg Dash. "Melting Below Zero." *Scientific American* (February 2000): 34–37.

World Meteorological Organization. *Preventing and Mitigating Natural Disasters: Working Together for a Safer World.* Geneva: World Meteorological Organization, 2006.

WEB SITES

Academy of Natural Sciences, Philadelphia. "Louis Agassiz (1807–1873)." University of California. Available online. URL: www.ucmp.berkeley.edu/history/agassiz.html. Accessed March 9, 2006.

Antarctic Connection. "Vostok Station." Available online. URL: http://www.antarcticconnection.com/antarctic/stations/vostok.shtml. Accessed February 24, 2006.

Argo. "Welcome to the Argo home page." Available online. URL: www.argo.uscd.edu/. Accessed September 26, 2005.

Arnett, Bill. "Mars." Available online. URL: www.nineplanets.org/mars.html. Last updated October 18, 2005.

Arrhenius, Svante. "On the Influence of Carbonic Acid in the Air upon the Temperature of the Ground." (excerpts) *Philosophical Magazine* 41, 237–276 (1896). Available online. URL: http://web.lemoyne.edu/~GIUNTA/Arrhenius.html. Accessed March 9, 2006.

Asahi Glass Foundation. "Profiles of the 1998 Blue Planet Prize Recipients." Asahi Glass Foundation. Available online. URL: www.af-info.or.jp/eng/honor/hot/enr-budyko.html. Accessed March 10, 2006.

Astronomical University, University of Uppsala. "Anders Celsius (1701–1744)". University of Uppsala. Available online. URL: www.astro.uu.se/history/Celsius_eng.html. Accessed March 10, 2006.

Beaty, Bill. "Ball Lightning Page." Available online. URL: www.amasci.com/tesla/ballgtn.html. Accessed February 13, 2006.

Beckman, John E. and Terence J. Mahoney. *The Maunder Minimum and Climate Change: Have Historical Records Aided Current Research?* Library and Information Services in Astronomy III. Astronomical Society of the Pacific. Available online. URL: http://www.stsci.edu/stsci/meetings/lisa3/beckmanj.html. Accessed January 9, 2006.

Blanchard, Duncan C. "The Snowflake Man." *Weatherwise* 23, 6, 260–269, 1970. Available online. URL: www.snowflakebentley.com/sfman.htm. Accessed March 10, 2006.

Cain, Dennis R. and Paul Kirkwood. "National Doppler Radar Sites." National Weather Service, NOAA. Available online. URL: www.crh.noaa.gov/radar/national.html. Last modified October 5, 2004.

Caracena, Fernando, Ronald H. Holle, and Charles A. Doswell III. "Microbursts: A Handbook for Visual Identification." NOAA. Available online. URL: www.cimms.ou.edu/~doswell/microbursts/Handbook.html. Last updated June 14, 2001.

Carver, Glenn. "The Ozone Hole Tour." Centre for Atmospheric Science, University of Cambridge. Available online. URL: www.atm.ch.cam.ac.uk/tour/psc.html. Accessed January 20, 2006.

Chung-Chieng, 'Aaron' Lai, and Zhen Huang. "Antarctic Circumpolar Wave and El Niño." Available online. URL: www.ees.lanl.gov/staff/cal/acen.html. Accessed February 13, 2006.

Climate Diagnostics Center. "El Niño/Southern Oscillation." NOAA-CIRES. Available online. URL: www.cdc.noaa.gov/ENSO/. Updated June 30, 2004.

Climate Prediction Center. "The ENSO Cycle." National Weather Service. Available online. URL: http://www.cpc.ncep.noaa.gov/products/analysis_monitoring/ensocycle/enso_cycle.shtml. Last modified December 19, 2005.

Climatic Research Unit, University of East Anglia. "Datasets/UK Climate/ Lamb Weather Types." Available online. URL: www.cru.uea.ac.uk/~mikeh/datasets/uk/lamb.htm. Accessed December 27, 2005.

Corrosion Doctors. "John Frederick Daniell (1790–1845)." Available online. URL: www.corrosion-doctors.org/Biographies/DaniellBio.htm. Accessed March 13, 2006.

Cotton, William R. "Weather Modification by Cloud Seeding: A Status Report 1989–1997." Department of Atmospheric Sciences, Colorado State University. Available online. URL: http://rams.atmos.colostate.edu/gkss.html. Accessed February 14, 2006.

Department of Atmospheric Science. "Coriolis Force: An Artifact of the Earth's Rotation." University of Illinois at Urbana–Champaign. Available online. URL: http://ww2010.atmos.uiuc.edu/(Gh)/guides/mtr/fw/crls.rxml. Accessed February 14, 2006.

Department of Energy. "Comparison of Global Warming Potentials" from the Second and Third Assessment Reports of the Intergovernmental Panel on Climate Change (IPCC). Department of Energy. Available online. URL: http://www.eia.doe.gov/oiaf/1605/gwp.html. Last updated August 12, 2002.

Dutch, Steven. "Channeled Scablands: Overview." Available online. URL: www.uwsp.edu/geo/projects/geoweb/participants/dutch/VTrips/Scablands0.HTM. Last update November 21, 2003.

Earth Observation Group. "Defense Meteorological Satellite Program (DMSP) Data Archive, Research, and Products." National Geophysical Data Center, NOAA. Available online. URL: www.ngdc.noaa.gov/dmsp/. Revised December 14, 2005.

European Science Foundation. "Greenland Icecore Project." European Science Foundation. Available online. URL: www.esf.org/esf_article.php?section=2&domain=3&activity=1&language=0&article=166. Last updated December 11, 2001.

EU-SEASED. "Eurocore Project Info." EU-SEASED. Available online. URL: www.eu-seased.net/eurocore/project_info.htm#seased1. Last updated January 31, 2003.

Federal Aviation Authority. "Density Altitude." FAA. Available online. URL: www.nw.faa.gov/ats/zdvartcc/high_mountain/density.html. Accessed February 14, 2006.

Fraser, Alistair B. "Bad Coriolis." Available online. URL: www.ems.psu.edu/~fraser/Bad/BadCoriolis.html. Accessed February 14, 2006.

Geerts, B., and E. Linacre. "Sunspots and Climate." Available online. URL: www-das.uwyo.edu/~geerts/cwx/notes/chap02/sunspots.html. Accessed February 1, 2006.

Geerts, B., and M. Wheeler. "The Madden–Julian Oscillation." University of Wyoming. Available online. URL: www-das.uwyo.edu/~geerts/cwx/notes/chap12/mjo.html. Posted May 1998; accessed January 6, 2006.

George Brown University. "Data-Model Comparisons." George Brown University. Available online. URL: www.geo.brown.edu/georesearch/esh/QE/Research/PaleoClm/DataMode/DataMode. htm. Accessed February 14, 2006.

Goulet, Chris M. "Magnetic Declination Frequently Asked Questions." Geocities. Available online. URL: www.geocities.com/magnetic_declination/. Last updated October 10, 2001.

Greenland Guide Index. Available online. URL: www.greenland-guide.dk/default.htm. Accessed January 17, 2006.

Hayes, Michael J. *What is Drought?: Drought Indices;* National Drought Mitigation Center. Available online. URL: www.drought.unl.edu/whatis/indices.htm. Accessed November 23, 2005.

Heaps, Andrew, William Lahoz, and Alan O'Neill. "The Quasi-Biennial zonal wind Oscillation (QBO)." University of Reading. Available online. URL: http://ugamp.nerc.ac.uk/hot/ajh/qbo.htm. Accessed January 24, 2006.

Higham, Thomas. "Radiocarbon Dating." Available online. URL: www.c14dating.com/int.html. Accessed January 25, 2006.

Hinds, Stacey. "Cloud Seeding." Denver Water Community Relations. Available online. URL: www.denverwater.org/cloud_seeding.html. Accessed February 14, 2006.

Hoffman, Paul F. and Daniel P. Schrag. "The Snowball Earth." Harvard University. Available online. URL: www-eps.harvard.edu/people/faculty/hoffman/snowball_paper.html. Accessed January 30, 2006.

IGBP. "International Geosphere–Biosphere Program." Available online. URL: www.igbp.kva.se/cgi-bin/php/frameset.php. Accessed December 22, 2005.

International Association for Aerobiology. Available online. URL: www.isao.bo.cnr.it/aerobio/iaa/. Accessed February 13, 2006.

Irish Identity. "Navan's Most Famous Son." Hoganstand.com. Available online. URL: www.hoganstand.com/general/identity/extras/famousgaels/stories/beaufort.htm#top. Accessed March 10, 2006.

Italian Association of Aerobiology. "Italian Aeroallergen Network." Available online. URL: www.isao.bo.cnr.it/aerobio/aia/AIANET.html. Accessed February 13, 2006.

JCOMM Voluntary Observing Ships' Scheme. "Port Meteorological Officers." Available online. URL: www.bom.gov.au/jcomm/vos/pmo.html. Accessed February 24, 2006.

Kaplan, George. "The Seasons and the Earth's Orbit—Milankovitch Cycles." U.S. Naval Observatory. Available online. URL: http://aa.usno.navy.mil/faq/docs/seasons_orbit.html. Last modified October 30, 2003.

King's College London. "History of the College: John Frederick Daniell." Available online. URL: www.kcl.ac.uk/college/history/people/daniell.html. Accessed March 13, 2006.

Kordic, Ruby. "Venus." Available online. URL: http://ruby.kordic.re.kr/~vr/CyberAstronomy/Venus/HTML/index.html. Accessed Fenruary 23, 2006.

Kovarik, W. "Leaded Gasoline Information." Available online. URL: http://www.runet.edu/~wkovarik/papers/leadinfo.html. Accessed February 13, 2006.

Landsea, Chris. "What may happen with tropical cyclone activity due to global warming?" NOAA. Available online. URL: www.aoml.noaa.gov/hrd/tcfaq/G3.html. Accessed February 21, 2006.

Maiden, M., R. G. Barry, R. L. Armstrong, J. Maslanik, T. Scambos, A. Brenner, S. T. X. Hughes, D. J. Cavalier, A. C. Fowler, J. Francis, and K. Jezek. "The Polar Pathfinders: Data Products and Science Plans." American Geophysical Union. Available online. URL: www.agu.org/eos_elec/96149e.html. Accessed December 15, 2005.

Major, Gene. "Global Change Master Directory, A Directory to Earth Science Data and Services." NASA Goddard Space Flight Center. Available online. URL: http://gcmd.nasa.gov/records/GCMD_CDIAC_NDP13.html. Last updated December 2005.

Mantua, Nate. "The Pacific Decadal Oscillation (PDO)." Available online. URL: http://tao.atmos.washington.edu/pdo/. Accessed July 14, 2005.

Martin, Tom. "The Tay Bridge Disaster." Available online. URL: www.tts1.demon.co.uk/tay.html. Accessed February 15, 2006.

McCabe, Greg, Mike Palecki, and Julio Betancourt. "Pacific and Atlantic Ocean Influences on Multidecadal Drought Frequency in the U.S." *Past Research Highlights*. U.S. Geological Survey and University of Arizona Desert Laboratory. Available online. URL: wwwpaztcn.wr.usgs.gov/rsch_highlight/articles/200404.html.) Last modified June 11, 2004. Accessed September 26, 2005.

McClung, Alex. "The Earth Observing System." Goddard Space Flight Center. Available online. URL: http://eospso.gsfc.nasa.gov/eos_homepage/description.php. Updated January 24, 2006.

McIntyre, David H. "Coriolis Force and Noninertial Effects." Corvallis: Department of Physics, Oregon State University. Available online. URL: www.physics.orst.edu/~mcintyre/coriolis/. Accessed February 14, 2006.

McPhaden, Michael J. "The Tropical Atmosphere Ocean Project." NOAA. Available online. URL: www.pmel.noaa.gov/tao/. Updated daily. Accessed February 21, 2006.

Mitchell, Todd. "High and low index Arctic Oscillation (AO)." University of Washington. Available online. URL: http://tao./atmos.washington.edu/analyses0500/ao_definition.html. Posted November 2000; accessed September 26, 2005.

NASA. "The Earth Radiation Budget Experiment." Available online. URL: http://asd-www.larc.nasa.gov/erbe/ASDerbe.html. Accessed February 14, 2006.

———. Goddard Space Flight Center. "Upper Atmosphere Research Satellite (UARS)." NASA Facts On Line. Available online. URL: www.gsfc.nasa.gov/gsfc/service/gallery/fact_sheets/earthsci/uars.htm. Posted January 1994, accessed February 23, 2006.

———. "SeaWinds Wind Report." Available online. URL: http://haifung.jpl.nasa.gov/. Accessed January 27, 2006.

———. "Tsunami: The Big Wave." TRW Inc. Available online. URL: http://observe.arc.nasa.gov/nasa/exhibits/tsunami/tsun_bay.html. Accessed February 21, 2006.

National Hurricane Center. "About the Tropical Prediction Center." NOAA. Available online. URL: www.nhc.noaa.gov/aboutintro.shtml. Last modified December 13, 2005.

National Physical Laboratory. "Frequently Asked Questions: How do I use a Fortin or Kew Pattern mercury barometer?" National Physical Laboratory. Available online. URL: www.npl.co.uk/pressure/faqs/usehgbaro.html. Accessed March 21, 2006.

National Snow and Ice Data Center. "Northern Hemisphere EASE-Grid Annual Freezing and Thawing Indices 1901–2002." NSIDC. Available online. URL: http://nsidc.org/data/docs/fgdc/ggd649_freeze_thaw_nh/index.html. Reviewed November 7, 2005.

National Weather Service. "Automated Surface Observing System." NOAA. Available online. URL: www.nws.noaa.gov/asos/. Accessed February 13, 2006.

———. "Cooperative Observer Program." NOAA. Available online. URL: www.nws.noaa.gov/om/coop/. Last updated February 8, 2006.

———. "National Weather Service." NOAA. Available online. URL: www.weather.gov/. Accessed January 11, 2006.

———. "NOAA History: A Science Odyssey." NOAA. Available online. URL: www.history.noaa.gov/legacy/nwshistory.html. Last updated April 21, 2004.

———. "NOAA Weather Radio All Hazards." NOAA. Available online. URL: www.weather.gov/nwr/. Last updated January 31, 2006; www.nws.noaa.gov/nwr/nwrbro.htm. Accessed March 2, 2006.

———. "Family of Services." NOAA. Available online. URL: www.nws.noaa.gov/datamgmt/fos/fospage.html. Accessed January 11, 2006.

Natural Environment Research Council. "Welcome to the Rapid Climate Change Home Page." NERC. Available online. URL: www.noc.soton.ac.uk/rapid/rapid.php. Last modified January 19, 2006.

NOAA. "Automated Surface Observations." Available online. URL: www.nws.noaa.gov/asos/. Last updated October 1999. Accessed February 13, 2006.

———. "Geostationary Operational Environmental Satellites." NOAA. Available online. URL: http://www.oso.noaa.gov/goes/. Accessed December 14, 2005.

———. "Global Systems Division." Available online. URL: www.fsl.noaa/gov/. Last modified December 16, 2005.

———. "GOES Project Science." NOAA. Available online. URL: http://rsd.gsfc.nasa.gov/goes/. Accessed December 14, 2005.

———. "Initial Joint Polar-Orbiting Operational Satellite System." Office of Systems Development, NOAA Satellite and Information Service. Available online. URL: http://projects.osd.noaa.gov/IJPS/. Accessed December 21, 2005.

———. "National Severe Storms Laboratory." Department of Commerce, National Oceanic and Atmospheric Administration Office of Oceanic and Atmospheric Research. Available online. URL: www.nssl.noaa.gov/. Last updated January 6, 2006.

———. National Weather Service Forecast Office, La Crosse, Wisc. "The Tornadoes of June 28, 1865." NOAA. Available online. URL: www.crh.noaa.gov/arx/events/tors_jun1865.php. Last modified November 8, 2005.

———. "The United States Voluntary Observing Ships Scheme." NOAA. Available online. URL: www.vos.noaa.gov/vos_scheme.shtml. Last modified September 29, 2003.

Nobel Foundation. "Svante Arrhenius—Biography." Nobel Foundation. Available online. URL: http://nobelprize.org/chemistry/laureates/1903/arrhenius-bio.html. Accessed March 9, 2006.

O'Connor, J. J., and E. F. Robertson. "Benoît Paul Emile Clapeyron." University of St. Andrews. Available online. URL: www-groups.dcs.st-and.ac.uk/~history/Mathematicians/Clapeyron.html. Accessed March 10, 2006.

———. "Daniel Bernoulli." St. Andrews University. Available online. URL: www-groups.dcs.st-and.ac.uk/~history/Mathematicians/Bernoulli_Daniel.html. Accessed March 10, 2006.

———. "Gaspard Gustave de Coriolis." University of St. Andrews. Available online. URL: www-groups.dcs.st-and.ac.uk/~history/Mathematicians/Coriolis.html. Accessed March 10, 2006.

———. "Vilhelm Friman Koren Bjerknes." St. Andrews University. Available online. URL: www-groups.dcs.st-and.ac.uk/~history/Mathematicians/Bjerknes_Vilhelm.html. Accessed March 10, 2006.

Poulsen, Erling. "Early Danish Thermometers: The Thermometers of Ole Rømer." Available online. URL: www.rundetaarn.dk/engelsk/observatorium/tempeng.htm. Accessed February 16, 2006.

Quade, Paul. "Density-Altitude." Available online. URL: http://futurecam.com/densityAltitude.html. Accessed February 14, 2006.

Qiu, B. and F. F. Jin. "Antarctic circumpolar waves: An indication of ocean–atmosphere coupling in the extratropics." Available online. URL: www.agu.org/pubs/abs/gl/97GL02694/97GL02694.html. Accessed February 13, 2006.

Salawitch, Ross J. "Polar Stratospheric Clouds." Jet Propulsion Laboratory. Available online. URL: http://remus.jpl.nasa.gov/info.htm. Accessed January 20, 2006.

Sample, Sharron. "Destination Earth, 40+ Years of Earth Science." NASA. Available online. URL: www.earth.nasa.gov/history/landsat/landsat.html. Last updated July 12, 2005.

Science Central. "Pan-American Aerobiology Association." ScienceCentral.com. Available online. URL: www.sciencecentral.com/site/480560. Accessed February 13, 2006.

Scripps Institution of Oceanography. "Indian Ocean Experiment: An International Field Experiment in the Indian Ocean." La Jolla, Calif.: Scripps Institution of Oceanography, University of California, San Diego. Available online. URL: http://www-indoex.ucsd.edu/. Posted August 29, 1999.

Sonic.net. "Dendrochronology." Available online. URL: www.sonic.net/bristlecone/dendro.html. Accessed February 21, 2006.

Soper, Davison E. "Atmosphere of Venus." University of Oregon. Available online. URL: http://zebu.uoregon.edu/~soper/Venus/atmosphere.html. Accessed February 23, 2006.

Spokane Outdoors. "Channeled Scablands Theory." Available online. URL: www.spokaneoutdoors.com/scabland.htm. Accessed February 13, 2006.

Steitz, David E. "Future Missions to Study Clouds, Aerosols, Volcanic Plumes." NASA. Available online. URL: http://liftoff.msfc.nasa.gov/home/news/article3.html. Updated December 22, 1998.

TED Case Studies. "Bhopal Disaster." Available online. URL: www.american.edu/TED/bhopal.htm. Accessed February 13, 2006.

Tornado and Storm Research Organisation. "TORRO: The Tornado and Storm Research Organisation." Available online. URL: www.torro.org.uk/TORRO/index.php. Accessed February 20, 2006.

Union Carbide Corporation. "Bhopal Information Center." Available online. URL: www.bhopal.com/. Accessed February 13, 2006.

United Nations. "United Nations System-Wide Earthwatch." United Nations Environment Programme. Available online. URL: http://earthwatch.unep.net/. Last updated November 7, 2005.

United States Environmental Protection Agency. "Ozone Science: The Facts Behind the Phaseout."

U.S. EPA. Available online. URL: www.epa.gov/docs/ozone/science/sc_fact.html. Last updated June 21, 2004.

University of Colorado. "National Snow and Ice Data Center." Boulder: University of Colorado. Available online. URL: http://nsidc.org/index.html. Accessed January 11, 2006.

———. "Topex/Poseidon home page." University of Colorado. Available online. URL: http://ccar.colorado.edu/research/topex/html/topex.html. Accessed February 20, 2006.

University of Montana. "Milankovitch Cycles and Glaciation." Available online. URL: www.homepage.montana.edu/~geol445/hyperglac/time1/milankov.htm. Accessed January 10, 2006.

University of Oxford. "G. M. B. Dobson (25 February 1889–11 March 1976)." Available online. URL: www-atm.physics.ox.ac.uk/user/barnett/ozoneconference/dobson.htm. Most recent revision February 1, 2006.

University of Texas. "Topex/Poseidon Educational Outreach." University of Texas. Available online. URL: www.tsgc.utexas.edu/topex/. Last modified January 29, 2001.

Utah Geological Survey. "Commonly Asked Questions About Utah's Great Salt Lake and Lake Bonneville." Available online. URL: www.ugs.state.ut.us/online/PI-39/. Accessed December 27, 2005.

Van Domelen, David J. "Getting Around The Coriolis Force." Physics Education Research Group, Department of Physics, Ohio State University. Available online. URL: www.physics.ohio-state.edu/~dvandom/Edu/newcor.html. Accessed February 14, 2006.

Villwock, Andreas. "The North Atlantic Oscillation." Available online. URL: www.clivar.org/publications/other_pubs/iplan/iip/pd1.htm. Last updated June 4, 1998, accessed January 13, 2006.

Viroqua. "The Viroqua Tornado June 28th, 1865". Available online. URL: www.wx-fx.com/viroqua.html. Accessed February 23, 2006.

Wang, Bin, and Xihua Xu. "Northern Hemisphere Summer Singularities and Climatological Intraseasonal Oscillation." *Journal of Climate,* 10, 5: 1071–85. Available online. URL: http://ams.allenpress.com/amsonline/?request=get-abstract...doi=10.1175%2F1520-04 42(1997)010%3C1071:NHSMSA%3E2.0.CO%3B2. Accessed January 10, 2006.

Watanabe, Susan. "NASA's CloudSat Mission, Revealing the Inner Secrets of Clouds." NASA. Available online. URL: http://www.nasa.gov/mission_pages/cloudsat/main/. Last updated January 10, 2006.

West, Peter. "Lake Vostok." National Science Foundation Office of Legislative and Public Affairs. Available online. URL: www.nsf.gov/od/lpa/news/02/fslakevostok.htm. Posted May 2002; accessed February 24, 2006.

White, Iain. "The Flandrian: The Case for an Interglacial Cycle." University of Portsmouth. Available online. URL: www.envf.port.ac.uk/geog/teaching/quatgern/q8b.htm. Last updated July 2002.

Wikipedia. "The 2004 Indian Ocean earthquake." Available online. URL: http://en.wikipedia.org/wiki/2004_Indian_Ocean_earthquake. Last modified February 21, 2006.

World Data Center for Paleoclimatology. "Ice Core Gateway Greenland Ice Sheet Project (GISP)." World Data Center for Paleoclimatology. Available online. URL: http://www.ncdc.noaa.gov/paleo/icecore/greenland/gisp/gisp.html. Accessed December 16, 2005.

World Glacier Inventory. The National Snow and Ice Data Center. Available online. URL: http://nsidc.org/data/g01130.html. Accessed December 15, 2005.

World Glacier Monitoring Service. Available online. URL: www.geo.unizh.ch/wgms/index.htm. Accessed December 15, 2005.

World Meteorological Organization. "Global Atmosphere Watch: Ozone Bulletins and Data." WMO. Available online. URL: www.wmo.ch/web/arep/ozone.html. Accessed January 16, 2006.

———. "THORPEX, A World Weather Research Programme." WMO. Available online. URL: www.wmo.int/thorpex/about.html. Accessed February 17, 2006.

———. "Tropical Cyclone Programme (TCP)." Available online. URL: www.wmo.ch/web/www/TCP/rsmcs.html. Last modified August 8, 2002.

———. "World Climate Programme." Available online. URL: www.wmo.ch/index-en.html. Accessed March 7, 2006.

———. "World Meteorological Organisation." Available online. URL: www.wmo.ch/index-en.html. Accessed March 7, 2006.

———. "World Weather Watch." Available online. URL: www.wmo.ch/index-en.html. Accessed March 7, 2006.

World Meteorological Organization, International Council for Science, and the Intergovernmental Oceanographic Commission of UNESCO. "Global Energy and Water Cycle Experiment." Available online. URL: www.gewex.org/index.html. Accessed January 10, 2006.

Wunderground.com. "Radar Frequently Asked Questions (FAQ)." The Weather Underground, Inc. Available online. URL: www.wunderground.com/radar/help.asp. Accessed January 12, 2006.

INDEX

air 11–12, 39, 40
Dobson spectrophotometer 149
Fabry, Charles 600
greenhouse effect 219
isoprene and perpenes 256
light detection and ranging (LIDAR) 278
Marum, Martinus van 625
nitrogen oxides (NO_x) 319
photochemical smog 351
polar stratospheric clouds (PSCs) 362
Schönbein, Christian 641
total ozone mapping spectrometer (TOMS) 403–404
World Meteorological Organization (WMO) 564
zenith angle 569
ozone depletion potential 342, *342t*
ozone hole 341
ozone layer (ozonosphere) **341–342**
CFCs 16, 73
Crutzen, Paul 593–594
Dobson, G. M. B. 598
Dobson spectrophotometer 149
Molina, Mario 628–629
Montreal Protocol 564, 674
Rowland, F. Sherwood 637–638
ozonosphere 40

P

Pa (pascal) 503
Pacific air 15
Pacific- and Indian-Ocean Common Water (PIOCW) **343**
Pacific Decadal Oscillation (PCO) **343**
Pacific highs **343**

Pacific Ocean 326, *327t,* 433. *See also* ocean currents
Pacific typhoons 492
pack ice 408
Paine, Halbert E. 573
Paleoarchean era 343
Paleocene-Eocene thermal maximum (PETM, Initial Eocene Thermal Maximum, IETM) **344**
Paleocene epoch 343–344
paleoclimatology 92, 94–95
beetle analysis **56**
Climate-Leaf Analysis Multivariate Program (CLIMAP) **82**
glaciomarine sediment 212
Globigerina ooze 215
ice cores. *See* ice cores
leaf margin analysis (LMA) **277**
oxygen isotopes 340
paleoclimatology 92, 94–95, 641–642
pollen analysis 363–364
radiocarbon dating 381–382
radiometric dating 381–382
Shackleton, Nicholas 641–642
solar-topographical theory 432
strand lines 446
tree rings 486–488, *487*
varves 509–510
wine harvest 562
Paleogene period **344**
paleoglaciology 211
Paleoproterozoic era **344**
Paleozoic era **344–345**
Palmer, W. C. 151
Palmer Drought Severity Index (PDSI) 151, *151t*
Palmer Hydrological Drought Index (PHDI) 151
palynology 363
pampero 283

PAN (peroxyactetyl nitrate) 18
panas oetara 289
pancake ice 408
Pangaea 124, *124m,* 359
pannus clouds 113
panspermia 578
Panthalassa 124, *124m*
papagaya (norte) 289
parabolic dunes 399
parameterization 345
parcel of air (air parcel) 345
adiabat 5–6
buoyancy 63
hydrostatic equation 236
lapse rates 272–274, *273*
Poisson's equation 360
thermodynamic diagrams 463, *464*
parhelion (sun dogs) 334, *334*
Paris Convention on the Regulation of Aerial Navigation (1919) 42
partial pressure 22
Dalton's law of partial pressures 594–595
gas laws 202
Raoult's law 388
particulates. *See* aerosols
partly cloudy 96
pascal (Pa) 503
Pascal, Blaise **631**
passive instruments 345
Pastonian interglacial 247
past weather **346**
Pat (typhoon) 665
path length 346
Paul (hurricane) 665
Pauline (hurricane) 665
PDSI (Palmer Drought Severity Index) 151, *151t*
PE (potential evapotranspiration) 169–170, 470, 471, *471t*
PE (precipitation efficiency) 308, 470, 471, *471t*, 646
peak gust 556
pearl-necklace lightning 280
Pearson, Allen 197, 606
pea soupers 416

pedestal rocks 165–166, *166*
Peggy (typhoon) 665
Peléean eruptions 516
penetrative convection 128
Penman formula 167
pennant flags 439–440, *440*
Pennsylvanian epoch 69
pentads 346
penumbra *346,* **346**
peppered moth *(Biston betularia)* 245
perched aquifers 34–35, *35*
perchloromethane (carbon tetrachloride) **70**
percolation 221, 348–349, *348t*
perfluorocarbons (PFCs) 219, *219t*
pergelisol (permafrost) **346–348**, *347m*
perhumid climates 88
periglacial climates 88
perihelion *346,* **346**
Peripatetic School 646
perlucidus clouds 113
permafrost (pergelisol) **346–348**, *347m*
permanent drought 152
permeability (hydraulic conductivity) **348–349**, *348t*
permeable membranes 338
Permian period **349–350**
Pérot, Albert 600
peroxyactetyl nitrate (PAN) 18
persistence 263, 545
persistence forecast 534
Peru Current (Humboldt Current) 162–163, 505, 686, *688m*
Peruvian dew (garúa) 282
PGF (pressure gradient force) **372–373**
Phaenomena (Aratus) 535
Phanerozoic eon **350**
phases *350,* **350**
PHDI (Palmer Hydrological Drought Index) 151
phenology 249, **350–351**
Philosophical Basis of Evolution, The (Croll) 593
phloem 486